HARCOURT
Math

CALIFORNIA EDITION

Harcourt School Publishers

Orlando • Boston • Dallas • Chicago • San Diego

www.harcourtschool.com

ISBN 0-15-315517-5

7 8 9 10 11 12 048 10 09 08 07 06 05

Mathematics Advisors

Jerome Dancis
Associate Professor of Mathematics
University of Maryland
College Park, Maryland

Tom Roby
Assistant Professor of Mathematics
California State University
Hayward, California

Senior Author

Evan M. Maletsky
Professor of Mathematics
Montclair State University
Upper Montclair, New Jersey

Authors

Angela Giglio Andrews
Math Teacher, Scott School
Naperville District #203
Naperville, Illinois

Grace M. Burton
Chair, Department of Curricular Studies
Professor, School of Education
University of North Carolina
  at Wilmington
Wilmington, North Carolina

Howard C. Johnson
Dean of the Graduate School
Associate Vice Chancellor
  for Academic Affairs
Professor, Mathematics and
  Mathematics Education
Syracuse University
Syracuse, New York

Lynda A. Luckie
Administrator/Math Specialist
Gwinnett County Public Schools
Lawrenceville, Georgia

Joyce C. McLeod
Visiting Professor
Rollins College
Winter Park, Florida

Vicki Newman
Classroom Teacher
McGaugh Elementary School
Los Alamitos Unified School District
Seal Beach, California

Janet K. Scheer
Executive Director
Create A Vision
Foster City, California

Karen A. Schultz
College of Education
Georgia State University
Atlanta, Georgia

Program Consultants and Specialists

Janet S. Abbott
Mathematics Consultant
California

Arax Miller
Curriculum Coordinator and English
  Department Chairperson
Chamlian School
Glendale, California

Lois Harrison-Jones
Education and Management
  Consultant
Dallas, Texas

Rebecca Valbuena
Language Development Specialist
Stanton Elementary School
Glendora, California

iii

Number Sense and Operations

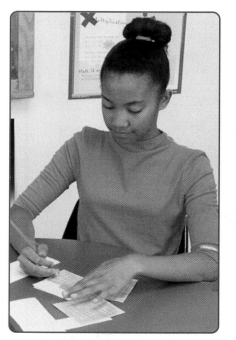

Chapter Features

Daily Review and Practice
  Quick Review
  Mixed Review and
  Test Prep
Intervention
  Troubleshooting,
  pp. H5-H7
Extra Practice
  pp. H36-H37

Technology Resources

**Harcourt Math
Newsroom Video:**
Chapter 5, p. 95

E-Lab:
Chapter 6, p. 129

Multimedia Glossary:
*The Learning Site at
www.harcourtschool.com/
glossary/math*

UNIT WRAPUP

Fraction Concepts and Operations

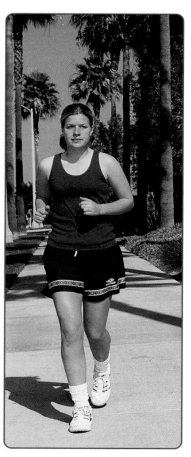

UNIT WRAPUP

Chapter Features

Daily Review and Practice
   Quick Review
   Mixed Review and
   Test Prep
Intervention
   Troubleshooting,
   pp. H12-H14
Extra Practice
   pp. H42-H44

Technology Resources

**Harcourt Math
Newsroom Video:**
Chapter 11, p. 234

E-Lab:
Chapter 12, p. 249

**Mighty Math Astro
Algebra:**
Chapter 13, pp. 258, 260

Multimedia Glossary:
The Learning Site at
**www.harcourtschool.com/
glossary/math**

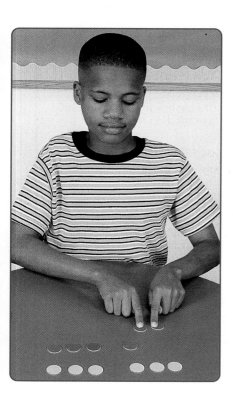

Chapter 13
MULTIPLY AND DIVIDE WITH INTEGERS . . 254

UNIT WRAPUP

Algebra: Expressions and Equations

Chapter 16
MULTIPLICATION AND
DIVISION EQUATIONS

UNIT WRAPUP

UNIT 6
CHAPTERS 17–19

Geometry and Spatial Reasoning

Chapter Features

Daily Review and Practice
 Quick Review
 Mixed Review and
  Test Prep

Intervention
 Troubleshooting,
  pp. H14, H18-H20

Extra Practice
 pp. H48-H50

Technology Resources

**Harcourt Math
Newsroom Video:**
Chapter 18, p. 342

E-Lab:
Chapter 19, p. 357

**Mighty Math Cosmic
Geometry:**
Chapter 17, p. 327
Chapter 18, pp. 338, 349

Multimedia Glossary:
The Learning Site at
**www.harcourtschool.com/
glossary/math**

Chapter 19

UNIT WRAPUP

Ratio, Proportion, Percent, and Probability

Chapter Features

Daily Review and Practice
Quick Review
Mixed Review and
Test Prep

Intervention
Troubleshooting,
pp. H9, H21–H24

Extra Practice
pp. H51–H54

Technology Resources

**Harcourt Math
Newsroom Video:**
Chapter 20, p. 380

E-Lab:
Chapter 20, p. 377
Chapter 21, p. 407
Chapter 22, p. 425

**Mighty Math Cosmic
Geometry:**
Chapter 20, p. 388

**Mighty Math Number
Heroes:**
Chapter 23, p. 438

Multimedia Glossary:
The Learning Site at
**www.harcourtschool.com/
glossary/math**

UNIT 8
CHAPTERS 24–27
Measurement

Chapter Features

Daily Review and Practice
  Quick Review
  Mixed Review and
  Test Prep

Intervention
  Troubleshooting,
  pp. H18, H20, H23,
  H25-H28

Extra Practice
  pp. H55-H58

Technology Resources

Harcourt Math Newsroom Video:
Chapter 25, p. 476

E-Lab:
Chapter 25, p. 475
Chapter 26, p. 491
Chapter 27, p. 510

Mighty Math Cosmic Geometry:
Chapter 27, p. 511

Multimedia Glossary:
The Learning Site at
www.harcourtschool.com/ glossary/math

Algebra: Patterns and Relationships

Chapter Features

Daily Review and Practice
  Quick Review
  Mixed Review and
    Test Prep
Intervention
  Troubleshooting,
    pp. H18, H29-H31
Extra Practice
  pp. H59-H61

Technology Resources

Harcourt Math Newsroom Video:
Chapter 28, p. 529

E-Lab:
Chapter 28, p. 531
Chapter 30, p. 571

Mighty Math Cosmic Geometry:
Chapter 29, p. 544

Mighty Math Astro Algebra:
Chapter 30, p. 559

Multimedia Glossary:
The Learning Site at
www.harcourtschool.com/ glossary/math

STUDENT HANDBOOK

WELCOME!

The authors of *Harcourt Math* want you to be a good mathematician, but most of all we want you to enjoy learning math and feel confident that you can do it. We invite you to share your math book with family members. Take them on a guided tour through your book!

THE GUIDED TOUR

Choose a chapter you are interested in. Show your family some of these things in the chapter that will help you learn.

✓ CHECK WHAT YOU KNOW ••••••••••••••••••••

Do you need to review any skills before you begin the next chapter? If you do, you will find help in the Handbook in the back of your book.

••• THE MATH LESSONS

☑ **Quick Review** to check the skills you need for the lesson.

☑ **Learn section** to help you study problems, models, examples, and questions that give you different ways to learn.

☑ **Check for Understanding** to make sure you understood the lesson.

☑ **Practice and Problem Solving** to help you practice what you have just learned.

☑ **Mixed Review and Test Prep** to keep your skills sharp and prepare you for important tests. Look back at the pages shown by each problem to get help if you need it.

STUDENT HANDBOOK ••••••••••••••••••••••••••••••

Now show your family the **Student Handbook** in the back of your book. The sections will help you in many different ways.

✓ **Troubleshooting** will help you review and remember skills from last year.

✓ **Extra Practice** will help you make sure that you are ready to move on to the next lesson.

✓ **Sharpen Your Test-Taking Skills** will help you feel confident that you can do well on a test.

✓ **Answers to Selected Exercises** will help you check your answers to part of an assignment.

California Standards tell you what you are expected to learn this year. Each lesson's standards are listed at the bottom of the lesson page.

Invite your family members to:

► talk with you about what you are learning.

► help you correct errors you have made on completed work.

► help you set a time and find a quiet place to do math homework.

► solve problems with you as you play, shop, and do household chores together.

► visit **The Learning Site** at www.harcourtschool.com

► have *FUN WITH MATH!*

Have a great year!

The Authors

FOCUS ON PROBLEM SOLVING

Analyze
Choose
Solve
Check

Good problem solvers need to be good thinkers. They also need to know the strategies listed on page 1. This plan can help you learn how to think through a problem.

Analyze the problem.

What are you asked to find?	Restate the question in your own words.
What information is given?	Look for numbers. Find how they are related.
Is there information you will not use? If so, what?	Decide whether you need all the information you are given.

Choose a strategy to solve.

What strategy will you use?	Think about some problem solving strategies you can use. Then choose one.

Solve the problem.

How can you use the strategy to solve the problem?	Follow your plan. Show your solution.

Check your answer.

Look back at the problem. Does the answer make sense? Explain.	Be sure you answered the question that is asked.
What other strategy could you use?	Solving the problem by another method is a good way to check your work.

Try It
Here's how you can use the problem-solving steps to solve a problem.

Draw a Diagram

During summer vacation, Kayla will visit her cousin and then her grandmother. She will be gone for 5 weeks and 2 days and will spend 9 more days with her cousin than with her grandmother. How long will she stay with each?

PROBLEM SOLVING STRATEGIES

▶ **Draw a Diagram or Picture**
Make a Model
Predict and Test
Work Backward
Make an Organized List
Find a Pattern
Make a Table or Graph
Solve a Simpler Problem
Write an Equation
Use Logical Reasoning

Analyze
Kayla will visit her cousin and grandmother for 5 weeks and 2 days this summer. I need to find out how long she will stay with each if she spends 9 more days with her cousin than with her grandmother.

Choose
I can *draw a diagram* to show how long she will stay. I can use boxes for the length of each stay. The length of the boxes will represent the lengths of the stays.

Solve
5 weeks and 2 days is 37 days in all. So, my diagram looks like this:

Step 1

cousin [9]
grandmother [ ]

Sum of all parts (days) = 37.
37 − 9 = 28
28 ÷ 2 = 14

Step 2

cousin [14 | 9] = 23
grandmother [14] = 14   So, each equal part = 14.

So, Kayla will stay with her cousin for 23 days and with her grandmother for 14 days.

Check
23 days is 9 days longer than 14 days. The total of the two stays is 23 + 14, or 37 days. This solution fits the description of Kayla's vacation trip.

Draw a Diagram

Eight basketball teams will play in a tournament. A team is out after it loses once. How many games does a team have to win in order to win the tournament?

Analyze There are 8 teams. If a team loses, it is out of the tournament. If a team wins, it keeps playing.

Choose You can *draw a diagram* to show which teams play in each game of the tournament. Use the diagram to show each pair of teams playing against each other.

Team 1
Team 2

Solve Use the diagram to pair up the 8 teams. Make up a winner for each game. Then show the next games until the tournament has a winner.

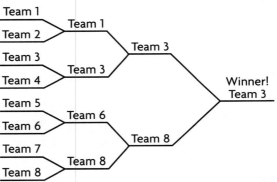

So, a team must win 3 games to win the tournament.

Check Follow Team 3 through the tournament listing. Count the number of times it wins.

Problem Solving Practice

1. **What if** half as many teams play in the tournament? How many games will have to be played to determine the winner?

2. After the game, the teams have sandwiches at their coaches' houses. You can build your own sandwich with lettuce, tomato, and ham. How many different ways can you stack the 3 fillings to make a sandwich?

Make a Model

A̲lice has three pieces of ribbon. Their lengths
are 7 in., 10 in., and 12 in. How can Alice use
these ribbons to measure a length of 15 in.?

Analyze

Each of the three ribbons is a different length. No ribbon is 15 in.
long, but you can combine the ribbon lengths to measure 15 in.

Choose

You can *make a model* of each piece of ribbon. Then you can
organize the ribbons to show new lengths.

Solve

When you put two pieces together, you can form lengths of 17, 19,
and 22 in. All of these are too long.

One Way
Alice can fold the 10-in. piece in half to model a 5-in. length.
Then she can put the folded piece together with the 10-in. piece
to make 15 in.

Another Way
Alice can place the 7-in. piece above the 12-in. piece to show that
the 12-in. piece extends 5 in. beyond it. Then that 5 in. and the
10-in. piece together will make 15 in.

Check

Check that you used the correct lengths for the ribbon pieces.
Then add the lengths of the two longer pieces and subtract the
length of the short one. The result is 15 in.

Problem Solving Practice

1. Draw pictures of models you could make to measure other
   lengths with the three pieces of ribbon.

2. **What if** Andy is building a stand by using four cubes? He stacks
   the cubes, one on top of the other, and paints the outside of
   each cube (not the bottom). How many faces of the cubes are
   painted?

Predict and Test

Margaret sells tickets to a pizza dinner at her school. An adult ticket costs $3.50. A student ticket costs $2.50. Margaret sells 10 more adult tickets than student tickets. She collects $155 in ticket sales. A total of 450 people attend the pizza dinner. How many student tickets has Margaret sold?

Analyze

Student tickets cost $2.50 and adult tickets cost $3.50. Margaret sells 10 more adult tickets than student tickets. Her sales are $155. The number of people at the dinner is given, but this information is not needed.

Choose

You can use *predict and test* to solve. Use number sense and the needed information to predict how many student tickets Margaret has sold. Then test your prediction and revise it if needed.

Solve

Make a table to show each prediction and its result. Be sure you have 10 more adult tickets than student tickets each time.

PREDICTION		TEST	
Adult	Student	Sales	
20	10	(20 × $3.50) + (10 × $2.50) = $95	too low, revise
40	30	(40 × $3.50) + (30 × $2.50) = $215	too high, revise
30	20	(30 × $3.50) + (20 × $2.50) = $155	✓ correct

So, Margaret has sold 20 student tickets.

Check

Check that each multiplication is correct: 30 × $3.50 = $105, and 20 × $2.50 = $50; $105 + $50 = $155. The answer checks.

Problem Solving Practice

1. The sum of Margaret's and her younger brother's ages is 38. The difference between their ages is 8. How old is Margaret's brother?

2. Tables for the pizza dinner are set up in 3 rooms. One room has 11 tables to seat 98 people. Some tables are for 8 people, and others are for 10 people. How many tables for 8 people are set up in this room?

Analyze
Choose
Solve
Check

Work Backward

Joe and his brother Tim go shopping. At the toy store, they use half of their money to buy a video game. Then they go to the pizza parlor and spend half of the money they have left on a pizza. Then they spend half of the remaining money to rent a video. After these stops, they have $4.50 left. How much money did they have at the start?

Analyze

You need to find how much money they had at the start. You know that they had $4.50 left and that they spent half of their money at each of three stops.

Choose

Start with the amount you know they have left—$4.50—and *work backward*. Knowing how much they have left will help you calculate first how much they spent to rent the video, then how much the pizza cost, then the price of the video game, and finally how much money they had at the start.

Solve

amount they had left → $4.50

$4.50 → twice that amount before a video → 2 × $4.50 = $9

$9 → twice that amount before a pizza → 2 × $9 = $18

$18 → twice that amount before a video game → 2 × $18 = $36

So, the amount they had at the start was $36.

Check

Put $36 in the original problem to check the amount they spent at each stop.

$36 ÷ 2 = $18, $18 ÷ 2 = $9, $9 ÷ 2 = $4.50

The amount they had left matches the amount given in the problem.

Problem Solving Practice

1. The Lauber family has 4 children. Joe is 5 years younger than his brother Mark. Tim is half as old as his brother Joe. Mary, who is 10, is 3 years older than Tim. How old is Mark?

2. If you divide this mystery number by 4, add 8, and multiply by 3, you get 42. What is the mystery number?

Analyze
Choose
Solve
Check

Make an Organized List

At Fun City Amusement Park you can throw 3 darts at a target to score points and win prizes. If each dart lands within the target area, how many different total scores are possible?

2
5
10

Analyze

If all 3 darts hit the center circle, you get 30 points. This is the highest score. If all 3 darts hit the outside circle, you get 6 points. This is the lowest score. If the 3 darts hit a different combination, other scores are possible.

Choose

Make an organized list to determine all possible hits and score totals. List the value of each dart and the total for all three darts.

Solve

First consider all 3 darts hitting the center circle. List the value of each dart and the total score. Then consider 2 darts hitting the center circle and the third dart hitting a different circle. List the value of each dart and the total scores. Do the same for 1 dart hitting the center circle and no darts hitting the center circle.

3 Darts Hit Center	2 Darts Hit Center	1 Dart Hits Center	0 Darts Hit Center
10 + 10 + 10 = 30	10 + 10 + 5 = 25	10 + 5 + 5 = 20	5 + 5 + 5 = 15
	10 + 10 + 2 = 22	10 + 5 + 2 = 17	5 + 5 + 2 = 12
		10 + 2 + 2 = 14	5 + 2 + 2 = 9
			2 + 2 + 2 = 6

So, there are 10 possible scores.

Check

Make sure that all possible combinations of scores are listed and that each set of scores in the list is different.

Problem Solving Practice

1. The Yogurt Store at Fun City sells 3 flavors of yogurt: chocolate, vanilla, and strawberry. You want to get a scoop of each flavor in a waffle cone. How many different ways can the scoops be arranged?

2. How many ways can you make change for a quarter by using dimes, nickels, and pennies?

Find a Pattern

Analyze
Choose
Solve
Check

A contractor can build stairways to a deck or patio with any number of steps. She uses the pattern at the right to build them. How many blocks are needed to build a stairway with 7 steps?

Analyze

As the number of steps increases, so does the number of blocks. You must find the number of blocks needed for a stairway with 7 steps.

Choose

You can *find a pattern* for the number of blocks needed. For each stairway, count the number of blocks to make each step and then find the total number of blocks. Use the pattern to find the number of blocks for 7 steps.

Solve

The pattern shows that the number of the step in the stairway is the same as the number of blocks needed to make it. The 2nd step is made with 2 blocks, the 3rd step is made with 3 blocks, and so on.

Number of Step	Side View	Number of Blocks	Cumulative Total
2		2	$2 + 1 = 3$
3		3	$3 + 2 + 1 = 6$
4		4	$4 + 3 + 2 + 1 = 10$

So, a stairway with 7 steps has 7 blocks in the 7th step, 6 blocks in the 6th step, and so on. The total number of blocks is $7 + 6 + 5 + 4 + 3 + 2 + 1 = 28$ blocks.

Check

Sketch a stairway with 7 steps. Check the number of blocks needed. The number in the sketch matches the answer.

$7 + 6 + 5 + 4 + 3 + 2 + 1 = 28$

Problem Solving Practice

1. A cereal company adds baseball cards to certain boxes of cereal. Cards are added to the 3rd box, the 6th box, the 11th box, and the 18th box of cereal. If this pattern continues, how many boxes will have baseball cards in them when a case of 40 is ready to be shipped? Explain.

Describe the pattern and find the missing numbers.

2. 1, 4, 16, 64, 256, ■ , ■ , 16,384

Make a Table

In math, Mrs. Laurence gave her class a 100-point test. Students who score 80 or above are eligible to join the Math Club. The test scores are in the box below. How many students scored 80 or above? How many students scored below 80?

90	83	80	77	78	91	92	73
62	83	79	88	72	85	93	84
75	68	82	75	94	70	98	82

Analyze

You have the test scores for the class. You need to find how many students scored below 80 and how many students scored 80 or above.

Choose

You can *make a table* and tally the test scores. This organizes the data and makes it easier to answer questions about the scores.

Solve

Make a two-column table for the range of scores. As you read each test score, place a tally mark across from the appropriate range. Be sure you end up with 24 tallies, one for each student.

SCORES	TALLIES
60–69	||
70–79	卌 |||
80–89	卌 |||
90–99	卌 |

Now use the table to answer the questions.

SCORES OF 80 OR ABOVE	SCORES BELOW 80
8 students scored 80–89.	8 students scored 70–79.
6 students scored 90–99.	2 students scored 60–69.

So, 14 students scored 80 or above, and 10 students scored below 80.

Check

Recount the tallies in each row and check your addition.

Problem Solving Practice

1. Starting at 3:00, the director will give each student 5 minutes to try out for the talent show. The students, in order, are Ben, Tarek, Jan, Ed, and Frank. At what time does Frank start?

2. Kelly reads stories to children at the library. There are three sessions, each lasting 45 minutes, with 30 minutes between sessions. If Kelly starts reading at 10:00 A.M., at what time does she finish?

Analyze
Choose
Solve
Check

Solve a Simpler Problem

Make a table of the first 2 odd numbers and their sum, the first 3 odd numbers and their sum, and so on. How does the table show a pattern? What is the sum of the first 20 odd numbers?

Analyze

You know how to begin finding sums for the first 2 odd numbers and the first 3 odd numbers. You need to find a pattern in the table to help you find the sum of the first 20 odd numbers.

Choose

You can begin by *solving a simpler problem*. Start with sums for the first 2 odd numbers, followed by sums for 3 odd numbers, 4 odd numbers, and 5 odd numbers.

Solve

Organize the data in a table and look for a pattern in the sums.

ODD NUMBERS	SUM	PATTERN
1 + 3	4	2^2
1 + 3 + 5	9	3^2
1 + 3 + 5 + 7	16	4^2
1 + 3 + 5 + 7 + 9	25	5^2

The table shows a pattern in the sums that relates to square numbers. Extend the pattern in the table to 20 odd numbers.

6 odd numbers $\rightarrow$ 36

7 odd numbers $\rightarrow$ 49

8 odd numbers $\rightarrow$ 64

20 odd numbers $\rightarrow$ 400

So, the sum of the first 20 odd numbers is 20^2, or 400.

Check

Extend the table a few more rows to check that there really is a pattern of square numbers. For example, find sums for the first 9, 10, 11, and 12 odd numbers. The answers check.

Problem Solving Practice

1. Martha has 5 pairs of slacks and 4 blouses for school. How many different outfits can she make with these items?

2. What is the least 5-digit number that can be divided by 50 with a remainder of 17?

Write an Equation

Analyze
Choose
Solve
Check

A school district can afford to build a gym with three basketball courts with an area of 18,700 square feet. The gym needs to be 110 feet wide to leave walking room around the courts. How long will the gym be?

Analyze

The gym is rectangular. To find the area a rectangle covers, you multiply the length by the width. You know the area and the width of the gym and need to find its length.

Choose

You can *write an equation* for the area of the rectangle. Begin with the formula for the area of a rectangle. Use the numbers you know to find the missing number.

Solve

You find the area of a rectangle by multiplying its length by its width.

$$A = l \times w$$ Write the formula.
$$18,700 = l \times 110$$ Use the numbers you know.
$$l = 18,700 \div 110$$ Write a related division equation.
$$l = 170$$

So, the length of the gym should be 170 feet.

Check

Place your answer in the original equation.

$$A = l \times w$$
$$A = 170 \times 110$$
$$A = 18,700$$

The product matches the information in the problem.

Problem Solving Practice

1. A rectangular box of crackers has a volume of 128 cubic inches. The area of its base is 16 square inches. What is the height of the box?

2. The perimeter of an isosceles triangle is 70 cm. The sides of equal length each measure 28 cm. What is the length of the unknown side?

Use Logical Reasoning

Christine, Elizabeth, and Sharon like different sports. One girl likes to snow ski, one likes to run track, and the other likes to swim. Elizabeth is a good friend of the girl who likes track. Sharon is shorter than the girl who likes to ski. Elizabeth does not like cold weather. Which girl likes which sport?

Analyze

You know that there are three girls and that each one likes a different sport. Clues will help you determine which girl likes which sport.

Choose

You can *use logical reasoning* to solve this problem.

Solve

Make a table. List the sports and the names of the friends. Work with the clues one at a time. Write "yes" in a box if the clue applies to that girl. Write "no" in a box if the clue does not apply. Only one box in each row and column can have a "yes" in it.

a. Elizabeth is a good friend of the girl who likes track, so Elizabeth does not run track. Write "no" in the appropriate box.

	ski	track	swim
Christine	yes	no	no
Elizabeth	no	no	yes
Sharon	no	yes	no

b. Sharon is shorter than the girl who likes to ski, so Sharon does not ski. Write "no" in the appropriate box.

c. Elizabeth does not like cold weather. Write "no" in the appropriate box. With the other sports eliminated, Elizabeth must be the person who likes to swim. Write "yes" in the appropriate box. For Christine and Sharon, write "no" in each box under *swim*.

So, Christine likes to snow ski, Sharon likes track, and Elizabeth likes to swim.

Check

Compare your answers to the clues in the problem. Make sure none of your conclusions conflict with the clues.

Problem Solving Practice

- Brent, Michael, and Tim always order their favorite pizzas: pepperoni, three-cheese, and sausage. Brent is allergic to pepperoni. Tim is going to see a movie with the boy who loves sausage. Mike does not like meat on his pizza. Which kind of pizza does each boy order?

Compare Strategies

Mrs. Hagen is eager to plant her spring garden. She wants to plant 24 tomato plants in a rectangular array. For each plant, there will be a 1-foot square of space. How many different arrays can she make?

Analyze There are 24 plants to be put into a rectangular array. You need to find how many arrays are possible.

Choose Lexi and Jake use different strategies to solve this problem. Lexi chooses to *draw a diagram*, and Jake chooses to *write an equation*.

Solve

Lexi's Way: Draw a Diagram

Lexi uses graph paper to draw diagrams of rectangles to show all possible arrays. Each rectangle represents a garden that covers 24 square feet. The rectangles are 1 × 24, 2 × 12, 3 × 8, and 4 × 6.

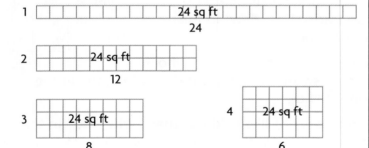

Jake's Way: Write an Equation

Jake uses the equation $A = l \times w$ for finding the area of a rectangle to identify the different rectangles that could be formed. Each length and width needs to be a whole number.

$A = l \times w$
$24 = 1 \times 24$
$24 = 2 \times 12$
$24 = 3 \times 8$
$24 = 4 \times 6$

Think: What are the factors of 24?

So, Lexi and Jake both find that Mrs. Hagen can choose from four rectangular arrays.

Check Lexi can do the problem Jake's way, and Jake can try it Lexi's way.

Problem Solving Practice

1. Mr. Sargent is buying math games for his class. One game costs $2.95, one costs $3.75, and one costs $6.00. He wants 6 of each game for his class. What is the total cost of these games?

2. How many triangles are in the figure at the right?

Multistep Problems

The Enrichment Program Committee at the Franklin School orders 500 T-shirts to sell at the school Field Day. Each T-shirt costs the school $4.20 and sells for $13.50. All 500 T-shirts are sold. After paying for the T-shirts, the school has reached its goal of raising $4,000. How much money has the school made?

FRANKLIN SCHOOL ENRICHMENT PROGRAM

Look at the facts in the problem. T-shirts cost $4.20 each and sell for $13.50. The committee orders and sells 500 T-shirts. You need to find how much money has been made after the costs are deducted. You know your answer should be at least $4,000.

One Way

First subtract to find how much the school has made on each T-shirt.
    money earned on 1 T-shirt: $13.50 − $4.20 = $9.30

Then multiply to find the total earned.
    money earned on 500 T-shirts: $9.30 × 500 = $4,650

Another Way

Find the total sales and subtract the total cost.

    Total sales: 500 × $13.50 = $6,750

    Total cost: 500 × $4.20 = $2,100

    Total earned = $6,750 − $2,100 = $4,650

So, the school has made $4,650 on the sale of the T-shirts.

Problem Solving Practice

1. At the school Field Day, frozen yogurt cones sell for $2.25. Lin buys one for herself and one for each of her 4 friends. How much change does she get back from $20?

2. Fran is making a cabinet for her shell collection. She buys 3 boards at $2.75 each and 4 hinges at $0.99 each. What is the total cost of the supplies?

3. Mrs. Smith bought bread for $0.98, eggs for $1.17, ground turkey for $3.18, and 3 bunches of carrots for $0.65 each. She gave the clerk $10.03. How much change should Mrs. Smith receive?

CHAPTER 1

Whole Number Applications

The first Space Shuttle mission, STS-1, went up in 1981 with two astronauts. Since then, NASA has flown nearly 100 Shuttle missions. To the nearest hour, the first mission, STS-1, lasted 30 hours. The STS-78 mission lasted 406 hours. About how many hours longer did the STS-78 mission last than the STS-1 mission?

SELECTED SPACE SHUTTLE MISSIONS

Number of Hours vs. Mission

- STS-1 (1981): 30
- STS-9 (1983): 248
- STS-34 (1989): 120
- STS-47 (1992): 191
- STS-78 (1996): 406

Check What You Know

Use this page to help you review and remember important skills needed for Chapter 1.

✅ Vocabulary

Choose the best term from the box.

1. In $8 \times 3 = 24$, the __?__ is 24.
2. In $48 \div 12 = 4$, the __?__ is 12.
3. In $40 - 30 = 10$, the __?__ is 10.
4. In $4\overline{)412}$, the __?__ is 412.
5. In $3 + 8 + 5 = 16$, the __?__ is 16.

> factor
> difference
> dividend
> divisor
> product
> sum

✅ Place Value of Whole Numbers (See p. H3.)

Name the place value of the digit 8.

6. 56,485,013
7. 2,403,815
8. 4,568,137
9. 423,859,100
10. 207,108,629
11. 813,593,005
12. 375,178,955,276
13. 351,480,994,133
14. 748,630,123,944

Give the value of the blue digit.

15. 321,497,580
16. 321,497,580
17. 321,497,580
18. 628,705,814,956
19. 628,705,814,956
20. 628,705,814,956

✅ Round Whole Numbers (See p. H4.)

Round to the nearest thousand.

21. 2,467
22. 5,609
23. 28,500
24. 299
25. 34,831
26. 19,089
27. 6,136
28. 43,712
29. 134,612
30. 217,501
31. 59,832
32. 539,513

Round to the nearest ten thousand.

33. 51,677
34. 228,260
35. 12,435
36. 78,562
37. 639,108
38. 40,500
39. 90,499
40. 381,810

LOOK AHEAD

In Chapter 1 you will

- estimate and compute with whole numbers
- evaluate expressions
- use mental math to compute and to solve equations

Estimate with Whole Numbers

Learn how to estimate sums, differences, products, and quotients of whole numbers.

Vocabulary

clustering
underestimate
overestimate

Remember that when rounding, you look at the digit to the right of the place to which you are rounding.
• If that digit is 5 or greater, round up.
• If that digit is less than 5, round down.

The longest throw of a flying disc was 656 ft 2 in., made by Scott Stokely in 1995. Carmen, David, and Leona each threw a disc. Is the total distance of the three throws close to the record distance?

You don't need an exact sum to answer the question, so estimate.

$$
\begin{array}{rcl}
134 & \to & 130 \\
148 & \to & 150 \\
+183 & \to & +180 \\
\hline
& & 460
\end{array}
$$
Round each number to the nearest ten.

NAME	DISTANCE
David	134 ft
Leona	148 ft
Carmen	183 ft

The estimate, 460, is not close to the record distance of 656 ft, so the total distance is not close either.

Use **clustering** to estimate a sum when the addends are about the same.

EXAMPLE 1

Estimate $1{,}802 + 2{,}182 + 1{,}999$.

$$
\begin{array}{r}
1{,}802 \\
2{,}182 \\
+1{,}999 \\
\end{array}
$$
The three addends all cluster around 2,000, so use 2,000 for each number.

$3 \times 2{,}000 = 6{,}000$ *Multiply.*

So, the sum is about 6,000.

You can use rounding to estimate a difference.

EXAMPLE 2

Estimate $31{,}928 - 20{,}915$.

Round to the nearest ten thousand.

$$
\begin{array}{r}
30{,}000 \\
-20{,}000 \\
\hline
10{,}000
\end{array}
$$

Round to the nearest thousand.

$$
\begin{array}{r}
32{,}000 \\
-21{,}000 \\
\hline
11{,}000
\end{array}
$$

So, both 10,000 and 11,000 are reasonable estimates.

CALIFORNIA STANDARDS ○━┓NS 2.0 Students calculate and solve problems involving addition, subtraction, multiplication, and division. **MR 2.0** Students use strategies, skills, and concepts in finding solutions. *also,* **MR 2.4, MR 2.5**

You can show an estimate by using the "approximately equal to" symbol, ≈.

5,125 − 1,920 ≈ 3,000   *Read: 5,125 − 1,920 is approximately equal to 3,000.*

An estimate that is less than the exact answer is an **underestimate**. An estimate that is greater than the exact answer is an **overestimate**.

$$\begin{array}{r} 366 \\ +198 \\ \hline 564 \end{array}$$ *Exact answer* $$\begin{array}{r} 370 \\ +200 \\ \hline 570 \end{array}$$ *Round up.* *Overestimate*

$$\begin{array}{r} 144 \\ \times 123 \\ \hline 17,712 \end{array}$$ *Exact answer* $$\begin{array}{r} 100 \\ \times 100 \\ \hline 10,000 \end{array}$$ *Round down.* *Underestimate*

EXAMPLE 3

Students set up 28 rows of 36 seats each for a talent show in the school cafeteria. About how many programs should the school print for the show?

Estimate 28 × 36. To make sure there are enough programs, find an overestimate.

$$\begin{array}{r} 28 \\ \times 36 \\ \hline \end{array} \rightarrow \begin{array}{r} 30 \\ \times 40 \\ \hline \end{array}$$ *Round each factor up to the nearest ten.*

$$\begin{array}{r} 30 \\ \times 40 \\ \hline 1,200 \end{array}$$ *Multiply. Because each factor is rounded up, the product is an overestimate.*

So, the school should print about 1,200 programs.

Remember that compatible numbers are numbers that divide without a remainder, are close to the actual numbers, and are easy to compute with mentally.

To estimate a quotient, use rounding or compatible numbers.

EXAMPLE 4

The employees of the Briar Creek office building have collected 1,545 lb of paper to be recycled. The building has 36 offices. About how many pounds of paper, on average, did the employees in each office collect?

Estimate 1,545 ÷ 36.

1,600 ÷ 40   *4 is compatible with 16, so use 1,600 ÷ 40.*
1,600 ÷ 40 = 40   *Divide.*

So, the employees in each office collected an average of about 40 lb.

- Is 1,488 ÷ 36 easier to estimate using rounding or compatible numbers? Explain.

Math **I**dea ▶ Some of the strategies you can use to estimate are rounding, clustering, and compatible numbers.

Think and Discuss ▶ Look back at the lesson to answer each question.

1. **Explain** why 6,000 + 1,500 is an overestimate or an underestimate of 6,108 + 1,524.

2. **Describe** how to estimate a quotient by using compatible numbers. Give an example to illustrate your answer.

Guided Practice ▶ Estimate the sum or difference.

3. 723	4. 2,940	5. 4,480	6. 5,449
+819	3,140	4,100	4,869
	+2,834	+3,967	+4,834

7. 667	8. 8,855	9. 34,855	10. 67,184
−133	−2,268	−11,268	−49,650

Estimate the product or quotient.

11. 36	12. 59	13. 48	14. 490
× 9	×33	×29	× 66

15. 321 ÷ 4   16. 1,544 ÷ 28   17. 4,156 ÷ 64   18. 8,429 ÷ 39

Independent Practice ▶ Estimate the sum or difference.

19. 1,700	20. 293	21. 5,765	22. 43,643
2,008	348	5,948	+84,211
+2,324	+343	+6,324	

23. 389	24. 3,556	25. 44,123	26. 667,184
− 43	−3,339	−29,512	−249,650

27. 17,809 − 2,145   28. 321,059 + 42,950

TECHNOLOGY LINK

To learn more about estimation with whole numbers, watch the **Harcourt Math Newsroom Video** *Weather Stacked Stadium.*

Estimate the product or quotient.

29. 364	30. 53	31. 482	32. 1,874
× 12	×41	×299	× 582

33. 1,844 ÷ 22   34. 3,575 ÷ 56   35. 6,435 ÷ 529   36. 21,416 ÷ 521

37. 4,135 × 784   38. 62,217 ÷ 4,889

Tell whether the estimate is an *overestimate* or an *underestimate*. Then show how the estimate was determined.

39. 352 + 675 ≈ 1,100   40. 4,134 + 47 ≈ 4,100   41. 96 × 19 ≈ 2,000

42. 709 + 151 ≈ 850   43. 291 × 28 ≈ 9,000   44. 25 × 29 ≈ 900

Use estimation to compare. Write < or > for ●.

45. 614×41 ● $21{,}119 + 1{,}899$ **46.** $18{,}391 \div 19$ ● 59×21

47. $4{,}012 - 3{,}508$ ● $3{,}624 \div 6$ **48.** $12{,}283 + 19{,}971$ ● $209{,}910 \div 7$

49. 711×63 ● $28{,}520 + 16{,}990$ **50.** 513×52 ● $29{,}190 - 1{,}986$

Problem Solving ▶ Applications

Use Data   For 51–52, use the table.

51. About how many more library books does Harvard have than the University of Illinois at Urbana?

52. Estimate the total number of library books at Yale and the University of California at Berkeley.

53. The theater of the Natural Science and History Museum was filled to capacity for 276 shows. The theater holds 36 people. How many people attended the shows?

54. ✎ **Write About It**   Is it easier to use rounding or compatible numbers to estimate $756 \div 74$? Explain.

55. Two numbers, each rounded to the nearest hundred, have a product of 60,000. What are two possible numbers?

LARGEST UNIVERSITY LIBRARIES IN THE UNITED STATES (1999)	
Library	Books
Harvard University	12,877,360
Yale University	9,485,823
University of Illinois–Urbana	8,474,737
University of California–Berkeley	8,078,685
University of Texas	7,019,508

MIXED REVIEW AND TEST PREP

For 56–58, find the perimeter and area.

56.
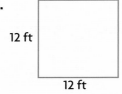
12 ft
12 ft

57.

12 in.
15 in.
9 in.

58.

10 cm
3 cm

59. TEST PREP   Ms. Cannon baked a total of 60 apple and blueberry pies. She baked a dozen more apple pies than blueberry pies. How many apple pies did she bake?

A 18   **B** 24   **C** 36   **D** 48

60. TEST PREP   Allen and Mei began working at the same time. It took Allen 50 minutes to mow the lawn, while Mei took $1\frac{1}{4}$ hr to paint the fence. How much longer did Mei work than Allen?

F 85 min   **G** 65 min   **H** 35 min   **J** 25 min

Extra Practice page H32, Set A)

19

Use Addition and Subtraction

Learn how to add and subtract whole numbers.

PROGRAM	YEARS	TOTAL HOURS IN SPACE
Mercury	1961–1963	54
Gemini	1965–1966	1,940
Apollo	1968–1972	7,506
Skylab	1973–1974	12,352

Remember that you add when joining groups of different sizes and subtract when taking away or comparing groups.

During the early years of space flight, the total time United States astronauts spent in space increased with each program.

EXAMPLE 1

How many hours did U.S. astronauts spend in space?

Find 54 + 1,940 + 7,506 + 12,352.

Estimate to check for reasonableness.

Round to the nearest thousand.   Find the sum.

```
     54              0           54
  1,940          2,000        1,940    Compare your estimate.
  7,506    →     8,000        7,506    21,852 is close to 22,000, so
+12,352       +12,000      +12,352    the sum is reasonable.
               22,000        21,852
```

So, U.S. astronauts spent 21,852 hours in space.

EXAMPLE 2

How many more hours did U.S. astronauts spend in Skylab than on the Gemini missions?

Find 12,352 − 1,940.

Estimate to check for reasonableness.   Find the difference.

```
 12,352         12,000       12,352    Compare your estimate.
- 1,940    →   - 2,000      - 1,940    10,412 is close to 10,000, so
               10,000        10,412    the difference is reasonable.
```

So, U.S. astronauts spent 10,412 more hours in Skylab than on the Gemini missions.

CALIFORNIA STANDARDS O—π NS 2.0 Students calculate and solve problems involving addition, subtraction, multiplication, and division. **MR 2.1** Use estimation to verify the reasonableness of calculated results. **MR 3.1** Evaluate the reasonableness of the solution in the context of the original situation. *also,* **MR 2.7**

Think and ▶
Discuss

Look back at the lesson to answer each question.

1. **Explain** how you know whether to add or subtract when solving a word problem.

2. **Explain** why it is a good idea to find an estimate before or after you find the exact answer.

Guided ▶
Practice

Find the sum or difference. Estimate to check.

3. 835 + 604   **4.** 6,901 + 342 + 67   **5.** 40,190 − 13,982

PRACTICE AND PROBLEM SOLVING

Independent ▶
Practice

Find the sum or difference. Estimate to check.

6. 9,500 − 289   **7.** 21,670 + 14,704   **8.** 31,227 + 56,995

9.   999,999   **10.**   987,654   **11.**   50,000,000
      +111,385   −456,789   − 3,604,381

Solve.

12. 1,485 + 2,019 + 1,310 + 3,665 + 798

13. 43,875 + 81,420 − 38,288 + 12,108 − 23,990

Problem Solving ▶
Applications

Use Data   For 14–15, use the graph.

14. How many more calories are in 1 cup of almonds than in 1 cup of carrots mixed with 1 cup of raisins? Estimate to check.

15. ✎ **Write a problem** that uses data from the graph and that can be solved by adding or subtracting.

16. When the international space station is completed, it will be 290 ft long, which is 206 ft longer than Skylab. Skylab was 41 ft longer than Mir, the Russian space station. How long is Mir?

International space station

CALORIES

Calories in 1 Cup

- Almonds 849
- Apples 73
- Carrots 46
- Raisins 419

Food

MIXED REVIEW AND TEST PREP

17. 6,785 + 4,521

18. Complete. 36 in. = ▇ ft

19. List the factors of 24.

20. List the first six multiples of 9.

21. **TEST PREP**   Which is a prime number?

   A 15   **B** 27   **C** 31   **D** 50

Use Multiplication and Division

Learn how to multiply and divide whole numbers.

Sometimes you can multiply to solve a word problem.

EXAMPLE 1

Remember that you multiply when joining equal-sized groups, and you divide when separating into equal-sized groups or when finding how many in each group.

Sixth-grade students sold 132 books of carnival ride coupons. How many ride coupons did they sell if there were 18 in each book?

Multiply. 132×18 Estimate. $130 \times 20 = 2,600$

$$
\begin{array}{r}
132 \\
\times\ 18 \\
\hline
1\ 056 \\
+1\ 32 \\
\hline
2,376
\end{array}
$$

Compare the exact product to your estimate. Since 2,376 is close to the estimate of 2,600, the exact product is reasonable.

So, the students sold 2,376 coupons.

You can omit the zero placeholders when you multiply. Just be careful to line up the products correctly.

Correct	Incorrect
132	132
$\times\ 24$	$\times\ 24$
528	528
$+264$	$+264$

EXAMPLE 2

Season tickets to an amusement park are on sale for $125 each. On the first day of the sale, the amusement park sold 12,383 tickets. How much money did the amusement park receive for season tickets that day?

Multiply. $12,383 \times 125$

Estimate. $12,000 \times 130 = 1,560,000$

$$
\begin{array}{r}
12,383 \\
\times\ \ \ \ 125 \\
\hline
61\ 915 \\
247\ 66 \\
+1\ 238\ 3 \\
\hline
1,547,875
\end{array}
$$

Compare the exact product to your estimate. Since 1,547,875 is close to the estimate of 1,560,000, the exact product is reasonable.

So, the amusement park received $1,547,875.

CALIFORNIA STANDARDS O—n**NS 2.0** Students calculate and solve problems involving addition, subtraction, multiplication, and division. **MR 2.1** Use estimation to verify the reasonableness of calculated results. **MR 3.1** Evaluate the reasonableness of the solution in the context of the original situation. *also,* **MR 1.3, MR 2.7**

Sometimes you have to use division to solve word problems.

EXAMPLE 3

Remember that the procedure for dividing is divide, multiply, subtract, compare, and bring down.

Mrs. Lopez is redesigning the company cafeteria to seat 540 employees. Each table in her design seats 12 employees. How many tables will she need?

Divide. 540 ÷ 12

Estimate. 480 ÷ 12 = 40

$$\begin{array}{r} 45 \\ 12\overline{)540} \\ -48\downarrow \\ \hline 60 \\ -60 \\ \hline 0 \end{array}$$

Compare the exact quotient to your estimate. Since 45 is close to the estimate of 40, the exact quotient is reasonable.

So, Mrs. Lopez will need 45 tables in her design.

• What if each table seats 10 employees? How many tables will Mrs. Lopez need?

Sometimes a division problem has a zero in the quotient.

EXAMPLE 4

A school collected 2,568 newspapers. The newspapers were bundled in packages of 25. How many packages of newspapers did the school bundle?

Divide. 2,568 ÷ 25

Estimate. 2,500 ÷ 25 = 100

$$\begin{array}{r} 102 \text{ r}18 \\ 25\overline{)2,568} \\ -25 \\ \hline 06 \\ -0 \\ \hline 68 \\ -50 \\ \hline 18 \end{array}$$

Compare the exact answer to your estimate. Since 102 r18 is close to the estimate of 100, the exact quotient is reasonable.

So, the school bundled 102 packages of newspapers.

In Example 4 there is a remainder. Some calculators allow you to show a whole-number remainder.

2,568 ⊡÷R⊡ 25 ⊡=⊡ | 102 R18 |

You can express a remainder with an *r*, or you can express it as a fractional part of the divisor or as a decimal. The quotient and remainder in Example 4 can also be expressed as $102\frac{18}{25}$ or as 102.72.

Think and Discuss ▶ Look back at the lesson to answer each question.

1. **Explain** why the school bundled 102 packages of newspapers instead of 103 packages of newspapers in Example 4.

2. **Tell** the different ways to express the remainder for the division problem $153 \div 6$.

Guided Practice ▶ Multiply or divide. Estimate to check.

3. $1,113 \times 712$ **4.** $2,115 \div 72$ **5.** $16,225 \times 219$

Multiply or divide.

6. $\begin{array}{r} 13 \\ \times 14 \\ \hline \end{array}$ **7.** $8\overline{)432}$ **8.** $12\overline{)144}$ **9.** $\begin{array}{r} 962 \\ \times\ 40 \\ \hline \end{array}$

10. 159×340 **11.** $7,658 \times 111$ **12.** $7,044 \div 14$ **13.** $1,068 \div 19$

Independent Practice ▶ Multiply or divide. Estimate to check.

14. $2,250 \div 18$ **15.** $4,904 \times 196$ **16.** $193,200 \div 46$
17. $7,021 \times 498$ **18.** $249,900 \div 49$ **19.** $24,587 \times 71$

TECHNOLOGY LINK

More Practice: Use **Mighty Math Calculating Crew,** *Captain Nick Knack,* Level V.

Multiply or divide.

20. $16\overline{)1,664}$ **21.** $\begin{array}{r} 298 \\ \times\ 89 \\ \hline \end{array}$ **22.** $\begin{array}{r} 5,233 \\ \times\ 238 \\ \hline \end{array}$ **23.** $52\overline{)728}$

24. $4\overline{)412}$ **25.** $\begin{array}{r} 380 \\ \times\ 55 \\ \hline \end{array}$ **26.** $\begin{array}{r} 2,382 \\ \times\ 12 \\ \hline \end{array}$ **27.** $24\overline{)626}$

28. $\begin{array}{r} 327 \\ \times 123 \\ \hline \end{array}$ **29.** $26\overline{)2,314}$ **30.** $68\overline{)24,820}$ **31.** $\begin{array}{r} 5,470 \\ \times\ 240 \\ \hline \end{array}$

32. $29\overline{)13,253}$ **33.** $\begin{array}{r} 6,378 \\ \times\ 291 \\ \hline \end{array}$ **34.** $\begin{array}{r} 2,009 \\ \times\ 562 \\ \hline \end{array}$ **35.** $120\overline{)10,080}$

Divide. Write the remainder as a fraction.

36. $5\overline{)49}$ **37.** $7,349 \div 20$ **38.** $386 \div 15$ **39.** $4\overline{)3,385}$

40. ⓐ⁺ᵇ⁄ᶜ **Algebra** What is the least whole number, n, for which it is true that $n \div 8 > 542 + 258$?

41. ⓐ⁺ᵇ⁄ᶜ **Algebra** What is the least whole number, n, for which it is true that $70 \times n > 29,000$?

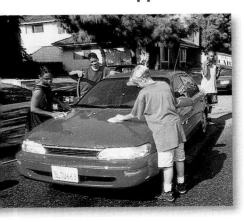

Problem Solving ▶ Applications

42. Yuji's parents bought an entertainment center for $1,176. They plan to pay for it with 14 equal monthly payments. How much will each payment be if a $3.50 service charge is added every month?

43. Lincoln Middle School had a car wash to raise money. The students charged $3.00 for every car and $5.00 for every van. If they washed 23 cars and 18 vans, how much did they earn?

44. ❓ **What's the Error?** Describe the error. Then solve the problem correctly.

$$49\overline{)30{,}395} \quad 62\ r15$$

45. There are 5 children in Brenda's family. Brenda is 16 years old. Brenda's twin sisters are 6 years old, and her brothers are 10 and 12 years old. What is the mean of the 5 children's ages?

MIXED REVIEW AND TEST PREP

46. How much more than 96,784 is 142,981? (p. 20)

47. Complete. 4 lb = ■ oz

48. Order 804, 824, 818, and 803 from least to greatest.

49. **TEST PREP** How many meters are in 1,800 millimeters?

   A 0.18 m   **B** 1.8 m   **C** 18 m   **D** 180 m

50. **TEST PREP** Teresa is buying perfume for her mother. The prices are $16.19, $15.89, $15.99, and $17.00. How much will she save by buying the least expensive instead of the most expensive?

   F $0.01   **G** $0.11   **H** $0.81   **J** $1.11

LINKUP to Reading

Strategy • Use Context Many word problems contain clues such as *more than, fewer than, twice as many,* and *total.* Be sure to interpret clue words within the context of the problem before you choose an operation.

Use Data **For each problem, write the clue words and the operation. Then solve the problem.**

1. How many more calories will you burn in 1 hr by skiing than by hiking?

2. On Saturday, Connie spent an hour in gymnastics class and then walked for 1 hr. How many total calories did she burn during these two activities?

3. Clara burned half as many calories while raking the lawn for 1 hr than she did while jogging for 1 hr. How many calories did she burn while raking the lawn?

CALORIES BURNED PER HOUR	
Activity	**Calories**
Walking (at 2 mi per hr)	112
Gymnastics	128
Hiking	191
Jogging	224
Cross-country skiing	326

PROBLEM SOLVING STRATEGY
Predict and Test

Analyze
Choose
Solve
Check

Learn how to use the strategy *predict and test* to solve problems with whole numbers.

You can solve some problems by using your number sense to predict a possible answer. You should then test your answer and revise your prediction if necessary.

There were 25 problems on a test. For each correct answer, 4 points were given. For each incorrect answer, 1 point was subtracted. Tania answered all 25 problems. Her score was 85. How many correct answers did she have?

Analyze   What are you asked to find?

What facts are given?

Is there any numerical information you will not use? If so, what?

Choose   What strategy will you use?

You can use the strategy *predict and test.* Use the given information and your number sense to predict about how many correct answers Tania had. Then test your prediction, and revise it if necessary.

Solve   How will you solve the problem?

Make a table to show your prediction, your test of it, and any revisions you need. Be sure that the total of correct and incorrect problems is 25.

PREDICTION		TEST	
Correct	Incorrect	SCORE	
20	5	$(20 \times 4) - 5 = 75$	*too low, so revise*
21	4	$(21 \times 4) - 4 = 80$	*too low, so revise*
22	3	$(22 \times 4) - 3 = 85$	← *correct*

So, Tania had 22 correct answers.

Check   How can you check your answer?

What if Tania's score were 65? How many incorrect answers would she have?

CALIFORNIA STANDARDS ○━┓NS 2.0 Students calculate and solve problems involving addition, subtraction, multiplication, and division. **MR 3.2** Note the method of deriving the solution and demonstrate a conceptual understanding of the derivation by solving similar problems. *also,* **AF 3.1, MR 1.1, MR 1.3, MR 2.0, MR 2.7**

Solve by predicting and testing.

1. Rodney bought a total of 40 oranges and apples. He bought 14 fewer apples than oranges. How many of each fruit did he buy?

2. The Mighty Tigers soccer team played a total of 25 games. They won 9 more games than they lost, and 2 games ended in ties. How many games did they win?

3. The perimeter of a rectangular garden is 40 ft. If the length is 6 ft more than the width, what are the length and width of the garden?

   A $l = 12$ ft, $w = 6$ ft

   B $l = 13$ ft, $w = 7$ ft

   C $l = 6$ ft, $w = 12$ ft

   D $l = 7$ ft, $w = 13$ ft

4. The perimeter of a rectangular lawn is 28 yd. If the length is 4 yd more than the width, what are the length and width of the lawn?

   F $l = 10$ yd, $w = 4$ yd

   G $l = 9$ yd, $w = 5$ yd

   H $l = 5$ yd, $w = 9$ yd

   J $l = 4$ yd, $w = 10$ yd

MIXED STRATEGY PRACTICE

5. Rosalia waters her tomato plants every other day. She waters her pepper plants every 3 days. If she waters both on April 20, what are the next three dates on which she will water both?

6. Stacy spent a total of $28.45. She bought a ticket for a basketball game for $8.50, food for $7.95, and some T-shirts for $6.00 each. How many T-shirts did she buy?

7. Sam has 98 baseball cards. This is 2 more than twice as many as Paul has. How many cards does Paul have?

8. Use the table below. If the pattern continues, how many miles in all will four runners run on the fifth day?

9. Melina and her two sisters collect stamps. Melina has twice as many as her older sister, who has 33 stamps. Melina has three times as many as her younger sister. How many stamps do they have in all?

10. ❓ **What's the Question?** The sum of the ages of Jeff, Elijah, and Stefan is 41. Jeff is 14 years old. Stefan is 3 years older than Elijah. The answer is 12 years old.

11. The train that leaves at 11:45 A.M. usually arrives in New York City 34 minutes after that. Today it arrived at 12:24 P.M. How late was the train?

ALGEBRA
Use Expressions

Learn how to identify, write, and evaluate expressions involving whole numbers.

In a basketball game, a team scored 27 points in the first half and 38 points in the second half. To represent the total points, you could use a numerical expression. A **numerical expression** is a mathematical phrase that includes only numbers and operation symbols.

$$27 + 38 \leftarrow \text{total points}$$

Here are other examples of numerical expressions.

$$60 + 25 \qquad 42 \div 7 \qquad 16 - 3 \qquad 51 \times 36 \qquad 30 + 12 + 41$$

If you didn't know how many points the team scored in the second half, you could use a variable to represent the points. A **variable** is a letter or symbol that can stand for one or more numbers. An expression that includes a variable is called an **algebraic expression**.

$27 + p \leftarrow$ *Use p to represent points scored in second half.*

Here are other examples of algebraic expressions.

$$5 + n \qquad 7 \times a \qquad k - 3 \qquad y \div 2 \qquad 6 \times 5 \times b$$

In an expression, there are several ways to show multiplication.

$$7 \times a \text{ can be written as } 7a, 7(a), \text{ or } 7 \cdot a.$$

Word expressions can be translated into numerical or algebraic expressions.

Vocabulary
numerical expression
variable
algebraic expression
evaluate

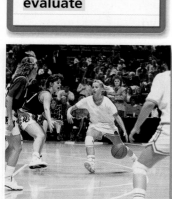

EXAMPLE 1

Write a numerical or algebraic expression for the word expression.

A. three dollars less than five dollars $5 - 3$

B. two times a distance, d $2 \times d$, $2(d)$, $2d$, or $2 \cdot d$

To **evaluate** a numerical expression, you find its value. To evaluate an algebraic expression, replace the variable with a number and then find the value.

EXAMPLE 2

Evaluate each expression.

A. $a + 150$, for $a = 18$

$a + 150$	*Replace a*
$18 + 150$	*with 18.*
168	*Add.*

B. $b \div 10 \times 3$, for $b = 120$

$b \div 10 \times 3$	*Replace b with*
$120 \div 10 \times 3$	*120.*
12×3	*Divide and*
36	*then multiply.*

CALIFORNIA STANDARDS O¬nNS 2.0 Students calculate and solve problems involving addition, subtraction, multiplication, and division. **AF 1.0** Students write verbal expressions and sentences as algebraic expressions and equations; they evaluate algebraic expressions, solve simple linear equations, and graph and interpret their results. *also*, **MR 2.4, MR 2.5**

Think and Discuss ▶ Look back at the lesson to answer each question.

1. **Explain** the difference between a numerical expression and an algebraic expression. Give some examples of each.

2. **Show** four different ways to write an algebraic expression for the product of the number 10 and the variable g.

Guided Practice ▶ Write a numerical or algebraic expression for the word expression.

3. forty-six less than one hundred twenty-five

4. one hundred seven more than y

5. y divided by fifteen

Evaluate each expression.

6. 21×15

7. $100 - g$, for $g = 54$

8. $s \div 8$, for $s = 720$

PRACTICE AND PROBLEM SOLVING

Independent Practice ▶ Write a numerical or algebraic expression for the word expression.

9. twenty-five \$20 bills

10. q more than two hundred fifteen

11. seventy-six decreased by k

12. x divided by fourteen

Evaluate each expression.

13. 15×31

14. $3,021 + 915$

15. $10,340 - 1,340$

16. $k - 65$, for $k = 95$

17. $\frac{d}{7} \times 2$, for $d = 490$

18. $100b$, for $b = 54$

19. $m \div n$, for $m = 1,230$ and $n = 410$

20. cd, for $c = 5$ and $d = 200$

Problem Solving Applications ▶

21. Let n represent the number of free throws Nathan scored. Bryan scored 12 more free throws than Nathan. Write an algebraic expression to show how many free throws Bryan scored.

22. ✎ **Write About It** Explain how to evaluate an algebraic expression when you know the value of each variable. Give an example.

23. **REASONING** Tiffany, Deidre, Luisa, Kendall, and Ann Marie are runners. Kendall can outrun Luisa and Tiffany, but Deidre can outrun Kendall. Luisa can outrun Ann Marie, but Deidre can outrun Luisa. Which one of the girls is the fastest runner?

MIXED REVIEW AND TEST PREP

24. 530×42 (p. 22)

25. $3,870 \div 18$ (p. 22)

26. $1,234 + 453$ (p. 20)

27. $8,000 - 357$ (p. 20)

28. **TEST PREP** Christina bought 3 pens and 1 notebook for \$5.85. A pen cost \$1.20. How much did the notebook cost?

A \$1.20   **B** \$1.25   **C** \$1.85   **D** \$2.25

Extra Practice page H32, Set D

ALGEBRA
Mental Math and Equations

Learn how to use mental math to solve equations.

QUICK REVIEW

Evaluate each expression.

1. 25×8
2. $80 + d$, for $d = 37$
3. $g \div 8$, for $g = 72$
4. $450 - 225$
5. Find the missing factor. $\blacksquare \times 8 = 24$

Vocabulary

equation

solution

An **equation** is a statement showing that two quantities are equal. These are equations:

$$6 + 7 = 13 \qquad 24 \div 3 = 8 \qquad k - 3 = 1 \qquad 2d = 18 \qquad a + b = 11$$

If an equation contains a variable, you can solve the equation by finding the value of the variable that makes the equation true. That value is the **solution**.

EXAMPLE 1

Remember that a variable is a letter or symbol that stands for one or more numbers.

Which of the numbers 8, 9, and 10 is a solution of the equation $12p = 108$?

Replace p with 8.	*Replace p with 9.*	*Replace p with 10.*
$12(8) \overset{?}{=} 108$	$12(9) \overset{?}{=} 108$	$12(10) \overset{?}{=} 108$
$96 = 108$ *false*	$108 = 108$ *true*	$120 = 108$ *false*

The solution is 9 because $12(9) = 108$.

• Which of the numbers 4, 5, and 6 is a solution of the equation $222 \div n = 37$?

Some equations with variables can be solved by using mental math. Think of the value of the variable that makes the equation true. Then check your answer.

EXAMPLE 2

The Statue of Liberty's hand is about 16 ft long. The index finger is 8 ft long. What is the length of the palm of her hand? Solve the equation $16 = c + 8$ by using mental math.

$16 = c + 8$	*What number added to 8 gives 16?*
$8 = c$	*The solution is 8.*

Check:

$16 = 8 + 8$	*Replace c with 8.*
$16 = 16$	*8 + 8 is equal to 16.*

• Solve the equation $m \times 7 = 56$ by using mental math.

CALIFORNIA STANDARDS AF 1.0 Students write verbal expressions and sentences as algebraic expressions and equations; they evaluate algebraic expressions, solve simple linear equations, and graph and interpret their results. **O—ɴAF 1.1** Write and solve one-step linear equations in one variable. *also,* **O—ɴNS 2.0, MR 2.5**

Think and Discuss ▸ Look back at the lesson to answer each question.

1. Tell whether 4 is a solution of the equation $x + 3 = 9$. If it is not, find the solution.

2. Give an example of an equation with a solution of 5.

Guided Practice ▸ Determine which of the given values is the solution of the equation.

3. $f \div 7 = 3$; $f = 19, 20,$ or 21 **4.** $t + 9 = 20$; $t = 10, 11,$ or 12

Solve each equation by using mental math.

5. $7 = x + 3$ **6.** $\dfrac{h}{9} = 3$ **7.** $4 \times k = 16$

PRACTICE AND PROBLEM SOLVING

Independent Practice ▸ Determine which of the given values is the solution of the equation.

8. $3h = 39$; $h = 11, 12,$ or 13 **9.** $17 - x = 12$; $x = 5, 6,$ or 7

10. $48 + s = 57$; $s = 8, 9,$ or 10 **11.** $3 = 54 \div k$; $k = 16, 17,$ or 18

Solve each equation by using mental math.

12. $p - 7 = 7$ **13.** $9m = 81$ **14.** $13 + r = 30$

15. $x - 16 = 4$ **16.** $h \div 8 = 7$ **17.** $14 = k - 15$

18. $87 = e \div 10$ **19.** $12 \times v = 240$ **20.** $t \div 6 = 125$

21. $12 + 4 + d = 25$ **22.** $3 \times 4 = c - 8$ **23.** $p + 14 = 32 - 12$

Problem Solving Applications ▸

24. The equation $w + 12 = 40$ describes the number of men and women riding the bus to a convention. If w is the number of women riding the bus, how many men are riding the bus?

25. Mr. Murakami teaches 5 classes of 25 students each. One hundred of his students are sixth graders. How many are not sixth graders?

26. ❓ **What's the Question?**   A roller coaster has 7 cars. Fifty-six people can ride the roller coaster at one time. The answer is 8.

MIXED REVIEW AND TEST PREP

27. Evaluate $a + 14$ for $a = 27$. (p. 28)

28. $525 \div 25$ (p. 22)

29. Find $4,310 - 1,900 + 3,450 - 870$. (p. 20)

30. Find the greatest common factor of 15 and 35.

31. TEST PREP   Andre left the house at 8:45 A.M. He arrived home $4\frac{1}{2}$ hours later. At what time did Andre arrive home?

A 11:45 A.M.   **B** 12:45 A.M.   **C** 12:45 P.M.   **D** 1:15 P.M.

Extra Practice page H32, Set E

1. **VOCABULARY** A way to estimate a sum when all of the addends are about the same is ___?___. (p. 16)

2. **VOCABULARY** A letter or symbol that can stand for one or more numbers is a(n) ___?___. (p. 28)

3. **VOCABULARY** A statement showing that two quantities are equal is a(n) ___?___. (p. 30)

Estimate. (pp. 16–19)

4.	593	5.	1,420	6.	935	7.	2,371	8.	43,816
	+724		+5,791		−549		−1,456		−39,972

9. 48×6 10. 308×67 11. $374 \div 7$ 12. $276 \div 42$ 13. $3,764 \div 591$

Find the sum or difference. (pp. 20–21)

14.	4,762	15. $9,724 - 286$	16.	50,031	17.	737	18.	678,040
	+39,038				− 9,352		4,650		−329,193
							+11,821		

Multiply or divide. (pp. 22–25)

19.	526	20. 123×12	21. $2,250 \div 18$	22.	189	23. $40\overline{)3,206}$
	$\times$ 42				$\times 108$	

Evaluate each expression for $a = 63$, $b = 150$, and $c = 7$. (pp. 28–29)

24. $a + 305$ 25. $b - 36$ 26. $300 \div b$ 27. $3c$ 28. $a \div 9$

29. $215 - b$ 30. $112 \div c$ 31. $a \times 5$ 32. $2a + 4$ 33. $a + b$

Solve each equation by using mental math. (pp. 30–31)

34. $3m = 27$ 35. $14 = q + 6$ 36. $20 = y - 9$ 37. $w \div 50 = 5$ 38. $74 + a = 85$

Solve. (pp. 26–27)

39. At school during spirit week, Ming sold a total of 36 red and blue ribbons. She sold 6 more red ribbons than blue ribbons. How many of each color did she sell?

40. Colton worked two days on a project for school. He worked a total of 195 minutes. If he worked 45 minutes longer on the first day, how long did Colton work each day?

Understand the problem.

See item **7**.

The words describe an algebraic expression. *Less* describes the relationship of the given number to the unknown number.

Also see problem **1**, p. H62.

Choose the best answer.

1. Which is the best estimate for this sum?
$6{,}204 + 5{,}893 + 6{,}028 + 5{,}991$

A 22,000   **C** 28,000

B 24,000   **D** 30,000

2. Which is the best estimate for this quotient?

$$4{,}923 \div 71$$

F 7

G 70

H 700

J 7,000

3. A new jogging track at a recreation center is 580 feet longer than the old track. The new track is 1,320 feet long. How long was the old track?

A 740 ft   **C** 1,900 ft

B 860 ft   **D** Not here

4. There are 9,800 beads to put in boxes. Each box holds 48 beads. How many boxes are needed to hold all of the beads?

F 48 boxes   **H** 205 boxes

G 204 boxes   **J** Not here

5. Which is the best estimate for this product?

$$81 \times 409$$

A 32,000

B 28,000

C 3,200

D 2,800

6. What is the value of $35 + b$ for $b = 5$?

F 7   **H** 40

G 30   **J** 175

7. Which algebraic expression represents the expression "28 less than a number, q"?

A $28 - q$

B $q + 28$

C $q \div 28$

D $q - 28$

8. What is the value of n for $8 \times n = 24$?

F 3

G 16

H 32

J 192

9. A bus can carry 16 people. There are 9 people on board and there is room for x more people. Which equation shows this relationship?

A $16 + x = 9$ **C** $9 + x = 16$

B $x - 9 = 16$ **D** $x - 16 = 9$

10. Which algebraic expression represents the expression "7 more than a number, t"?

F $7 \times t$

G $t - 7$

H $7 - t$

J $t + 7$

11. $1{,}225 \times 37$

A 12,250   **C** 45,325

B 34,225   **D** 48,390

12. $115{,}309 - 67{,}899$

F 47,410   **H** 58,410

G 57,410   **J** 58,510

CHAPTER 2

Operation Sense

A bird can reach almost any part of its body with its beak because it has an extremely flexible neck. The bones in the neck are called cervical vertebrae. Humans have 7 cervical vertebrae. Most birds have more cervical vertebrae than humans or other mammals. Flamingos have $4^2 + 8 \times 2 - 13$ cervical vertebrae. About how many times as many cervical vertebrae does a flamingo have than a human?

CERVICAL VERTEBRAE

Animal	Number of Cervical Vertebrae
Human	
Owl	
Pigeon	
Swan	
Calif. Condor	

Number of Cervical Vertebrae
0 2 4 6 8 10 12 14 16 18 20 22 24 26

Check What You Know

Use this page to help you review and remember important skills needed for Chapter 2.

 Repeated Multiplication (See p. H15.)

Find the product.

1. $3 \times 3 \times 3$
2. $2 \times 2 \times 2 \times 2$
3. $4 \times 4 \times 4$
4. $5 \times 5 \times 5 \times 5$
5. $10 \times 10 \times 10 \times 10$
6. $9 \times 9 \times 9$
7. $8 \times 8 \times 8$
8. $6 \times 6 \times 6$
9. $7 \times 7 \times 7$
10. $10 \times 10 \times 10$
11. $4 \times 4 \times 4 \times 4$
12. $5 \times 5 \times 5$

 Properties (See p. H2.)

Name the property illustrated.

13. $48 + 13 + 5 = 13 + 48 + 5$
14. $8 \times (3 + 1) = (8 \times 3) + (8 \times 1)$
15. $0 \times 999 = 0$
16. $15 \times 1 = 15$
17. $(9 + 5) + 10 = 9 + (5 + 10)$
18. $27 + 36 = 36 + 27$
19. $(4 + 2) \times 7 = (4 \times 7) + (2 \times 7)$
20. $7 \times 9 \times 2 = 2 \times 9 \times 7$
21. $1 \times 148 = 148$
22. $(2 \times 9) \times 5 = 2 \times (9 \times 5)$
23. $8 \times (20 + 8) = (8 \times 20) + (8 \times 8)$
24. $6 \times 3 = 3 \times 6$

✔ **Use of Parentheses** (See p. H15.)

Solve the equation.

25. $5 \times (4 + 3) = (a \times 4) + (a \times 3)$
26. $7 + (3 + 9) = (7 + m) + 9$
27. $4 + 6 + 3 = 4 + (6 + r)$
28. $(9 \times 4) + (9 \times 2) = h \times (4 + 2)$
29. $4 \times (8 - 4) = (4 \times t) - (4 \times 4)$
30. $(5 + 6) + (4 + 7) = (5 + 4) + (r + 7)$

Evaluate the expression.

31. $(4 + 7) + 9$
32. $5 \times (6 + 2)$
33. $(1 + 7) \times 4$
34. $(7 + 8) + 3$
35. $4 + (7 - 3)$
36. $6 \times (9 - 3)$
37. $2 \times (3 + 5 + 8)$
38. $(8 + 5) + (3 + 1)$
39. $4 + (5 + 7) + 2$
40. $2 \times (3 \times 1)$

LOOK AHEAD

In Chapter 2 you will

- use properties and mental math to find sums, differences, products, and quotients
- use exponents
- use order of operations

MENTAL MATH
Use the Properties

Learn how to use properties and mental math to find sums, differences, products, and quotients.

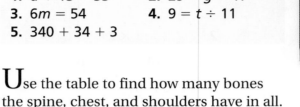

Vocabulary

compensation

Remember these properties:

Commutative Property
$4 + 8 = 8 + 4$
$6 \times 7 = 7 \times 6$

Associative Property
$(6 + 8) + 5 = 6 + (8 + 5)$
$(2 \times 9) \times 5 = 2 \times (9 \times 5)$

Distributive Property
$12 \times 32 = 12 \times (30 + 2) = (12 \times 30) + (12 \times 2)$

Use the table to find how many bones the spine, chest, and shoulders have in all. Mentally find the sum by first reordering using the Commutative Property.

$26 + 25 + 4 = 26 + 4 + 25$ *Commutative Property*

$\qquad\qquad = 30 + 25$ *Use mental math.*
$\qquad\qquad = 55$

Mentally find the sum by regrouping using the Associative Property.

$(26 + 25) + 4 = 26 + (25 + 4)$ *Associative Property*

$\qquad\qquad = 26 + 29$ *Use mental math.*
$\qquad\qquad = 55$

BONES IN THE HUMAN BODY	
Part	**Number of Bones**
Head	28
Spine	26
Throat	1
Chest	25
Shoulders	4
Arms	6
Hands	54
Legs	10
Feet	52

So, the spine, chest, and shoulders have a total of 55 bones.

You can use the Distributive Property to mentally solve a problem.

| **EXAMPLE 1** |

How many bones are in 5 models of the human spine?
$5 \times 26 = 5 \times (20 + 6)$ *Break 26 into parts.*

$\qquad\quad = (5 \times 20) + (5 \times 6)$ *Use the Distributive Property. Multiply mentally.*

$\qquad\quad = 100 + 30$ *Add the products.*
$\qquad\quad = 130$

So, there are 130 bones in 5 models.

You can also use the Commutative and Associative Properties.

| **EXAMPLE 2** |

James has 8 storage boxes on each of 6 shelves. Each box contains 5 items. How many items are there altogether?

Commutative Property   Associative Property
$(8 \times 6) \times 5 = (8 \times 5) \times 6$ $(8 \times 6) \times 5 = 8 \times (6 \times 5)$
$\qquad\qquad = 40 \times 6$ $\qquad\qquad = 8 \times 30$
$\qquad\qquad = 240$ $\qquad\qquad = 240$

A strategy you can use for some addition and subtraction problems is **compensation**. For addition, change one number to a multiple of 10 and then adjust the other number to keep the balance.

EXAMPLE 3

Mr. Forge and his friends play basketball for an hour on Fridays and Saturdays. On Friday they scored a total of 44 points, and on Saturday they scored 57 points. Use compensation to find the total points scored for both days.

44 + 57 = (44 + 6) + (57 − 6)   *Add 6 to 44 and subtract 6*
                                  from 57.

        = 50 + 51   *Use mental math to add.*

        = 101

So, the total points scored is 101.

When you use compensation to subtract, you have to do the same thing to each number.

EXAMPLE 4

Use compensation to find 128 − 56.

128 − 56 = (128 + 4) − (56 + 4)   *Add 4 to 128 and to 56 before*
                                    subtracting.

         = 132 − 60

         = 72

So, the difference is 72.

You can sometimes divide mentally by breaking a number into smaller parts that are each divisible by the divisor.

EXAMPLE 5

Use mental math to find 396 ÷ 4.

396 = 360 + 36   *Break 396 into parts.*

360 ÷ 4 = 90 and 36 ÷ 4 = 9   *Divide each part by 4 mentally.*

90 + 9 = 99   *Add the parts of the quotient.*

So, 396 ÷ 4 = 99.

• Tell another way to break 396 into two parts to divide by 4.

Math **I**dea ▶ Using the number properties and other mental math strategies will help you add, subtract, multiply, and divide mentally.

CHECK FOR UNDERSTANDING

Think and ▶
Discuss

Look back at the lesson to answer each question.

1. **Tell** how using the Associative Property in Example 2 made the problem easier to solve.

2. **Explain** two different ways to use compensation to find 349 + 138 mentally.

Use mental math to find the value.

3. 12×17 **4.** $45 + 9 + 15$ **5.** $124 + 17 + 16$ **6.** 9×36

7. $(6 + 37) + 13$ **8.** $2 \times 9 \times 50$ **9.** 5×29 **10.** 11×43

11. $39 + 16$ **12.** $83 + 38$ **13.** $426 \div 3$ **14.** $16 + 35$

15. $279 \div 3$ **16.** $137 - 51$ **17.** $65 - 22$ **18.** $567 \div 7$

PRACTICE AND PROBLEM SOLVING

Use mental math to find the value.

19. 24×7 **20.** $73 - 27$ **21.** 45×11 **22.** 12×35

23. $87 + 98$ **24.** $(12 + 23) + 8$ **25.** 4×27 **26.** $18 + 26$

27. 4×53 **28.** $64 - 29$ **29.** $24 + 32 + 16$ **30.** 19×14

31. $126 + 118$ **32.** $293 - 137$ **33.** $765 \div 9$

34. $32 + 36$ **35.** $19 + 26$ **36.** $4 \times 6 \times 50$

37. $25 \times 30 \times 2$ **38.** $172 \div 4$ **39.** $1,526 - 498$

40. $40 \times 15 \times 2$ **41.** $(4 \times 33) + (4 \times 7)$ **42.** $(6 \times 24) + (6 \times 6)$

43. $192 \div n$ for $n = 3$ **44.** $c \times 9 \times 5$ for $c = 8$

45. $p \div 12$ for $p = 624$ **46.** $a + 19 + 32$ for $a = 18$

Name each missing reason.

47. $80 \times 3 = (8 \times 10) \times 3$ 80 means 8×10.

$\qquad = 8 \times (10 \times 3)$ Associative Property of Multiplication

$\qquad = 8 \times (3 \times 10)$ ___?___

$\qquad = (8 \times 3) \times 10$ ___?___

$\qquad = 24 \times 10$ ___?___

$\qquad = 240$ ___?___

48. What if the product of three whole numbers is 210? Without using 1 as a factor, what are the possible choices for the numbers?

Use Data   **For 49–51, use the data below.**

49. Use mental math to find how many CDs were bought in all. Explain how you got your answer.

50. If Nick and Selena each gave 12 CDs to Brenda, how many would Brenda have then?

51. How many CDs would Brenda, Selena, Ricardo, and Nick each have if they shared their CDs equally?

CDs Bought
Brenda 12
Selena 17
Nick 25
Ricardo 18

52. A toy store has 7 boxes on each of 4 shelves. Each box has 25 items in it. How many items altogether are on the 4 shelves?

53. Ann needs 250 signatures on a petition. On Monday she got 23 signatures, on Tuesday she got 3 times as many as on Monday, and on Wednesday and Thursday she got 45 each. How many more signatures does she need?

54. Jocelyn has $253.47. Her aunt gives her $87.95 more. Jocelyn buys a pair of shoes for $39.99, three T-shirts for $7.77 each, and two pairs of jeans for $60.22. Use estimation to find about how much money Jocelyn has left.

55. 📕 **Write About It** Explain how to use compensation to add two numbers. Give an appropriate example to support your explanation.

MIXED REVIEW AND TEST PREP

56. Use mental math to solve. $a \div 7 = 21$ (p. 30)

57. Multiply. 732×46 (p. 22)   **58.** Divide. $64,270 \div 35$ (p. 22)

59. TEST PREP Rob changes 4 quarts of oil in his car every 3,000 miles. How many quarts of oil will Rob have used after driving 9,000 miles? (p. 22)

A 3   **B** 9   **C** 12   **D** 36

60. TEST PREP Joe bought 3 basketballs for $22.99 each and a net for $5.99. Which number sentence can be used to find the total cost of the basketballs and net? (p. 28)

F $3 \times (22.99 + 5.99)$ **H** $(3 \times 5.99) + 22.99$

G $(3 \times 22.99) + 5.99$ **J** $(3 + 22.99) + 5.99$

Thinker's CORNER

Math Fun Practice using mental math strategies to solve this puzzle.

1. Copy the diagram. Place the values of the expressions below in the circles so that every sum of three numbers in a line is the same.

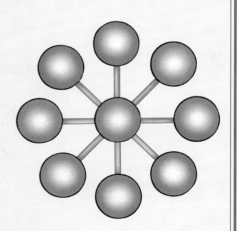

$84 \div 2$
$36 + 8$
$28 + 16 + 2$
$3 \times 8 \times 2$ $8 + 14 + 32$
$28 + 22$ $448 \div 8$
4×13 $4 + 38 + 16$

2. Use mental math. What is the sum of each row of three numbers?

Extra Practice   page H33, Set A   **39**

ALGEBRA
Exponents

Learn how to represent numbers by using exponents.

Vocabulary

exponent

base

1. 3×3 **2.** 6×6

3. $4 \times 4 \times 4$ **4.** $2 \times 2 \times 2 \times 2$

5. $9 \times 9 \times 9$

Some football stadiums can seat over 100,000 people. Large numbers can be hard to understand. On the right are four ways to write 100,000 using smaller numbers.

$10 \times 10,000$
$10 \times 10 \times 1,000$
$10 \times 10 \times 10 \times 100$
$10 \times 10 \times 10 \times 10 \times 10$

Another way to write 100,000 is by using exponents. An **exponent** shows how many times a number called the **base** is used as a factor.

$$\overset{exponent}{10^5} = 10 \times 10 \times 10 \times 10 \times 10 = 100,000$$

base   equal factors

EXPONENT FORM	READ	VALUE
10^1	The first power of ten	10
$10^2 = 10 \times 10$	Ten squared, or the second power of ten	100
$10^3 = 10 \times 10 \times 10$	Ten cubed, or the third power of ten	1,000

Remember that when you multiply two or more numbers to get a product, the numbers multiplied are called factors.
$8 \times 3 \times 4 = 96$
The numbers 8, 3, and 4 are factors of 96.

EXAMPLE 1

Find the values of 2^4, 4^2, and 6^3.

$2^4 = 2 \times 2 \times 2 \times 2$
    $= 16$
2 is a factor four times.

$4^2 = 4 \times 4$
    $= 16$
4 is a factor two times.

$6^3 = 6 \times 6 \times 6$
    $= 216$
6 is a factor three times.

Note: The first power of any number equals that number.

$6^1 = 6$ $9^1 = 9$ $10^1 = 10$

The zero power of any number, except zero, is defined to be 1.

$6^0 = 1$ $19^0 = 1$ $10^0 = 1$

EXAMPLE 2

Write 125 using an exponent and the base 5.

$125 = 5 \times 25 = 5 \times 5 \times 5$ *Find the equal factors.*
      $= 5^3$ *Write the base and the exponent.*

So, $125 = 5^3.$

CALIFORNIA STANDARDS ○━┓ NS 2.0 Students calculate and solve problems involving addition, subtraction, multiplication, and division. *also,* MR 2.4, MR 2.5

CHECK FOR UNDERSTANDING

Think and Discuss ▶ Look back at the lesson to answer each question.

1. **Tell** how many zeros there are in the standard form of 10^7.

2. **Explain** how to write $6 \times 6 \times 6 \times 6$ using exponents.

Guided Practice ▶ Write the equal factors. Then find the value.

3. 2^3 4. 5^2 5. 3^4 6. 9^3 7. 1^4

PRACTICE AND PROBLEM SOLVING

Independent Practice ▶ Write the equal factors. Then find the value.

8. 4^5 9. 7^3 10. 1^{12} 11. 5^3 12. 2^5

13. 34^2 14. 13^2 15. 10^8 16. 20^2 17. 2^{10}

18. 10^4 19. 3^0 20. 15^2 21. 25^1 22. 90^2

Write in exponent form.

23. $12 \times 12 \times 12$ 24. $1 \times 1 \times 1 \times 1 \times 1$ 25. $4 \times 4 \times 4 \times 4$

26. $2 \times 2 \times 2 \times 2 \times 2$ 27. $n \times n$ 28. $y \times y \times y \times y$

Express with an exponent and the given base.

29. 64, base 8   30. 216, base 6   31. 10,000; base 10

32. Write 64 using a base of 8, a base of 4, and a base of 2.

Problem Solving Applications ▶

33. **Use Data** When did California have a population less than 10^7? Explain.

Population of California

34. Ben saves ticket stubs from the movies he sees. He puts them in an album that has 27 pages, each with spaces for 24 stubs. How many stubs can the album hold?

35. **? What's the Question?** Scott has 3^2 video games and Aaron has 2^3 games. The answer is 1.

MIXED REVIEW AND TEST PREP

36. Use mental math to find the value of $279 \div 3$. (p. 30)

37. Round 45,621 to the nearest thousand. (p. 16)

38. $943,012 - 57,806$ (p. 20)   39. $32,047 \div 43$ (p. 22)

40. **TEST PREP** Which expression is equivalent to $(3 + 5) \times 2$? (p. 28)

A $(7 - 3) + 10$ **B** $(4 \times 2) + 8$ **C** $(2 \times 3) + 5$ **D** $10 + (7 \times 10)$

Extra Practice page H33, Set B

Explore Order of Operations

Explore how to evaluate expressions by using order of operations.

You need a calculator.

QUICK REVIEW

1. $16 + 24$ **2.** $(24 \div 6) \div 2$

3. 14×10 **4.** $(75 - 21) + 8$

5. $4 \times (6 + 2)$

Vocabulary

order of operations

algebraic operating system (AOS)

Erin's older brother, Todd, was working on his homework. At the top of the page, he wrote, "**P**lease **E**xcuse **M**y **D**ear **A**unt **S**ally." Erin didn't understand. She said, "We don't have an aunt named Sally."

Todd said, "You'll see why I wrote this on my paper."

Activity 1

• Use paper and pencil to find the value of $23 + 12 \times (6 - 2)^2$.

• How does your answer compare with the answer of a classmate?

Math Idea ▶ When you find the value of an expression with more than one operation, you need to use the **order of operations**.

     1. Perform operations in parentheses.
     2. Clear exponents.
     3. Multiply and divide from left to right.
     4. Add and subtract from left to right.

• Use the order of operations to evaluate $(41 + 31) \div 2^2 - 8$.

Think and Discuss

• What do the underlined letters in "**P**lease **E**xcuse **M**y **D**ear **A**unt **S**ally" represent?

• List the order of operations you would use to find the value of $3^3 + 5 - 3 \times (21 \div 7)$. Explain why the order is important.

Practice

Tell the order in which you would perform the operations in each expression. Then find the value of the expression.

1. $(120 - 14) + 4^2 \times 3$

2. $3 + 5^2 \times 2 \div 10 - 4$

3. $9 \times 1 + 12 \times 2 \div 8 + 4$

4. $5 + (7 - 4)^2 - 8 \div 2$

5. $16 - 2^3 - (9 - 7) \times 4$

CALIFORNIA STANDARDS **AF 1.3** Apply algebraic order of operations and the commutative, associative, and distributive properties to evaluate expressions; and justify each step in the process. **AF 1.4** Solve problems manually by using the correct order of operations or by using a scientific calculator. *also,* **NS 2.0, MR 2.4, MR 2.5, MR 3.2**

You can use a calculator to evaluate expressions with more than one operation. Some calculators use an **algebraic operating system (AOS)** that automatically follows the order of operations.

Activity 2

- Use your calculator to find the value of $8 \div 2 + 6 \times 3 - 4$.

- Following the order of operations, use paper and pencil to find the value of $8 \div 2 + 6 \times 3 - 4$.

- Exchange papers with a classmate, and check each other's work.

Think and Discuss
- How does the calculator value for $8 \div 2 + 6 \times 3 - 4$ compare with the value you got by using paper and pencil? Does your calculator use an AOS?

To find the value of an expression with a calculator that does not use an AOS, follow the order of operations or use the memory keys.

Follow the order of operations to find the value of the expression $2 + 6 \times 3^2 - 4$.

3   6 2 4   52.

Use the memory keys to find the value of the expression $9^2 + 6 \div 2 \times 4$.

9   6 2 4   93.

- When you enter values into a calculator that does not have an AOS, how do you know which values to enter first?

TECHNOLOGY LINK

More Practice: Use E-Lab, *Order of Operations.*
www.harcourtschool.com/
elab2002

Practice
1. A calculator shows the display 9 as the value of $12 + 15 \div 3$. Does the calculator use an AOS? Explain.

2. How could you use memory keys to evaluate the expression $12 + 15 \div 3$?

3. How could you use the order of operations to evaluate the expression $12 + 15 \div 3$?

Use a calculator to find the value.

4. $12 + 8 \times 4^2$ 5. $9 + (6 - 2) \times 5$ 6. $18 \div (6 - 4) + 5$

MIXED REVIEW AND TEST PREP

7. Find the value for 6^3. (p. 40)

8. $34{,}056 + 2{,}207$ (p. 20)

9. Evaluate $6p - q$ for $p = 7$ and $q = 11$. (p. 28)

10. $807 \div 45$ (p. 22)

11. **TEST PREP** Write $5 \times 5 \times 5 \times 5$ in exponent form. (p. 40)

A 4^5 B 5^3 C 5^4 D 10^2

43

ALGEBRA
Order of Operations

Learn how to use the order of operations.

To evaluate an expression that contains more than one operation, you use rules called the order of operations. The first letters of these words help you remember the order of operations.

Please **E**xcuse **M**y **D**ear **A**unt **S**ally

Parentheses
Exponents
Multiplication or **D**ivision
Addition or **S**ubtraction

EXAMPLE 1

Tell the operations used to evaluate the expression.

$35 \div 7 + 5 \times 3^2$	*Clear exponents.*
$35 \div 7 + 5 \times 9$	*Divide.*
$5 + 5 \times 9$	*Multiply.*
$5 + 45$	*Add.*
50	

EXAMPLE 2

Find the value of the expression
$285 + 93 \div (3 - 2) \times 3 \times 4^2$.

$285 + 93 \div (3 - 2) \times 3 \times 4^2$	*Operate inside parentheses.*
$285 + 93 \div 1 \times 3 \times 4^2$	*Clear exponents.*
$285 + 93 \div 1 \times 3 \times 16$	*Divide.*
$285 + 93 \times 3 \times 16$	*Multiply twice.*
$285 + 4{,}464$	*Add.*
$4{,}749$	

CHECK FOR UNDERSTANDING

Think and Discuss ▶ **Look back at the lesson to answer each question.**

1. **Show** where to insert parentheses to make this equation true.
   $420 - 100 \div 40 = 8$

2. **Tell** which operation you would do last to evaluate the expression $7 + 8 - 2^3 \div 4$.

Guided Practice ▶ **Evaluate the expression.**

  3. $30 - 15 \div 3$ **4.** $5^2 - (40 \div 4) \div 2$ **5.** $5^2 + 10^2 \div 25 - 1$

CALIFORNIA STANDARDS **AF 1.3** Apply algebraic order of operations and the commutative, associative, and distributive properties to evaluate expressions; and justify each step in the process. **AF 1.4** Solve problems manually by using the correct order of operations or by using a scientific calculator. *also,* **NS 2.0, MR 1.0, MR 3.2**

Independent ▶ Practice

Evaluate the expression.

6. $45 \div 15 + 2 \times 3$

7. $7 \times 2^2 + 6 - 9$

8. $3 + 4 \times 250$

9. $12 + (36 \div 4)^2 - 25$

10. $2^6 - (27 - 8) + (5^2 - 21)$

11. $(43 + 57) \times (9 - 6)^0$

12. $4^4 - (5^3 - 7^2) + (3^3 - 25)^3$

13. $(6^2 + 3^2)^2 \div 5 \times 3 + 3$

14. $(24 + 1^8) \times (7 - 5)^3$

15. $(7 \times 4)^2 - (34 + 1^8) \times 2^3$

$\frac{a+b}{c}$ Algebra Evaluate the expression for $a = 4$ and $b = 7$.

16. $21 \div b + 8$

17. $a \times 31 - 8^2$

18. $(8 - a) \div 2 + 7$

19. $b^2 \div 7 \times (6 + 5)$

Problem Solving ▶ Applications

For 20–21, write and evaluate an expression to solve.

20. Heather mailed 3 packages that cost $2.50 each and 2 packages that cost $1.50 each. How much did she spend on postage?

21. Minh bought a watermelon for $3.25 and 3 cantaloupes for $1.29 each. He gave the clerk $20. How much change did he get in return?

22. **? What's the Error?** Joe and Brett found the value of $2 + 6 \times 3^2 - 4$. Joe said the answer is 68 and Brett said the answer is 52. Decide who made the error and describe what the error is.

23. The Island Theater holds 236 people. It was filled to capacity for each of its 43 shows last week. This week 8,299 people attended shows at the theater. How many people attended shows at the Island Theater during these two weeks?

24. What are the equal factors and the value for 7^3? (p. 40)

25. Use mental math to find the value. 34×6 (p. 36)

26. Subtract. $2,500 - 1,646$ (p. 20)

27. Divide. $163 \div 5$ (p. 22)

28. **TEST PREP** Ricky's math class collected a total of 1,364 pennies and quarters. They collected 234 more pennies than quarters. How much money did Ricky's class collect? (p. 26)

A $149.24

C $229.40

B $158.34

D $237.12

PROBLEM SOLVING SKILL
Sequence and Prioritize Information

Analyze
Choose
Solve
Check

Learn how to solve problems by sequencing and prioritizing information.

Mrs. Rucki gets paid on the first workday of each month. She deposits her check in the bank and uses the money to pay her bills. Of the money that's left she uses part for spending money and puts the rest into savings.

On March 1, the following occurred:

- Mrs. Rucki had bills of $740, $85, $102, $33, and $52 to pay.
- She received her monthly paycheck for $1,570.
- She wanted to save at least $300 from her check.
- She needed at least $250 spending money.

The sequence, or order, in which Mrs. Rucki does these things is important. For example, she cannot use the money until the bank gives her credit for her deposit. That may take three days.

Mrs. Rucki made the list at the left. The list shows the things to do and the order she planned to do them.

- Why does it make sense for Mrs. Rucki to write "Put the rest into my savings account" last?

Suppose Mrs. Rucki follows the sequence above. How much will she have left to put into her savings account?

$740 + $85 + $102 + $33 + $52 = $1,012 *Find the total of her bills.*

$1,570 - $1,012 = $558 *Find the amount left after she pays her bills.*

$558 - $250 = $308 *Find what is left after she withdraws spending money.*

So, Mrs. Rucki could put $308 into her savings account.

Math Idea ▶ The order in which parts of an activity are carried out is often important to success. Some parts of an activity may be more important than others.

 CALIFORNIA STANDARDS MR 1.0 Students make decisions about how to approach problems. **MR 1.1** Analyze problems by identifying relationships, distinguishing relevant from irrelevant information, identifying missing information, sequencing and prioritizing information, and observing patterns. also, **MR 2.0, MR 3.2**

PROBLEM SOLVING PRACTICE

For 1–3, use the schedule at the right. Each show is 50 minutes long.

1. Jennifer and her family will visit Ocean World Park from 9:30 to 4:00. They want to see the Water Skiing show at 10:00. Name the order in which they can see all the other shows.

2. The Jackson family wants to see the Underwater Acrobats at 12:00 and then take 45 minutes to eat lunch. What other show could they see before leaving the park at 3:00?

   A Aquarium Tour
   B Animal Acts
   C Whale Acts
   D Water Skiing

Ocean World Park Show Times	
9:00, 12:00	Underwater Acrobats
9:00, 3:00	Whale Acts
10:00, 2:00	Animal Acts
10:00, 1:00	Water Skiing
11:00, 4:00	Aquarium Tour

3. Michele wants to go on the Aquarium Tour at 11:00. Before she leaves the park, she wants to see the Animal Acts show at 2:00. At which other time could she see a show after the Aquarium Tour?

   F 9:00 **H** 1:00
   G 10:00 **J** 2:00

MIXED APPLICATIONS

4. Ed wants to buy 3 boxes of cereal at $2.99 each, 2 melons at $1.29 each, and 5 cans of juice at $0.79 each. Ed has $12.80. How much more money does he need to buy all of the items?

5. At the Discount Book Barn, books cost $5.00 and magazines cost $1.50. José bought some books and magazines for $21. How many of each did he buy?

6. **Use Data** Use the graph below. The highest waterfall in the world is Angel Falls in Venezuela. About how much higher is Angel Falls than Ribbon Falls?

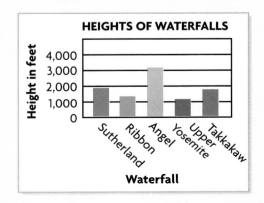

7. Tim has saved $1,475 to make a down payment on a new car. He wants to save a total of three times that amount plus $275. How much money does Tim want to save for a down payment?

8. Ron and his two cousins own videotapes. Ron has twice as many as his older cousin, who has 27 videos. Ron has three times as many as his younger cousin. How many videotapes do they have in all?

9. Harry hammered nails into a board to make a circular pegboard. The nails were the same distance apart, and the sixth nail was directly opposite the eighteenth nail. How many nails formed the circle?

10. ✍ **Write About It** Make a list of the steps you would follow to make a scrambled egg.

1. **VOCABULARY**   In the expression 2^3, the number 2 is the ___?___ . (p. 40)

2. **VOCABULARY**   To find the value of an expression that has more than one operation, you need to use the ___?___ . (p. 42)

3. **VOCABULARY**   In the expression 8^4, the number 4 is the ___?___ . (p. 40)

Use mental math to find the value. (pp. 36–39)

4. $19 + 43$

5. $76 - 37$

6. $32 + (48 + 83)$

7. 26×12

8. $5,986 \times 1$

9. $4 \times 6 \times 25$

Write the equal factors. Then find the value. (pp. 40–41)

10. 6^2

11. 9^2

12. 3^5

13. 5^4

14. 2^0

15. 10^6

16. 7^5

17. 25^1

18. 4^3

19. 12^2

Evaluate the expression. (pp. 44–45)

20. $4 \times 5 - 6 \times 3$

21. $(12 \times 7) + 9^2$

22. $13 + 4 \times (20 + 35)$

23. $4^3 - 4 \times 12$

24. $36 \div (24 - 18) + 9$

25. $45 - (12 \times 3) \div 6$

26. $16 \times 9 \div 2^3$

27. $(100 - 28) \div 3^2$

28. $43 + (6^2 - 3^3)^2$

29. $(19 + 9^2)^2 \div 5^2 + 94$

Use Data   For 30–31, use the table at the right.

30. Michael wants to surprise his family by having dinner ready when they get home. The table shows the length of time each of the foods needs to cook. If Michael wants to have everything ready to eat at the same time, in what order should he start cooking the food? (pp. 46–47)

31. Suppose Michael decides to cook baked potatoes instead of rice. The potatoes take 75 minutes to bake. In what order should he start cooking the food? (pp. 46–47)

32. Doreen has $40.00. She wants to buy 3 pairs of socks for $2.75 each, gloves for $9.99, and 2 T-shirts for $8.79 each. How much money will she have left? (pp. 42–43)

33. Each of three pyramids of Egypt is made up of about 2.5 million large stones. Is the total number of stones greater than or less than 10^8? (pp. 40–41)

Cooking Times for Food

Type of Food	Cooking Time
Chicken	55 Minutes
Peas	5 Minutes
Rice	20 Minutes
Stuffing	15 Minutes
Dinner Rolls	12 Minutes

TIP!

Look for important words.
See item **5**.

An important word is **not**. The words **not** true mean you need to find the one expression among the answer choices that is **not** an equality.

Also see problem **2**, p. H62.

Choose the best answer.

1. Which can be expressed as 3^5?

   A $3 \times 3 \times 3 \times 3 \times 3$

   B $5 \times 5 \times 5$

   C 5×3

   D 3×3

2. Lori has 79 stickers. This is three more than twice as many as Beth has. How many stickers does Beth have?

   F 34 **H** 38

   G 36 **J** 40

3. Calvin spent $43 on a video game and a CD. The video game cost $13 more than the CD. How much did the video game cost?

   A $13 **C** $30

   B $28 **D** $56

4. Kim's class has been saving pennies to go on a field trip to the Gem and Mineral Museum. The class has collected 10,000 pennies. How is 10,000 written using an exponent with the base 10?

   F 1^4 **H** 10^4

   G 10^2 **J** 10^5

5. Which number sentence is **not** true?

   A $30 \times 12 \times 8 = 12 \times 30 \times 8$

   B $30 + 8 + 12 = 30 + 12 + 8$

   C $30 - 12 + 8 = 30 + 8 - 12$

   D $30 + 8 \times 12 = 30 \times 12 + 8$

6. Which expression is equivalent to $(51 + 32) + 71$?

   F $(51 + 71) + (32 + 71)$

   G $51 + (32 + 71)$

   H $93 + 51$

   J $19 + 71$

7. Which expression is equivalent to $42 + 76$?

   A $40 + 78$ **C** $42 + 72$

   B $40 + 79$ **D** $142 + 67$

8. Which expression is equivalent to $(3^2 + 7) \div (4 - 2)$?

   F $(3 + 7)^2 - 2$

   G $(15 + 1) \div 2$

   H $(3 \times 5)^2 \div 15$

   J $100 - (4 \div 2)$

9. What is the value of $42 - d$ for $d = 6$?

   A 7 **C** 48

   B 36 **D** Not here

10. Which algebraic expression represents the expression "13 less than a number, t"?

    F $13 \times t$ **H** $13 - t$

    G $t + 13$ **J** $t - 13$

11. Which is the best estimate for this sum?
    $$15{,}192 + 3{,}751 + 1{,}551$$

    A 18,000 **C** 20,000

    B 19,000 **D** 21,000

12. Which is another way to write $y \times y \times y$?

    F $3 \times y$ **H** y^3

    G $y \times 3y$ **J** y^4

CHAPTER 3

Decimal Concepts

In 1990 the United States Food and Drug Administration required food manufacturers to put complete nutrition labels on all foods. Possible foods for a trail mix snack are listed in the table along with their nutritional value (Daily Value). To make a low-fat trail mix, which cereal would be better to use? Use > or < to compare the fat content.

Trail Mix Item	Serving Size	Calories	Daily Value		
			Fat	Fiber	Carbohydrate
Peanuts	3 tbsp	240	0.30 or 30%	0.12 or 12%	0.03 or 3%
Sunflower Seeds	3 tbsp	170	0.22 or 22%	0.08 or 8%	0.08 or 8%
Raisins	$\frac{1}{4}$ cup	130	0	0.09 or 9%	0.10 or 10%
Dried Cranberries	$\frac{1}{4}$ cup	110	0.01 or 1%	0.08 or 8%	0.09 or 9%
Toasted Oat Cereal	1 cup	110	0.03 or 3%	0.11 or 11%	0.09 or 9%
Bran Flakes Cereal	$\frac{1}{2}$ cup	60	0.01 or 1%	0.52 or 52%	0.08 or 8%

Check What You Know

Use this page to help you review and remember important skills needed for Chapter 3.

✓ Represent Decimals (See p. H2.)

Write the decimal that is modeled.

1.
2.
3.
4.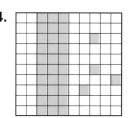

✓ Write and Read Decimals (See p. H3.)

Write the decimal.

5. 836 and 23 hundredths
6. 364 and 2 tenths
7. 93 thousand, 450 and 38 hundredths
8. 595 thousand, 821 and 9 tenths
9. 306 thousand, 7 and 6 hundredths

Complete to show how to read the numbers.

10. 1,463.05 1 thousand, 463 and __?__
11. 204.7 204 and __?__ tenths
12. 32,617.45 32 __?__ , 617 and 45 __?__
13. 4,382.1 4 thousand, 382 and __?__

✓ Compare Whole Numbers (See p. H3.)

Compare the numbers. Write <, >, or = for ●.

14. 143 ● 140
15. 808 ● 880
16. 716 ● 716
17. 691 ● 961
18. 94 ● 49
19. 405 ● 305
20. 383 ● 383
21. 5,937 ● 397
22. 4,062 ● 4,206
23. 689 ● 648
24. 6,098 ● 6,908
25. 4,801 ● 4,108

✓ Round Decimals (See p. H4.)

Round to the nearest whole number.

26. 3.64
27. 1.49
28. 6.938
29. 41.8
30. 18.70
31. 72.06

Round to the nearest tenth.

32. 69.64
33. 26.37
34. 52.489
35. 8.630
36. 9.479
37. 14.507
38. 90.63
39. 55.58
40. 26.397

LOOK AHEAD

In Chapter 3 you will
- represent, compare, and order decimals
- estimate with decimals
- use decimals and percents

Represent, Compare, and Order Decimals

Learn how to use place value to express, compare, and order decimals.

Remember that when reading a number with a decimal point, read the decimal point as "and." Read 8.2 as "eight and two tenths."

Genna read that it costs $0.03 to use her hair dryer for 30 minutes, $0.08 to use her clock for a month, and $0.53 to wash and dry a load of laundry.

Place value helps you understand numbers. The digits and the position of each digit determine a number's value. Read each number on the place-value chart. These numbers are part of the decimal system. Notice that 3 has a value of 3 thousandths, 3 tens, or 3 ten-thousandths, depending on its position in the number.

	Ten Thousands	Thousands	Hundreds	Tens	Ones	Tenths	Hundredths	Thousandths	Ten-Thousandths
PLACE VALUE									
0.053					0	0	5	3	
32.4				3	2	4			
8.0023					8	0	0	2	3

When you read and write numbers, you are using place value.

EXAMPLE 1

A. Standard form: 0.053
   Expanded form: 0.05 + 0.003
   Word form: *fifty-three thousandths*

B. Standard form: 32.4
   Expanded form: 30 + 2 + 0.4
   Word form: *thirty-two and four tenths*

C. Standard form: 8.0023
   Expanded form: 8 + 0.002 + 0.0003
   Word form: *eight and twenty-three ten-thousandths*

Math Idea ▶ Knowing the place value of digits will help you read, write, and calculate numbers correctly, including decimal numbers.

CALIFORNIA STANDARDS O—n NS 1.0 Students compare and order positive and negative fractions, decimals, and mixed numbers. Students solve problems involving fractions, ratios, proportions, and percentages. O—n NS 1.1 Compare and order positive and negative fractions, decimals, and mixed numbers and place them on a number line. *also* MR 2.5, MR 3.2

Mark notices that one jar of cinnamon contains 2.6 oz and another contains 2.3 oz. He wants to buy the jar with the greater amount of cinnamon.

You can use a number line to compare 2.6 and 2.3.

2.1	2.2	2.3	2.4	2.5	2.6	2.7	2.8

TECHNOLOGY LINK

More Practice: Use **Mighty Math Calculating Crew**, *Nautical Number Line*, Levels O and P.

Since 2.6 is to the right of 2.3 on the number line, 2.6 is greater than 2.3.	Since 2.3 is to the left of 2.6 on the number line, 2.3 is less than 2.6.
2.6 > 2.3 ↓ greater than	2.3 < 2.6 ↓ less than

So, the 2.6-oz jar has more cinnamon.

You can also use place value to compare decimal numbers.

EXAMPLE 2

Compare 7.28 and 7.2. Use < or >.

Start at the left.

7.**28**   7.**2**   *Compare the ones digits. They are the same.*

7.2**8**   7.**2**   *Compare the tenths digits. They are the same.*

7.2**8**   7.2**0**   *Add a zero so both numbers have the same number of places. Compare the hundredths digits. 8 is greater than 0.*

So, 7.28 > 7.2, and 7.2 < 7.28.

- Which is greater, 7.2 or 7.08? Explain.

Remember that you can add a zero to the right of a decimal without changing its value.
7.2 = 7.20
2 tenths is the same as 20 hundredths.

You can use place value to order two or more decimal numbers.

EXAMPLE 3

The prices for the same kind of CD player in four different stores are $132.95, $132.50, $130.25, and $135.25. Order the prices of the CD players from least to greatest.

Compare two numbers at a time.

$132.95 > $132.50 $132.95 > $130.25 $132.95 < $135.25

$132.50 > $130.25 $132.50 < $135.25 $130.25 < $135.25

So, the list of numbers in order from least to greatest is $130.25, $132.50, $132.95, $135.25.

- List 9.365, 9.271, 9.356, and 9.065 in order from greatest to least.

Think and ▶ Discuss

Look back at the lesson to answer each question.

1. **Name** the number that is 4 hundredths greater than 2.0369.

2. **Tell** the place immediately to the right of hundred-thousandths.

Guided ▶ Practice

Read the number. Write the value of the blue digit.

3. 15,425.007   4. 2,654,000.25   5. 550.76

Write the number in expanded form.

6. 0.605   7. 0.00103   8. 12.0089   9. 342.046

Compare the numbers. Write <, >, or = for ●.

10. 1.15 ● 1.14   11. 92.3 ● 92.30   12. 0.82 ● 0.84

Write the numbers in order from least to greatest.

13. 1.361, 1.351, 1.363   14. 125.3, 124.32, 125.33

PRACTICE AND PROBLEM SOLVING

Independent ▶ Practice

Read the number. Write the value of the blue digit.

15. 5.0547   16. 827.142   17. 345.79456

Write the number in expanded form.

18. 46.00105   19. 0.0362   20. 1,500.1   21. 2.456

Compare the numbers. Write <, >, or = for ●.

22. 99.06 ● 99.6   23. 133.2 ● 133.32   24. 707.07 ● 707.07

25. 32.630 ● 32.63   26. 457.3685 ● 457.5683   27. 49.302 ● 49.203

28. 1 + 0.1 + 0.05 ● 1 + 0.1 + 0.04

29. 5 + 0.2 + 0.003 ● 5 + 0.3 + 0.02

Write the numbers in order from least to greatest.

30. 1.41, 1.21, 1.412, 1.12   31. 1.45, 1.05, 0.405, 1.25, 1.125

32. 35.2, 35.72, 35.171, 35.7   33. 9.82, 9.082, 8.91, 9.285, 9.85

Write the numbers in order from greatest to least.

34. 5.004, 5.040, 5.4   35. 125.33, 125.3, 125.35, 125.4

36. $3\frac{1}{10}$, 3.001, $3\frac{1}{100}$, 3   37. 14.01, $14\frac{1}{10}$, 41.01, $14\frac{3}{100}$

**Problem Solving ▶
Applications**

38. The world's smallest cut diamond is 0.0009 inch in diameter and weighs 0.0012 carat. Write both numbers in words.

39. A movie studio announced that the box office sales for its new release reached nine million, four hundred fifty-six thousand, three hundred two dollars in the first week. Write ten times that amount in standard and word forms.

40. **?** **What's the Error?** Jon says 8.01 and 8.10 are equal because each number has the same digits. Explain his error.

Use Data For 41–42, use the graph at the right.

41. Estimate the total amount of rainfall during one year in Seattle.

42. Gene would like to visit Seattle to do some outdoor sightseeing. He would like to visit when the normal amount of rainfall is below 2 inches. When might be the best time for Gene to visit?

MONTHLY RAINFALL FOR SEATTLE, WASHINGTON

Rainfall (in inches) — Months: Jan Feb Mar Apr May Jun Jul Aug Sep Oct Nov Dec

---

MIXED REVIEW AND TEST PREP

43. $12 + 3^3 \times 8$ (p. 42) **44.** 234×38 (p. 22) **45.** Evaluate $406 \div c$ for $c = 14$. (p. 28)

46. TEST PREP Which shows the value of 4^4? (p. 40)

A 16 **B** 124 **C** 256 **D** 2,414

47. TEST PREP If you do homework for 35 to 45 minutes a day, which is a reasonable estimate of the number of minutes you do homework for 8 days? (p. 16)

F less than 200 min **H** between 500 and 700 min
G between 300 and 500 min **J** more than 700 min

Thinker's CORNER

Spin a Decimal Practice comparing and ordering decimals as you play this game.

Materials: a spinner numbered 0–9, a place-value chart

- In a small group, decide whether the player with the greatest decimal or least decimal will win. Taking turns, spin the pointer six times. After each spin, write the digit on your place-value chart. Once you have written it on your chart, the digit cannot be moved or erased.

- Take turns reading your six-digit numbers aloud. The player with the greatest or the least decimal wins the round and receives a point.

- Continue playing until a player has five points.

Extra Practice page H34, Set A

PROBLEM SOLVING STRATEGY
Make a Table

Analyze
Choose
Solve
Check

Learn how to solve problems by organizing data in a table.

QUICK REVIEW

Order the numbers from greatest to least.

1. 263, 224, 276, 265

2. 364, 356, 375, 325

3. 8,658; 8,586; 8,459; 8,903

4. 2,961; 2,906; 2,609; 2,169

5. 3,764; 3,456; 3,765; 3,425

Andrew chooses new in-line skates from the models below.

RC-204 $99.95; PRX-100 $78.50; D-500 $99.99; OP-1000 $78.99; ZZ-2 $91.50; ZA-45 $99.25

Andrew wants to buy the second most expensive model. His parents want him to buy the second least expensive model. Which models are these? What is the difference in price of the two models?

Analyze

What are you asked to find?

What information is given?

Choose

What strategy will you use?

You can use the strategy *make a table* to show the prices of the in-line skates in order from least to greatest.

Solve

How will you solve the problem?

Compare the prices and order them in a table. Then find the second most expensive model, the second least expensive model, and the difference in their prices.

$78.50 < $78.99 < $91.50 < $99.25 < $99.95 < $99.99

MODEL	PRICE	
PRX-100	$78.50	
OP-1000	$78.99	← *second least expensive*
ZZ-2	$91.50	
ZA-45	$99.25	
RC-204	$99.95	← *second most expensive*
D-500	$99.99	

Subtract: $99.95 − $78.99 = $20.96

So, the difference in price of models RC-204 and OP-1000 is $20.96.

Check

How can you check your answer?

What if Andrew chose model ZZ-2? How much more would he spend than if he bought model OP-1000?

CALIFORNIA STANDARDS **MR 2.0** Students calculate and solve problems involving addition, subtraction, multiplication, and division. **MR 3.2** Note the method of deriving the solution and demonstrate a conceptual understanding of the derivation by solving similar problems. *also* O━┓ NS 1.0, MR 1.3

PROBLEM SOLVING PRACTICE

PROBLEM SOLVING STRATEGIES

Draw a Diagram or Picture
Make a Model
Predict and Test
Work Backward
Make an Organized List
Find a Pattern
► **Make a Table or Graph**
Solve a Simpler Problem
Write an Equation
Use Logical Reasoning

Solve the problem by making a table.

Below are the fractions of games won by 8 baseball teams.

Hawks	0.650	Bulldogs	0.725
Tigers	0.750	Lions	0.490
Angels	0.675	Flames	0.700
Dolphins	0.550	Giants	0.695

1. Which team is in second place?

  A Tigers   **B** Flames   **C** Bulldogs   **D** Angels

2. How many teams are behind the Giants?

  F 2   **G** 3   **H** 4   **J** 5

Use Data For 3–4, reorder the data in the table at the right from greatest to least.

3. Which appliance uses the greatest amount of electricity?

4. Which appliance uses the least amount of electricity?

Electricity Used By Appliances

Appliance	Electricity (In Kilowatts)
Refrigerator	0.6
Air conditioner	1.5
Color TV	0.33
Iron	1.2
Coffeepot	0.9
Toaster	1.2

MIXED STRATEGY PRACTICE

5. A calculator, pen, and notebook cost $14.00 altogether. The calculator costs $9.00 more than the pen and $8.50 more than the notebook. How much does each item cost?

6. Kelly left the house with $16.00. She had $4.50 left after buying a movie ticket for $6.75, buying two snacks for $1.75 each, and paying for a bus ride. How much did she pay for the bus ride?

7. Carlene makes greeting cards. It costs $0.20 to make each card. She then sells them for $0.75 each. How many cards does she need to sell in order to make a profit of $33.00?

8. Frank walks 5 blocks to school for every 3 blocks Robert walks. They walk a total of 24 blocks. How many blocks from school does Robert live?

9. In a survey, teens prefer the Boomer portable stereo over the Blaster, but not as much as the Soundmaster. The Tekesound was preferred above all the others. Which portable stereo was least preferred?

10. **? What's the Question?** Glen read 69 pages each day for 6 days. He then read 23 pages each day for 4 days. The answer is about 500.

Estimate with Decimals

Learn how to estimate decimal sums, differences, products, and quotients.

QUICK REVIEW

Round each number to the nearest 100.

1. 346   **2.** 2,506   **3.** 54,067

4. 30,463   **5.** 156,556

You can estimate sums, differences, products, and quotients of decimals. To estimate with decimal numbers, use the same methods you used to estimate with whole numbers.

EXAMPLE 1

A long-distance phone company charges the rates shown at the right for calls from the United States. DeAnn made one-minute calls to India, Jordan, and Pakistan. About how much did the three calls cost?

Estimate $1.79 + $1.87 + $2.17.

WIRED WORLD PHONE COMPANY	
Country	**Rate per Minute**
Argentina	$0.39
China	$0.49
France	$0.13
Germany	$0.09
India	$1.79
Ireland	$0.15
Jordan	$1.87
Pakistan	$2.17

$1.79   *The three addends*
$1.87   *cluster around $2.00.*
+$2.17   *So, multiply $2.00 by 3.*

3 × $2.00 = $6.00   *Multiply.*

So, the three calls cost about $6.00.

EXAMPLE 2

Estimate.

A. 36.4 × 18.25

Round to the nearest ten.

$$\begin{array}{r} 36.4 \\ \times 18.25 \end{array} \rightarrow \begin{array}{r} 40 \\ \times 20 \\ \hline 800 \end{array}$$

So, 36.4 × 18.25 ≈ 800.

Remember that the symbol ≈ means "is approximately equal to."

B. 162.8 ÷ 8.16

Use compatible numbers.

$$8.16\overline{)162.8} \rightarrow 8\overline{)160}^{\,20}$$

So, 162.8 ÷ 8.16 ≈ 20.

CHECK FOR UNDERSTANDING

Think and ▶ Discuss

Look back at the lesson to answer each question.

1. Tell how you could estimate the sum of 4.79, 18.99, and 3.09.

2. Explain how to use compatible numbers to estimate 423.2 ÷ 2.7.

CALIFORNIA STANDARDS O—n NS 2.0 Students calculate and solve problems involving addition, subtraction, multiplication, and division. **MR 2.0** Students use strategies, skills, and concepts in finding solutions. *also* **MR 2.4, MR 2.5**

Estimate.

3. $18.7 + 23.1$ **4.** $123.76 \div 9$ **5.** $185.32 - 101.99$

6. 39.83×36 **7.** $67.8 + 66.1 + 71.7$ **8.** 817.3×11

PRACTICE AND PROBLEM SOLVING

**Independent ▶
Practice**

Estimate.

9. $6.7 + 9.4 + 15.82$ **10.** 12.2×8.3 **11.** $82.5 \div 9.3$

12. $\$266.08 - \$97.30$ **13.** $9.8 + 38.2$ **14.** 31.5×2.8

15. 6.8×18.2 **16.** $103.08 - 45.32$ **17.** 56.20×30.7

18.
$$\begin{array}{r} 689.89 \\ -\ \ 98.5 \\ \hline \end{array}$$

19.
$$\begin{array}{r} 1{,}038.54 \\ \times\ \ \ 26.12 \\ \hline \end{array}$$

20. $234.91 \div 5.79$

21.
$$\begin{array}{r} \$7{,}805.90 \\ +\ 9{,}158.43 \\ \hline \end{array}$$

22. $81.5 \times 23.1 \div 3.9$ **23.** $18.2 \times (7.2 - 4.9)$

Estimate to compare. Write < or > for each ●.

24. 4.32×8.56 ● 40 **25.** 25 ● $81.27 \div 4.1$ **26.** $34.6 - 12.4$ ● 14

**Problem Solving ▶
Applications**

Use Data   **For 27–29, the table shows the types of waste that make up a typical 100 pounds of garbage in the United States.**

27. About how many pounds of newspapers are included in every 300 pounds of garbage thrown away?

28. About how many more pounds of food waste than glass are thrown away for every 500 pounds of garbage?

What We Throw Away	
Type of Garbage	**Weight (in lb)**
Food Waste	8.6
Yard Waste	17.4
Newspapers	7.9
Other Paper	30.8
Metals	9.2
Glass	5.2
Other	20.9

29. ✎ **Write a problem** involving estimation. Use the data about yard waste, other paper, and metals, shown in the table.

MIXED REVIEW AND TEST PREP

30. Write the value of the digit 8 in the number 342.285. (p. 52)

31. Write 2.523, 2.325, 2.532, and 2.235 in order from least to greatest. (p. 52)

32. $20{,}817 - 19{,}805$ (p. 20)

33. $25{,}801 \div 23$ (p. 22)

34. TEST PREP Evaluate $a \times 32$ for $a = 426$. (p. 28)

     A 472   **B** 2,130   **C** 13,522   **D** 13,632

Decimals and Percents

Learn how to write a decimal as a percent and a percent as a decimal.

QUICK REVIEW

1. 2,457 + 4,541   2. 3,470 − 350   3. 6 × 5

4. 50 ÷ 5   5. Write the decimal for $\frac{2}{100}$.

Vocabulary

percent

The graph at the right shows the responses to a question about what people in the United States like to eat for breakfast.

Percent means "per hundred" or "hundredths." The symbol used to write a percent is %.

40 percent: $40\% = \frac{40}{100}$

So, 40 out of 100 have toast or a roll.

26 percent: $26\% = \frac{26}{100}$

So, 26 out of 100 have cold cereal.

BREAKFAST FOODS

Activity

You need: two 10 × 10 grids (decimal squares)

- On one grid, shade complete squares to make the first letter of your first name. On the other grid, shade complete squares to make the first letter of your last name. Make each letter as large as possible. Some examples are shown below.

- Since 30 out of 100 squares are shaded for the letter *Q*, you can write 0.30 or 30%. What decimals and percents can be written for the squares shaded for *X* and *Z*?

- Count the number of complete squares you shaded on each of your grids. What percent of the squares are shaded?

Remember that when you read a decimal, you name the place with the least value. For 0.93, the place with the least value is hundredths. The number is read as "93 hundredths."

You can think about place value when you change decimals to percents or percents to decimals.

EXAMPLES

A. Write 0.08 as a percent.
0.08 is 8 hundredths.
So, 0.08 = 8%.

B. Write 32% as a decimal.
32% is 32 hundredths.
So, 32% = 0.32.

CALIFORNIA STANDARDS ○—¬NS 1.0 Students compare and order positive and negative fractions, decimals, and mixed numbers. Students solve problems involving fractions, ratios, proportions, and percentages. *also* MR 1.1, MR 2.4

Think and ▶
Discuss

Look back at the lesson to answer the question.

1. **Discuss** how you know that Sharon has 18 red cars if 18% of her 100 model cars are red.

Guided ▶
Practice

Write the decimal and percent for the shaded part.

2.   3.   4.

Write the corresponding percent or decimal.

5. 70%   6. 0.20   7. 0.03   8. 84%   9. 50%

Independent ▶
Practice

Write the decimal and percent for the shaded part.

10.   11.   12.

Favorite Types of Music

Alternative	21
Rhythm and Blues	14
Rap	21
Rock	20
Other or none	24

Write the corresponding percent or decimal.

13. 62%   14. 0.05   15. 28%   16. 45%   17. 53%

18. 0.63   19. 0.85   20. 33%   21. 0.4   22. 7%

Problem Solving ▶
Applications

23. **Use Data** The table shows how 100 teens responded to a survey. Write a decimal and a percent to show the number of teens who did not choose Alternative music.

24. **Write About It** Explain how to write 0.6 as a percent.

MIXED REVIEW AND TEST PREP

25. Order 27.8, 27.5, 27.82 from least to greatest. (p. 52)

26. Evaluate. $35 + 17 \times 3^2 - 16$ (p. 44)   27. 168×92 (p. 22)   28. $3,470 \div 42$ (p. 22)

29. **TEST PREP** Which shows the sum $34,904 + 15,456 + 6,943$? (p. 20)

  A 55,920   **B** 56,920   **C** 68,870   **D** 72,780

1. **VOCABULARY** A word that means "per hundred" is ___?___ . (p. 60)

Write the value of the blue digit. (pp. 52–55)

2. 3.2497 3. 14.5805 4. 0.09003

5. 628.0402 6. 1.81738 7. 78.05124

Write the numbers in order from least to greatest. (pp. 52–55)

8. 2.365, 2.305, 2.3, 2.35, 2.035 9. 125.3, 124.32, 125.33, 12.245, 120.4

Estimate. (pp. 58–59)

10. 27.6 + 135.2 11. 4.8 × 2.3 12. 30.7 − 6.25 13. 89.75 ÷ 8

14. 8.45 + 8.99 + 9.2 15. 219.48 − 107.43 16. 416.2 × 31 17. 40.02 ÷ 6.3

Write the decimal and percent for the shaded part. (pp. 60–61)

18.   19.   20.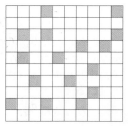

Write the corresponding percent or decimal. (pp. 60–61)

21. 74% 22. 0.07 23. 39% 24. 0.61

25. 0.6 26. 3% 27. 0.04 28. 84%

Solve.

29. The county library charges a fine of $0.10 a day for overdue books. The university library charges a fine of $0.50 for the first day and $0.05 for each additional day. On what day would overdue books have the same fine at both libraries? (pp. 56–57)

30. Planes leave Sacramento for San Francisco every 45 min. The first plane leaves at 5:45 A.M. What is the departure time closest to 4:30 P.M.? (pp. 56–57)

31. Donna is on the decoration committee. She spent $15.90 on streamers, $12.15 on balloons, $6.84 on tape, $19.98 on banner paper, and $13.22 on banner paint. What is a reasonable estimate of the amount she spent? (pp. 58–59)

32. Frank earns $6.25 per hour. One week he worked 16 hours. About how much did Frank earn that week? (pp. 58–59)

33. Kirk has to list 125.3, 124.32, 125.33, 12.345, 120.4 in order from greatest to least. Which number should he list third? (pp. 52–55)

Get the information you need.

See item **8**.

Recall that a percent is the ratio of a number to 100. Write the ratio as a decimal.

Also see problem **3**, p. H63.

Choose the best answer.

1. Which of these numbers rounds to 450 when rounded to the nearest ten and to 500 when rounded to the nearest hundred?

   A 415   **C** 452

   B 428   **D** 478

2. Earth's orbit is more than 100,000,000 kilometers from the sun. How is this number written in exponential notation?

   F 10^7 km   **H** 10^9 km

   G 10^8 km   **J** 10^{10} km

3. What is the value of $2 + 3 \times 4$?

   A 11   **C** 20

   B 14   **D** Not here

4. Each of the 42 members of the band contributed $3.75 toward a gift for the band director. Which is a reasonable estimate of the total amount collected?

   F Between $80 and $90

   G Between $90 and $110

   H About $120

   J About $160

5. Kate has 4 bags of birdseed that have masses of 2.5 kilograms, 1.25 kilograms, 1.9 kilograms, and 2.15 kilograms. Which shows the bags in order from least mass to greatest?

   A 1.9 kg, 2.15 kg, 2.5 kg, 1.25 kg

   B 2.5 kg, 2.15 kg, 1.9 kg, 1.25 kg

   C 2.15 kg, 1.25 kg, 2.5 kg, 1.9 kg

   D 1.25 kg, 1.9 kg, 2.15 kg, 2.5 kg

6. A restaurant manager bought a total of 73 apples and oranges. She bought 11 more oranges than apples. How many of each kind of fruit did she buy?

   F 31 apples and 42 oranges

   G 27 apples and 46 oranges

   H 36 apples and 37 oranges

   J 42 apples and 31 oranges

7. With tax, a CD player costs $77.75. Ken saves $4.85 each week. Which is a reasonable estimate of the number of weeks he must save for the CD player?

   A More than 20 weeks

   B Between 18 and 20 weeks

   C Between 15 and 17 weeks

   D Less than 15 weeks

8. How is 6% written as a decimal?

   F 6.0   **H** 0.06

   G 0.6   **J** 0.006

9. How is 0.36 written as a percent?

   A 0.36%   **C** 36%

   B 3.6%   **D** 360%

10. Which is greater than 16.30?

    F 16.45   **H** 16.03

    G 16.23   **J** 1.730

11. Which is the value of $(3 - 2) \times 5 + 6^2$?

    A 29   **C** 41

    B 40   **D** 122

12. Which of these is in order from least to greatest?

    F 0.0310, 0.301, 0.310

    G 0.310, 0.0310, 0.301

    H 0.301, 0.0310, 0.310

    J 0.0310, 0.310, 0.301

Decimal Operations

Today's computers use optical laser writing technology to store billions of bytes of information. A Jaz® disk holds 1 gigabyte of information. A gigabyte is about 1,000,000,000 (10^9) bytes. Sometimes capacity is given in megabytes. A megabyte is about 1,000,000 (10^6) bytes. A 3.5-inch floppy disk holds just 1.44 megabytes of information. About how many bytes is that?

STORAGE CAPACITY IN GIGABYTES

Type of Disk: Zip®, CD-ROM, Jaz®, DVD-RAM

Number of Gigabytes: 0 1 2 3 4 5 6

Check What You Know

Use this page to help you review and remember important skills needed for Chapter 4.

 Whole-Number Operations (See p. H4.)

Add or subtract.

1. $7 + 28 + 12$ **2.** $45 - 15$ **3.** $63 - 19$ **4.** $19 + 41 + 27 + 23$

5. $34 - 17 - 7$ **6.** $27 + 56 + 100$ **7.** $143 + 79$ **8.** $213 - 88$

Multiply.

9. $\begin{array}{r} 63 \\ \times\ 4 \\ \hline \end{array}$ **10.** $\begin{array}{r} 49 \\ \times\ 9 \\ \hline \end{array}$ **11.** $\begin{array}{r} 19 \\ \times 76 \\ \hline \end{array}$ **12.** $\begin{array}{r} 88 \\ \times 32 \\ \hline \end{array}$

13. $\begin{array}{r} 80 \\ \times 50 \\ \hline \end{array}$ **14.** $\begin{array}{r} 75 \\ \times 11 \\ \hline \end{array}$ **15.** $\begin{array}{r} 200 \\ \times\ 15 \\ \hline \end{array}$ **16.** $\begin{array}{r} 340 \\ \times\ 20 \\ \hline \end{array}$

Divide.

17. $4\overline{)96}$ **18.** $5\overline{)127}$ **19.** $9\overline{)423}$ **20.** $7\overline{)760}$

21. $32\overline{)448}$ **22.** $20\overline{)3,660}$ **23.** $37\overline{)1,073}$ **24.** $23\overline{)4,715}$

 Multiply Decimals by 10, 100, and 1,000 (See p. H13.)

Multiply.

25. 4.3×10 **26.** $9.61 \times 1,000$ **27.** 8.4×100

28. $25.397 \times 1,000$ **29.** 194.05×100 **30.** 408.08×10

✔ **Remainders** (See p. H5.)

Divide. Write the remainder as a decimal.

31. $4\overline{)35}$ **32.** $5\overline{)56}$ **33.** $8\overline{)100}$

34. $12\overline{)243}$ **35.** $15\overline{)2,412}$

Divide. Write the remainder as a fraction.

36. $6\overline{)45}$ **37.** $8\overline{)77}$ **38.** $12\overline{)134}$

39. $14\overline{)550}$ **40.** $18\overline{)459}$

> ### LOOK AHEAD
>
> **In Chapter 4 you will**
> - add, subtract, multiply, and divide decimals
> - evaluate decimal expressions and solve equations

Add and Subtract Decimals

Learn how to add and subtract decimals.

0.17 m
0.60 m
1.40 m
1.26 m
2.00 m

The tallest known mammal was a 6.1-m giraffe named George. Born in Kenya, George spent most of his life in the Chester Zoo in England.

To find the height of the giraffe shown, you must add five partial heights.

Math Idea ▶ When you add or subtract decimals, align the decimal points first and then add or subtact the digits, one place at a time.

2.00	*Align the decimal points.*
1.26	
1.40	
0.60	
+0.17	*Place the decimal point.*
5.43	*Then add.*

So, the total height is 5.43 m.

You can use estimation to check for reasonableness.

EXAMPLE 1

Jamie has $85.75. Running shoes cost $68.45. How much money will Jamie have left after buying the running shoes?

Estimate.

$85.75   → $86   *Round to the nearest dollar.*
− 68.45   − 68
                  $18

Find the answer.

$85.75
− 68.45   *Align the decimal points.*
$17.30   *Place the decimal point. Then subtract.*

Use your estimate to check the reasonableness of the answer. Compare the two. Since $17.30 is close to the estimate of $18, the answer is reasonable.

So, Jamie will have $17.30 left.

CALIFORNIA STANDARDS O—NS 2.0 Students calculate and solve problems involving addition, subtraction, multiplication, and division. **MR 2.1** Use estimation to verify the reasonableness of calculated results. *also* **MR 2.4**

Some decimal numbers that you add or subtract do not have the same number of decimal places.

The samples below show how Courtney and Jamie found 17.06 + 5.493. Think about which sum is reasonable.

Courtney

17.06
+5.493
7.199

Jamie

17.06
+5.493
22.553

Use estimation to find which sum is reasonable.

17.06 + 5.493 → 17 + 5 = 22   Jamie's sum is reasonable.

Remember that you can add zeros to the right of a decimal without changing its value. Use zeros to make all decimals have the same number of decimal places.

EXAMPLE 2

Add. 3.45 + 7 + 0.835
Estimate.

3.45		3	*Round to the nearest whole number.*
7	→	7	
+0.835		+1	
		11	

Find the answer.

3.45		3.450	*Align the decimal points.*
7	or	7.000	*Use zeros as placeholders.*
+0.835		+0.835	
		11.285	*Place the decimal point. Add.*

Compare the answer to your estimate. 11.285 is close to the estimate of 11. The answer is reasonable.

So, 3.45 + 7 + 0.835 = 11.285.

EXAMPLE 3

Find the difference. 351.4 − 65.25
Estimate.

351.4	→	350	
− 65.25		− 70	*Round to the nearest ten.*
		280	

Find the answer.

351.4		351.40	*Align the decimal points.*
− 65.25	or	− 65.25	*Use a zero as a placeholder.*
		286.15	*Place the decimal point. Subtract.*

Compare the answer to your estimate. 286.15 is close to the estimate of 280. The answer is reasonable.

So, 351.4 − 65.25 = 286.15.

Think and ▶
Discuss

Look back at the lesson to answer each question.

1. **Tell** why it is important to align the decimal points when you add or subtract.

2. **Explain** the steps you would use to find $67 - 34.58$.

Guided ▶
Practice

Add or subtract. Estimate to check.

3. $6.18
     $- 5.55$

4.   0.45
     0.5
     $+1.349$

5.   6
     5.43
     1.4
     $+5.755$

6.   10.72
     $- 1.3$

7. $3.2 + 2.68 + 15.043$

8. $142.108 - 63.8$

Copy the problem. Place the decimal point correctly in the answer.

9. $37.5 - 0.19 = 3731$

10. $0.431 + 1.549 + 2.017 = 3997$

11. $6 + 118.59 + 0.35 = 12494$

12. $9.7 - 3.01 = 669$

Independent ▶
Practice

Add or subtract. Estimate to check.

13.   50.28
      $+37.52$

14.   153.95
      $+434.16$

15.   805.41
      $+633.25$

16.   31.62
      $- 5.8$

17. $3.2 - 2.6$

18. $735.1 + 37 + 105.73$

19. $370.92 - 83.247$

20. $275.2 - 86.05$

21. $123.1 + 140 + 225.45$

22. $\$8 + \$215.49 + \$0.75$

23. $620.87 - 91.386$

24. $56.60 - 8.476$

Copy the problem. Place the decimal point correctly in the answer.

25. $23.64 + 233.5 = 25714$

26. $\$25.67 + \$7.16 + \$0.35 = \$3318$

27. $11.2 - 1.78 = 942$

28. $4.98 - 3.235 = 1745$

Estimate to determine if the given sum is reasonable. Write *yes* or *no*.

29. $14.78 + 122.4 = 137.18$

30. $\$32.76 + \$8.09 + \$0.49 = \$41.34$

31. $58.02 - 9.473 = 3.671$

32. $427.7 - 39.27 = 388.43$

a+b/c Algebra Evaluate each expression for $d = 4.3$.

33. $d - 3.05$

34. $1 + d + 0.7$

35. $8 + d$

36. $37.60 - d$

37. $d - 2.084$

38. $(d + 16.05) - 4.5$

Problem Solving ▶ Applications

39. Jake's batting average is 0.325. Last year it was 0.235. What is the difference between his average last year and this year?

40. Karen is saving $15.75 every week to buy a camera that costs $165.45. In how many weeks will Karen have enough money to buy the camera?

41. Estimate 30.53 + 95.7 + 75.12. Is the sum more than or less than your estimate? Explain how you know.

42. ✎ **Write About It** Why is it important to estimate the answer when you add and subtract decimals?

43. The computer club collected 3,905 lb of paper. Other clubs collected 3,950; 3,590; and 3,509 lb each. Write the weights of paper in order from greatest to least.

MIXED REVIEW AND TEST PREP

44. Write 0.05 as a percent. (p. 60)

45. Write the decimal for 46%. (p. 60)

46. Evaluate $a \div c$ for $a = 4,602$ and $c = 37$. (p. 28)

47. TEST PREP Which shows the value of 7^4? (p. 40)

   A 283   **B** 343   **C** 2,381   **D** 2,401

48. TEST PREP Tanya bought milk for $2.09, two loaves of bread for $1.05 each, cheese for $4.50, and three bottles of juice for $3.00 each. How much did Tanya pay for the items? (p. 20)

   F $12.04   **G** $12.64   **H** $17.69   **J** $23.04

LINKUP to Science

Microbiologist Microbiologists are scientists who specialize in the study of the tiny cells that make up every living thing. While cells vary widely in size, most plant and animal cells are so small that they can be seen only with a microscope. Microbiologists often must use decimals when measuring the sizes of cells and their structures, as in these examples:

Average plant cell	0.000035 meter
Small bacterium	0.0000002 meter
Cell wall or membrane	0.0000000075 meter

• Use your school library to find the sizes of blood cells, skin cells, and nerve cells in the human body. How much larger or smaller is each of these cell types than the average plant cell?

Multiply Decimals

Learn how to multiply decimals.

You need colored pencils and decimal squares.

MATH LAB

QUICK REVIEW

1.	42	2.	83	3.	52	4.	93	5. 36 × 4
	× 7		× 4		× 6		× 5	

You can use a model to find the product of a decimal number and a whole number.

Activity

- To find 3 × 0.14, shade 0.14, or 14 small squares, three times. Use a different color and shade a different group of 14 small squares each time.

- Count the number of shaded squares. What is 3 × 0.14?

0.14

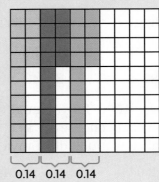
0.14 0.14 0.14

- Use a decimal square to find 5 × 0.18.

- Describe how you shaded your decimal square.

- What is 5 × 0.18?

Sometimes, when the factors are greater, as in 8 × 1.52, it is easier to compute the product by not using decimal squares.

EXAMPLE 1

Ed buys 9 sports cards at $1.12 each. How much does he spend?

Multiply. $1.12 × 9

Estimate. $1.12 × 9 → $1 × 9 = $9

Find the answer.

$$\begin{array}{r} \$1.12 \\ \times 9 \\ \hline \$10.08 \end{array}$$

Multiply as with whole numbers. Since the estimate is $9, place the decimal point after the 10.

Since the estimate is $9, the answer is reasonable.

So, Ed pays $10.08 for sports cards.

CALIFORNIA STANDARDS O—NS 2.0 Students calculate and solve problems involving addition, subtraction, multiplication, and division. **MR 2.1** Use estimation to verify the reasonableness of calculated results. *also* **MR 1.3, MR 2.4, MR 2.5, MR 3.1**

Multiply a Decimal by a Decimal

You can use a decimal square or paper and pencil to find the product of two decimals.

 EXAMPLE 2

Multiply. 0.2×0.6

Shade 6 columns blue for 0.6.

Shade 2 rows yellow for 0.2.

The area in which the shading overlaps shows the product, or 0.2 of 0.6.

So, $0.2 \times 0.6 = 0.12$.

Place the decimal point in a product by estimating or by adding the number of decimal places in the factors.

$$
\begin{array}{r}
0.2 \leftarrow \text{1 decimal place} \\
\underline{\times 0.6} \leftarrow \text{1 decimal place} \\
0.12 \leftarrow \textit{1 + 1, or 2, decimal places}
\end{array}
$$

EXAMPLE 3

Mr. Ponti works 37.5 hr per week. He earns $8.70 an hour. How much does he earn in a week?

Multiply. $\$8.70 \times 37.5$

Estimate. $\$8.70 \times 37.5 \rightarrow \$9 \times 38 = \$342$

Find the answer.

$$
\begin{array}{r}
\$8.70 \leftarrow \text{2 decimal places} \\
\underline{\times \ 37.5} \leftarrow \text{1 decimal place} \\
4350 \\
6090 \\
\underline{+2610} \\
326.250 \leftarrow \text{3 decimal places}
\end{array}
$$

Multiply as with whole numbers.
Place the decimal point in the product.

Since the estimate is $342, the answer is reasonable.

So, Mr. Ponti earns $326.25.

When you multiply decimals, you sometimes have to insert zeros in the answer.

EXAMPLE 4

Multiply. 0.037×0.062

$$
\begin{array}{r}
0.037 \leftarrow \text{3 decimal places} \\
\underline{\times 0.062} \leftarrow \text{3 decimal places} \\
74 \\
\underline{+222} \\
0.002294
\end{array}
$$

Multiply as with whole numbers.
Place the decimal point in the product.
The answer must have 6 decimal places, so place 2 zeros to the left of 2.

So, $0.037 \times 0.062 = 0.002294$.

You can also use the Distributive Property to multiply with decimals.

EXAMPLE 5

Multiply. 6 × 5.9

Use estimation to place the decimal point in the product.
Estimate. 6 × 5.9 → 6 × 6 = 36

Find the answer.
6 × 5.9 = (6 × 5) + (6 × 0.9)   *Use the Distributive Property.*

$\quad\quad = 30 + 5.4$ *Since the estimate is 36, place the*
$\quad\quad = 35.4$ *decimal point after the 35.*

Since the estimate is 36, the answer is reasonable.

So, 6 × 5.9 = 35.4.

CHECK FOR UNDERSTANDING

Think and Discuss ▶ Look back at the lesson to answer the question.

1. **Explain** how you would place the decimal point in the product 0.27 × 0.476.

Guided Practice ▶ Use the decimal square shown to help you multiply.

2.

4 × 0.12

3.

0.7 × 0.5

4.

0.6 × 0.4

Tell the number of decimal places there will be in the product.

5. 3.62 × 7

6. 2.15 × 8.18

7. 4.04 × 5.2

Copy the problem. Place the decimal point in the product.

8. 9 × 5.4 = 486

9. 0.7 × 4.1 = 287

10. 2.2 × 0.55 = 1210

Multiply. Estimate to check.

11. 0.42 × 2.9

12. 1.25 × 0.4

13. 3.23 × 8

PRACTICE AND PROBLEM SOLVING

Independent Practice ▶ Use the decimal square shown to help you multiply.

14.

5 × 0.18

15.

0.3 × 0.8

16.

0.7 × 0.4

Tell the number of decimal places there will be in the product.

17. 3.79×8.2 **18.** 0.876×0.2 **19.** 1.3842×0.91

Copy the problem. Place the decimal point in the product.

20. $6.37 \times 2.91 = 185367$ **21.** $20.4 \times 9.52 = 194208$

22. $7.32 \times 3 = 2196$ **23.** $0.82 \times 0.5 = 410$ **24.** $32.5 \times 0.06 = 1950$

Multiply. Estimate to check.

25. 3×4.6 **26.** 9×2.5 **27.** 7.3×5

28. 8.2×5 **29.** 1.2×4.1 **30.** 0.9×6.3

31. 0.2×0.4 **32.** 12.6×0.45 **33.** 0.21×2.1

34. 6.15×2.4 **35.** 4.08×1.35 **36.** 6.21×0.95

37. 24.63×1.09 **38.** 29.147×5.61 **39.** 0.189×2.09

40. 118.001×0.37 **41.** 148.9×0.006 **42.** $1{,}200.5 \times 8.2$

Problem Solving ▶
Applications

43. The recipe for a pie calls for 2.25 lb of apples and 0.75 lb of walnuts. Apples are $1.60 per pound and walnuts are $4.95 per pound. How much will the apples and walnuts cost in all?

44. Keith has a wall that is 5.2 m wide. He has 3 bookcases that are each 1.9 m wide. Is there enough room for all of the bookcases to be placed against the wall? Explain.

45. ✎ **Write a problem** that uses multiplication of two decimals to find the answer. The product must have four decimal places.

46. **Use Data**   Look at the graph at the right. Rochelle had to earn 134 points in the first four rounds to advance in a competition. Rochelle says that she did advance. Is this a reasonable statement? Explain.

ROCHELLE'S SCORES

Round	Points
6	75
5	60
4	77
3	47
2	58
1	68

MIXED REVIEW AND TEST PREP

47. Add. $46.2 + 3.45 + 16$ (p. 66)

48. Subtract. $604.5 - 76.38$ (p. 66)

49. Evaluate the expression $a \div c$ for $a = 14{,}067$ and $c = 47$. (p. 28)

50. TEST PREP   Which shows the value of m when $8 = m \div 9$? (p. 36)

A 70   **B** 72   **C** 720   **D** 7,200

51. TEST PREP   Which is the best estimate for $21{,}563 \div 43$? (p. 16)

F 500   **G** 700   **H** 5,000   **J** 7,000

Extra Practice   page H35, Set B

LESSON **4.3**

Explore Division of Decimals

Explore how to use a model to divide decimals.

You need decimal squares, colored pencils, scissors.

Remember

1 tenth (0.1) = 1 column

1 hundredth (0.01) = 1 small square

QUICK REVIEW

1. 6)372   **2.** 8)427   **3.** 5)524

4. 4)123   **5.** 252 ÷ 6

You can shade and cut apart decimal squares to divide a decimal by a whole number.

Activity 1

Find 3.66 ÷ 3.

• Shade 3.66 decimal squares.

• Divide the shaded wholes into 3 equal groups. Divide the 66 hundredths into 3 equal groups.

• What decimal names each group? What is the quotient?

• Use decimal squares to find 1.32 ÷ 4.

Think and Discuss

• Find 366 ÷ 3. How is the quotient the same as for 3.66 ÷ 3? How is it different?

• Find 132 ÷ 4. How is the quotient the same as for 1.32 ÷ 4? How is it different?

Practice

Use decimal squares to find the quotient.

1. 4.04 ÷ 4

2. 3.25 ÷ 5

3. 1.35 ÷ 3

CALIFORNIA STANDARDS O━nNS 2.0 Students calculate and solve problems involving addition, subtraction, multiplication, and division. **MR 2.4** Use a variety of methods, such as words, numbers, symbols, charts, graphs, tables, diagrams, and models, to explain mathematical reasoning.

You can shade and cut apart decimal squares to divide a decimal by a decimal.

TECHNOLOGY LINK

More Practice:
Use E-Lab, *Exploring Division of Decimals*.
www.harcourtschool.com/elab2002

Activity 2

Find 3.6 ÷ 1.2.

• Shade 3.6 decimal squares.

• Cut apart the 6 tenths.

• Divide the shaded squares and shaded tenths into equal groups of 1.2. How many groups of 1.2 are in 3.6? What is the quotient?

• Use decimal squares to find 3.2 ÷ 1.6.

• How many groups of 1.6 are in 3.2?

Think and Discuss

• Find 36 ÷ 12. How is the quotient the same as for 3.6 ÷ 1.2? How is the problem different from 3.6 ÷ 1.2?

• You know that 3.6 ÷ 12 = 0.3 and 3.6 ÷ 1.2 = 3. What do you think 3.6 ÷ 0.12 equals?

Practice
Use decimal squares to find the quotient.

1. 7.8 ÷ 1.3   **2.** 5.6 ÷ 0.8   **3.** 1.56 ÷ 0.52

4. 5.5 ÷ 1.1   **5.** 3.6 ÷ 0.9   **6.** 1.8 ÷ 0.45

7. 0.42 ÷ 0.14   **8.** 10.8 ÷ 2.7   **9.** 0.64 ÷ 0.08

MIXED REVIEW AND TEST PREP

10. Multiply. 64.7 × 3.6 (p. 70)

11. Write the decimal for 57%. (p. 60)

12. Write the value of 3^4. (p. 40)

13. Divide. 3,759 ÷ 42 (p. 22)

14. **TEST PREP** Evaluate the expression. 12 × 25 ÷ (12 + 18) (p. 44)

    A 8   **B** 10   **C** 43   **D** 300

Divide with Decimals

Learn how to divide a decimal by a whole number and how to divide a decimal by a decimal.

Remember that compatible numbers are numbers that divide without a remainder, are close to the actual numbers, and are easy to compute mentally.

Jacki wants to transfer photographs onto a CD to use with her computer. She bought a box of 5 blank CDs for $26.45. What was the cost of each CD?

Dividing a decimal by a whole number is like dividing whole numbers.

Divide. $26.45 \div 5$

Use compatible numbers to estimate.

$26.45 \div 5 \quad \rightarrow \quad 25 \div 5 = 5$

Find the answer.

$$
\begin{array}{r}
5.29 \\
5\overline{)26.45} \\
-25\downarrow \\
\hline
14 \\
-10\downarrow \\
\hline
45 \\
-45 \\
\hline
0
\end{array}
$$

Place a decimal point above the decimal point in the dividend.

Divide.

Since the estimate is $5, the answer is reasonable. So, each CD cost $5.29.

Sometimes you have to place a zero in the quotient when the dividend is less than the divisor.

EXAMPLE 1

Divide. $0.285 \div 3$

$$
\begin{array}{r}
0.095 \\
3\overline{)0.285} \\
-0\downarrow \\
\hline
28 \\
-27\downarrow \\
\hline
15 \\
-15 \\
\hline
0
\end{array}
$$

Place a decimal point above the decimal point in the dividend.

Divide. Since 2 tenths is less than 3 ones, place a zero in the tenths place in the quotient.

So, $0.285 \div 3 = 0.095$.

CALIFORNIA STANDARDS ○━ィNS **2.0** Students calculate and solve problems involving addition, subtraction, multiplication, and division. **MR 2.1** Use estimation to verify the reasonableness of calculated results. *also* ○━ィMG **1.1**

- Use a calculator to find the first three quotients in each set.

- Look for a pattern. Try to predict the last quotient in each set.

Set A	Set B
0.48 ÷ 0.03 = ■	0.621 ÷ 0.023 = ■
4.8 ÷ 0.3 = ■	6.21 ÷ 0.23 = ■
48 ÷ 3 = ■	62.1 ÷ 2.3 = ■
480 ÷ 30 = ■	621 ÷ 23 = ■

- Describe the pattern that helped you predict the last quotient in each set.

- Look at 4.8 ÷ 0.3. Multiply both numbers by 10. How do the quotients for 4.8 ÷ 0.3 and 48 ÷ 3 compare? How does multiplying the divisor and the dividend by 10 affect the quotient?

To divide a decimal by a decimal, first multiply the divisor and the dividend by a power of 10 to change the divisor to a whole number.

$$0.7\overline{)62.44} \quad \rightarrow \quad 7\overline{)624.4}$$

THINK: $0.7 \times 10 = 7$

$62.44 \times 10 = 624.4$

EXAMPLE 2

Divide. 22.8 ÷ 0.8

$0.8\overline{)22.8}$ *Make the divisor a whole number by multiplying the divisor and dividend by 10.*

$0.8 \times 10 = 8$ $22.8 \times 10 = 228$

Place the decimal point in the quotient. Divide.

$$\begin{array}{r} 28.5 \\ 8\overline{)228.0} \\ -16 \\ \hline 68 \\ -64 \\ \hline 40 \\ -40 \\ \hline 0 \end{array}$$

Since there is a remainder, place a zero in the tenths place in the dividend, and continue to divide.

So, 22.8 ÷ 0.8 = 28.5.

- Think about 63.7 ÷ 0.24. To change the divisor to a whole number, you multiply by 100. What does the dividend become?

Sometimes there aren't enough places in the dividend to move the decimal to the right.

EXAMPLE 3

Divide. 158.4 ÷ 0.12

$0.12\overline{)158.40}$

Multiply the divisor and dividend by 100.
0.12 × 100 = 12   158.4 × 100 = 15,840
Write a zero in the dividend.

$$
\begin{array}{r}
1320 \\
12\overline{)15840} \\
-12 \\
\hline
38 \\
-36 \\
\hline
24 \\
-24 \\
\hline
00 \\
\end{array}
$$

Divide.

Since the remainder is zero, the quotient is a whole number. You do not need to put the decimal point in the answer.

So, 158.4 ÷ 0.12 = 1,320.

CHECK FOR UNDERSTANDING

Think and Discuss ▶ Look back at the lesson to answer each question.

1. **Explain** how to change the divisor and the dividend before you solve the problem 55.8 ÷ 0.18.

2. **Compare** the quotients 4.5 ÷ 1.5 and 45 ÷ 15.

Guided Practice ▶ Rewrite the problem so that the divisor is a whole number.

3. 9.6 ÷ 1.6   **4.** 73.6 ÷ 0.5   **5.** 48.24 ÷ 2.4

Copy the problem. Place the decimal point in the quotient.

6. 28.50 ÷ 2.50 = 1140   **7.** 34.178 ÷ 2.3 = 1486   **8.** 62.44 ÷ 7 = 892

Divide. Estimate to check.

9. 7.88 ÷ 4   **10.** 33.66 ÷ 11   **11.** $0.55\overline{)2.42}$

PRACTICE AND PROBLEM SOLVING

Independent Practice ▶ Rewrite the problem so that the divisor is a whole number.

12. 48.4 ÷ 0.4   **13.** 8.19 ÷ 0.09   **14.** 3.7 ÷ 2.1

15. 2.39 ÷ 0.05   **16.** 45.218 ÷ 0.23   **17.** 233.58 ÷ 10.2

Copy the problem. Place the decimal point in the quotient.

18. 325.8 ÷ 3 = 1086   **19.** 53.07 ÷ 8.7 = 61   **20.** 10.2 ÷ 2 = 51

21. 84.87 ÷ 12.3 = 69   **22.** 274.89 ÷ 1.5 = 18326

Divide. Estimate to check.

23. $11\overline{)109.01}$

24. $90\overline{)10.8}$

25. $60\overline{)12.6}$

26. $0.38\overline{)13.3}$

27. $6.41\overline{)135.892}$

28. $38.2\overline{)469.86}$

29. $44.28 \div 5.4$

30. $80.1 \div 9$

31. $90.3 \div 6$

32. $1.26 \div 0.2$

33. $13.2 \div 0.06$

34. $42.5 \div 0.05$

Evaluate each expression.

35. $3 - 1.6 \times 0.4$

36. $5 - 2.4 \div 8$

37. $0.09 \div 3 + 3$

Problem Solving ▶ Applications

38. Emily rents 5 DVD movies for $28.75. Is the price for one movie closer to $5 or to $6? Explain.

39. Jonelle saves $4.95 every week to buy a video that costs $29.70. She has already saved $9. For how many more weeks does she need to save money to have enough to buy the video?

40. **? What's the Error?** Michael divided 4.25 by 0.25 and got a quotient of 0.17. Explain the error. What is the correct quotient?

41. Andrew reads 22 pages the first day. He plans to increase the number of pages he reads by 4 a day until he finishes the book. How many pages will he be reading at the end of 3 days?

MIXED REVIEW AND TEST PREP

42. Multiply. 2.051×8.6 (p. 70)

43. Is 208.605 greater than or less than 206.605? (p. 52)

44. Use mental math to find $210 \div 42$. (p. 36)

45. **TEST PREP** Which shows the value of r when $18 + r = 52$? (p. 30)

    A 34   **B** 30   **C** 24   **D** 14

46. **TEST PREP** Jocelyn works 12 hours per week. She earns $7.25 an hour. How much does she earn in 4 weeks? (p. 22)

    F $87   **G** $174   **H** $290   **J** $348

Thinker's CORNER

REASONING The ancient Greeks discovered that there is one particular number that can be used to relate the dimensions of any circle. They called this particular number π (pi) and found that it is the same number regardless of the size of the circle.

- Determine the approximate value of π by dividing the distance around each circle described below (the circumference) by the distance across that circle (the diameter). Is the answer the same for each?

		Circumference	Diameter
1.	Circle 1	6.28 in.	2 in.
2.	Circle 2	9.42 in.	3 in.

PROBLEM SOLVING SKILL
Interpret the Remainder

Analyze
Choose
Solve
Check

Learn how to interpret the remainder in a division problem.

The sixth-grade class at Hightown Middle School has an annual class picnic. With the help of some students, Ms. Gordon is planning for the picnic this year.

Ms. Gordon needs 163 boxed drinks for lunch. A package holds 6 boxes. How many packages of drinks will Ms. Gordon need to buy?	$27 \text{ r}1$ $6\overline{)163}$ -12 43 -42 1	Since 27 packages hold only 162 total boxes, increase the quotient by 1. So, Ms. Gordon needs to buy 28 packages of drinks.
Ms. Gordon has 158 ft of ribbon to use for games. She will cut the ribbon into 8-ft lengths. How many 8-ft lengths of ribbon will Ms. Gordon have?	$19 \text{ r}6$ $8\overline{)158}$ -8 78 -72 6	The remainder is not enough for another 8-ft length of ribbon. Drop the remainder. So, Ms. Gordon will have 19 pieces of ribbon.
Students collected 158 prizes. The prizes were put in packages of 5 each. The remaining prizes were given to another class in the school. How many prizes were given to the other class?	$31 \text{ r}3$ $5\overline{)158}$ -15 8 -5 3	The remainder is your answer. So, 3 prizes were given to the other class.

Talk About It ▶ • What does the remainder in the first problem mean?

• **What if** Ms. Gordon buys 28 packages of boxed drinks? How many extra boxed drinks will Ms. Gordon buy?

• **What if** Ms. Gordon has a total of 165 ft of ribbon? How many 8-ft lengths of ribbon will she have? How long is the ribbon that's too short?

Solve the problem by interpreting the remainder.

A total of 39 students and adults tour the science museum to see an exhibit about the shaping of the Earth's surface. The tour director can take groups of up to 5 on each tour. All 39 students and adults need to see the exhibit.

1. How many complete groups of 5 people can the tour director take?

   A 5 groups   **B** 6 groups   **C** 7 groups   **D** 8 groups

2. What is the least number of tours the director will have to give to accommodate all 39 students?

   F 6 groups   **G** 7 groups   **H** 8 groups   **J** 9 groups

3. James has $4.39 to buy magnets at the science museum. Each magnet costs $0.95. He wants to buy as many magnets as he can. How many magnets can James buy?

4. Sharon bought a package of 15 postcards at the museum. She gave the same number of postcards to each of her 4 teachers and kept the ones left over. How many postcards did Sharon keep?

MIXED APPLICATIONS

5. A train that is scheduled to arrive at 5:15 P.M. arrives 20 minutes late. If the train left at 9:30 A.M., how long was the trip?

6. A total of 51 students and teachers are using cars to go on a field trip. Six people ride in each car. How many cars are needed for the field trip?

7. Mark estimates that he needs 1 minute to solve a short homework problem and 5 minutes for each long problem. If he has 25 short and 8 long problems, how long should his homework take?

8. Tina makes bracelets for her friends. She uses 3 red beads for every 7 yellow beads to make a pattern. For one bracelet, she uses a total of 50 beads. How many of each color does she use?

9. A board game has 3 times as many red playing pieces as blue. It has 5 times as many green pieces as blue. There are 12 blue pieces. How many playing pieces are there in all?

10. Edgar has twice as many library books as his brother. If Edgar has 10 books in all, how many library books must he return to have the same number as his brother?

11. **? What's the Question?** Use the table at the right. Each Presidential term in office is 4 years. The answer is that he served more than one term but less than two.

United States Presidents	
President	Years in Office
Richard Nixon	1969-1974
Gerald Ford	1974-1977
Jimmy Carter	1977-1981
Ronald Reagan	1981-1989
George Bush	1989-1993

ALGEBRA
Decimal Expressions and Equations

Learn how to evaluate expressions and solve equations with decimals.

Different groups of friends eat lunch together at a cafe. The cost of the lunch is $6.45 per person. Write an expression to find the total cost of a lunch for a group of friends.

Let a be the number of friends buying lunch.

$a \times 6.45$ or $6.45a$ *Write the expression.*

The number of friends varies. What is the total cost for 7 friends?

$a \times 6.45$ *Write the expression.*

7×6.45 *Replace a with 7.*

45.15 *Multiply.*

So, the total cost is $45.15.

• Evaluate the expression $w \div 3 + 9.3$ for $w = 4.8$.

You solved equations with whole numbers by using mental math. You can use the same methods to solve some equations with decimals.

EXAMPLE

Solve the equation $h \div 6 = 0.6$ by using mental math.

$h \div 6 = 0.6$ *What number divided by 6 = 0.6?*

$h = 3.6$ THINK: *6 × 0.6 = 3.6.*

$3.6 \div 6 = 0.6$ *Check your answer. Replace h with 3.6.*
$0.6 = 0.6$

So, $h = 3.6$.

• Solve. $t + 4.24 = 9.48$

CALIFORNIA STANDARDS **AF 1.0** Students write verbal expressions and sentences as algebraic expressions and equations; they evaluate algebraic expressions, solve simple linear equations, and graph and interpret their results. **○━ AF 1.1** Write and solve one-step linear equations in one variable. *also*

Think and Discuss ▶ Look back at the lesson to answer each question.

1. **What if** 12 friends went to lunch at the cafe? Show how you would find the total cost of the lunch.

2. **Explain** how you would solve the equation $3.2 = a \div 2$.

Guided Practice ▶ Evaluate each expression.

3. $a + 3.4$
   for $a = 8.3$

4. $1.6 \div b$
   for $b = 0.4$

5. $9.16 - a$
   for $a = 4.08$

Solve each equation by using mental math.

6. $\frac{4.8}{k} = 8$

7. $m - 12.7 = 6.3$

8. $3t = 21.9$

PRACTICE AND PROBLEM SOLVING

Independent Practice ▶ Evaluate each expression.

9. $2h$
   for $h = 2.3$

10. $9.6 \div a$
    for $a = 3$

11. $j + 7.1$
    for $j = 6.9$

12. $4.17 - c$
    for $c = 1.09$

13. $m \div 6 + 3.6$
    for $m = 1.8$

14. $g + h - 3.2$
    for $g = 4.1$ and $h = 2.3$

Solve each equation by using mental math.

15. $r + 8.1 = 15.8$

16. $1.7 = \frac{d}{4}$

17. $4a = 32.8$

18. $x - 2.4 = 8.6$

19. $p + 11.1 = 28.7$

20. $3.2r = 7.1 + 5.7$

Problem Solving Applications ▶

21. Let n represent the number of miles Jeremy rides his bicycle to attend 6 baseball practices. Write an expression to show how far he travels for one practice. Evaluate the expression for $n = 28.8$ mi.

22. (?) **What's the Error?** Explain the error at the right. Give the correct solution.

$$24.8 + x = 30$$
$$x = 54.8$$

23. David jogs 3 mi each weekday and 7 mi each Saturday. Ray jogs 18 mi each week. How much farther does David jog in a week?

MIXED REVIEW AND TEST PREP

24. $4.38 \div 7.5$ (p. 76)

25. $18 \div 6 + (16 \times 3) - 14$ (p. 44)

26. $6,045 - 973$ (p. 20)

27. Order from least to greatest. 3.58, 3.08, 3.85, 3.508 (p. 52)

28. **TEST PREP** The sum of two numbers is 35. Their difference is less than 10. Which is not a possible pair? (p. 26)

   A 15, 20   **B** 17, 18   **C** 10, 25   **D** 22, 13

Add or subtract. Estimate to check. (pp. 66–69)

1.
$$\begin{array}{r} 3.9 \\ 4 \\ +5.91 \\ \hline \end{array}$$

2.
$$\begin{array}{r} 7.6 \\ -0.95 \\ \hline \end{array}$$

3.
$$\begin{array}{r} 3.02 \\ 0.17 \\ +4.338 \\ \hline \end{array}$$

4. $19.3 - 2.56$

5. $0.126 + 5.3 + 3.04$

6. $245 - 39.05$

Tell the number of decimal places there will be in the product. Then multiply. (pp. 70–73)

7.
$$\begin{array}{r} 8.3 \\ \times 12.9 \\ \hline \end{array}$$

8.
$$\begin{array}{r} 7.82 \\ \times\ 4.5 \\ \hline \end{array}$$

9.
$$\begin{array}{r} 0.13 \\ \times 2.07 \\ \hline \end{array}$$

10.
$$\begin{array}{r} 53.6 \\ \times 1.23 \\ \hline \end{array}$$

11. 20.01×8.2

12. 6.9×17.4

13. 4.91×6.2

14. 7.02×5.5

Divide. Estimate to check. (pp. 76–79)

15. $1.4\overline{)35}$

16. $3.7\overline{)5.92}$

17. $0.45\overline{)1.08}$

18. $0.25\overline{)85}$

19. $22.8 \div 3$

20. $9.72 \div 2.7$

21. $33.33 \div 1.1$

22. $6.9 \div 0.3$

Evaluate each expression. (pp. 82–83)

23. $(2.3 + c) + 1.7$ for $c = 8$

24. $3 \times d + b$ for $d = 1.7$ and $b = 5.4$

25. $c \div 2 - a$ for $c = 8$ and $a = 2.3$

26. $(5.4 - a) + d$ for $a = 2.3$ and $d = 1.7$

27. $4 \times a - c$ for $a = 2.3$ and $c = 8$

28. $(c - 5.4) \times c$ for $c = 8$

29. $(d + c) - 5$ for $d = 1.7$ and $c = 8$

30. $(a + d) \div 4$ for $a = 2.3$ and $d = 1.7$

Solve each equation by using mental math. (pp. 82–83)

31. $c + 14.07 = 32.97$

32. $6a = 24.78$

33. $8d = 6.4$

34. $7.14 - g = 3.24$

35. $2.4 + f = 4.76$

36. $4y = 29.6$

37. $1.86 \div r = 6.2$

38. $t + 3.6 = 4.5 + 3.3$

Solve. (pp. 80–81)

39. There are 9 golf balls in a box. How many boxes does Hector need if he wants to give away 500 golf balls as souvenirs?

40. On Tuesday, Sabrina gets an invitation to a party that is in 20 days. She can go to the party if it does not fall on the weekend. Can she go to the party? Explain.

Choose the answer.
See item **4**.

If your answer doesn't match one of the choices, check your computation and the placement of the decimal point. If your computation is correct, mark the letter for Not here.

Also see problem **6**, p. H64.

Choose the best answer.

1. Elise bought 4 balloons. Each balloon cost $0.98, including tax. Which is the total cost of the 4 balloons?

   A $3.62   **C** $3.96

   B $3.92   **D** $4.00

2. At the end of a trip, the odometer of a car read 4,572.1 miles. The odometer read 2,998.7 miles at the beginning of the trip. How far did the car go?

   F 7,570.8 mi

   G 7,460.8 mi

   H 2,684.4 mi

   J 1,573.4 mi

3. 0.27 + 1.098

   A 1.368   **C** 1.125

   B 1.255   **D** Not here

4. 1.65 ÷ 0.5

   F 0.33   **H** 33

   G 3.3   **J** Not here

5. Stu wants to buy a poster that costs $5.99, a desk lamp that costs $22.18, and bookends that cost $18.98. What is the cost of the 3 items he wants to buy, before tax is added?

   A $46.15   **C** $47.15

   B $47.00   **D** $48.15

6. What is the value of $x + 16.714$ for $x = 20.3$?

   F 37.014   **H** 3.614

   G 36.014   **J** 0.3614

7. A sweatshirt costs $9 more than a T-shirt. Together, the shirts cost $23. What is the cost of each type of shirt?

   A T-shirt: $4, sweatshirt: $19

   B T-shirt: $16, sweatshirt: $7

   C T-shirt: $12, sweatshirt: $11

   D T-shirt: $7, sweatshirt: $16

8. What is the value of s for $s + 1.1 = 5.9$?

   F 3.7   **H** 6.0

   G 4.8   **J** 7.0

9. In which pair of numbers is there a 5 in the hundredths place of both numbers?

   A 553.621; 8,516.2

   B 87.35; 62,531.9

   C 36.157; 4,062.058

   D 513.26; 792.56

10. 7.8×0.3

    F 2.34   **H** 8.1

    G 3.24   **J** Not here

11. Which number sentence is true?

    A $0.8 > 8.0$

    B $0.8 \geq 8.0$

    C $0.8 \leq 8.0$

    D $0.8 < 8.0$

12. $5^2 - 3 + 2 \times 8$

    F 15   **H** 40

    G 38   **J** 160

MATH DETECTIVE

Who Am I?

REASONING   Use the given information plus your knowledge of division to find the mystery number. Be prepared to explain how you solved the mystery.

Mystery Number 1

I'm the least whole number that gives a remainder of 4 when divided by 5 and when divided by 9. Who am I?

Mystery Number 2

I'm the first whole number greater than 240 that gives a remainder of 5 when divided by 9. Who am I?

Mystery Number 3

I'm the first whole number greater than 100 that gives a remainder of 3 when divided by 8 and when divided by 7. Who am I?

Mystery Number 4

I'm the least whole number that when divided by 4, by 5, and by 7 gives a remainder of 3. Who am I?

Think It Over!

- **Write About It**   Explain how you found each mystery number.

- **Stretch Your Thinking**   I'm the least number that when divided by 5 and when divided by 7 produces a remainder of 3. How much do you have to add to me to get a remainder of 0 when you divide by 5 and when you divide by 7?

Scientific Notation

The weight of one type of passenger jet is about 820,000 lb. This number, 820,000, is written using *standard notation*.

You can also write the weight using **scientific notation**. Scientific notation is a shorthand method for writing large numbers.

A number written in scientific notation has two parts separated by a multiplication symbol.

$$8.2 \times 10^5$$

The first part is a number that is at least 1 but less than 10.   The second part is a power of 10.

To write the number 820,000 in scientific notation:

Count the number of places the decimal point must be moved to the left to form a number that is at least 1 but less than 10.

$$820,000 \quad \rightarrow \quad 8.2$$

5 places

Since the decimal point moved 5 places, the power of 10 is 5.

$$820,000 = 8.2 \times 10^5$$

The airplane's weight written in scientific notation is 8.2×10^5 lb.

EXAMPLE

About 32,500,000 passengers used Newark International Airport in 1998. Write the number of passengers in scientific notation.

$$32,500,000 \quad \rightarrow \quad 3.25 \times 10^7$$

7 places

The number of passengers was about 3.25×10^7.

TALK ABOUT IT

- Is 1.2×10^9 greater than 9.98×10^8? Explain.

- Show how to correctly write 782.5×10^8.

TRY IT

Write the number in scientific notation.

1. 602,000   2. 199   3. 3,400,000
4. 8,540   5. 5,010,000   6. 113,000
7. 72   8. 48,900   9. 26,200,000

820,000 pounds lifting off!

VOCABULARY

1. An expression that includes a variable is a(n) __?__ expression. (p. 28)

2. When you __?__ a decimal by a power of 10, the decimal point moves one place to the right for each factor of 10. (p. 87)

EXAMPLES

EXERCISES

Chapter 1

- **Add and subtract whole numbers.**
  (pp. 20–21)

$$
\begin{array}{r}
3{,}921 \\
68 \\
+\ 205 \\
\hline
4{,}194
\end{array}
\qquad
\begin{array}{r}
3{,}000 \\
-1{,}650 \\
\hline
1{,}350
\end{array}
$$

Find the sum or difference.

3. $756 + 902$
4. $4{,}293 + 256 + 19$
5. $3{,}511 - 1{,}345$
6. $729 + 8 + 3{,}996$
7. $16{,}092 - 5{,}618$
8. $25{,}080 - 19{,}387$

- **Multiply and divide whole numbers.**
  (pp. 22–25)

$$
\begin{array}{r}
273 \\
\times\ 86 \\
\hline
1\ 638 \\
+21\ 84 \\
\hline
23{,}478
\end{array}
$$

$$
\begin{array}{r}
203\ \text{r}16 \\
41\overline{)8{,}339} \\
-8\ 2 \\
\hline
13 \\
-\ 0 \\
\hline
139 \\
-123 \\
\hline
16
\end{array}
$$

Find the product or quotient.

9. $\begin{array}{r} 57 \\ \times\ 38 \end{array}$
10. $\begin{array}{r} 430 \\ \times\ 17 \end{array}$
11. $\begin{array}{r} 276 \\ \times\ 809 \end{array}$

12. $489 \div 26$
13. $9{,}671 \div 42$
14. $9{,}538 \div 19$

- **Use mental math to solve equations.**
  (pp. 30–31)

  Solve. $w \div 6 = 7$ THINK: *What number divided by 6 equals 7?*

  $w = 42$ $42 \div 6 = 7$

Solve each equation by using mental math.

15. $b + 5 = 12$
16. $a - 3 = 5$
17. $3y = 18$
18. $k \div 5 = 4$

Chapter 2

- **Use the order of operations to evaluate expressions.** (pp. 44–45)

  Evaluate. $25 \div 5 + (8 - 4)^2 \times 3$

  $25 \div 5 + (4)^2 \times 3$ *Operate in parentheses.*
  $25 \div 5 + 16 \times 3$ *Clear exponents.*
  $5 + 16 \times 3$ *Divide.*
  $5 + 48$ *Multiply.*
  53 *Add.*

Evaluate the expression.

19. $3 \times 6 + 7^2 - 9$
20. $(8 + 7) \div 3 + (5 - 3)^3$
21. $(15 - 3 \times 4) + 8 \div 2$
22. $9^2 - 20 \times 2 + 5$
23. $32 + (8^2 - 50) \times 2$
24. $16 \div 2^3 + 4 \times 3$

Chapter 3

- **Compare and order decimals.** (pp. 52–55)

  Compare 2.8 and 2.83.

2.8	2.83	*Start at the left.*
2.8	2.83	*← same number of ones*
2.8	2.83	*← same number of tenths*
2.80	2.83	*Add a zero to compare.*

  2.8 < 2.83, or 2.83 > 2.8

Compare the numbers. Write <, > or = for ●.

25. 3.72 ● 3.7   **26.** 5.02 ● 5.021

Write the numbers in order from least to greatest. Use <.

27. 1.67, 1.76, 1.607, 1.706, 1.076

28. 0.0014, 0.4001, 0.0401, 0.0041, 0.014

- **Write decimals as percents and percents as decimals.** (pp. 60–61)

  Write 0.05 as a percent.
  0.05 is 5 hundredths.
  So, 0.05 is 5%.

Write the percent or decimal.

29. 0.28   **30.** 7%

31. 47%   **32.** 0.6

Chapter 4

- **Add and subtract decimals.** (pp. 66–69)

  Find the difference. 8.2 − 6.391

8.200	*Align the decimal points.*
−6.391	*Use zeros as placeholders.*
1.809	*Place the decimal point. Subtract.*

Find the sum or difference.

33. 29.6 + 0.935   **34.** 26.53 + 5.238

35. 76.03 − 58.94   **36.** 347.31 − 48.896

- **Multiply and divide decimals.** (pp. 70–79)

  Find the quotient. 2.46 ÷ 0.6

  ```
        4.1
  0.6)2.46        Make the divisor a
     -2 4         whole number.
       06           0.6 × 10 = 6
       -6         2.46 × 10 = 24.6
        0         Divide.
  ```

Find the product or quotient.

37. 75.8 × 6   **38.** 4.83 × 0.9

39. 3.92 × 0.58   **40.** 68.49 × 3.6

41. 36.48 ÷ 12   **42.** 43.5 ÷ 0.3

43. 59.04 ÷ 4.8   **44.** 8.094 ÷ 0.95

- **Evaluate expressions with decimals.**
  (pp. 82–83)

  Evaluate $r + s + 3$ for $r = 6.9$ and $s = 4.8$.

  | $r + s + 3$ | *Replace r with 6.9 and s* |
  | | *with 4.8.* |

  $6.9 + 4.8 + 3 = 14.7$

Evaluate each expression for $c = 2.5$, $d = 3.2$, and $f = 4.1$.

45. $(c \times 3) + d$ **46.** $(10 - d) + f$

47. $\dfrac{d}{8}$ **48.** $f - c + d$

PROBLEM SOLVING APPLICATIONS

49. Jake has half as many pennies as his sister has. Together their pennies are worth 72 cents. How many pennies does Jake have? (p. 26)

50. Movie tickets cost $3.75 each for the matinee. How many tickets can Ms. Hamil buy with $20? How much money will she have left over? (p. 76)

Wind Energy

Wind energy has been used for centuries to pump water, grind grain, and power sailboats. Today, windmills called wind turbines are used to generate electricity. There are more wind turbines in California than anywhere else in the world.

These wind turbines are located at California's San Gorgonio Pass. When the blades spin, they drive generators, which in turn produce electricity.

Use Data For 1–7, use the table.

1. Order the locations from greatest number of turbines to least.

WIND TURBINES IN CALIFORNIA	
Location	Number of Turbines
Altamont Pass	6,000
San Gorgonio Pass	3,500
Tehachapi Pass	5,000
Other CA areas	500

The total number of wind turbines in the world is about 8.5 times the number at Altamont Pass.

2. About how many wind turbines are there in the world?

3. About how many more wind turbines are there outside of California than there are inside California?

Denmark has 3,600 wind turbines. This is the second greatest total in the world.

4. How many times as many wind turbines are there in California as there are in Denmark? Write the answer as a fraction and as a mixed number.

5. How many more turbines are there in California than there are in Denmark?

Electricity is measured in kilowatt-hours (kWh). The wind turbines at Tehachapi Pass together generate 25 million kWh of electricity every day.

6. It costs about $0.035 to produce 1 kWh of wind turbine energy. How much does it cost to produce the wind turbine energy at Tehachapi Pass each day?

7. On average, how many kWh are generated daily by each turbine at Tehachapi Pass?

Solar Energy

Each day, huge amounts of free energy strike the Earth in the form of radiation from the sun. This *solar* energy can be converted to heat and electricity for human use. California is one of the world's largest producers and users of solar energy.

1. In 1997, solar power cells produced about 126 million watts of electricity throughout the world. Cells made in California produced about 0.3 of the world total. How much electricity was produced by cells made in California?

2. It costs $0.045 to produce 1 kilowatt of electricity by natural gas. Leah's clothes dryer uses 4 kilowatts of electricity per hour. How much does it cost her to run her dryer for 2 hours?

Use Data   **For 3–5, use the table.**

Solar panels produce energy without polluting the environment.

3. About how many times as much electricity is generated weekly in California by natural gas as by wind and solar power combined?

4. Write an equation you can use to find how much more electricity is generated each week in California by coal power than is generated by solar power. Then solve the equation.

5. The total amount of electricity generated each week in California is 5,156 gWh. How much of the total comes from sources that are not listed in the table?

ELECTRICITY GENERATED WEEKLY IN CALIFORNIA	
Energy Source	**Energy (gWh)***
Natural Gas	1,672
Coal	1,030
Wind	53
Solar	16
One gigawatt-hour (gWh) equals one million kilowatt-hours.	

CHAPTER 5

Collect and Organize Data

Hubbard Glacier
in Alaska

If you could look at Earth from space, you would see that most of its surface is covered with water. However, most of that water is salt water which is unusable for drinking, watering crops, or manufacturing. What information about Earth's water supply can you find in the graph?

WATER ON THE EARTH

 Fresh Water

 Ice

Salt Water

If you could put all the water on Earth into 100 buckets, 97 buckets would hold the salt water of the oceans and seas, and 2 buckets would hold the frozen fresh water of glaciers and icecaps. Only 1 bucket would hold liquid fresh water.

Check What You Know

Use this page to help you review and remember important skills needed for Chapter 5.

✓ Vocabulary

mean
median
mode
range

Choose the best term from the box.

1. The middle number in a group of numbers arranged in numerical order is the __?__ .

2. The sum of a group of numbers divided by the number of addends is the __?__ .

3. The difference between the greatest number and the least number in a set of data is the __?__ .

✓ Reading a Table (See p. H5.)

STUDENTS' FAVORITE SPORTS				
	Gymnastics	**Basketball**	**Baseball**	**Football**
Boys	4	29	16	13
Girls	17	14	9	12

Use the data in the table above to answer the questions.

4. How many boys like baseball the best?

5. How many more girls than boys like gymnastics the best?

6. Which sport was selected by more boys than any other?

7. How many girls like gymnastics or football the best?

8. How many students were surveyed?

✓ Mean, Median, and Mode (See p. H6.)

Find the mean for each set of data.

9. 7, 6, 11, 7, 9

10. 84, 73, 92, 77, 90

11. 8.9, 8.8, 7.9, 8.3

12. 4.6, 5.1, 6.2, 5.4, 4.8, 5.7

Find the median and the mode for each set of data.

13. 16, 32, 24, 10, 48, 32, 28

14. 10, 15, 7, 8, 12, 13, 13, 19, 21

15. 12.1, 10.8, 11.6, 10.8, 11

16. 1.4, 0.9, 3.6, 2.5, 0.9, 6.2, 2.4

✓ Range (See p. H6.)

Find the range for each set of data.

17. 6, 3, 7, 5, 9

18. 105, 102, 116, 103, 96

LOOK AHEAD

In Chapter 5 you will

- identify different kinds of samples and determine if surveys are biased
- organize data in frequency tables and line plots
- work with and analyze measures of central tendency

Samples

Learn how to identify different types of samples and to determine if a sample is representative of the population.

Vocabulary

survey

population

sample

convenience sample

random sample

systematic sample

1. 240
   +360

2. 94
   −60

3. 60
   ×20

4. $30\overline{)270}$

5. $10\overline{)850}$

Sample   All Teenagers

A **survey** is a method of gathering information about a group. Surveys are usually made up of questions or other items that require responses.

You can survey the **population**, the entire group of individuals or objects, such as all teenagers. Or, if the population is large, you can survey a part of the group, called a **sample**.

EXAMPLE 1

Suppose Jenna wants to find out the favorite game among students in her math class. Describe the population. Should Jenna survey the population or use a sample? Explain.

The population consists of all the students in Jenna's math class. Jenna should survey the population since it is small.

• **What if** the population is changed to all of the 1,800 students in Jenna's school?

Surveying the population is not always possible, so samples are used.

EXAMPLE 2

Renee asked students at Roosevelt Middle School to indicate their favorite sport. She surveyed two samples and then the entire population of 600 students. How do the results from the two samples compare with the results from the population?

SAMPLE SIZE	NUMBER CHOOSING BASKETBALL	PERCENT OF SAMPLE
50	18	36%
100	33	33%

POPULATION	NUMBER CHOOSING BASKETBALL	PERCENT OF POPULATION
600	192	32%

The results from the samples are different from, but close to, the results from the population.

CALIFORNIA STANDARDS **SDAP 2.0** Students use data samples of a population and describe the characteristics and limitations of the samples. **SDAP 2.1** Compare different samples of a population with the data from the entire population and identify a situation in which it makes sense to use a sample. **O—ᴨ SDAP 2.2** Identify different ways of selecting a sample and which method makes a sample more representative for a population. *also* **O—ᴨ SDAP 2.4**

Types of Samples

There are many types of sampling methods. The table below shows three different sampling methods.

TECHNOLOGY LINK

To learn more about samples, watch the **Harcourt Math Newsroom Video** *What is in a Poll?*

TYPE	DEFINITION	EXAMPLE
convenience sample	The most available individuals or objects in the population are selected to obtain results quickly.	Choose a specific location, such as the cafeteria or the library, and survey students as they walk by you.
random sample	Every individual or object in the population has an equal chance of being selected. This produces the sample that is most representative of the population.	Assign a number to each student and then choose students by randomly selecting numbers with a computer.
systematic sample	An individual or object is randomly selected and then others are selected using a pattern.	Randomly choose a student from a list of students, and then choose every fourth student.

EXAMPLE 3

Identify each type of sample.

A. A bicycle shop owner randomly surveys a customer and then surveys every tenth customer after that to determine when customers expect to buy their next new bike.

This is a systematic sample since individuals are selected according to a pattern.

B. A sixth-grade middle school math teacher randomly selects 50 students in her math classes and asks them if they have a home computer.

This is a random sample since everyone has an equal chance of being selected.

C. The owner of a movie rental shop asks customers in the shop on a given day to fill out a card naming their favorite movie.

This is a convenience sample since the most available individuals are selected.

Math Idea ▶ The results from a sample can depend on how the sample is chosen. It is important to select a sample that is representative of the population. For example, if the population includes men and women, then the sample must include both men and women.

Think and ▶ Discuss

Look back at the lesson to answer each question.

1. **Explain** why a sample is often used rather than a population when conducting a survey.

2. **Tell** when you might survey a population rather than a sample.

3. **Discuss** why a random sample is more likely to represent a population than a convenience sample.

Guided ▶ Practice

Describe the population.

4. Suppose a random sample of 1,250 women in a city was taken in order to determine the favorite type of music of all women in the city.

Tell whether you would survey the population or use a sample. Explain.

5. You want to find out where the students in your class want to go for a field trip.

6. You want to know the favorite mountain bike of teenagers in your state.

For 7–9 determine the type of sample. Write *convenience, random,* or *systematic.*

7. A camp director was asked to conduct a survey to find which team game the campers liked best. She randomly selected one camper and then surveyed every fifth person on her list of campers.

8. The names of one hundred high school students are drawn from a box containing all student names.

9. The owners of a sports store ask shoppers in the store a question about the kind of running shoes they like.

Independent ▶ Practice

Describe the population.

10. To find out the average salary of female architects, a researcher randomly selected 100 female architects.

Tell whether you would survey the population or use a sample. Explain.

11. You want to find out the daily exercise habits of the varsity basketball team.

12. You want to know the favorite kinds of food of all teens in a large city.

For 13–15, determine the type of sample. Write *convenience*, *random*, or *systematic*.

13. A cafeteria worker randomly selected a student and then asked every tenth student who entered the cafeteria a question.

14. When conducting a survey, Rachel selected 50 people by drawing names out of a hat without looking.

15. To find out the favorite movie of students at her school, a teacher asked students in her math class to complete a survey.

Problem Solving ▶
Applications

16. Jim asked Lake Middle School students to indicate their favorite pet. The results are in the tables at the right. He surveyed two samples and then the population of 202 students. How do the results from the two samples compare with the results from the population?

SAMPLE SIZE	NUMBER CHOOSING DOGS	PERCENT OF SAMPLE
10	2	20%
50	24	48%

POPULATION	NUMBER CHOOSING DOGS	PERCENT OF POPULATION
320	160	50%

Dalton wants to conduct a survey to find out which computer game is most popular among the students in his school.

17. How can Dalton get a random sample?

18. How can Dalton get a convenience sample?

19. **Write About It** Choose a topic you could find out about by conducting a survey. Then tell how you would choose a random sample.

20. A target has three circles, one inside of the other. The bull's-eye is 10 points, the middle circle is 6 points, and the outer circle is 4 points. If 3 darts are thrown and only 2 hit the target, what are the possible scores?

MIXED REVIEW AND TEST PREP

Evaluate for *a* = 3.1 and *b* = 6.2. (p. 82)

21. $a + b$

22. $a \times b$

23. $b \div a$

24. **TEST PREP** Which is the product of 5.25 and 1.9? (p. 70)

   A 9.975   **B** 99.75   **C** 997.5   **D** 9,975

25. **TEST PREP** Martha saved $0.50 one week. Then she saved $0.50 more each week than she had the week before. How many weeks did it take to save a total of $10.50? (p. 66)

   F 5 weeks   **G** 6 weeks   **H** 7 weeks   **J** 8 weeks

Extra Practice page H36, Set A

Bias in Surveys

Learn how to determine whether a sample or question in a survey is unbiased.

Vocabulary

unbiased sample
biased sample
biased question

Does the earth revolve around the sun or does the sun revolve around the earth? In a recent survey of over 1,000 adults, 79% knew the correct answer.

When you collect data from a survey, your sample should represent the whole population. Every individual in that population should have an equal chance of being selected. Then the sample is an **unbiased sample**.

If individuals or groups from the population are not represented in the sample, then the sample is a **biased sample**. If the sample for the survey above included males only, it would be biased since the survey was for all adults.

EXAMPLE 1

Latosha wants to find out how many hours Polk Middle School students spend on the Internet. If she surveys students from this school, which samples below will be biased? Explain.

A. 200 girls

B. 200 athletes

C. 200 randomly selected students

D. Students who ride their bikes to school

Choices A, B, and D are biased. Choice A excludes boys, choice B only includes athletes, and choice D excludes students who don't ride their bikes to school.

Sometimes, questions are biased. A **biased question** leads to a specific response or excludes a certain group.

EXAMPLE 2

Is the following question biased?

Do you agree with a well-known movie critic that movies longer than two hours are boring?

This question is biased since it leads you to agree with the movie critic.

CALIFORNIA STANDARDS SDAP 2.3 Analyze data displays and explain why the way in which the question was asked might have influenced the results obtained and why the way in which the results were displayed might have influenced the conclusions reached. ○━ SDAP 2.4 Identify data that represent sampling errors and explain why the sample (and the display) might be biased. *also* MR 2.0, MR 2.5

CHECK FOR UNDERSTANDING

Think and ▶ Discuss

Look back at the lesson to answer the question.

1. **Write** the question in Example 2 so that it is not biased.

Guided ▶ Practice

Tell whether the sample is *biased* or *unbiased*. Explain.

The local gym wants to find out how its members feel about the new exercise equipment.

2. Female members

3. Members under age 20

4. Members who used the gym in August

5. 50 randomly selected members

PRACTICE AND PROBLEM SOLVING

Independent ▶ Practice

Tell whether the sample is *biased* or *unbiased*. Explain.

The Middletown Mall is conducting a survey of its shoppers to find out which days they prefer to shop.

6. 400 teenage shoppers

7. 400 randomly selected shoppers

8. 400 female shoppers

9. Shoppers who enter the record store

Determine whether the question is biased. Write *biased* or *unbiased*.

10. Is basketball your favorite sport?

11. Do you agree with the fruit industry president that green apples taste better than red apples?

Problem Solving ▶ Applications

12. **REASONING**   Students at Memorial Middle School are being surveyed to find out their choice of a new school mascot. If 300 students are randomly selected, do equal numbers of boys and girls have to be chosen? Explain.

13. 📖 **Write About It**   Is this question biased? Explain.
I think pizza is the best choice for lunch, don't you?

MIXED REVIEW AND TEST PREP

14. A store owner randomly selects a customer and then surveys every tenth customer. What type of sample is this? (p. 98)

Find the quotient. (p. 76)

15. $3.6 \div 0.9$

16. $1.44 \div 0.12$

17. $270 \div 0.03$

18. **TEST PREP**   Find the value of $3^2 \times (6 + 4)$. (p. 44)

   A 22   **B** 30   **C** 58   **D** 90

PROBLEM SOLVING STRATEGY
Make a Table

Analyze
Choose
Solve
Check

Learn how to solve problems by displaying information in a table.

Mrs. Donovan asked the students in her class to tell the number of movies they had seen since the start of the school year. The results are shown at the right.

Number of Movies Seen

1 5 3 2 1 6 1 2 3 3 4 5
1 1 6 3 4 2 2 2 4 3 2 1
2 2 4 5 1 3 2 1

How many students saw at least 3 movies? How many students saw fewer than 3 movies?

Analyze

What are you asked to find?

What information is given?

Is there information you will not use? If so, what?

Choose

What strategy will you use?

You can make a tally table to organize the data. Then you can use the table to answer the questions.

Solve

How can you organize the tally table?

Make a row for each of the numbers of movies the students have seen. Then read the data and make a tally mark for each piece of information next to the appropriate number.

Next, use the table to answer the questions.

There are $6 + 4 + 3 + 2$, or 15, students who saw at least 3 movies and $8 + 9$, or 17, students who saw fewer than 3 movies.

NUMBER OF MOVIES SEEN	
1	ⅢⅢ Ⅲ
2	ⅢⅢ ⅢⅠ
3	ⅢⅢ Ⅰ
4	ⅢⅠ
5	Ⅲ
6	Ⅱ

Check

How can you check your answer?

What if two students were absent that day and they each had seen 5 movies? How many students would have seen at least 3 movies? fewer than 3 movies?

CALIFORNIA STANDARDS MR 2.0 Students use strategies, skills, and concepts in finding solutions. **MR 3.2** Note the method of deriving the solution and demonstrate a conceptual understanding of the derivations by solving similar problems. *also* **MR 3.3**

Solve the problem by displaying the data in a table.

Members of the math club took a survey to find out how each person gets to school. These are the results.

skateboard	bike	in-line skates	bike
walk	bus	bike	walk
skateboard	bike	bike	bus
walk	skateboard	in-line skates	bike
bus	walk	bus	walk
skateboard	bike	bus	walk

Draw a Diagram or Picture
Make a Model
Predict and Test
Work Backward
Make an Organized List
Find a Pattern
► **Make a Table or Graph**
Solve a Simpler Problem
Write an Equation
Use Logical Reasoning

1. Make a table using the data above. How many categories of data are in your table?

   A 2   **B** 3   **C** 4   **D** 5

2. How many of the math club members walk or ride a skateboard to school?

   F 8   **G** 9   **H** 10   **J** 11

3. How many of the math club members ride a bike or the bus to school?

4. How many more math club members walk to school than ride in-line skates to school?

MIXED STRATEGY PRACTICE

5. Julia receives a $50.00 gift certificate to a music store. If she wants to buy three $13.99 CDs and two $7.99 cassettes, how much of her own money will Julia have to add to the $50.00 gift certificate?

6. A basketball team scores 23 points, including six 1-point foul shots. How many 2-point and 3-point baskets could they have made? Make a list of all the possibilities.

7. An automobile club sells books of 10 movie tickets for $45.00. If tickets usually cost $8.75 each, how much do you save on 10 tickets by buying the book?

8. A fence separating two gardens is 24 ft long. If there is a post in the ground every 3 ft, how many posts are in the fence?

9. On a wildlife outing, Enrique spots 3 more seagulls than egrets and 5 more egrets than geese. If he spots 8 geese, how many birds does he spot in all?

10. If today is Tuesday, what day of the week will it be 200 days from today?

11. ✏️ **Write a problem** using the data from the math club survey at the top of the page.

Frequency Tables and Line Plots

Learn how to record and organize data collected in a survey.

Vocabulary

frequency table

cumulative frequency

In 2000, the largest iceberg on record contained enough ice to provide everyone on earth with 4 gallons of water per day for 40 years! Most of an iceberg is below the surface of the water. The data in the table below show the number of icebergs a scientist observed every day for 20 days.

NUMBER OF ICEBERGS OBSERVED				
12	15	11	16	15
16	16	15	13	16
16	14	12	14	16
15	13	15	14	15

You can use a line plot to record data.

EXAMPLE 1

Use the data above to make a line plot.

Step 1: Draw a horizontal line.

Step 2: On your line, write the numerical values for the number of icebergs, using vertical tick marks.

Step 3: Plot the data.

Each X represents the number of icebergs the scientist observed during one day.

Number of Icebergs Observed

TECHNOLOGY LINK

More Practice: Use *Data ToolKit* to make frequency tables and line plots.

A line plot helps you see if there are any gaps or extremes in the data. You can also see where data cluster. The line plot in Example 1 shows that the scientist observed 15–16 icebergs per day more often than she observed 11–14 icebergs per day.

CALIFORNIA STANDARDS SDAP 1.1 Compute the range, mean, median, and mode of data sets. *also* MR 2.4, SDAP 1.0

Frequency Tables

A **frequency table** shows the total for each category or group. You can add a cumulative frequency column to the table. **Cumulative frequency** is a running total of the frequencies.

EXAMPLE 2

The local mall surveyed 75 teens regarding the overall quality of food at the mall's food court. Use the data from the survey to make a frequency table. Do you think most teens like the quality of food at the food court?

Quality of Food at Food Court	
Quality	Tally
Very Good	![tally 20]
Good	![tally 30]
Fair	![tally 15]
Poor	![tally 10]

List the categories of quality in one column. Record the total for each category in the frequency column. Record the running total in the cumulative frequency column.

QUALITY OF FOOD AT FOOD COURT		
Quality	Frequency	Cumulative Frequency
Very Good	20	20
Good	30	50
Fair	15	65
Poor	10	75

← 20 + 30 = 50
← 50 + 15 = 65
← 65 + 10 = 75

So, most teens liked the quality of food at the food court: 50 out of 75 teens surveyed rated the food as good or very good.

A local mall took a survey to find the number of times people go to the mall in one year. The survey results are shown below.

NUMBER OF TRIPS TO THE MALL										
12	20	5	7	13	11	3	9	19	16
8	17	17	16	1	6	8	5	14	9

EXAMPLE 3

Remember that the range of a set of data is the difference between the greatest number and the least number in the set of data.

Use the data about the number of trips to the mall to make a cumulative frequency table with intervals.

20 − 1 = 19   *Find the range.*

Since 4 × 5 = 20, which is close to 19, make 4 intervals that include 5 consecutive numbers.   *Use the range to determine intervals.*

Complete the cumulative frequency table.

TRIPS TO THE MALL		
Number of Trips	Frequency	Cumulative Frequency
1-5	4	4
6-10	6	10
11-15	4	14
16-20	6	20

Cumulative frequency tables with intervals can also be used to organize data about different age groups. The data in the table below show the ages of joggers at the Get Fit exercise trail.

AGES OF JOGGERS									
27	31	11	22	25	42	44	36	33	24
19	34	48	17	28	33	32	30	40	31
38	18	22	23	24	27	36	19	26	32

EXAMPLE 4

Use the data about the ages of joggers at the Get Fit exercise trail to make a cumulative frequency table with intervals.

$48 - 11 = 37$ *Find the range.*

Since $4 \times 10 = 40$, which is close to 37, make 4 intervals that include 10 consecutive numbers.   *Use the range to determine intervals.*

Complete the cumulative frequency table.

AGES OF JOGGERS		
Age	Frequency	Cumulative Frequency
10–19	5	5
20–29	10	15
30–39	11	26
40–49	4	30

CHECK FOR UNDERSTANDING

Think and Discuss ▶ **Look back at the lesson to answer each question.**

1. **REASONING** **Name** the intervals you could have in Example 3 if you made 2 intervals. Explain.

2. **Explain** how to find the size of the survey sample by using the frequency table in Example 4.

Guided Practice ▶ **For 3–5, use the data in the table.**

3. Make a line plot.

4. Make a cumulative frequency table.

5. Find the range.

AGES OF POP MUSIC LISTENERS						
12	13	17	16	15	13	25
28	27	18	15	12	14	23
13	15	17	13	22	18	15

PRACTICE AND PROBLEM SOLVING

Independent Practice ▶ **For 6–8, use the data in the table.**

6. Make a line plot.

7. Make a cumulative frequency table.

8. Find the range.

NUMBER OF KILOMETERS BIKED						
6	11	7	8	8	5	15
14	12	10	11	6	7	9
15	15	8	9	12	17	10

Kim asked people in her neighborhood about their favorite type of movie. Use the data for 9–10.

FAVORITE MOVIES	
Movies	Tally
Drama	$\cancel{||||}$ $\cancel{||||}$
Action	$\cancel{||||}$ $|||$
Comedy	$\cancel{||||}$ $\cancel{||||}$ $|||$
Mystery	$\cancel{||||}$ $||||$

9. Organize the data into a frequency table. Include a cumulative frequency column.

10. What is the size of the sample?

For 11–14, use the data in the box at the right.

11. Make a line plot.

12. Find the range.

WEIGHTS OF STUDENTS' PET DOGS (LB)					
40	10	22	33	44	40
41	42	12	35	20	10
30	25	20	18	46	35
34	35	40	40	35	38

13. How many numbers would be in each interval if you used the range to make 4 intervals?

14. Make a cumulative frequency table using 4 intervals for the weights.

Problem Solving ▶ Applications

15. **REASONING** Survey your classmates about their favorite hobbies. Organize the data in a tally table and a frequency table. Can you make a line plot of the data? Explain.

16. Decide on appropriate intervals and make a cumulative frequency table using intervals for the data in the table below.

POUNDS OF VEGETABLES CONSUMED YEARLY											
12	16	18	12	9	21	14	17	18	5	23	25	32
17	16	8	9	22	21	14	18	7	23	24	19	13

17. **? What's the Question?** In a set of data, the greatest value is 54 and the least value is 30. If the answer is 24, what's the question?

MIXED REVIEW AND TEST PREP

Tell whether the question is *biased* or *unbiased*. (p. 98)

18. Do you feel that sixth graders should do two hours of homework each night instead of watching TV?

19. Whom do you intend to vote for in the upcoming Student Government election?

20. Tell if the expression is numerical or algebraic. $a + 3.9$ (p. 28)

21. **TEST PREP** Which is the value of 4^5? (p. 40)

   A 9 **B** 20 **C** 625 **D** 1,024

22. **TEST PREP** Sharon rode her bike a total of 36 km at a rate of 9 km per hr. How long did she ride? (p. 22)

   F 2 hr **G** 4 hr **H** 6 hr **J** 9 hr

Extra Practice page H36, Set C

Measures of Central Tendency

QUICK REVIEW

1. 420 ÷ 5   **2.** 392 ÷ 7   **3.** 320 ÷ 8   **4.** 540 ÷ 60   **5.** 720 ÷ 9

Three measures of central tendency are mean, median, and mode. Measures of central tendency can help you to describe a set of data.

In April 1998, amateur rocket builder Gib Reynolds set an American record for 7- to 13-year-olds. His model rocket soared to a height of 47 meters. Data for heights some model rockets traveled, including Gib's, are shown at the right. Find the mean, median, and mode for the data.

ROCKET HEIGHTS (M)				
47	28	33	35	28

Mean: (47 + 28 + 33 + 35 + 28) ÷ 5 = 171 ÷ 5 = 34.2 m

Median: 28 28 33 35 47; 33 m   **Mode:** 28 m

Sometimes a line plot can help you to find the mode and the median.

EXAMPLE 1

Remember that the mean is the sum of a group of numbers divided by the number of addends.

The median is the middle number in a group of numbers arranged in numerical order.

The mode is the number that occurs most often.

Use the data to make a line plot. Find the mode and the median.

DAILY TEMPERATURES (°F)										
72	82	83	78	81	78	73	74	75	73	76
71	75	80	83	72	72	78	81	79	82	76

```
                X                       X
                X   X       X   X       X
        X   X   X   X   X   X   X   X   X   X   X   X   X
    +---+---+---+---+---+---+---+---+---+---+---+---+---+---+
   70  71  72  73  74  75  76  77  78  79  80  81  82  83  84
```

Mode: The modes are 72 and 78 since each occurs three times.

Median: Since there are 22 temperatures, the median is the mean of the 11th and 12th temperatures.   (76 + 78) ÷ 2 = 77

Math Idea ▶ When you want to summarize a set of data as one value, you can use one of the three measures of central tendency.

EXAMPLE 2

Pedro jogged 6 mi, 5 mi, 2 mi, 2 mi, and 4 mi over 5 days. Which measure of central tendency is most useful to describe the data?

Mean: (6 + 5 + 2 + 2 + 4) ÷ 5 = 19 ÷ 5 = 3.8

Median: 2 2 4 5 6; 4   **Mode:** 2

The mean, or 3.8, and the median, 4, are close to most of the data while the mode, or 2, is closer to the lower end of the data. So, the mean or median is most useful to describe the data.

CALIFORNIA STANDARDS SDAP 1.1 Compute the range, mean, median, and mode of data sets. SDAP 1.4 Know why a specific measure of central tendency (mean, median, mode) provides the most useful information in a given context. *also* MR 2.4, O¬ NS 2.0, SDAP 1.0

Sometimes the mean is not the best measure of central tendency to describe a set of data.

EXAMPLE 3

The bald eagle, America's national bird, was once a greatly endangered species. In recent decades, the species has made a dramatic recovery. The map at the right shows the number of nests in five zones of Montana. Which measure of central tendency is most useful to describe the data?

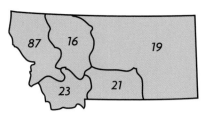

Numbers of nests: 87, 16, 19, 21, 23

Mean: $(87 + 16 + 19 + 21 + 23) \div 5 = 166 \div 5 = 33.2$

Median: 16 19 21 23 87 → 21

Mode: No value occurs more than any other, so there is no mode.

The median, 21, is close to most of the data. The high value of 87 makes the mean greater than any of the four other values. So, the median is most useful to describe the data.

CHECK FOR UNDERSTANDING

Think and Discuss ▶ **Look back at the lesson to answer the question.**

1. **What if,** in Example 3, the region with 87 nests had only 27 nests? Which of the measures would change? Which measure would be most useful to describe the data?

Guided Practice ▶ **For 2–4, use the table below.**

Day	Sun	Mon	Tue	Wed	Thu	Fri	Sat
Hours of Sleep	8	6	7	7	4	10	7

2. Find the mean.   3. Find the median.   4. Find the mode.

Find the mean, median, and mode.

5. 22, 24, 22, 29, 33, 14   6. 124, 120, 132, 133, 119, 90, 87

PRACTICE AND PROBLEM SOLVING

Independent Practice ▶ **For 7–9, use the table below.**

Game	1	2	3	4	5	6	7	8
Points Scored	10	12	10	17	18	12	20	24

7. Find the mean.   8. Find the median.   9. Find the mode.

Find the mean, median, and mode.

10. 76, 63, 40, 52, 52, 40, 6, 15   11. 365, 180, 360, 720, 59

12. Use the data below to make a line plot. Use your line plot to find the median and the mode. Then use the data to find the mean. Which measure of central tendency would be most useful to describe the data?

YEARLY RAINFALL (IN.)									
47	11	8	14	15	16	13	13	17	22

13. Over the past 8 days, Fernando spent $5, $6, $9, $8, $10, $15, $15, and $2. Which measure of central tendency would be most useful to describe the data?

14. Algebra Reggie's average score on five math tests is 92. On his first four tests, Reggie's scores were 88, 97, 93, and 82. What was Reggie's score on his fifth test?

15. ❓ **What's the Error?** Tim wrote $(7 + 3 + 10 + 4) \div 3 = 8$ to find the mean of 7, 3, 10, and 4. What is his error? What is the correct mean?

MIXED REVIEW AND TEST PREP

Use the table for 16–17. (p. 102)

16. How many named blue or red as their favorite color?

17. What is the size of the sample?

FAVORITE COLOR	FREQUENCY	CUMULATIVE FREQUENCY
Blue	32	32
Red	17	49
Green	22	71

Write the decimal as a percent. (p. 60)

18. 0.65   **19.** 0.9

20. TEST PREP Which is the product of 0.56 and 2.5? (p. 70)

   A 0.14   **B** 1.4   **C** 14   **D** 140

Spinnermeania Play the game below to practice finding the mean, median, and mode of a set of data.

Materials: spinner showing 0–9 for each player

- Each player spins four times, recording the results.

- Each player determines the mean, median, and mode of the numbers he or she spun. The player with the greatest number in each category gets a point. If there is a tie, each player with that number gets a point.

- Repeat for four more rounds, increasing the number of spins in each round to five, eight, ten, and twelve. The player with the greatest total points at the end of the five rounds wins.

Extra Practice page H36, Set D

Outliers and Additional Data

Learn how outliers and additional data affect the mean, median, and mode.

Vocabulary

outlier

When you add data to a data set, some measures of central tendency for the set change. How they change depends on how the new data are related to the original data.

EXAMPLE 1

Gina saved dimes for 8 weeks. She recorded the number of dimes in a table. In the ninth week, her parents gave her 25 dimes. In the tenth week, her grandparents gave her 52 dimes. Find the mean, median, and mode for Gina's data, using the values for the 8 weeks of her savings. Then find the measures for all 10 weeks.

DIMES GINA SAVED								
Week	1	2	3	4	5	6	7	8
Dimes	17	18	30	15	1	6	19	6

Data for 8 Weeks
Mean: $(17 + 18 + \ldots + 19 + 6) \div 8 = 14$ **Mode:** 6
Median: 1 6 6 15 17 18 19 30; $(15 + 17) \div 2 = 16$

Data for 10 Weeks
Mean: $(17 + 18 + \ldots + 25 + 52) \div 10 = 18.9$ **Mode:** 6
Median: 1 6 6 15 17 18 19 25 30 52; $(17 + 18) \div 2 = 17.5$

EXAMPLE 2

At age 61, Lev Sarkisov climbed Mount Everest. Suppose he climbed with a group whose ages were 30, 31, 33, 30, 29, 35, 33, 30, and 28. Which measure of central tendency is most affected by adding Sarkisov's age to the group?

Data without Sarkisov's Age

Mean: $(30 + 31 + \ldots + 30 + 28) \div 9 = 31$ **Mode:** 30

Median: 28 29 30 30 30 31 33 33 35; 30

Data with Sarkisov's age

Mean: $(30 + 31 + \ldots + 61) \div 10 = 34$ **Mode:** 30

Median: 28 29 30 30 30 31 33 33 35 61; $(30 + 31) \div 2 = 30.5$

The mean increases by 3, the median increases by 0.5, and the mode remains the same. So, the mean is the measure most affected by adding Sarkisov's age.

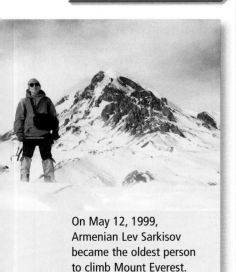

On May 12, 1999, Armenian Lev Sarkisov became the oldest person to climb Mount Everest.

CALIFORNIA STANDARDS SDAP 1.2 Understand how additional data added to data sets may affect these computations of measures of central tendency. **SDAP 1.3** Understand how the inclusion or exclusion of outliers affects measures of central tendency. *also* **SDAP 1.0, SDAP 1.1, MR 2.0**

109

Bethany used a line plot to record the number of concert tickets she sold on 10 different days. Most of the data are between 11 and 16. The data values at 32 and 34 are outliers, or extreme values. An **outlier** is a data value that stands out from other data values in a set. Outliers can significantly affect measures of central tendency.

```
                X
        X   X           X
        X   X           X   X                                   X       X
    ┼───┼───┼───┼───┼───┼───┼───┼───┼───┼───┼───┼───┼───┼───┼───┼───┼───┼───┼───┼───┼───┼───┼───┼───┼───┼───┼
    10  11  12  13  14  15  16  17  18  19  20  21  22  23  24  25  26  27  28  29  30  31  32  33  34  35
```

EXAMPLE 3

A. Find the measures of central tendency for the 8 data values in the line plot if the outliers are not included.

Mean: $(11 + 11 + 12 + 12 + 12 + 15 + 15 + 16) \div 8 = 104 \div 8 = 13$
Median: 11 11 12 12 12 15 15 16; 12   **Mode:** 12

B. Find the measures of central tendency for all the data values with the outliers included.

Mean: $(104 + 32 + 34) \div 10 = 170 \div 10 = 17$
Median: 11 11 12 12 12 15 15 16 32 34; $(12 + 15) \div 2 = 13.5$
Mode: 12

C. Suppose Bethany sold 1 ticket the next day and 0 tickets the final day. Include these data values with the data in the line plot and find the measures of central tendency for all 12 data values.

Mean: $(170 + 1 + 0) \div 12 = 171 \div 12 = 14.25$
Median: 0 1 11 11 12 12 12 15 15 16 32 34; 12   **Mode:** 12

• How do the outliers in part B affect the mean, median, and mode?

CHECK FOR UNDERSTANDING

Think and Discuss ▶ **Look back at the lesson to answer the question.**

1. **What if** Gina's grandparents had given her 32 dimes in Example 1? How would the measures of central tendency be affected?

Guided Practice ▶ 2. Find the mean, median, and mode for the ages in the table.

3. Find the mean, median, and mode if a 28-year-old joins the club.

AGES OF MEMBERS OF MODEL AIRPLANE CLUB			
15	12	9	13
22	19	12	14

Jeff received these scores on his science quizzes: 75, 70, 70, 45, 100, 70, 70, 80, 75, 80.

4. **a.** Use the scores to make a line plot. Circle the outliers.

   b. Find the mean, median, and mode with and without the outliers.

   c. Why did these outliers have so little effect on the mean, median, and mode?

Independent ▶ Practice

The table shows the points scored by two basketball players in the first six games of the season.

POINTS SCORED		
Game	Abe	Bart
1	7	15
2	5	0
3	4	0
4	20	12
5	7	13
6	2	5

5. a. Find the mean, median, and mode of each player's points for six games.

b. In the seventh game, Abe scored 20 points and Bart scored 0 points. Find the mean, median, and mode of each player's points for seven games.

c. How are the measures of central tendency for Abe's and Bart's scores affected by the points scored in the seventh game?

The table shows the number of pets owned by children in a pet club.

6. a. Draw a line plot for the data about the number of pets children own. Circle the outliers.

NUMBER OF PETS OWNED			
1	1	2	1
3	12	4	4
3	3	10	1

b. Find the mean, the median, and the mode for the data with and without the outliers.

c. How does including the outliers affect the mean? the median? the mode?

Problem Solving ▶ Applications

7. a. Emily scored 75, 85, 35, 85, 70, 10 on her first 6 math quizzes. What score does Emily need on her seventh quiz so that after 7 quizzes she will have a median score of 75?

b. *REASONING* The greatest score Emily can make on a math quiz is 100. Explain whether it is possible for her to have a mean score of 70 after taking the seventh quiz.

8. *REASONING* The high temperatures in °F over 7 days in town were 72°, 73°, 70°, 68°, 70°, 71°, and 39°. Explain why the mean would not be a good measure to describe the temperatures.

MIXED REVIEW AND TEST PREP

Find the mean, median, and mode of the data. (p. 106)

9. 5, 6, 8, 6, 5, 7, 5, 4

10. 42, 56, 44, 38, 10

11. 10, 12, 12, 10, 8, 5

12. A company claims that its 8-year warranty lasts 4 times as long as any other warranty. What is the longest of the other warranties? (p. 22)

13. TEST PREP There are 180 players in a baseball league. Each team has 12 players. How many teams are in the league? (p. 22)

A 12 teams **B** 15 teams **C** 16 teams **D** 18 teams

Data and Conclusions

Learn how to decide whether conclusions based on data are valid.

Would you take $1,000 to give up watching television for three months? In a recent survey of adults, 87% answered "Yes!"

Kendra used the same question to survey students at her middle school. She randomly selected 60 of these students to participate in the survey. She found that 90% of the students she surveyed answered "Yes." So, Kendra concluded that about 90% of all students at her middle school would take $1,000 to give up watching television for three months.

EXAMPLE 1

Is Kendra's conclusion valid?

To determine if Kendra's conclusion is valid, use the checklist below.

Remember that when you want to collect data by asking a question, how you state the question is very important. The question must be unbiased.

• Who are the people you are interested in studying (the population)?	✓ The population Kendra is interested in is the students at her middle school.
• Was the sample selected from the correct population?	✓ Yes. Kendra surveyed students from her middle school.
• Was the sample randomly selected?	✓ Yes. The sample was randomly selected.
• Was the question unbiased?	✓ Yes. The question was unbiased.

So, Kendra's conclusion is valid.

• **What if** Kendra had surveyed 60 students who participate in school athletics? Would her conclusion be valid? Explain.

Math **I**dea ▶ In order to draw a valid conclusion from a set of data, you must select a random sample from the correct population and ask an unbiased question.

CALIFORNIA STANDARDS ⊶ SDAP 2.5 Identify claims based on statistical data and, in simple cases, evaluate the validity of the claims. *also* MR 1.0, MR 2.4

Barry read the results of a survey conducted ten years ago at his school. A random sampling of sixth graders produced the results displayed in the circle graph at the right.

Barry wanted to see if a survey of sixth graders at his school today would produce the same results. He randomly selected 200 sixth graders from his school and asked them the same question.

What is your favorite type of television program?

Sitcoms 70%

Dramas 12%

Movies 18%

> **EXAMPLE 2**

What conclusions can Barry draw from the results of his survey shown in the circle graph below?

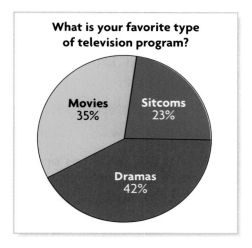

What is your favorite type of television program?

Movies 35%

Sitcoms 23%

Dramas 42%

Barry can conclude the following:

Type of Show	Conclusion
Sitcoms	Interest in sitcoms decreased.
Movies	Interest in movies increased.
Dramas	Interest in dramas increased.

• How do you know Barry's conclusions are valid?

CHECK FOR UNDERSTANDING

Think and Discuss ▶ Look back at the lesson to answer each question.

1. **Tell** if Kendra's conclusions in Example 1 would still be valid if she had surveyed students from her first two classes of the day. Explain.

2. **What if,** in Example 2, Barry had asked, "Do you agree with me that a drama is the best kind of television show to watch?" Would he be able to make the same conclusions? Explain.

For 3, write *yes* or *no* to tell whether the conclusion is valid. Explain your answer.

3. Five hundred randomly selected teenagers from Fresno are asked to name their favorite subject. Eighty percent of these teenagers respond that their favorite subject is math. You conclude that math is the most popular subject among teenagers in Fresno.

Use Data For 4–5, use the data in the graph.

4. The graph at the right shows the results of a survey of 600 randomly selected middle school students. Jen concludes that more middle school students are interested in football than in any other sport. Is her conclusion valid? Explain.

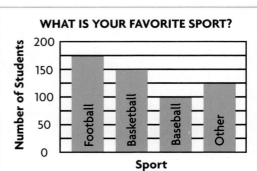

WHAT IS YOUR FAVORITE SPORT?

5. Robert concludes that more high school students are interested in football than in any other sport. Is his conclusion valid? Explain.

PRACTICE AND PROBLEM SOLVING

For 6–7, write *yes* or *no* to tell whether the conclusion is valid. Explain your answer.

6. All the members of the science club were asked if they wanted to take more science classes. Each of them answered yes. You decide that science is becoming more popular among all students.

7. A sample of your friends shows that their favorite color is green. You decide that the same is probably true for the entire school.

Use Data For 8–9, use the data in the graph.

8. The graph at the right shows the results of a survey of 1,000 randomly selected teenagers from California. José concludes that corn is the favorite vegetable of teenagers from New York. Is his conclusion valid? Explain.

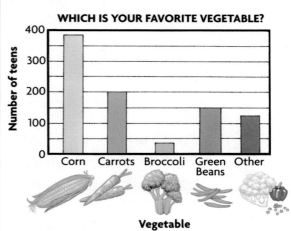

WHICH IS YOUR FAVORITE VEGETABLE?

9. Jerome concludes that carrots are the second favorite vegetable among California teenagers. Is his conclusion valid? Explain.

Problem Solving ▸ Applications

For a survey completed five years ago, middle school students from San Diego were asked, "What is your favorite hobby?"

FAVORITE HOBBIES	
Reading	36%
Crafts	7%
Coin Collecting	2%
Stamp Collecting	2%
Building Models	12%
Trading Cards	39%
Other	2%

A research company randomly selected 750 current middle school students from San Diego and asked them the same question. The results are in the table.

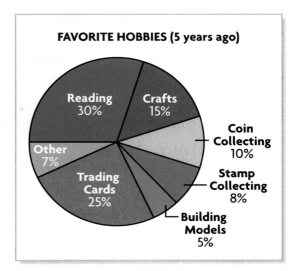

FAVORITE HOBBIES (5 years ago)

Reading 30%
Crafts 15%
Coin Collecting 10%
Stamp Collecting 8%
Building Models 5%
Trading Cards 25%
Other 7%

10. What conclusion can you draw about student interest in reading?

11. What conclusion can you draw about student interest in crafts?

12. What conclusion can you draw about student interest in coin collecting?

13. What conclusion can you draw about student interest in trading cards?

14. What conclusion can you draw about student interest in models?

15. How can you tell that your conclusions are valid?

16. For which of the hobbies did interest increase the most? decrease the most?

17. **? What's the Error?** A survey of middle school athletes showed that their favorite exercise is jogging. Cathy looked at the survey results and concluded that all students prefer jogging. Explain Cathy's error.

18. Robert spent $60 at the sports equipment store. He bought a glove for $35 and 5 baseballs. What was the price for each baseball?

MIXED REVIEW AND TEST PREP

Find the mean, median, and mode of the data. (p. 106)

19. 2, 4, 4, 3, 6, 8, 1, 6

20. 22, 44, 31, 46, 22

21. 6, 8, 11, 10, 2, 5

22. **TEST PREP** Which is the solution of $x + 3 = 10$? (p. 30)

   A $x = 3$ **B** $x = 7$ **C** $x = 10$ **D** $x = 13$

23. **TEST PREP** Which is the solution of $x \div 3 = 6$? (p. 30)

   F $x = 2$ **G** $x = 3$ **H** $x = 9$ **J** $x = 18$

For 1–2, determine the type of sample. Write *convenience*, *random*, or *systematic*. (pp. 94–97)

1. A cereal company surveys a person in every third apartment in an apartment building.

2. Alicia wants to find out what snack foods students in her school prefer. She randomly surveys 100 students.

3. **VOCABULARY**   In a survey, if individuals or groups in the population are not represented by the sample, the sample is __?__ .
   (p. 98)

Tell whether the sample is biased or unbiased. (pp. 98–99)

4. A store randomly surveys 10 out of every 100 customers about the quality of its service.

5. A teacher surveys girls about the best day to give a test.

For 6–9, use the table at the right. (pp. 102–105)

6. What is the sample size?

7. Find the range.

8. Make a line plot.

9. Make a frequency table.

STUDENTS' HEIGHTS (in inches)									
62	72	63	62	69	70	60	64	66	63
71	62	65	68	63	70	64	62	67	70

Find the mean, median, and mode. (pp. 106–109)

10. 17, 12, 23, 19, 23

11. 6.2, 5.5, 8.4, 5.5

12. 265, 235, 171, 253

13. Ten people in Ana's neighborhood formed a bike club. The table shows the ages of the members. (pp. 106–109)

   a. Find the mean, median, and mode for the ages of the ten members.

   b. Find the mean, median, and mode if the adult advisor for the club is 27 years old and is a club member. Compare with part *a*.

AGES OF MEMBERS OF BIKE CLUB				
13	14	10	11	13
12	13	9	11	10

14. Janyce wanted to find out where visitors to the local mall come from. She surveyed visitors, tallied her results, and came up with the following percents: California—34%; Idaho—14%; Nevada— 11%; Oregon—16%; Montana—8%; Washington—11%; other states—6%. Make a table to organize the information from greatest to least percent. (pp. 100–101)

15. If 50% of the people who visit the mall in Exercise 14 come from within 20 miles of the mall, which is a reasonable conclusion? Why? (pp. 100–101)

   A The mall where Janyce took her survey is in southern California.

   B The mall where Janyce took her survey is in eastern Nevada.

   C The mall where Janyce took her survey is in southern Washington.

   D The mall where Janyce took her survey is in northern California.

Eliminate choices.

See item **5.**

Keep the question in mind as you read each answer choice. Look for the most important characteristic of a survey sample to decide what was wrong with the survey.

Also see problem **5**, p. H64.

Choose the best answer.

1. Which symbol makes this true?

   8.34 ● 8.342

   A <   **C** =

   B >   **D** ÷

2. Which of the following is equivalent to 27 + 43?

   F 25 + 50

   G 30 + 40

   H 27 + 40 + 12

   J 20 + 60

3. Which of the following numbers is between 8.07 and 8.3?

   A 8.06

   B 8.017

   C 8.29

   D 8.312

4. The student council wants to conduct a survey to get an idea of what other students think about changing the school colors. Which group would be the best sample?

   F 25 friends

   G 25 football players

   H 25 teachers

   J 25 students selected at random

5. Before Warren ran for class president, he surveyed 20 of his friends, asking "Would you vote for me for class president?" Of those surveyed, 100% answered "yes." Warren was very surprised when he lost the election. What was wrong with his survey?

   A Nothing; his friends changed their minds and didn't vote for him.

   B His survey sample did not represent the student body.

   C He should have included teachers in his survey.

   D His survey did not ask the right question.

For 6–7, use the information below.

Jon had the following scores for 5 rounds of golf: 78, 80, 69, 75, 73.

6. What is the mean of the scores?

   F 50   **H** 74

   G 71   **J** 75

7. What is the range of the scores?

   A 11   **C** 80

   B 75   **D** Not here

8. The ages of students in the Hiking Club are 13, 11, 15, 14, 12, 13, 13. What is the mode of this set of data?

   F 11   **H** 13

   G 12   **J** 14

9. What is $2 \times 2 \times 2 \times 2 \times 2 \times 2 \times 2$ expressed in exponential notation?

   A 2^2 **C** 2^7

   B 7^2 **D** 2^6

10. 883.03 ÷ 22.7

   F 389   **H** 3.89

   G 38.9   **J** Not here

CHAPTER 6

Graph Data

In July 1994, fragments of Comet Shoemaker-Levy bombarded Jupiter. One impact left a crater about the same diameter as Earth, which is approximately 12,756 km wide. Evidence of past impacts of meteors, asteroids, or comets also appears on Earth and the Moon. One newly discovered crater on the floor of Chesapeake Bay has a diameter of about 80 km. On a bar graph, about how many times as long would the bar for the Jupiter crater be than the bar for the Chesapeake Bay crater?

EARTH'S AND MOON'S LARGEST CRATERS	
Location	**Approximate Diameter (km)**
Popigri, Siberia	97
Manicouagan, Canada	100
Chicxulub, Mexico	180
Orientale basin, Moon	1,300
Ibrium basin, Moon	1,800
Aitken basin, Moon	2,500

Check What You Know

Use this page to help you review and remember important skills needed for Chapter 6.

 Read Bar Graphs (See p. H6.)

For 1–4, use the bar graph at the right.

1. List the four launch vehicles in order of size from tallest to shortest.

2. About how tall is the Saturn V?

3. About how much taller is the Saturn V than the Ariane IV?

4. About how much taller is the tallest launch vehicle than the shortest?

HEIGHTS OF LAUNCH VEHICLES

For 5–8, use the bar graph at the left.

5. Name two Great Lakes whose combined area is less than the area of Lake Superior.

6. What is the area of the third-largest Great Lake?

7. How much larger is the area of Lake Michigan than the area of Lake Ontario?

8. Which Great Lake has an area of about 22,000 sq mi more than the area of Lake Erie?

AREAS OF THE GREAT LAKES

✓ **Read Stem-and-Leaf Plots** (See p. H7.)

For 9–15, use the stem-and-leaf plot at the right.

9. What is the score shown by the third stem and fourth leaf?

10. What is the median of Michael's golf scores?

11. What is the range of Michael's golf scores?

12. What is the mode of Michael's golf scores?

13. How many scores of 90 will Michael need in order to have a mode of 90 for all of his rounds?

14. How many rounds of golf did Michael play?

15. If Michael plays two more rounds and has scores of 86 and 81, what would be the median of the scores?

MICHAEL'S GOLF SCORES

Stems	Leaves
7	6 7 8 9 9
8	0 2 3 5 5 6 8 9
9	0 0 1 2 3 8

> ### LOOK AHEAD
>
> **In Chapter 6 you will**
> - make and analyze stem-and-leaf plots, box-and-whisker graphs, histograms, and line plots
> - find unknown values
> - analyze graphs

LESSON 6.1

Make and Analyze Graphs

Learn how to display and analyze data in bar graphs, line graphs, and circle graphs.

QUICK REVIEW

Compare. Use < or >.
1. 5 ● 9   2. 25 ● 20   3. 87 ● 88
4. 19 ● 23   5. 309 ● 304

When data are grouped in categories, a bar graph is a good way to display the data. Look at the bar graph below.

Vocabulary

multiple-bar graph

multiple-line graph

HEAVIEST MARINE MAMMALS

The data in the table below show the weights of other mammals.

WEIGHTS OF SOME MAMMALS				
Mammal	Lion	Gorilla	Tiger	Grizzly Bear
Male (lb)	400	450	420	500
Female (lb)	300	200	300	400

A **multiple-bar graph** shows two or more sets of data.

EXAMPLE 1

Use the data in the table above to make a double-bar graph. How do the weights of males compare to the weights of females?

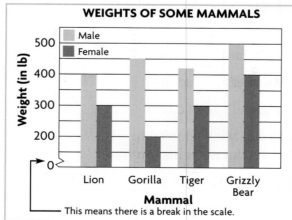

Determine an appropriate scale.

Use bars of equal width. Use the data to determine the heights of the bars.

Title the graph and both axes. Include a key.

For each of the mammals in the graph above, the males weigh more than the females.

CALIFORNIA STANDARDS MR 1.1 Analyze problems by identifying relationships, distinguishing relevant from irrelevant information, identifying missing information, sequencing and prioritizing information, and observing patterns. *also* **MR 2.2, MR 2.4**

Line Graphs

A bar graph and a line graph are often used to display the same data. However, a line graph is the better choice when the data show change over time. The line graph below shows the change in the price of a stock every five years from 1970 to 2000.

Like a multiple-bar graph, a **multiple-line graph** can show two or more different sets of data on one graph.

MONEY IN SAVINGS ACCOUNTS					
	March	**April**	**May**	**June**	**July**
Bob	$163	$172	$151	$138	$102
Jan	$43	$55	$76	$79	$96

EXAMPLE 2

Use the data in the table above to make a double-line graph. If the trends continue, how would you describe the amount of money that Bob is saving? that Jan is saving?

Determine an appropriate scale.

Mark a point for each amount saved for Bob and connect the points.

Mark a point for each amount saved for Jan and connect the points.

Title the graph and both axes. Include a key.

The money in Bob's savings account is decreasing. The money in Jan's savings account is increasing.

• Use the graph to predict what will happen to each savings account in August.

Graphs show pictures of data. Bar graphs and line graphs use axes to help you analyze data. A circle graph helps you compare parts to the whole or to other parts.

EXAMPLE 3

TECHNOLOGY LINK
More Practice: Use *Data Toolkit* to make bar graphs, line graphs, and circle graphs.

In 1987, compact discs (CDs) were new. The circle graph shows the way recorded music was sold in the United States in 1987. About how many cassettes were sold for every CD sold that year?

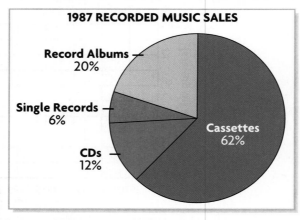

1987 RECORDED MUSIC SALES

Record Albums — 20%
Single Records — 6%
CDs — 12%
Cassettes 62%

Find the parts that represent cassettes and CDs.

Compare the percent of recorded music sold on cassettes to the percent sold on CDs.

Cassettes made up 62% of recorded music sales, while CDs made up 12%. $62 \div 12 \approx 60 \div 12$, or 5.

So, about 5 cassettes were sold for every CD sold.

- By 1997, the percent of recorded music sold in the form of CDs was about 6 times as great as it had been in 1987. About what percent of recorded music was sold on CDs in 1997?

CHECK FOR UNDERSTANDING

Think and ▶ Discuss

Look back at the lesson to answer each question.

1. **Explain** why the graphs in Example 1 and Example 2 have a key.

2. **Describe** the kind of data you would display in a bar graph, a line graph, and a circle graph.

Guided ▶ Practice

Tell if you would use a bar, line, or circle graph to display the data.

3. The average monthly rainfall in your city over two years

4. A family budget divided into types of expenses

PRACTICE AND PROBLEM SOLVING

Independent ▶ Practice

Tell if you would use a bar, line, or circle graph to display the data.

5. The heights of five different students

6. The price of a stock over a period of several months

7. The way you spend your weekly allowance

8. The population of six different cities in your state

9. Make a multiple-bar graph using the data in the table at the right.

NUMBER OF RAINY DAYS				
	April	**May**	**June**	**July**
1999	12	3	13	5
2000	6	7	9	11

10. Make a multiple-line graph using the data in the table at the right.

AVERAGE STOCK PRICES				
	Sep	**Oct**	**Nov**	**Dec**
Stock A	$26	$29	$32	$27
Stock B	$14	$10	$8	$19

Problem Solving ▶ Applications

Use Data **For 11–13, use the double-line graph below.**

11. How do the comedy video rentals compare to the action video rentals?

12. Between which two months did the number of action video rentals decrease the most?

13. *REASONING* If the trends continue, what will happen to the number of comedy and action video rentals in July?

Use Data **For 14–16, use the circle graph at left.**

14. The number of people aged 19–64 is about how many times the number who are 65 or older?

15. *REASONING* Which of the three age groups do you think spends the most money? Explain.

16. ❓ **What's the Error?** Jay looked at the circle graph at the left and said that 61% of the U.S. population is aged 19 or older. What is his error?

U.S. POPULATION BY AGE (1997)

19-64 61%

26%

13%

18 or younger

65 or older

MIXED REVIEW AND TEST PREP

17. Fifteen students in the gym are asked what their favorite sport is. All of them say "basketball." Ron concludes that the favorite sport of all students is basketball. Is Ron's conclusion valid? Explain. (p. 112)

Find the product. (p. 70)

18. 6×0.4

19. 0.8×0.7

20. TEST PREP Which is 0.27×0.4? (p. 70)

    A 0.108   **B** 0.675   **C** 1.08   **D** 10.8

21. TEST PREP Find the value of 21^2. (p. 40)

    F 21   **G** 2^3 **H** 42   **J** 441

Find Unknown Values

Learn how to estimate unknown values from a graph and solve for the values by using logic, arithmetic, and algebra.

For fitness training this week, Karen wants to walk a total of 24 mi. She likes to walk at a rate of 4 mi per hr. The table above shows how long it will take her to walk certain distances.

Time (hr)	1	2	3	4	5
Distance (mi)	4	8	12	16	20

EXAMPLE 1

Make a line graph with the data in the table and use it to estimate how long it will take Karen to walk 24 mi.

WALKING AT 4 MI PER HR

Look at the graph. If the pattern continues, it looks as if it will take Karen 6 hr to walk 24 mi.

EXAMPLE 2

Use logical reasoning and arithmetic to find how long it will take Karen to walk 24 mi.

Look at the data in the table. Notice that whenever the time increases by 1, the distance increases by 4 mi.

5 + 1 = 6, or 6 hr   *Add 1 to the last time in the table.*

20 + 4 = 24, or 24 mi   *Add 4 to the last distance in the table.*

So, it will take Karen 6 hr to walk 24 mi.

You can also use the formula $d = rt$ to solve problems about distance (d), rate (r), and time (t). The distance traveled is a product of the rate of speed and the amount of time.

EXAMPLE 3

Use the formula $d = rt$ to find how long it will take Karen to walk 32 mi if she walks at a rate of 4 mi per hr.

$d = rt$ *Write the formula.*

$32 = 4 \times t$ *Replace d with 32 and r with 4. What number multiplied by 4 gives 32?*

$8 = t$ *The solution is t = 8.*

So, it will take Karen 8 hr to walk 32 mi.

CALIFORNIA STANDARDS MR 2.3 Estimate unknown quantities graphically and solve for them by using logical reasoning and arithmetic and algebraic techniques. *also AF 2.3, MR 3.2*

Think and ▶
Discuss

Look back at the lesson to answer the question.

1. **Explain** how you used the graph in Example 1 to predict how long it would take Karen to walk 24 mi.

Guided ▶
Practice

Ty averaged 50 mi per hr on his car trip. Use the table for 2–4.

2. Make a line graph. Use the line graph to find how long it will take Ty to drive 250 mi.

Time (hr)	1	2	3	4
Distance (mi)	50	100	150	200

3. Use logical reasoning and arithmetic to find how long it will take Ty to drive 250 mi.

4. Use the formula $d = rt$ to find how long it will take Ty to drive 350 mi.

PRACTICE AND PROBLEM SOLVING

Independent ▶
Practice

Juan averaged 8 mi per hr while biking. Use the table for 5–7.

5. Use the data in the table to make a line graph. Use the line graph to estimate how long it will take Juan to bike 48 mi.

Time (hr)	1	2	3	4
Distance (mi)	8	16	24	32

6. Use logical reasoning and arithmetic to find how long it will take Juan to bike 48 mi.

7. Use the formula $d = rt$ to find how long it will take Juan to bike 64 mi.

Problem Solving ▶
Applications

Calvin is training for a marathon. The total distance he has run each week is shown in the table.

8. Make a line graph using the data in the table.

9. If the trend in Calvin's training continues, about how far will he run in the sixth week of training?

10. ✎ **Write a problem** using Calvin's training data. Explain your solution.

Distance Run

Week	1	2	3	4	5
Miles	14	20	26	33	39

MIXED REVIEW AND TEST PREP

Find the mean, median, and mode. (p. 106)

11. 9, 6, 4, 4, 5

12. 12, 14, 32, 28

13. 2, 9, 9, 9, 6, 5, 8, 4

14. Order 0.76, 0.765, and 0.076 from least to greatest. (p. 52)

15. **TEST PREP** Find the value of $56 - 32 \div (4 \times 2)$. (p. 44)

    A 3   **B** 12   **C** 52   **D** 96

Stem-and-Leaf Plots and Histograms

Learn how to display and analyze data in stem-and-leaf plots and histograms.

Vocabulary

stem-and-leaf plot
histogram

You can use a **stem-and-leaf plot** to organize data when you want to see each item in the data. For a stem-and-leaf plot, choose the stems first and then write the leaves.

EXAMPLE 1

Bryan Berg built a 19 ft $16\frac{1}{2}$ in. high card tower with 102 levels.

The table shows the number of levels reached, without cards falling, at a card-tower building competition. Use the data to make a stem-and-leaf plot.

CARD-TOWER COMPETITION					
21	18	32	47	50	33
19	21	11	54	31	18
33	42	21	29	16	12

11 12 16 18 18 19
21 21 21 29
31 32 33 33
42 47
50 54

First, group the data by tens digits. Then, order the data from least to greatest.

Card Tower Competition

Stems	Leaves
1	1 2 6 8 8 9
2	1 1 1 9
3	1 2 3 3
4	2 7
5	0 4

Use the tens digits as stems. Use the ones digits as leaves. Write the leaves in increasing order.

The line 4 | 2 7 means 42 and 47.

EXAMPLE 2

Use the data from a domino stacking competition to make a stem-and-leaf plot. Then use the stem-and-leaf plot to help find the mode and the median.

NUMBER OF DOMINOES STACKED									
97	88	74	96	98	58	68	90	80	90
	72	86	69	78	93	84	99	92	85

Stems	Leaves
5	8
6	8 9
7	2 4 8
8	0 4 5 6 8
9	0 0 2 3 6 7 8 9

90 occurs more than any other number.

There are 19 scores. The median is the 10th score.

Mode: 90   Median: 86

CALIFORNIA STANDARDS MR 2.4 Use a variety of methods, such as words, numbers, symbols, charts, graphs, tables, diagrams, and models, to explain mathematical reasoning. *also* **SDAP 1.0, SDAP 1.1**

A **histogram** is a bar graph that shows the frequency, or the number of times, data occur within intervals. The bars in a histogram are connected, rather than separated.

BAR GRAPH

HISTOGRAM

R**emember** that you can use the range of a set of data to help determine intervals.

This bar graph is used to show information about individual movie customers. The histogram is used to show information about groups of movie customers.

EXAMPLE 3

The table below shows the number of sit-ups students in gym class did in one minute. Make a histogram for the data.

NUMBER OF SIT-UPS									
28	19	32	45	44	12	24	32	35	47
55	59	24	25	37	36	38	36	42	41

First, make a frequency table with intervals of 10. Start with 10.

Interval	10-19	20-29	30-39	40-49	50-59
Frequency	2	4	7	5	2

Title the graph and label the scales and axes.

Graph the number of students who did sit-ups within each interval.

CHECK FOR UNDERSTANDING

Think and ▶
Discuss

Look back at the lesson to answer each question.

1. **Tell** how the stem-and-leaf plot in Example 1 is useful in showing how well people did in the competition.

2. **Explain** how data displayed in a histogram are different from data displayed in a bar graph.

127

Guided ▶ Practice

3. Make a stem-and-leaf plot of the data 32, 24, 44, 57, 31, 25, 41, 26.

Tell whether a bar graph or a histogram is more appropriate.

4. number of customers at different intervals of time

5. populations of five different states

PRACTICE AND PROBLEM SOLVING

Independent ▶ Practice

6. Make a stem-and-leaf plot of the data 89, 74, 63, 65, 68, 74, 71, 80.

Tell whether a bar graph or a histogram is more appropriate.

7. heights of the tallest mountains in the United States

8. ages of 75 ice-skating competitors at a competition

Problem Solving ▶ Applications

Use the data in the table for 9.

9. a. Make a stem-and-leaf plot of the data.

PLANT HEIGHTS IN INCHES					
28	36	25	24	20	32
15	18	28	12	19	26

b. Use the stem-and-leaf plot to help find the median and mode.

c. REASONING Can you tell what the mean is for this set of data by looking at the stem-and-leaf plot? Explain.

AGES OF MARIA'S MUSIC CUSTOMERS									
10	25	33	14	54	62	29	44	41	40
11	31	41	24	65	16	39	50	51	55
19	22	17	26	31	42	17	18	42	37

10. Use the data in the table above to make a histogram.

11. **Write About It** Write a question that can be answered using the data from Exercise 10. Explain your answer.

MIXED REVIEW AND TEST PREP

12. Thirty students in the gym are asked their favorite sport. They all respond by saying "basketball." Is this valid for all students? (p. 112)

13. What type of sample does Ron get if he randomly surveys 175 out of 500 students? (p. 94)

14. Evaluate $3k$ for $k = 6.5$. (p. 82)

15. Write 0.73 as a percent. (p. 60)

16. **TEST PREP** Mt. McKinley has a height of 20,320 ft. Mt. Whitney has a height of 14,494 ft. How much higher is Mt. McKinley than Mt. Whitney? (p. 20)

A 34,814 ft   **B** 5,826 ft   **C** 5,800 ft   **D** 3,286 ft

   Extra Practice page H37, Set C

Explore Box-and-Whisker Graphs

Explore how to make a box-and-whisker graph and understand its parts.

You need at least eleven 3 in. × 5 in. cards, marker.

Vocabulary
box-and-whisker graph
lower extreme
upper extreme
lower quartile
upper quartile

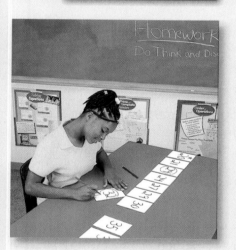

A **box-and-whisker graph** shows how far apart and how evenly data are distributed.

Activity

* Write each of the data in the table on a separate card.

NUMBER OF TROPHIES WON									
35	21	24	32	36	20	24	29	27	30

* Order the data from least to greatest. Draw a star on the card with the least value, or **lower extreme**, and the card with the greatest value, or **upper extreme**.

* Find the median of the data. If the median is not one of the numbers already written, write it on a card, put it in the middle of the data, and circle it.

* Find the median of the lower half of the data. This median is called the **lower quartile.** Circle it. Separate the data to the left of the lower quartile from the rest of the data.

* Find the median of the upper half of the data. This median is the **upper quartile**. Circle it. Separate the data to the right of the upper quartile from the rest of the data.

Think and Discuss

* Look at your cards. Into how many parts do the lower quartile, the median, and the upper quartile separate the data?

* What fraction of the data are to the left of the lower quartile? to the right of the upper quartile? What fraction of the data are between the lower quartile and the upper quartile?

You have found all you need to make this box-and-whisker graph.

TECHNOLOGY LINK
More Practice: Use E-Lab, *Exploring Box-and-Whisker Graphs.*
www.harcourtschool.com/elab2002

Practice

* Make a box-and-whisker graph of the data.

POINTS SCORED									
10	12	9	22	17	7	14	8	11	19

CALIFORNIA STANDARDS SDAP 1.0 Students compute and analyze statistical measurement for data sets. **SDAP 1.1** Compute the range, mean, median, and mode of data sets.

Box-and-Whisker Graphs

Learn how to analyze a box-and-whisker graph.

The only actual data you can identify from the data set in a box-and-whisker graph are the extremes.

EXAMPLE 1

The data in the box-and-whisker graph represent the diameters in kilometers of some asteroids observed by a scientist. What was the diameter of the largest asteroid? of the smallest asteroid?

0 1 2 3 4 5 6 7 8 9 10 11 12

The upper extreme is 12 and the lower extreme is 4. So, the diameter of the largest asteroid was 12 km, and the diameter of the smallest asteroid was 4 km.

• Look again at the graph. Which of the following can you determine—mean, median, mode, range?

EXAMPLE 2

Data about the number of meteors seen during a particular hour at different locations during the 1999 Leonids meteor shower are shown in the box-and-whisker graph. What does the graph show about how the data are distributed?

80 82 84 86 88 90 92 94 96 98 100

The data in the lowest $\frac{1}{4}$ of the data set are very close together. The data in each part of the middle $\frac{1}{2}$ are farther apart. They are closer to the lower extreme than to the upper extreme. The data in the highest $\frac{1}{4}$ are farther apart than those in the middle. At least one person saw 82 meteors, and at least one person saw 100 meteors.

CHECK FOR UNDERSTANDING

Think and ▶ Discuss

Look back at the lesson to answer the question.

1. **Explain** why you can't find the mean and the mode when looking at a box-and-whisker graph.

CALIFORNIA STANDARDS SDAP 1.0 Students compute and analyze statistical measurements for data sets. **SDAP 1.1** Compute the range, mean, median, and mode of data sets. *also* **MR 2.5**

For 2–3, use the box-and-whisker graph.

2. What are the
median, lower
quartile, and
upper quartile?

3. What are the lower and upper extremes and the range?

PRACTICE AND PROBLEM SOLVING

For 4–5, use the box-and-whisker graph.

4. What are the
median, lower
quartile, and
upper quartile?

5. What are the lower and upper extremes and the range?

For 6–8, use the data in the table below.

6. What are the median,
lower quartile, and
upper quartile?

WEIGHTS OF PANTHERS FOOTBALL PLAYERS					
85	102	89	86	104	92
103	97	91	100	100	93

7. What are the lower
and upper extremes and the range?

8. Make a box-and-whisker graph and a histogram. Compare them.

The box-and-whisker graph shows data about the numbers of
points scored by the basketball team each game.

9. What are the
least and
greatest
numbers of
points?

10. What does the graph show about how the data are distributed?

11. ❓ What's the Question? Use the box-and-whisker graph for
9–10. The answers are 88 and 94.

MIXED REVIEW AND TEST PREP

12. If you are displaying data about the number of cars that cross an
intersection during different intervals of time, is a bar graph or a
histogram more appropriate? (p. 126)

13. A grocer wants to know if customers like the new store hours. She
surveys 50 male customers. Is the sample biased? (p. 98)

Write the value of the blue digit. (p. 52)

14. 9.64

15. 124.024

16. **TEST PREP**   Which is $3 \times 3 \times 3 \times 3$ in exponent form? (p. 40)

   A 3×4 **B** 3^2 **C** 3^3 **D** 3^4

Extra Practice   page H37, Set D)

131

Analyze Graphs

Learn how to analyze data displays and determine how results and conclusions may have been influenced.

Data can be displayed in many different ways. Sometimes, the way a question is asked can influence the results that are displayed.

Rita took a survey asking the following question: Do you agree with me that George Washington was the greatest U.S. President, or would you choose Thomas Jefferson or Abraham Lincoln?

EXAMPLE 1

The results of Rita's survey are displayed in the bar graph shown at the right. Could the way the question was asked have influenced the results? Explain.

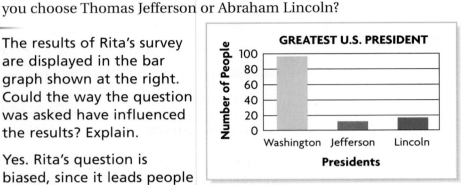

Yes. Rita's question is biased, since it leads people to agree with her that George Washington was the greatest U.S. President. As a result, the graph is misleading.

EXAMPLE 2

The results of two other surveys are shown below. Which graph is more likely to come from which question? Explain your reasoning.

Question 1:   Would you rather visit the Grand Canyon, Mount Rushmore, or the Statue of Liberty?

Question 2:   Would you rather visit the spectacular Grand Canyon, or would you rather visit Mount Rushmore or the Statue of Liberty?

Graph A   **Graph B**

Graph A probably goes with Question 2, since the question is biased and leads people to choose the Grand Canyon. Graph B probably goes with Question 1, since that question is not biased.

CALIFORNIA STANDARDS ⚬━ SDAP 2.3 Analyze data displays and explain why the way in which the question was asked might have influenced the results obtained and why the way in which the results were displayed might have influenced the conclusions. *also* MR 1.0, MR 1.1, MR 2.4

Graphs can communicate information quickly. That's why they are used by advertisers on television and in magazines and newspapers. Some graphs can be misleading and influence conclusions that are drawn.

EXAMPLE 3

Jon looked at the bar graph below and concluded that the Mississippi River is twice as long as the Missouri River. Explain Jon's mistake and tell why his conclusion is wrong.

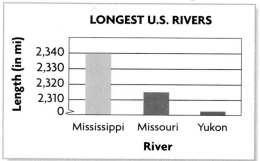

LONGEST U.S. RIVERS

The bar for the Mississippi is twice as long as the bar for the Missouri. However, if you look at the scale, you see that the rivers are about the same length. Because the lower part of the scale is missing, the differences are exaggerated.

When two graphs with different scales show two similar sets of data, comparing the graphs can sometimes be misleading.

EXAMPLE 4

The weekly ticket sales for the Mississippi River tour boat cruise are shown in the graphs below. Deb looked at the graphs and concluded that more tickets were sold in April than in March. Explain Deb's mistake.

It appears that the April ticket sales were greater than in March since the line for April is higher than the line for March. However, if you look at the scale for each graph, you can see that ticket sales were much greater in March than in April.

CHECK FOR UNDERSTANDING

Think and ▶ Discuss

Look back at the lesson to answer each question.
1. **Tell** how you could rewrite the question in Example 1 so it would not influence the results of the survey.

133

2. **Explain** how you could change the graph in Example 3 so it is not misleading.

Guided ▶ Practice

Rosa took a survey, asking the following question: Don't you think that the Mustangs are the best baseball team, or would you choose the Wildcats or Cougars? The results of her survey are displayed in the bar graph at the right.

3. Could the way the question was asked have influenced the results? Explain.

For 4–6, use the graph at the right. The graph is misleading.

4. About how many times as high is the bar for Angel Falls than for Tugela Falls?

5. Is Angel Falls 3 times as high as Tugela Falls? Explain.

6. How could you change the graph so it is not misleading?

PRACTICE AND PROBLEM SOLVING

Independent ▶ Practice

Miguel took a survey, asking the following question: What is your favorite fruit—apples, bananas, or delicious, juicy Florida navel oranges? The results of Miguel's survey are displayed in the circle graph.

7. Could the way the question was asked have influenced the results? Explain.

For 8–10, use the graph at the right. The graph is misleading.

8. About how many times as high is the bar for Brand B than for Brand A?

9. Does Brand B cost twice as much as Brand A? Explain.

10. How can you change the graph so it is not misleading?

Problem Solving ▶
Applications

11. Jeff looked at the bar graph at the right and concluded that Los Angeles has three times the population of Chicago. Explain Jeff's mistake and tell why his conclusion is wrong.

CITY POPULATIONS

12. How could you fix the graph so Jeff would not make a mistake?

13. Lin looked at the graphs below and concluded that from July to October, the temperatures in San Francisco are about the same as the temperatures in San Diego. Explain Lin's mistake.

SAN DIEGO TEMPERATURES

SAN FRANCISCO TEMPERATURES

MIXED REVIEW AND TEST PREP

For 14–15, use the box-and-whisker graph. (p. 129)

14. What is the median?

15. What are the least and greatest values?

16. Find the mean of the data set 18, 12, 10, 8, 9, 14, and 6. (p. 106)

17. TEST PREP   Which is the value of $2n + 3.7$ for $n = 2.9$? (p. 82)

A 8.6   **B** 9.5   **C** 10.4   **D** 21.46

18. TEST PREP   Choose the correct value. $(3 + 24) \div 3 \times 2 - 1$ (p. 44)

F 14   **G** 17   **H** 18   **J** 21

LiNKUP to Reading

Strategy • Classify and Categorize
To classify information means to group together similar information. To categorize the information, label the groups.

Shape	Color
rectangular	red, blue, white, or green
circular	red, blue, white, or green
oval	red, blue, white, or green

Paige wants to make a tablecloth. She has rectangular, circular, and oval patterns. She can choose from red, blue, white, or green fabric. By classifying and categorizing the data, you can see that there are 12 ways to make the tablecloth.

• The soccer coach is choosing new uniforms. The shorts are blue, black, or gray. The jerseys are white, yellow, green, blue, or turquoise. How many different uniforms are possible?

1. **VOCABULARY** A bar graph that shows frequencies within intervals is a(n) __?__ . (p. 127)

2. **VOCABULARY** A graph that shows how far apart and how evenly data are distributed is a(n) __?__ . (p. 129)

3. What type of graph would best show the highest and lowest temperatures for each of the last four years? (pp. 120–123)

4. What type of graph would best show high and low temperatures for a week? (pp. 120–123)

5. Make a double-line graph with the data at the right. (pp. 120–123)

6. Make a double-bar graph with the data at the right. (pp. 120–123)

END-OF-MONTH STOCK PRICES					
	Sep	Oct	Nov	Dec	Jan
Stock A	$80	$74	$45	$50	$52
Stock B	$50	$52	$52	$50	$45

For 7–9, use the graph at the right. (pp. 124–125)

7. About how much did the stock price decrease from Monday to Tuesday?

8. Describe the pattern in the graph.

9. If the trend continues, what do you think the stock price will be on Friday?

UH-OH STOCK

10. Make a stem-and-leaf plot for the data. (pp. 126–128)

11. Make a histogram for the data. (pp. 126–128)

POINTS SCORED					
33	52	45	47	34	52
34	58	48	52	46	59

HEIGHTS OF BUILDINGS (IN FT)						
20	50	80	20	40	45	85
25	30	80	60	70	75	55

For 12–17, use the following data: 14, 16, 9, 21, 35, 2, 26, 8, 17.
(pp.130–131)

2 · 8 9 14 ~~16~~ 17 21 26 35

12. Find the upper extreme.

13. Find the lower extreme.

14. Find the upper quartile.

15. Find the lower quartile.

16. Find the median.

17. Make a box-and-whisker graph.

For 18–20, use the graph at the right. (pp. 132–135)

18. About how many times as high is the bar for green peppers as the bar for carrots?

19. Were three times as many pounds of carrots sold as pounds of green peppers? Explain.

20. How could you change the graph so that it is not misleading?

AMOUNT OF PRODUCE SOLD

Check your work.

See item **6**.

Think about the different kinds of graphs and the most appropriate data for them. Check that your answer choice could be made into a bar graph.

Also see problem **7**, p. H65.

Choose the best answer.

1. Which type of graph would be the best to show the favorite sports of a sample of teenagers?

   A Stem-and-leaf plot   **C** Line graph

   B Histogram   **D** Bar graph

2. The owner of a music shop made a line graph to show the number of CDs sold during a 4-week period. Which trend does the graph show?

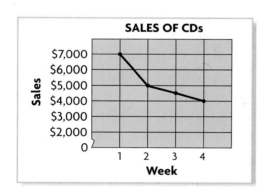

   F Sales are increasing.

   G Sales are even.

   H Sales are decreasing.

   J No trend is shown.

3. What is the value of $\frac{12 - 2^3}{4}$?

   A $^-5$ **C** 11

   B 1   **D** Not here

4. Which of the following is best suited for display in a multiple-bar graph?

   F Average temperatures a town recorded during a 1-year period

   G Number of CDs owned by sixth-grade students

   H Number of hours sixth- and seventh-grade students spend reading each week

   J Changes in water temperature over a 24-hour period

5. Which kind of graph is shown?

   A Circle graph

   B Histogram

   C Line graph

   D Stem-and-leaf plot

6. A bar graph would be best to display which data?

   F Heights of the tallest buildings in a city

   G Frequency of cars stopping at a tollbooth

   H Changes in a person's weight over a year

   J Part of a day a person spends reading

7. $22.8 - 3.11$

   A 18.79   **C** 19.79

   B 19.69   **D** Not here

MATH DETECTIVE

Measure by Measure

REASONING Use the clues and your knowledge of the measures of central tendency to find the set of numbers described. Be prepared to explain how you solved the mystery.

Mystery Number 1

Clues:
1. There are three numbers in the set.
2. The mean is 10.
3. The mode is 8.
What is the set of numbers?

Mystery Number 2

Clues:
1. There are three numbers in the set.
2. One number in the set is 5.
3. The median is 9.
4. The range is 16.
What is the set of numbers?

Mystery Number 3

Clues:
1. There are four numbers in the set.
2. The median is 12.
3. Two numbers in the set are 3 and 10.
4. The range is 15.
What is the set of numbers?

Think It Over!

- **Write About It** If you know that the sum of a set of six numbers is 150, which of the following measures could you find: mean, median, mode, range? Explain.

- **Stretch Your Thinking** A set of numbers forms the pattern 1, 3, 5, 7, ... If the median of the set is 15, what is the range?

Explore Scatterplots

Learn how to read and interpret a scatterplot.

Do you think there is any relationship between the number of people who paint a large building and the number of hours it takes them to finish the job?

The *scatterplot* displays data for 11 large buildings that were painted. It shows that as the number of painters increased, the time it took them to finish the job tended to decrease.

A scatterplot shows the relationship between two variables.

| Positive | Negative | None |

When the values of the two variables increase or decrease together, there is a **positive correlation**.

When the values of one variable increase while the others decrease, there is a **negative correlation**.

When the data points show no pattern of increase or decrease, there is **no correlation**.

TALK ABOUT IT

• Tell if the relationship between the speed of a car and the number of hours needed to drive 500 miles has a positive correlation, a negative correlation, or no correlation.

TRY IT

Sketch a scatterplot that could represent the situation. Then identify the type of correlation between the variables.

1. amount of time walking *and* total distance that you walk

2. number of rooms in house *and* street address of house

Write *positive correlation, negative correlation,* **or** *no correlation* **to describe the relationship shown in the scatterplot.**

3.

4.

VOCABULARY

1. Everyone in the population has the same chance of being selected in a(n) __?__ . (p. 95)

2. A bar graph that shows the frequency at which data occur within intervals is a(n) __?__ . (p. 127)

3. Individuals in the population are not represented in the sample if the sample is __?__ . (p. 98)

EXAMPLES

EXERCISES

Chapter 5

- **Identify the type of sample.** (pp. 94–97)

  Determine the sampling method used if workers on an assembly line check every tenth tire.

  This is a systematic sample.

Determine the type of sample. Write *convenience, random,* or *systematic.*

4. From a computer list of students, each with an equal chance of being selected, 100 students are chosen.
5. Every tenth person walking down the street is surveyed about the President.

- **Record and organize data.** (pp. 102–105)

  What is the size of the sample?

  SCORES ON MATH TEST

Score Interval	Frequency	Cumulative Frequency
91–100	5	5
81–90	8	13
71–80	6	19
Below 71	5	24

  There are 24 people in the sample.

For 6–7, use the following data.

NUMBER OF MOVIES SEEN IN ONE YEAR			
8	12	16	12
9	10	9	8
9	15	20	14
12	7	9	15

6. Make a line plot.
7. Make a cumulative frequency table with intervals.

- **Find the mean, median, and mode.**
  (pp. 106–111)

  24, 20, 24, 21, 26
  Mean: 24 + 20 + 24 + 21 + 26 = 115;
  　　　115 ÷ 5 = 23

  Median: 24

  Mode: 24

Find the mean, median, and mode.

8. 2, 7, 9, 4, 6, 8, 6
9. 5.3, 8.8, 4.7, 6.5, 4.7
10. 79, 87, 90, 100, 96, 89, 92, 87
11. Suppose you added the number 10.6 to the data in Exercise 9. Which measure(s) of central tendency would change?

Chapter 6

- **Display data in graphs.** (pp. 120–123)

TEACHERS AT CITY MIDDLE SCHOOLS

■ Female
■ Male

For 12, use the graph at the left.

12. Which schools have more female teachers than male teachers?
13. Would you use a bar graph, line graph, or circle graph to display a city's temperature readings for 1 month? Explain.
14. Make a multiple-line graph with the data below.

HIGH AND LOW TEMPERATURES					
	Mon	**Tues**	**Wed**	**Thurs**	**Fri**
Highs	45°	53°	41°	48°	50°
Lows	34°	39°	35°	40°	41°

- **Use stem-and-leaf plots and histograms.** (pp. 126–128)

In a stem-and-leaf plot, the tens digits of the data are stems, and the ones digits are leaves.

Ages

Stem	Leaves
1	1 3 3
2	0 3 5 5
3	7 7 8

Some students are participating in a jump-rope benefit. Their ages are 10, 15, 18, 20, 15, 11, 13, 10, 14, 14, 18, 19, 22, and 21.

15. Use the data to make a stem-and-leaf plot.
16. Use the data to make a histogram.

- **Make a box-and-whisker graph.** (pp. 129–131)

Heights of Students (in cm)
135 168 148 160 159 148 163 165 167

Draw a box-and-whisker graph.

130 135 140 145 150 155 160 165 170

For 17–20, use the table below.

HEIGHTS OF TREES (in ft)								
8	12	6	9	15	9	16	20	24

17. What is the median?
18. What are the lower and upper quartiles?
19. What are the lower and upper extremes?
20. Make a box-and-whisker graph.

PROBLEM SOLVING APPLICATIONS

21. Roberto has to pay $27 for his CDs at the record store. In how many ways can he pay, using only bills of $10, $5, and $1? (pp. 100–101)
23. The machines in the exact change lane of a tollway accept any combination of coins that total exactly 75¢, but they do not accept pennies or half dollars. In how many different ways can a driver pay the toll in an exact change lane? (pp. 100–101)

22. The sixth-grade students are having a car wash. They charge $4 for cars and $6 for SUVs. In how many ways can they earn $100? (pp. 100–101)
24. The debate club has 10 members. Each member will debate each of the other members only once. How many debates will they have? (pp. 100–101)

Water Resources

Water is a critical natural resource in California. Densely populated Southern California is relatively dry. More sparsely populated Northern California has more water.

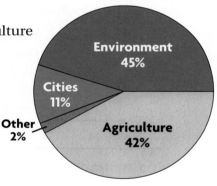

Water is stored in reservoirs and carried along aqueducts to the places where it is needed most.

Rainwater that flows into lakes, rivers, and streams is called *runoff.*

Use Data   For 1–3, use the circle graph.

1. About how many times as much runoff water is used by agriculture as is used by cities?

2. Which category uses about the same amount of runoff as agriculture?

3. Would you say that agriculture and "other" use a little less than half or a little more than half of the runoff water?

Use Data   For 4, use the table at the right.

4. The California Department of Water Resources has predicted the supply and demand for water in the year 2020 for both average conditions and drought conditions. Make and label a double-bar graph of the data.

Use Data   For 5, use the table below.

5. The United States Geological Survey measures the flow of hundreds of California streams. Make a line plot of the data.

WATER RUNOFF USE IN CALIFORNIA

- Environment 45%
- Cities 11%
- Other 2%
- Agriculture 42%

STREAMFLOW OF 18 CALIFORNIA STREAMS (cubic ft per second)					
5	7	8	5	7	4
6	7	6	9	6	5
5	7	4	6	7	7

WATER SUPPLY AND DEMAND IN 2020	Millions of Acre-feet*
Average Conditions	
Supply	63
Demand	71
Drought Conditions	
Supply	50
Demand	60

*One acre-foot equals the amount of water needed to cover 1 acre of land to a depth of 1 foot.

Dams

Dams, reservoirs, and aqueducts help to collect, store, and transport water in California. The state has about 1,400 dams of all types and sizes. Some of California's water comes from dams and reservoirs outside the state.

California's 230-ft-tall Oroville Dam (above) is the tallest dam in the United States. Hoover Dam, on the Arizona-Nevada border, also supplies some of California's water needs.

Use Data For 1–3, find the measure of central tendency for the dams listed in the table.

1. mean storage capacity

2. median length

3. range of storage capacities

4. Is the mean a good measure for comparing the storage capacities of the dams? Explain.

DAMS IMPORTANT TO CALIFORNIA			
Name	**Location**	**Storage Capacity**	**Length**
Parker Dam	California	600,000 acre-ft	856 ft
Imperial Dam	California	1,000 acre-ft	3,475 ft
Oroville Dam	California	3,500,000 acre-ft	6,920 ft
San Luis Dam	California	2,000,000 acre-ft	18,600 ft
Hoover Dam	Arizona/Nevada	28,500,000 acre-ft	1,244 ft

5. Make and label a bar graph showing the storage capacities of Parker, Oroville, San Luis, and Hoover Dams. Why would it be difficult to include Imperial Dam on your graph?

6. **Mental Math** One acre-foot is equivalent to 43,560 cubic feet of water. Mentally calculate the storage capacity of Imperial Dam in cubic feet.

7. **REASONING** Nora said that the longer a dam is, the greater is its storage capacity. Do you agree? Explain, using examples from the table to illustrate your answer.

8. Make and label a bar graph showing the lengths of the five dams in the table.

CHAPTER 7

Number Theory

Thousands of years ago, marbles were made of flint, stone, or baked clay. Today, marbles are made of different types of materials, and glass is the most popular. The glass is melted in a furnace and poured out. It is cut into cylinders, and these cylinders are rounded off by rollers and then cooled. Suppose 1,200 cooled marbles are put into packages of 10. How many packages could be made? Would there be any left over?

PACKAGES OF MARBLES		
Marbles per Package	Complete Package? Yes or No	Reason
3	Yes	1,200 ÷ 3 = 400 packages
4	Yes	1,200 ÷ 4 = 300 packages
5	Yes	1,200 ÷ 5 = 240 packages
6	Yes	1,200 ÷ 6 = 200 packages
7	No	1,200 ÷ 7 = 171 r3 packages
8		
9		

Check What You Know

Use this page to help you review and remember important skills needed for Chapter 7.

 Vocabulary

Choose the best term from the box.

1. A number whose only factors are 1 and itself is a __?__ .

2. A number that has more than two factors is a __?__ .

> composite
> number
> multiple
> prime number

 Prime Numbers (See p. H7.)

Decide whether the number is a prime number. Write _yes_ or _no_.

3. 2	**4.** 5	**5.** 4	**6.** 9
7. 11	**8.** 21	**9.** 37	**10.** 26
11. 13	**12.** 7	**13.** 45	**14.** 70

 Composite Numbers (See p. H7.)

Decide whether the number is a composite number. Write _yes_ or _no_.

15. 6	**16.** 15	**17.** 19	**18.** 81
19. 24	**20.** 53	**21.** 3	**22.** 25

Multiples (See p. H8.)

Write the next three multiples.

23. 4   4, 8, 12, ▦, ▦, ▦   **24.** 10   10, 20, 30, ▦, ▦, ▦   **25.** 12   12, 24, 36, ▦, ▦, ▦

26. 8   8, 16, 24, ▦, ▦, ▦   **27.** 5   5, 10, 15, ▦, ▦, ▦   **28.** 11   11, 22, 33, ▦, ▦, ▦

Write the first five multiples of each number.

29. 6	**30.** 22	**31.** 30
32. 7	**33.** 9	**34.** 13

 Factors (See p. H8.)

Write all of the factors of each number.

35. 8	**36.** 9	**37.** 11
38. 18	**39.** 54	**40.** 32

> **LOOK AHEAD**
>
> **In Chapter 7 you will**
> - use divisibility rules
> - find prime factors
> - determine the LCM and GCF of whole numbers

Divisibility

Learn how to tell if a number is divisible by 2, 3, 4, 5, 6, 8, 9, or 10.

Remember that a number is divisible by another number if the remainder is zero.

Chris knows that a number is divisible by 2 if the last digit is 0, 2, 4, 6, or 8. He also knows that a number is divisible by 3 if the sum of the digits is divisible by 3. He wonders if there is a rule for numbers that are divisible by 6.

Activity

You need: hundred chart

• Shade all the numbers divisible by 2.

• Circle all the numbers that are divisible by 3.

• Look at the numbers that are both shaded and circled. Divide these numbers by 6. What do you notice?

• What rule can Chris write about numbers divisible by 6?

Math Idea ▶ You can use divisibility rules to help you decide if a number is divisible by another number.

A number is divisible by	Divisible	Not Divisible
2 if the last digit is even (0, 2, 4, 6, or 8).	11,994	2,175
3 if the sum of the digits is divisible by 3.	216	79
4 if the last two digits form a number divisible by 4.	1,024	621
5 if the last digit is 0 or 5.	15,195	10,007
6 if the number is divisible by 2 and 3.	1,332	44
8 if the last three digits form a number divisible by 8.	5,336	3,180
9 if the sum of the digits is divisible by 9.	144	33
10 if the last digit is 0.	2,790	9,325

EXAMPLES ▶ Tell whether each number is divisible by 2, 3, 4, 5, 6, 8, 9, or 10.

A. 610 is divisible by
2; the last digit is even.
5; the last digit is 0 or 5.
10; the last digit is 0.

B. 459 is divisible by
3; the sum of the three digits is divisible by 3.
9; the sum of the digits is divisible by 9.

CALIFORNIA STANDARDS ⊙━NS 2.0 Students calculate and solve problems involving addition, subtraction, multiplication, and division. **MR 1.1** Analyze problems by identifying relationships, distinguishing relevant from irrelevant information, identifying missing information, sequencing and prioritizing information, and observing patterns. *also* **MR 2.2, MR 3.0**

Think and ▶
Discuss

Look back at the lesson to answer each question.

1. **Compare** the divisibility rules for 3 and 9.

2. **Tell** the advantages of knowing the divisibility rules.

Guided ▶
Practice

Tell whether each number is divisible by 2, 3, 4, 5, 6, 8, 9, or 10.

3. 56 **4.** 200 **5.** 784 **6.** 2,345 **7.** 3,009

PRACTICE AND PROBLEM SOLVING

Independent ▶
Practice

Tell whether each number is divisible by 2, 3, 4, 5, 6, 8, 9, or 10.

8. 75 **9.** 324 **10.** 45 **11.** 812 **12.** 501

13. 615 **14.** 936 **15.** 744 **16.** 5,188 **17.** 4,335

18. 1,407 **19.** 48,006 **20.** 7,064 **21.** 12,111 **22.** 1,044

For 23–25 write *T* or *F* to tell whether each statement is true or false. If it is false, give an example that shows it is false.

23. All even numbers are divisible by 2.

24. All odd numbers are divisible by 3.

25. Some even numbers are divisible by 5.

Problem Solving ▶
Applications

26. A number is between 80 and 100 and is divisible by both 5 and 6. What is the number?

27. *REASONING*   Use the divisibility rules you know to write a divisibility rule for 15.

28. Kirk has 20 trading cards. He has 10 more hockey cards than baseball cards. How many does Kirk have of each type of card?

29. Amber charges $8.25 per hour to baby-sit. One Saturday she baby-sat for 3.5 hr. How much did she earn baby-sitting on Saturday? Round your answer to the nearest cent.

MIXED REVIEW AND TEST PREP

30. What is the lower quartile of the data 24, 26, 28, 29, 30, 32, 34, 36, 37? (p. 129)

31. Write the percent for 0.05. (p. 60)

32. $36.4 \div 0.28$ (p. 76)

33. 858×19 (p. 22)

34. **TEST PREP**   Which is the value of $3^2 + 2 \times 5 - 6$? (p. 44)

A 13 **B** 24 **C** 34 **D** 49

Extra Practice   page H38, Set A

Prime Factorization

Learn how to write a composite number as the product of prime numbers.

Vocabulary

prime factorization

A prime number is a whole number greater than 1 whose only factors are itself and 1. Here are the prime numbers less than 50.

2, 3, 5, 7, 11, 13, 17, 19, 23, 29, 31, 37, 41, 43, 47

A composite number, like 104, has more than two factors. You can write a composite number as the product of prime factors. This is called the **prime factorization** of the number.

You can divide to find the prime factors of a composite number.

EXAMPLE 1

Find the prime factorization of 104.

```
2|104
2|52
2|26
13|13
   1
```

Repeatedly divide by the smallest possible prime factor until the quotient is 1.

$2 \times 2 \times 2 \times 13$

List the prime numbers you divided by. These are the prime factors.

So, the prime factorization of 104 is $2 \times 2 \times 2 \times 13$, or $2^3 \times 13$.

Use a factor tree to find the prime factors of a composite number.

EXAMPLE 2

Find the prime factorization of 156.

Choose any two factors of 156. Continue until only prime factors are left.

```
        156                           156
       /   \                         /   \
      2  ×  78   ← 2 is prime. 78    3  ×  52   ← 3 is prime. 52
           /\      is not prime.          /\      is not prime.
          2 × 39  ← 39 is not prime.     2 × 26  ← 26 is not prime.
              /\                             /\
             3 × 13 ← prime factors        2 × 13 ← prime factors
```

So, the prime factorization of 156 is $2 \times 2 \times 3 \times 13$, or $2^2 \times 3 \times 13$.

Math **I**dea ▶ Every composite number can be written as a product of two or more prime factors. No matter how you find the prime factors, you will get the same factors, but maybe in a different order.

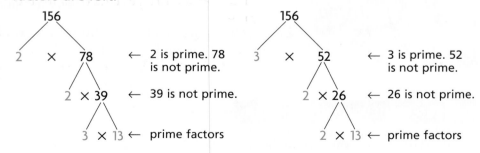

Think and ▶ Discuss

Look back at the lesson to answer each question.

1. **Tell** what the prime factorization would be in Example 2 if you started with 4 and 39.

2. **Tell** how you know when you have finished the prime factorization of a number.

Guided ▶ Practice

Use division or a factor tree to find the prime factorization.

3. 12   **4.** 65   **5.** 16   **6.** 42

Write the prime factorization in exponent form.

7. 21   **8.** 28   **9.** 254   **10.** 908

PRACTICE AND PROBLEM SOLVING

Independent ▶ Practice

Use division or a factor tree to find the prime factorization.

11. 128   **12.** 50   **13.** 76   **14.** 108

Write the prime factorization in exponent form.

15. 18   **16.** 302   **17.** 49   **18.** 217

19. 532   **20.** 45   **21.** 746   **22.** 99

Solve for n to complete the prime factorization.

23. $2 \times n \times 5 = 20$ **24.** $44 = 2^2 \times n$ **25.** $75 = 3 \times 5 \times n$

Problem Solving ▶ Applications

26. *REASONING*   The prime factorization of 50 is 2×5^2. Without dividing or using a factor tree, tell the prime factorization of 100.

27. *REASONING*   A number, c, is a prime factor of both 12 and 60. What is c?

28. Chris bathed his pet every eighth day in June, beginning on June 8. On what other dates did he bathe his pet in June?

29. ✏ **Write About It**   Do the prime factors of a number differ depending on which factors you choose first? Explain.

MIXED REVIEW AND TEST PREP

30. Is 3,543 divisible by 2, 3, or 9? (p. 146)

31. Write the percent for 0.6. (p. 60)

32. $18.3 + 22.6 + 17.03 + 21.99$ (p. 66)

33. Find the median for the data 13, 8, 9, 16, 18. (p. 106)

34. **TEST PREP** A sign is placed in a toy store, asking shoppers to complete a survey about their favorite video game. Which type of sampling is this? (p. 94)

    A convenience   **C** systematic

    B random   **D** compatible

LESSON 7.3

Least Common Multiple and Greatest Common Factor

Learn how to find the LCM and GCF of numbers and use them to solve problems.

QUICK REVIEW

List the first five multiples of each number.

1. 4   2. 2   3. 3

4. 9   5. 6

Vocabulary

least common multiple (LCM)

greatest common factor (GCF)

Kirk and Amber volunteer during December. Kirk volunteers every fourth day beginning December 4. Amber volunteers every third day beginning December 3. Find the first day they will volunteer together by listing the multiples of 4 and 3.

multiples of 4:   4, 8, 12, 16, 20, 24, 28

multiples of 3:   3, 6, 9, 12, 15, 18, 21, 24, 27, 30

The multiples that appear in blue are called common multiples. The smallest of the common multiples is called the **least common multiple**, or **LCM**. The LCM of 4 and 3 is 12, the product of the two numbers. So, the first day they volunteer together is December 12.

Examples 1 and 2 show two ways to find the LCM.

EXAMPLE 1

Find the LCM of 12 and 8.

12: 12, 24, 36, 48, 60, 72, 84, 96   *List the first eight multiples.*
 8: 8, 16, 24, 32, 40, 48, 56, 64   *Find the common multiples.*

So, the LCM of 12 and 8 is 24.   *Find the LCM.*

EXAMPLE 2

Find the LCM of 6, 9, and 18.

Write the prime factorizations.
$6 = 2 \times 3$ $9 = 3 \times 3 = 3^2$ $18 = 2 \times 3 \times 3 = 2 \times 3^2$

Write a product using each prime factor only once.
2×3

For each factor, write the greatest exponent used with that factor in any of the prime factorizations. Multiply.
$2 \times 3^2 = 18$

So, the LCM of 6, 9, and 18 is 18.

   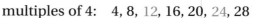 **CALIFORNIA STANDARDS** O⊸NS 2.4 Determine the least common multiple and the greatest common divisor of whole numbers; use them to solve problems with fractions. *also* **MR 3.2, MR 3.3**

Greatest Common Factor

Factors shared by two or more numbers are called common factors. The largest of the common factors is called the **greatest common factor**, or **GCF**.

To find the GCF of two or more numbers, list all the factors of each number, find the common factors, and then find the greatest common factor.

45: 1, 3, 5, 9, 15, 45   *The common factors are 1, 3, and 9.*

27: 1, 3, 9, 27   *The GCF of 45 and 27 is 9.*

The GCF can be used to solve problems.

EXAMPLE 3

Carlyn has 12 pens and 36 pencils. She is making packages with the same number of each item. What is the greatest number of packages she can make without any items left over? How many of each item will be in each package?

You can find the greatest number of packages by finding the GCF of 12 and 36.

12: 1, 2, 3, 4, 6, 12   *List the factors.*
36: 1, 2, 3, 4, 6, 9, 12, 18, 36   *Find the common factors.*

The GCF of 12 and 36 is 12.   *Find the GCF.*

So, Carlyn can make 12 packages without any items left over.

To find the number of each item, divide the number of pens and the number of pencils by the number of packages.

Pens: $12 \div 12 = 1$ Pencils: $36 \div 12 = 3$

So, there will be 1 pen and 3 pencils in each package.

To find the GCF of two numbers, you can also use their prime factors. List the prime factors, find the common prime factors, and then find their product.

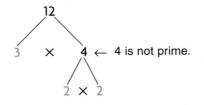

The prime factors of 12 are $2 \times 2 \times 3$.

The prime factors of 36 are $2 \times 2 \times 3 \times 3$.

The common prime factors are 2, 2, and 3.

Find the product of the common factors: $2 \times 2 \times 3$, or $2^2 \times 3 = 12$.

The GCF of 12 and 36 is 12.

151

EXAMPLE 4

Use prime factors to find the GCF of 48 and 72.

48: $2 \times 2 \times 2 \times 2 \times 3$ *Find the prime factors.*
72: $2 \times 2 \times 2 \times 3 \times 3$

2, 2, 2, and 3   *Find the common prime factors.*

$2 \times 2 \times 2 \times 3 = 24$ *Multiply the common factors.*

So, the GCF of 48 and 72 is 24.

CHECK FOR UNDERSTANDING

Think and Discuss ▶ **Look back at the lesson to answer each question.**

1. **Explain** why you are able to find a greatest common factor but not a greatest common multiple of 2 or more numbers.

2. **Tell** the number of pens and pencils there would be in each package in Example 3 if Carlyn made 6 packages.

Guided Practice ▶ **List the first five multiples of each number.**

3. 3   **4.** 7   **5.** 11   **6.** 15

Find the LCM of each set of numbers.

7. 3, 7   **8.** 2, 3   **9.** 6, 9   **10.** 5, 8, 20

Find the GCF of each set of numbers.

11. 6, 9   **12.** 4, 20   **13.** 9, 24   **14.** 12, 16, 20

PRACTICE AND PROBLEM SOLVING

Independent Practice ▶ **List the first five multiples of each number.**

15. 4   **16.** 8   **17.** 16   **18.** 55

19. 10   **20.** 27   **21.** 14   **22.** 39

Find the LCM of each set of numbers.

23. 3, 8, 24   **24.** 32, 128   **25.** 12, 20   **26.** 40, 105

27. 24, 30   **28.** 18, 21, 36   **29.** 12, 27   **30.** 48, 116

Find the GCF of each set of numbers.

31. 16, 18   **32.** 15, 18   **33.** 21, 306   **34.** 16, 24, 40

35. 25, 33   **36.** 200, 215   **37.** 24, 32, 40   **38.** 630, 712

Find a pair of numbers for each set of conditions.

39. The LCM is 36. The GCF is 3.

40. The LCM is 24. The GCF is 2.

Problem Solving ▶
Applications

41. Peter will distribute cereal samples with pamphlets about good nutrition. The samples come in packages of 15. The pamphlets come in packages of 20.
  a. What is the least number of cereal samples and pamphlets needed to have equal amounts?
  b. How many packages of each does he need?

42. Ruth has 36 markers and 48 erasers. She will put them in bags with the same number of each item. What is the greatest number of bags she can make?

43. **?** **What's the Error?** Jan says the LCM of 10 and 15 is 5. Find her error and the correct answer.

MIXED REVIEW AND TEST PREP

44. Write the prime factorization for 45 in exponent form. (p. 148)

45. Evaluate $a \div c$, for $a = 4{,}602$ and $c = 37$. (p. 28)

46. Find the value of 5^4. (p. 40)

47. **TEST PREP** Which is the exponential notation for
$3 \times 3 \times 3 \times 4 \times 4 \times 5 \times 5$? (p. 40)

  A $3 \times 4 \times 5^7$ **B** $3^3 \times 2^4 \times 2^5$ **C** $3^3 \times 4^2 \times 5^2$ **D** $3^3 \times 4^4 \times 5^5$

48. **TEST PREP** Which shows the decimal for 46%? (p. 60)

  F 0.046   **G** 0.46   **H** 4.6   **J** 46

LiNKUP to Careers

NASA Scientist When astronauts service and repair the Hubble Space Telescope, NASA scientists must set the orbit of the space shuttle so it will meet the Hubble Space Telescope in its orbit. The Hubble Space Telescope takes about 95 minutes to make one orbit around Earth. The space shuttle takes about 90 minutes to go around Earth.

One way to find when the two objects will meet is to find the LCM of the orbit times.

90: 90, 180, 270, 360, 450, 540, … 1,350; 1,440; 1,530; 1,620; 1,710

95: 95, 190, 285, 380, 475, 570, … 1,330; 1,425; 1,520; 1,615; 1,710

So, the space shuttle will meet the Hubble Space Telescope about every 1,710 minutes, or $28\frac{1}{2}$ hours.

• The Russian Mir space station orbits Earth about every 93 minutes. How often would Mir and the space shuttle meet?

PROBLEM SOLVING STRATEGY
Make an Organized List

Learn how to solve problems by making an organized list.

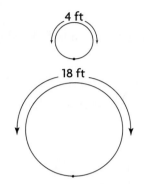

4 ft

18 ft

QUICK REVIEW

List the first four multiples of each number.

1. 3   **2.** 5   **3.** 9

4. 11   **5.** 20

Like many early bicycles, the Boneshaker had different sized wheels. Suppose the circumferences of the wheels were 4 ft and 18 ft. How many revolutions would each wheel make before marks on their rims would both be in the same positions again?

The Boneshaker was invented in 1865. The wooden wheels made for an uncomfortable ride.

Analyze   What are you asked to find?

What facts are given?

Choose   What strategy will you use?

You can use the strategy *make an organized list*. List the total distance traveled by each wheel for every complete turn.

Solve   How will you solve the problem?

Make a list of multiples of 4 and 18.
  multiples of 4: 4, 8, 12, 16, 20, 24, 28, 32, 36, 40
  multiples of 18: 18, 36, 54

The least common multiple is 36. When the wheels have traveled 36 ft, the marks will both be in the same positions again.

36 is the ninth multiple of 4, so the small wheel makes 9 revolutions.

36 is the second multiple of 18, so the large wheel makes 2 revolutions.

Check   How can you check your answer?

What if the smaller wheel had a circumference of 8 ft? How many revolutions would each wheel make before the marks were both in the same positions again?

CALIFORNIA STANDARDS O—ɴNS 2.4 Determine the least common multiple and the greatest common divisor of whole numbers; use them to solve problems with fractions. **MR 3.2** Note the method of deriving the solution and demonstrate a conceptual understanding of the derivation by solving similar problems. *also* **MR 2.0**

PROBLEM SOLVING PRACTICE

PROBLEM SOLVING STRATEGIES

Draw a Diagram or Picture

Make a Model

Predict and Test

Work Backward

▶ **Make an Organized List**

Find a Pattern

Make a Table or Graph

Solve a Simpler Problem

Write an Equation

Use Logical Reasoning

Solve the problem by making an organized list.

1. Justin and Amy help with the shopping for their individual families. Justin goes to the store every 3 days and Amy goes every 5 days. They see each other at the store on September 30. On what date will Justin and Amy see each other at the store again?

For 2–3, use the information below.

Susanna buys one bag with 40 snacks and another with 32. She wants to make up small snack packs for a party. All snack packs must have the same number of each snack.

2. If you make lists to find the greatest number of snack packs Susanna can make, what will you include in the lists?

   A addends   **B** factors   **C** multiples   **D** fractions

3. What is the greatest number of snack packs Susanna can make?

   F 2 packs   **G** 4 packs   **H** 8 packs   **J** 15 packs

MIXED STRATEGY PRACTICE

4. Ray begins an eight-week exercise program. He exercises 30 min during the first week, 45 min the second week, and 60 min the third week. If this pattern continues, how many hours and minutes will Ray exercise the sixth week?

5. DeAnn has a total of 74 in. of yarn to make gifts. She uses 13 in. of yarn for each of the 5 gifts she is making. Find how many inches of yarn DeAnn will have left after making the 5 gifts.

6. A total of 316 middle school teachers are going to a meeting. There are 30 more male teachers than female. How many teachers are male?

7. Christi rides her bicycle 7 blocks south, 3 blocks east, 5 blocks north, and 8 blocks west. How many blocks has she ridden when she crosses her own path?

8. Use the graph at the right. Look at the area in square miles for each lake. How much larger is Lake Michigan than Lake Erie and Lake Ontario together?

9. **? What's the Question?** Laura wants to buy the same number of plums and kiwis. Plums are sold in bags of 4 and kiwis are sold in bags of 7. The answer is 28 plums.

AREAS OF THE GREAT LAKES

Lake	Square Miles
Superior	31,820
Huron	23,010
Michigan	22,400
Erie	9,940
Ontario	7,540

1. **VOCABULARY** Writing a composite number as the product of prime factors is called __?__. (p. 148)

2. **VOCABULARY** The largest of the common factors of two or more numbers is the __?__. (p. 150)

3. **VOCABULARY** The smallest of the multiples of two or more numbers is the __?__. (p. 150)

Tell whether each number is divisible by 2, 3, 4, 5, 6, 8, 9, or 10. (pp. 146–147)

4. 42
5. 64
6. 96
7. 225

8. 330
9. 963
10. 450
11. 2,385

**Use division or a factor tree to find the prime factorization.
Write the prime factorization in exponent form.** (pp. 148–149)

12. 9
13. 8
14. 14
15. 18

16. 80
17. 12
18. 33
19. 50

20. 49
21. 98
22. 504
23. 891

24. List the first twenty multiples of 4 and the first twenty multiples of 10. What is the least common multiple? What multiples greater than 80 do they have in common? (pp. 150–153)

Find the LCM and the GCF of each set of numbers. (pp.150–153)

25. 3, 9
26. 2, 6
27. 6, 4
28. 10, 15

29. 8, 12
30. 9, 27
31. 15, 25
32. 25, 115

33. 27, 189
34. 6, 8, 12
35. 6, 9, 12
36. 8, 16, 20

Solve. (pp. 146–155)

37. A number is between 60 and 70. It is divisible by 3 and 9. What is the number?

38. Meat patties are sold in packages of 12. Buns are sold in packages of 8. What is the least number of meat patties and buns needed to have an equal number of each?

39. Cashews are sold in 8-oz jars, almonds in 12-oz jars, and peanuts in 16-oz jars. What is the least number of ounces of each type of nuts you can buy to make mixed nuts with equal amounts of each? How many jars of each would you need?

40. Alissa jogs in the park every 3 days. Erin jogs in the park once a week on Saturday. If they met in the park on Saturday, April 30, when will they meet in the park again?

Get the information you need.
See item 2.

You know that juice boxes are sold in packages of 8 and rice snacks are sold in packages of 10. Use the least common multiple to find an equal number of treats before you determine how many packages to buy.

Also see problem 3, p. H63.

1. What is the prime factorization of 90?

   A $2 \times 3^2 \times 5$ **C** $2^2 \times 3 \times 5^2$

   B $2 \times 3 \times 5$ **D** $2^2 \times 3^2 \times 5^2$

2. Hu is buying treats for a party. Juice boxes are sold in packages of 8. Rice snacks are sold in packages of 10. What is the least number of packages of juice and rice snacks he should buy to have an equal number of juice boxes and rice snacks?

   F 4 juice packs, 5 boxes of rice snacks

   G 5 juice packs, 4 boxes of rice snacks

   H 9 juice packs, 9 boxes of rice snacks

   J 10 juice packs, 8 boxes of rice snacks

3. What is the least common multiple of 18 and 27?

   A 3   **C** 54

   B 9   **D** 486

4. What is the prime factorization of 24?

   F 2×2^3 **H** $2^4 \times 3$

   G $2^3 \times 3$ **J** 6×2^2

5. Which number is divisible by both 6 and 9?

   A 45   **C** 243

   B 216   **D** 768

6. Carni earned the following amounts baby-sitting: $17, $15, $12, $17, $14. What was the mean amount that Carni earned?

   F $14   **H** $16

   G $15   **J** Not here

7. What is the greatest common factor of 16 and 24?

   A 4   **C** 48

   B 8   **D** 384

8. Marcus has a baseball game every fifth day in April. His first game is on April 5. How many games will there be in April?

   F 3   **H** 5

   G 4   **J** 6

9. A theater holds 381 people. All the available tickets for 18 shows were sold. Which is a good estimate for the number of people who attended the shows?

   A Less than 4,000 people

   B Between 4,000 and 6,000 people

   C Between 6,000 and 8,000 people

   D More than 8,000 people

10. Eleni bakes brownies for a bake sale at school. She wants to make 4 batches of brownies and package the brownies in boxes of 8. How many boxes will she need if there are 32 brownies in each batch?

    F 12   **H** 20

    G 16   **J** 32

CHAPTER 8

Fraction Concepts

Around the world, there are more than 30 different species of cats living in forests, grasslands, deserts, and mountains. The body lengths of cats range from 2 feet to 10 feet. The average tiger is $9\frac{1}{4}$ feet long, and the average lion is $6\frac{1}{2}$ feet long. The average bobcat is about $3\frac{1}{4}$ feet long. Compare the average length of a tiger to that of a lion.

AVERAGE BODY LENGTHS OF CATS

Average Body Length (in feet)

Type of Cat

Cheetah $4\frac{1}{2}$ ft, Snow Leopard 5 ft, Tiger $9\frac{1}{4}$ ft, Jaguar 6 ft, Lion $6\frac{1}{2}$ ft, Bobcat $3\frac{1}{4}$ ft

Check What You Know

Use this page to help you review and remember important skills needed for Chapter 8.

✅ Vocabulary

Choose the best term or symbol from the box.

denominator
<
numerator
>

1. The symbol for "is greater than" is __?__ .

2. The top number of a fraction is called the __?__ .

3. The bottom number of a fraction is called the __?__ .

✅ Compare and Order Whole Numbers

(See p. H3.)

Write <, >, or = for each ●.

4. 408 ● 480

5. 4,279 ● 4,277

6. 30 tens ● 3 hundreds

7. 9,315 ● 9,351

8. 18,808 ● 18,880

9. 356,782 ● 356,482

Order the numbers from greatest to least.

10. 3,400; 3,439; 3,399

11. 61,060; 62,000; 61,600

12. 98,450; 98,405; 98,540

✅ Model Fractions (See p. H8.)

Write the fraction for the shaded part.

13.

14.

15.

16.

✅ Model Percents (See p. H9.)

Write the percent for the shaded part.

17.

18.

19.

20.

LOOK AHEAD

In Chapter 8 you will

- find equivalent fractions
- write fractions in simplest form
- write fractions as mixed numbers and mixed numbers as fractions
- compare and order fractions
- find relationships among fractions, decimals, and percents

Equivalent Fractions and Simplest Form

Learn how to identify and write equivalent fractions, and how to write fractions in simplest form.

Vocabulary

equivalent fractions

simplest form

Fractions that name the same amount or the same part of a whole are called **equivalent fractions**. The figures show that the fractions $\frac{1}{4}$ and $\frac{2}{8}$ are equivalent, because they name the same part of a whole circle.

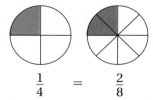

$$\frac{1}{4} = \frac{2}{8}$$

There are several ways to find equivalent fractions.

Activity

You need: fraction bars

Find how many eighths are equivalent to $\frac{1}{2}$.

- Place $\frac{1}{8}$ fraction bars along the $\frac{1}{2}$ bar until the lengths are equal. How many $\frac{1}{8}$ bars are there?

- Complete: $\frac{1}{2} = \frac{\blacksquare}{8}$
- Use fraction bars. How many fourths are equivalent to $\frac{1}{2}$?
- Complete: $\frac{1}{2} = \frac{\blacksquare}{4}$

Another way to find an equivalent fraction is to multiply or divide the numerator and denominator of a fraction by the same number, except 0. Doing this does not change the fraction's value, because this is the same as multiplying or dividing by 1.

EXAMPLE 1

Complete: $\frac{2}{4} = \frac{\blacksquare}{12}$

THINK: To get the denominator 12, I need to multiply the denominator 4 by 3. So, to get the missing numerator, I should multiply the numerator 2 by 3.

$$\frac{2 \times 3}{4 \times 3} = \frac{6}{12}$$

$\frac{3}{3} = 1$, so the product is still equal to $\frac{2}{4}$.

$$\frac{2}{4} = \frac{6}{12}$$

CALIFORNIA STANDARDS O⟶NS 2.4 Determine the least common multiple and the greatest common divisor of whole numbers; use them to solve problems with fractions (e.g., to find a common denominator to add two fractions or to find the reduced form for a fraction). *also* O⟶NS 2.0, MR 2.2, MR 2.5

EXAMPLE 2

Complete: $\dfrac{6}{10} = \dfrac{\blacksquare}{5}$

THINK: I can get the denominator 5 by dividing the denominator 10 by 2. So, to get the missing numerator, I should divide the numerator 6 by 2.

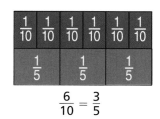

$$\dfrac{6}{10} = \dfrac{3}{5}$$

$\dfrac{6 \div \boxed{2}}{10 \div \boxed{2}} = \dfrac{3}{5}$ *$\dfrac{2}{2} = 1$, so the quotient is still equal to $\dfrac{6}{10}$.*

Math Idea ▶ When the numerator and denominator of a fraction have no common factors other than 1, the fraction is in **simplest form**.

$\dfrac{9}{16}$ **is** in simplest form because 9 and 16 have no common factors other than 1.

$\dfrac{9}{15}$ **is not** in simplest form because 9 and 15 have the common factor 3.

EXAMPLE 3

The sun is much farther from the Earth than the moon is. So, even though the sun is much larger than the moon, the sun's tide-raising force is only $\dfrac{12}{30}$ of the moon's force. Write $\dfrac{12}{30}$ in simplest form.

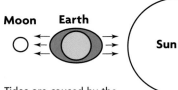

Tides are caused by the gravitational pull of the sun and the moon on Earth's oceans.

12: 1, 2, 3, 4, 6, 12   *Find the common factors of 12 and 30.*
30: 1, 2, 3, 5, 6, 10, 15, 30

$\dfrac{12}{30} = \dfrac{12 \div 3}{30 \div 3} = \dfrac{4}{10}$ *Divide the numerator and denominator by a common factor.*

$\dfrac{4}{10} = \dfrac{4 \div 2}{10 \div 2} = \dfrac{2}{5}$ *Repeat until the fraction is in simplest form.*

So, $\dfrac{2}{5}$ is the simplest form of $\dfrac{12}{30}$.

Bay of Fundy during low tide

In Example 3, the numerator and denominator were divided by common factors twice to find the simplest form. You can find it in just one step if you divide by the greatest common factor.

EXAMPLE 4

The world's highest tides occur in Canada's Bay of Fundy. Tides in Seattle, Washington, average only about $\dfrac{12}{42}$ of tidal heights in the Bay of Fundy. Write $\dfrac{12}{42}$ in simplest form.

12: 1, 2, 3, 4, 6, 12
42: 1, 2, 3, 6, 7, 14, 21, 42   *Find the GCF of 12 and 42.*
GCF = 6

$\dfrac{12}{42} = \dfrac{12 \div 6}{42 \div 6} = \dfrac{2}{7}$ *Divide the numerator and denominator by the GCF.*

Bay of Fundy during high tide

So, $\dfrac{2}{7}$ is the simplest form of $\dfrac{12}{42}$.

161

Think and ▶ Look back at the lesson to answer each question.
Discuss

1. **Describe** what you must do to find a fraction equivalent to $\frac{2}{3}$ if you first multiply the numerator by 5. What is the equivalent fraction?

2. **Explain** how you know that $\frac{5}{12}$ is in simplest form.

Guided ▶ Complete.
Practice

3. $\frac{3}{5} = \frac{\blacksquare}{20}$ 4. $\frac{3}{24} = \frac{\blacksquare}{8}$ 5. $\frac{8}{12} = \frac{\blacksquare}{6}$ 6. $\frac{3}{4} = \frac{\blacksquare}{24}$

7. $\frac{4}{8} = \frac{\blacksquare}{2}$ 8. $\frac{9}{24} = \frac{\blacksquare}{8}$ 9. $\frac{8}{18} = \frac{\blacksquare}{36}$ 10. $\frac{12}{54} = \frac{\blacksquare}{9}$

Write the factors common to the numerator and denominator.

11. $\frac{4}{8}$ 12. $\frac{9}{24}$ 13. $\frac{8}{18}$ 14. $\frac{12}{54}$

Write the fraction in simplest form.

15. $\frac{4}{32}$ 16. $\frac{14}{21}$ 17. $\frac{9}{54}$ 18. $\frac{48}{54}$

19. $\frac{22}{8}$ 20. $\frac{18}{5}$ 21. $\frac{9}{30}$ 22. $\frac{48}{32}$

Independent ▶ Complete.
Practice

23. $\frac{1}{2} = \frac{\blacksquare}{10}$ 24. $\frac{10}{15} = \frac{\blacksquare}{3}$ 25. $\frac{16}{20} = \frac{\blacksquare}{5}$ 26. $\frac{9}{27} = \frac{\blacksquare}{3}$

27. $\frac{2}{12} = \frac{1}{\blacksquare}$ 28. $\frac{\blacksquare}{36} = \frac{2}{9}$ 29. $\frac{21}{24} = \frac{7}{\blacksquare}$ 30. $\frac{40}{\blacksquare} = \frac{5}{8}$

31. $\frac{9}{\blacksquare} = \frac{3}{4}$ 32. $\frac{3}{\blacksquare} = \frac{12}{16}$ 33. $\frac{16}{28} = \frac{4}{\blacksquare}$ 34. $\frac{25}{40} = \frac{\blacksquare}{8}$

Write the factors common to the numerator and denominator.

35. $\frac{1}{7}$ 36. $\frac{9}{30}$ 37. $\frac{6}{27}$ 38. $\frac{9}{63}$

39. $\frac{10}{35}$ 40. $\frac{16}{40}$ 41. $\frac{3}{5}$ 42. $\frac{8}{10}$

Write the fraction in simplest form.

43. $\frac{4}{24}$ 44. $\frac{9}{12}$ 45. $\frac{6}{48}$ 46. $\frac{10}{15}$

47. $\frac{10}{18}$ 48. $\frac{20}{15}$ 49. $\frac{18}{90}$ 50. $\frac{28}{42}$

51. $\frac{22}{33}$ 52. $\frac{24}{30}$ 53. $\frac{60}{42}$ 54. $\frac{24}{28}$

55. $\frac{16}{64}$ 56. $\frac{8}{12}$ 57. $\frac{18}{90}$ 58. $\frac{48}{18}$

59. $\frac{4^2}{32}$ 60. $\frac{3^2}{12}$ 61. $\frac{2^3}{12}$ 62. $\frac{3^2}{6^2}$

TECHNOLOGY LINK

More Practice: Use E-Lab,
Equivalent Fractions.
www.harcourtschool.com/
elab2002

AMERICAN METEOR SOCIETY FIREBALL REPORT, 1998	
Month	Number
January	13
February	8
March	20
April	18
May	8
June	13

63. Use Data Look at the table at the left. A "fireball" is an extremely bright meteor occasionally seen streaking across the night sky. What fraction of the fireballs reported to the American Meteor Society during the first 6 months of 1998 occurred in March?

64. Technology Some calculators have a $\boxed{\text{SIMP}}$ key that can be used to simplify fractions. What fraction would this key sequence give? 10 $\boxed{/}$ 15 $\boxed{\text{SIMP}}$ $\boxed{=}$

65. ? **What's the Question?** Esteban has 6 apple muffins, 2 corn muffins, and 4 bran muffins. The answer is $\frac{1}{3}$ of the muffins.

66. ✎ **Write a problem** about everyday life that involves finding the simplest form of a fraction.

MIXED REVIEW AND TEST PREP

Evaluate each expression.

67. $(6 \div 3)^3 + 2^4$ (p. 44)

68. $(6^2 \div 3^2) + 1$ (p. 44)

69. 3.5×0.01 (p. 70)

70. TEST PREP 0.5×1.2 (p. 70)

    A 0.06   **B** 0.6   **C** 6   **D** 6.2

71. TEST PREP An adult's movie ticket costs $7.50, and a child's ticket costs $5.50. Find the total cost for 2 adults and 3 children. (p. 70)

    F $13.00   **G** $20.50   **H** $24.00   **J** $31.50

Thinker's CORNER

Algebra You can use what you know about equivalent fractions to solve some types of algebraic equations.

A. $\frac{2}{3} = \frac{x - 5}{27}$

THINK: I need to find a fraction equivalent to $\frac{2}{3}$ that has 27 as its denominator. I'll multiply by 9.

$\frac{2}{3} = \frac{\blacksquare}{27}$ $\frac{2 \times 9}{3 \times 9} = \frac{18}{27}$

So, $x - 5 = 18$. If 5 less than some number is 18, that number must be 23. So, $x = 23$.

B. $\frac{3}{s + 2} = \frac{18}{24}$

THINK: I need to find a fraction equivalent to $\frac{18}{24}$ that has 3 as its numerator. I'll divide by 6.

$\frac{3}{\blacksquare} = \frac{18}{24}$ $\frac{3}{4} = \frac{18 \div 6}{24 \div 6}$

So, $s + 2 = 4$. If some number plus 2 equals 4, that number must be 2. So, $s = 2$.

Solve the equation.

1. $\frac{x + 3}{10} = \frac{1}{2}$

2. $\frac{5}{8} = \frac{c - 5}{40}$

3. $\frac{3}{b - 4} = \frac{9}{15}$

4. $\frac{1}{(2 + y)} = \frac{6}{24}$

5. $\frac{4}{9} = \frac{(w + 4)}{18}$

6. $\frac{(k - 5)}{8} = \frac{12}{32}$

Mixed Numbers and Fractions

Learn how to write fractions as mixed numbers and mixed numbers as fractions.

A total solar eclipse occurs when the moon passes between the sun and the Earth, "eclipsing" the sun's light.

Total solar eclipses are rare, with only three visible in most of the U.S. since 1963. Eclipses last different times. In the graph, each quarter-section of a circle represents $\frac{1}{4}$ minute. How long will the 2017 eclipse last? It is represented by 11 sections, or $\frac{11}{4}$ minutes.

The graph shows that 11 sections equal 2 whole minutes plus $\frac{3}{4}$ of another minute. So, $\frac{11}{4} = 2\frac{3}{4}$. The 2017 eclipse will last $2\frac{3}{4}$ minutes.

The fraction $\frac{11}{4}$ has a value greater than 1 because the numerator is greater than the denominator. Sometimes a fraction such as $\frac{11}{4}$ is called an "improper fraction." Any such fraction can be written as a **mixed number**, like $2\frac{3}{4}$.

APPROXIMATE LENGTH OF U.S. TOTAL SOLAR ECLIPSES

1963	⊕ ◖
1970	⊕ ⊕ ⊕ ◖
1979	⊕ ⊕ ⊕
2017	⊕ ⊕ ◖

◹ = $\frac{1}{4}$ minute

Math **I**dea ▶ A mixed number has a whole-number part that is not 0 and a fraction part.

EXAMPLE 1

The longest total solar eclipse in the next 200 years will take place in 2186. It will last about $\frac{15}{2}$ minutes. Write $\frac{15}{2}$ as a mixed number.

$$\frac{15}{2} \rightarrow \begin{array}{r} 7\frac{1}{2} \\ 2\overline{)15} \\ -14 \\ \hline 1 \end{array}$$

Divide the numerator by the denominator. For the fraction part of the quotient, use the remainder as the numerator and the divisor as the denominator. Write the fraction in simplest form.

So, $\frac{15}{2} = 7\frac{1}{2}$. The 2186 eclipse will last $7\frac{1}{2}$ minutes.

You can also write a mixed number as a fraction.

EXAMPLE 2

Write the mixed number $3\frac{2}{5}$ as a fraction.

$$3\frac{2}{5} = \frac{3 \times 5}{5} + \frac{2}{5} = \frac{(3 \times 5) + 2}{5} = \frac{17}{5}$$

So, $3\frac{2}{5} = \frac{17}{5}$.

Multiply the whole number by the denominator. Add the numerator. Use the same denominator.

CALIFORNIA STANDARDS NS 1.0 Students compare and order positive and negative fractions, decimals, and mixed numbers. Students solve problems involving fractions, ratios, proportions, and percentages. *also* O⌐∎NS 2.0, MR 2.2, MR 2.5

Think and ▶ Look back at the lesson to answer the question.
Discuss

1. Tell how you know that a given number is a mixed number.

Write the fraction as a mixed number or a whole number.

2. $\frac{5}{3}$ **3.** $\frac{7}{2}$ **4.** $\frac{15}{5}$ **5.** $\frac{11}{3}$ **6.** $\frac{13}{4}$

Guided ▶ Write the mixed number as a fraction.
Practice

7. $1\frac{1}{4}$ **8.** $1\frac{3}{5}$ **9.** $2\frac{2}{3}$ **10.** $3\frac{4}{5}$ **11.** $5\frac{2}{7}$

PRACTICE AND PROBLEM SOLVING

Independent ▶ Write the fraction as a mixed number or a whole number.
Practice

12. $\frac{7}{4}$ **13.** $\frac{9}{2}$ **14.** $\frac{11}{2}$ **15.** $\frac{23}{4}$ **16.** $\frac{27}{3}$

17. $\frac{31}{6}$ **18.** $\frac{18}{11}$ **19.** $\frac{90}{7}$ **20.** $\frac{104}{13}$ **21.** $\frac{150}{9}$

22. $\frac{x}{y}$ for $x = 18$ and $y = 12$ **23.** $\frac{a}{b}$ for $a = 55$ and $b = 15$

Write the mixed number as a fraction.

24. $3\frac{2}{3}$ **25.** $6\frac{1}{2}$ **26.** $5\frac{1}{3}$ **27.** $1\frac{9}{10}$ **28.** $4\frac{1}{9}$

29. $9\frac{1}{4}$ **30.** $2\frac{3}{8}$ **31.** $4\frac{9}{11}$ **32.** $11\frac{4}{9}$ **33.** $18\frac{3}{5}$

Problem Solving ▶ **34.** ❓ **What's the Error?** Marti changed $3\frac{1}{4}$ to $12\frac{1}{4}$.
What mistake did she make? What is the correct answer?

35. 📓 **Write About It** Can any fraction be written as a mixed number? Explain.

36. Astronomy On June 20, 1955, a total solar eclipse lasted 7 min 7 sec. On June 20, 1974, a total solar eclipse lasted 5 min 8 sec. Which lasted longer? How much longer?

MIXED REVIEW AND TEST PREP

Write the prime factorization of each number. (p.148)

37. 36 **38.** 42 **39.** 23

40. Is the following question biased? If so, rewrite it so that it is unbiased: The film *Time Warp* is great, isn't it? (p. 98)

41. TEST PREP Noella is hiking a 25-km trail. She has hiked 3.8 km to the first overlook and another 6.5 km to the second overlook. How many kilometers does she have left to hike? (p. 66)

A 9.3 km **B** 10.3 km **C** 14.7 km **D** 15.7 km

Compare and Order Fractions

Learn how to compare and order fractions.

Remember that values increase as you move to the right on a number line. Values decrease as you move left.

increase ⟶

⟵ decrease

1. $\frac{1}{2} = \frac{\blacksquare}{10}$ 2. $\frac{2}{3} = \frac{\blacksquare}{9}$ 3. $\frac{4}{5} = \frac{\blacksquare}{20}$ 4. $\frac{3}{8} = \frac{\blacksquare}{24}$

5. Write the first four multiples of 12.

If two fractions have the same denominator, the fraction with the greater numerator is greater. So, $\frac{7}{12} > \frac{5}{12}$ because $7 > 5$.

If fractions do not have common denominators, you can use a number line to compare and order the fractions. The number line shows that $\frac{1}{4} < \frac{3}{8} < \frac{1}{2}$. From least to greatest the order is $\frac{1}{4}, \frac{3}{8}, \frac{1}{2}$.

You can also use the least common multiple (LCM) to compare and order fractions.

EXAMPLE

George Washington Carver, one of America's most honored scientists, was born a slave in 1864.

George Washington Carver made over 500 useful agricultural products using peanuts, sweet potatoes, and pecans. About $\frac{7}{10}$ of the products used peanuts and about $\frac{1}{4}$ used sweet potatoes. Did Carver make more products with sweet potatoes or with peanuts?

To compare $\frac{7}{10}$ and $\frac{1}{4}$, find the LCM of the denominators, 10 and 4.

10: 10, 20, 30, 40   *Write multiples of 10.*

  4: 4, 8, 12, 16, 20, 24, 28   *Write multiples of 4.*

LCM = 20

$\frac{7}{10} = \frac{7 \times 2}{10 \times 2} = \frac{14}{20}$ *Rewrite the fractions, using the LCM as a common denominator.*

$\frac{1}{4} = \frac{1 \times 5}{4 \times 5} = \frac{5}{20}$

$\frac{14}{20} > \frac{5}{20}$, so $\frac{7}{10} > \frac{1}{4}$.   *Compare $\frac{14}{20}$ and $\frac{5}{20}$.*

So, Carver made more products with peanuts.

CHECK FOR UNDERSTANDING

Think and ▶ Discuss

Look back at the lesson to answer each question.

1. **Explain** how to compare $\frac{4}{9}$ and $\frac{5}{9}$.

CALIFORNIA STANDARDS O—π NS 1.1 Compare and order positive and negative fractions, decimals, and mixed numbers and place them on a number line. O—π NS 2.4 Determine the least common multiple and the greatest common divisor of whole numbers; use them to solve problems with fractions. *also* O—π NS 1.0, MR 1.0, MR 2.4, MR 2.5

Compare the fractions. Write <, >, or = for each ●.

2. $\frac{13}{20}$ ● $\frac{9}{20}$ **3.** $\frac{1}{4}$ ● $\frac{9}{20}$ **4.** $\frac{5}{6}$ ● $\frac{2}{3}$ **5.** $\frac{3}{8}$ ● $\frac{6}{16}$

Use the number line to order the fractions from least to greatest.

0   $\frac{1}{12}$ $\frac{1}{6}$ $\frac{1}{4}$ $\frac{1}{3}$ $\frac{5}{12}$ $\frac{1}{2}$ $\frac{7}{12}$ $\frac{2}{3}$ $\frac{3}{4}$ $\frac{5}{6}$ $\frac{11}{12}$ 1

6. $\frac{3}{4}, \frac{1}{3}, \frac{11}{12}$ **7.** $\frac{2}{3}, \frac{1}{4}, \frac{6}{12}$ **8.** $\frac{1}{3}, \frac{5}{12}, \frac{2}{4}$

PRACTICE AND PROBLEM SOLVING

Compare the fractions. Write <, >, or = for each ●.

9. $\frac{6}{7}$ ● $\frac{4}{7}$ **10.** $\frac{3}{11}$ ● $\frac{4}{11}$ **11.** $\frac{4}{12}$ ● $\frac{1}{3}$ **12.** $\frac{17}{20}$ ● $\frac{3}{5}$

13. $\frac{5}{6}$ ● $\frac{15}{18}$ **14.** $\frac{7}{9}$ ● $\frac{11}{12}$ **15.** $\frac{3}{4}$ ● $\frac{5}{8}$ **16.** $\frac{11}{15}$ ● $\frac{2}{3}$

Use the number line to order the fractions from least to greatest.

0   $\frac{1}{12}$ $\frac{1}{6}$ $\frac{1}{4}$ $\frac{1}{3}$ $\frac{5}{12}$ $\frac{1}{2}$ $\frac{7}{12}$ $\frac{2}{3}$ $\frac{3}{4}$ $\frac{5}{6}$ $\frac{11}{12}$ 1

17. $\frac{9}{12}, \frac{1}{2}, \frac{2}{6}$ **18.** $\frac{4}{12}, \frac{4}{6}, \frac{7}{12}$ **19.** $\frac{7}{12}, \frac{5}{6}, \frac{1}{2}$

Order the fractions from least to greatest.

20. $\frac{5}{8}, \frac{1}{2}, \frac{3}{4}$ **21.** $\frac{2}{5}, \frac{3}{10}, \frac{1}{2}$ **22.** $\frac{11}{16}, \frac{3}{4}, \frac{5}{8}$

23. $\frac{1}{3}, \frac{1}{6}, \frac{1}{2}$ **24.** $\frac{2}{3}, \frac{2}{9}, \frac{2}{6}$ **25.** $\frac{3}{4}, \frac{1}{12}, \frac{5}{8}$

26. During a physical education class, $\frac{1}{3}$ of the students chose to play basketball, $\frac{4}{15}$ chose flag football, and $\frac{2}{5}$ chose tetherball. Which activity was chosen by the most students?

27. ❓ **What's the Error?** For dinner, a mushroom pizza is cut into eighths and a cheese pizza into twelfths. After the meal there are 3 pieces of each left. Pablo tells his mother that the same amount of each pizza is left. What mistake did he make?

28. *REASONING* Find a fraction that has a denominator of 15 and is between $\frac{2}{3}$ and $\frac{4}{5}$.

MIXED REVIEW AND TEST PREP

Evaluate each expression for $a = 2.3$, $b = 0.7$, and $c = 5.4$. (p. 82)

29. $a - b + c$ **30.** $(b \times 5) + a$

31. Find the LCM of 8 and 12. (p. 150)   **32.** Find the GCF of 16 and 40. (p. 150)

33. **TEST PREP** There are 12 cans of soup in 1 case. How many cases should you order if you need 132 cans of soup? (p. 22)

     A 10   **B** 11   **C** 1,200   **D** 1,584

Explore Fractions and Decimals

Explore how to convert fractions to decimals.

You need graph paper, scissors, colored pencils.

You can use decimal squares to help you convert fractions to decimals.

Activity

- Cut out a 10 × 10 grid from graph paper. Fold it into 2 equal parts. Then use a colored pencil to shade one of the equal parts. What fraction of the grid is shaded?

- How many small squares are in the whole grid? How many of these squares are shaded? What fraction compares these shaded squares to those in the whole grid? What decimal can you write for this fraction?

- How many columns are in the whole grid? How many columns are shaded? What fraction compares the shaded columns to the whole grid? What decimal can you write for this fraction?

- Are the fractions $\frac{1}{2}$, $\frac{5}{10}$, and $\frac{50}{100}$ equivalent? How do you know? What are two ways to write the fraction $\frac{1}{2}$ as a decimal?

- Cut out another 10 × 10 grid. Fold the grid into 5 equal parts and shade one part. How many rows or columns are shaded? How can you write $\frac{1}{5}$ as a decimal?

Think and Discuss

- How can you show tenths in a 10 × 10 grid?

- How can you show hundredths in a 10 × 10 grid?

- What fractions are easiest to write as decimals?

- How is the 10 × 10 grid helpful for writing fractions as decimals?

TECHNOLOGY LINK

More Practice: Use E-Lab, *Equivalent Fractions, Decimals, and Mixed Numbers.*
www.harcourtschool.com/ elab2002

Practice

Write the fraction as a decimal. Use decimal squares.

1. $\frac{3}{10}$ 2. $\frac{60}{100}$ 3. $\frac{7}{10}$ 4. $\frac{7}{100}$ 5. $\frac{90}{100}$

6. $\frac{2}{10}$ 7. $\frac{40}{100}$ 8. $\frac{6}{10}$ 9. $\frac{85}{100}$ 10. $\frac{8}{10}$

CALIFORNIA STANDARDS MR 2.0 Students use strategies, skills, and concepts in finding solutions. MR 2.4 Use a variety of methods, such as words, numbers, symbols, charts, graphs, tables, diagrams, and models, to explain mathematical reasoning. *also* MR 2.5

Fractions, Decimals, and Percents

Learn how to convert fractions to decimals, decimals to fractions, and fractions to percents.

Vocabulary

terminating decimal

repeating decimal

QUICK REVIEW

1. $24 \div 4$
2. $144 \div 6$
3. $155 \div 5$
4. Write 0.73 in words.
5. Write 0.026 in words.

There are many ways to write numbers. Some ways are as fractions, decimals, and percents.

Sometimes you may have to rewrite a given number in a different form. The easiest conversion is from a decimal to a fraction.

EXAMPLE 1

Remember that *percent* means "out of one hundred." For example, 25% means "25 out of 100."

Write each decimal as a fraction.
A. 0.7　　　　　　　　　　　　　**B.** 0.29

Use the decimal's place value to write each fraction.

$0.7 = \frac{7}{10}$ **THINK:** *"seven tenths"*　　　　　$0.29 = \frac{29}{100}$ **THINK:** *"twenty-nine hundredths"*

To rewrite a fraction as a decimal, use long division or a calculator.

EXAMPLE 2

A newborn koala is about 19 mm long. This is about $\frac{3}{4}$ in. Change $\frac{3}{4}$ to a decimal.

Use long division.

$$\begin{array}{r} 0.75 \\ 4)\overline{3.00} \end{array}$$

Divide the numerator by the denominator.

Use a calculator.

3 $\boxed{\div}$ 4 $\boxed{=}$

0.75

So, $\frac{3}{4} = 0.75$.

The decimal 0.75 is an example of a **terminating decimal**. The decimal comes to an end at 5. You know that a decimal terminates if you reach a remainder of zero when you are using long division.

The decimal for the fraction $\frac{4}{11}$ does not terminate. When you divide 4 by 11, you never reach a remainder of zero. This decimal is a **repeating decimal** because it shows a pattern of repeating digits.

$$\begin{array}{r} 0.3636 \\ 11)\overline{4.0000} \\ -3\,3 \\ \hline 70 \\ -66 \\ \hline 40 \\ -33 \\ \hline 70 \\ -66 \\ \hline 4 \end{array}$$

To write a repeating decimal, show three dots or draw a bar over the repeating part.

$\frac{4}{11} = 0.363636\ldots$　　　$\frac{4}{11} = 0.\overline{36}$

CALIFORNIA STANDARDS ⊶ **NS 1.0** Students compare and order positive and negative fractions, decimals, and mixed numbers. Students solve problems involving fractions, ratios, proportions, and percentages. *also* ⊶ **NS 2.0, MR 2.2, MR 2.4, MR 2.5**

169

To compare a fraction and a decimal, you can first rewrite the fraction as a decimal. Then compare the decimals.

EXAMPLE 3

A newborn panda weighs about $\frac{1}{4}$ lb. A newborn cocker spaniel weighs about 0.4 lb. Which animal weighs less at birth?

Solve by using long division.

$$\begin{array}{r} 0.25 \\ 4\overline{)1.00} \\ -\underline{8} \\ 20 \\ -\underline{20} \\ 0 \end{array}$$

Divide the numerator by the denominator.

$0.25 < 0.4$, so $\frac{1}{4} < 0.4$.

So, a newborn panda weighs less than a newborn cocker spaniel.

Solve by using a calculator.

1 ÷ 4 =   0.25

To write a fraction as a percent, first convert the fraction to a decimal. Then write the decimal as a percent.

EXAMPLE 4

The barrow ground squirrel of Point Barrow, Alaska, is the world's longest-hibernating animal. The squirrel hibernates $\frac{9}{12}$ of the year. What percent of the year does it hibernate?

$\frac{9}{12} = 0.75$ *Use long division or a calculator to rewrite the fraction as a decimal.*

$0.75 = \frac{75}{100}$ THINK: *75 hundredths. Write the decimal as a fraction.*

$= 75\%$ THINK: *Percent means "out of one hundred." So, 75 hundredths is 75 percent.*

So, the barrow ground squirrel hibernates 75% of the year.

CHECK FOR UNDERSTANDING

Think and Discuss ▶ **Look back at the lesson to answer each question.**

1. **Explain** how to use place value to change 0.026 to a fraction.

2. **Compare** a repeating decimal with a terminating decimal.

Guided Practice ▶ **Write the decimal as a fraction.**

3. 0.7   **4.** 0.39   **5.** 0.105   **6.** 0.007

Write as a decimal. Tell whether the decimal terminates or repeats.

7. $\frac{1}{4}$ **8.** $\frac{7}{20}$ **9.** $\frac{2}{3}$ **10.** $\frac{8}{11}$

Compare. Write $<$, $>$, or $=$ for each ●.

11. $0.62 ● \frac{1}{2}$ **12.** $\frac{12}{20} ● 0.9$ **13.** $\frac{1}{8} ● 0.125$

Write the fraction as a percent.

14. $\frac{7}{10}$ **15.** $\frac{1}{5}$ **16.** $\frac{1}{4}$ **17.** $\frac{40}{100}$

Independent ▶ Practice

Write the decimal as a fraction.

18. 0.4   **19.** 0.06   **20.** 0.35   **21.** 0.61

22. 0.115   **23.** 0.205   **24.** 0.079   **25.** 0.009

Write as a decimal. Tell whether the decimal terminates or repeats.

26. $\frac{1}{5}$ **27.** $\frac{1}{6}$ **28.** $\frac{1}{15}$ **29.** $\frac{5}{8}$

30. $\frac{11}{20}$ **31.** $\frac{3}{10}$ **32.** $\frac{5}{12}$ **33.** $\frac{1}{9}$

34. $\frac{11}{12}$ **35.** $\frac{9}{25}$ **36.** $\frac{17}{33}$ **37.** $\frac{15}{99}$

TECHNOLOGY LINK

More Practice: Use **Mighty Math Calculating Crew**, *Nautical Number Line*, Level Q

Compare. Write <, >, or = for each ●.

38. $\frac{1}{10}$ ● 0.04   **39.** 0.15 ● $\frac{3}{20}$ **40.** $\frac{1}{2}$ ● 0.52

41. 0.65 ● $\frac{3}{4}$ **42.** $\frac{1}{20}$ ● 0.1   **43.** 0.58 ● $\frac{7}{12}$

Write the fraction as a percent.

44. $\frac{9}{10}$ **45.** $\frac{3}{4}$ **46.** $\frac{1}{2}$ **47.** $\frac{6}{100}$

48. $\frac{3}{5}$ **49.** $\frac{25}{50}$ **50.** $\frac{3}{2}$ **51.** $\frac{1}{200}$

Problem Solving ▶ Applications

52. The goal of the East Side Animal Shelter is to have 0.8 of its animals adopted. One week, the shelter found homes for 20 of its 24 animals. Did the shelter reach its goal? Explain.

Use Data   For 53–55, use the table.

53. Write Brian's math test score as a decimal.

54. Did Juan get a higher score on the math or science test?

55. Which student got a higher score on the math test than on the science test?

STUDENT	MATH SCORE	SCIENCE SCORE
Brian	$\frac{18}{25}$	0.95
Juan	$\frac{21}{25}$	0.85
Sabina	$\frac{17}{25}$	0.75
Megan	$\frac{23}{25}$	0.90

56. ✎ **Write About It**   The decimal for $\frac{1}{9}$ is $0.\overline{1}$; for $\frac{2}{9}$, $0.\overline{2}$; and for $\frac{3}{9}$, $0.\overline{3}$. Explain how you could use this information to predict the decimal for $\frac{8}{9}$. Use division to check your method.

MIXED REVIEW AND TEST PREP

57. Write $4\frac{1}{2}$ as a fraction. (p. 164)

58. Write $\frac{36}{5}$ as a mixed number. (p. 164)

59. 79.02 − 2.13 (p. 66)

60. 48.541 + 11 (p. 66)

61. TEST PREP   On his last seven math quizzes, Rashard scored 86, 88, 92, 88, 33, 96, and 84. Find the mean of his scores with and without the outlier. (p. 109)

A 81, 89   **B** 84, 89   **C** 88, 88   **D** 89, 81

1. **VOCABULARY** When the numerator and denominator of a fraction have no common factors other than 1, the fraction is in __?__. (p. 161)

2. **VOCABULARY** A number that is made up of a whole number and a fraction is called a __?__. (p. 164)

Write the fraction in simplest form. (pp. 160–163)

3. $\frac{6}{12}$

4. $\frac{12}{16}$

5. $\frac{25}{30}$

Complete. (pp. 160–163)

6. $\frac{3}{5} = \frac{\blacksquare}{20}$

7. $\frac{2}{\blacksquare} = \frac{10}{35}$

8. $\frac{24}{32} = \frac{\blacksquare}{8}$

Write the fraction as a mixed number or a whole number. (pp. 164–165)

9. $\frac{7}{3}$

10. $\frac{30}{6}$

11. $\frac{19}{4}$

Write the mixed number as a fraction. (pp. 164–165)

12. $1\frac{5}{6}$

13. $3\frac{1}{3}$

14. $5\frac{7}{8}$

Compare the fractions. Write <, >, or = for each ●. (pp. 166–167)

15. $\frac{7}{8}$ ● $\frac{5}{8}$

16. $\frac{2}{3}$ ● $\frac{8}{12}$

17. $\frac{1}{3}$ ● $\frac{1}{2}$

18. $\frac{1}{2}$ ● $\frac{11}{20}$

19. $\frac{3}{4}$ ● $\frac{3}{8}$

20. $\frac{7}{25}$ ● $\frac{1}{5}$

Write the decimal as a fraction. (pp. 169–171)

21. 0.27

22. 0.1

23. 0.089

Write as a decimal. Tell whether the decimal terminates or repeats.
(pp. 169–171)

24. $\frac{1}{4}$

25. $\frac{5}{6}$

26. $\frac{7}{20}$

Write the fraction as a percent. (pp. 169–171)

27. $\frac{3}{4}$

28. $\frac{9}{100}$

29. $\frac{11}{25}$

30. In the election for class president, Marcus received $\frac{5}{12}$ of the votes, Denise received $\frac{1}{4}$ of the votes, and Alonzo received $\frac{1}{3}$ of the votes. Who won the election? (pp. 166–167)

31. **Use Data** Use the table to find the fraction of the new November films that are rated PG-13. Write your answer in simplest form. What percent is this? (pp. 160–163)

NOVEMBER FILMS	
Rating	**Number**
G	3
PG-13	8
PG	5
R	4

32. Of all U.S. car tunnels longer than 1 mile, $\frac{3}{8}$ are in Pennsylvania. Change $\frac{3}{8}$ to a decimal. (pp. 169–171)

33. On Library Day, $\frac{13}{20}$ of the students at Pine Street School checked books out of the library. What percent of the students checked out books? (pp. 169–171)

Understand the problem.

See item **8.**

To compare fractions and decimals, change them to the same form. Use the form that will be easier to work with. Then find the numbers that are in the proper order.

Also see problem **1**, p. H62.

Choose the best answer.

1. Which group contains fractions that are all equivalent to $\frac{1}{4}$?

   A $\frac{2}{8}, \frac{4}{20}, \frac{11}{44}$ **C** $\frac{6}{24}, \frac{15}{60}, \frac{50}{200}$

   B $\frac{3}{6}, \frac{20}{80}, \frac{3}{12}$ **D** $\frac{3}{12}, \frac{25}{100}, \frac{5}{9}$

2. Which is equivalent to 0.36?

   F $\frac{18}{100}$

   G $\frac{9}{25}$

   H $\frac{3}{6}$

   J $\frac{36}{10}$

3. Which pair contains numbers that are equivalent?

   A $3\frac{2}{3}, \frac{11}{3}$ **C** $4\frac{2}{5}, \frac{21}{5}$

   B $\frac{13}{4}, 2\frac{3}{4}$ **D** $\frac{16}{7}, 3\frac{1}{2}$

4. What is $\frac{135}{144}$ in simplest form?

   F $\frac{7}{8}$ **H** $\frac{15}{16}$

   G $\frac{11}{12}$ **J** Not here

5. Eldora measured the length of a remote control sailboat course as $\frac{5}{8}$ of a mile. What is the decimal equivalent of $\frac{5}{8}$?

   A 0.5

   B 0.58

   C 0.625

   D 0.8

6. In which pair are both numbers equivalent to $\frac{3}{4}$?

   F 0.25; 25% **H** 80%; $\frac{12}{15}$

   G 30%; $\frac{3}{10}$ **J** 75%; $\frac{12}{16}$

7. Evan has a bag of fruit. He has 9 apples, 4 oranges, and 3 bananas. What fraction represents the pieces of fruit that are oranges?

   A $\frac{4}{18}$

   B $\frac{1}{4}$

   C $\frac{5}{16}$

   D $\frac{3}{4}$

8. Which numbers are in order from least to greatest?

   F $0.3, \frac{3}{8}, \frac{2}{5}$

   G $\frac{2}{5}, 0.3, \frac{3}{8}$

   H $\frac{3}{8}, \frac{2}{5}, 0.3$

   J $\frac{2}{5}, \frac{3}{8}, 0.3$

9. If 5 packages of hot dogs cost $9.25, what is the cost of 1 package?

   A $0.92

   B $1.15

   C $1.85

   D $2.10

10. $\frac{5}{6} < \blacksquare$

   F $\frac{7}{12}$ **H** $\frac{7}{9}$

   G $\frac{7}{10}$ **J** $\frac{7}{8}$

CHAPTER 9

Add and Subtract Fractions and Mixed Numbers

In 1992, the United States Department of Agriculture and the Department of Health and Human Services introduced the Food Guide Pyramid. Bob used the information in the pyramid to make the diagram shown. He shades the number of servings he has each day to make sure he has a balanced diet. The shaded squares show what he ate for breakfast. How many cups of food were in his breakfast?

Dairy and Protein
5 Servings

Limit Fats, Oils, Sweets

Fruits and Vegetables
7 Servings

Grains, Rice, & Pasta
9 Servings

1 serving is about $\frac{1}{2}$ cup

Check What You Know

Use this page to help you review and remember important skills needed for Chapter 9.

✓ Vocabulary

Choose the best term from the box.

> divide
> equivalent
> estimate
> mixed number
> simplest form

1. A number with a whole-number part and a fraction part is called a(n) __?__ .

2. A fraction in which the numerator and denominator have no common factors other than 1 is in __?__ .

3. To write a fraction greater than 1 as a mixed number, __?__ the numerator by the denominator.

4. Fractions that name the same number or part are called __?__ fractions.

✓ Simplify Fractions (See p. H9.)

Write each fraction in simplest form.

5. $\frac{6}{8}$ 6. $\frac{5}{10}$ 7. $\frac{4}{12}$ 8. $\frac{18}{27}$ 9. $\frac{12}{9}$

10. $\frac{16}{20}$ 11. $\frac{12}{8}$ 12. $\frac{9}{15}$ 13. $\frac{10}{20}$ 14. $\frac{15}{18}$

15. $\frac{26}{39}$ 16. $\frac{12}{16}$ 17. $\frac{15}{9}$ 18. $\frac{6}{32}$ 19. $\frac{17}{51}$

20. $\frac{48}{54}$ 21. $\frac{100}{200}$ 22. $\frac{25}{10}$ 23. $\frac{80}{64}$ 24. $\frac{84}{96}$

✓ Add and Subtract Like Fractions (See p. H10.)

Find the sum or difference. Write each answer in simplest form.

25. $\frac{1}{3} + \frac{1}{3}$ 26. $\frac{3}{5} + \frac{1}{5}$ 27. $\frac{3}{8} + \frac{1}{8}$ 28. $\frac{5}{6} + \frac{1}{6}$

29. $\frac{3}{4} - \frac{1}{4}$ 30. $\frac{5}{8} - \frac{3}{8}$ 31. $\frac{2}{3} - \frac{1}{3}$ 32. $\frac{4}{5} - \frac{2}{5}$

33. $\frac{3}{6} + \frac{2}{6}$ 34. $\frac{7}{8} - \frac{1}{8}$ 35. $\frac{3}{4} + \frac{1}{4}$ 36. $\frac{7}{8} - \frac{3}{8}$

37. $\frac{4}{9} - \frac{2}{9}$ 38. $\frac{1}{12} + \frac{7}{12}$ 39. $\frac{13}{14} - \frac{3}{14}$ 40. $\frac{1}{15} + \frac{8}{15}$

LOOK AHEAD

In Chapter 9 you will
- estimate sums and differences of fractions and mixed numbers
- add and subtract fractions and mixed numbers

Estimate Sums and Differences

Learn how to estimate sums and differences of fractions and mixed numbers.

Estimate the sum or difference.

1.	823	2.	364	3.	736	4.	589	5.	6,755
	+116		−232		−381		+ 42		− 482

You can decide whether a fraction is closest to 0, $\frac{1}{2}$, or 1.

One Way Use a number line.

Look at the number line below. Is $\frac{3}{8}$ closest to 0, $\frac{1}{2}$, or 1?

0   $\frac{1}{8}$ $\frac{2}{8}$ $\frac{3}{8}$ $\frac{1}{2}$ $\frac{5}{8}$ $\frac{6}{8}$ $\frac{7}{8}$ 1

$\frac{3}{8}$ is closest to $\frac{1}{2}$.

Another Way Compare the numerator to the denominator.

| $\frac{1}{8}$ | $\frac{1}{6}$ | $\frac{2}{9}$ |

| $\frac{4}{9}$ | $\frac{5}{8}$ | $\frac{2}{3}$ |

| $\frac{7}{9}$ | $\frac{5}{6}$ | $\frac{7}{8}$ |

The numerators are much less than half the denominators. So, the fractions are close to 0.

The numerators are about one half of the denominators. So, the fractions are close to $\frac{1}{2}$.

The numerators are about the same as the denominators. So, the fractions are close to 1.

Sometimes when you add and subtract fractions, you do not need an exact answer.

EXAMPLE 1

The California section of the Pacific Crest Trail, which runs from Mexico to Canada, is 1,680 miles long. The average time it takes to hike the California section of the trail is 3–5 months. The Tanner family hiked $\frac{3}{10}$ of this section in June and July. Then they hiked $\frac{3}{8}$ of it in August and September. About how much of the trail did the Tanners hike from June through September?

Estimate $\frac{3}{10} + \frac{3}{8}$.

$$
\begin{array}{ll}
\frac{3}{10} \rightarrow \frac{1}{2} & \frac{3}{10} \text{ is between 0 and } \frac{1}{2}, \text{ but closer to } \frac{1}{2}. \\
+\frac{3}{8} \rightarrow +\frac{1}{2} & \frac{3}{8} \text{ is between 0 and } \frac{1}{2}, \text{ but closer to } \frac{1}{2}. \\
\hline
\qquad 1 & \text{The sum is greater than } \frac{1}{2}, \text{ but less than 1.}
\end{array}
$$

So, the Tanners hiked more than $\frac{1}{2}$ of the trail, but not all of it.

CALIFORNIA STANDARDS NS 2.1 Solve problems involving addition, subtraction, multiplication, and division of positive fractions and explain why a particular operation was used for a given situation. *also,* **NS 2.0, MR 2.4**

EXAMPLE 2

Estimate $\frac{7}{8} - \frac{5}{6}$.

$\frac{7}{8} \rightarrow \quad 1$ *$\frac{7}{8}$ is between $\frac{1}{2}$ and 1, but closer to 1.*

$\underline{-\frac{5}{6} \rightarrow -1}$ *$\frac{5}{6}$ is between $\frac{1}{2}$ and 1, but closer to 1.*

        0 *Subtract.*

The fractions are close to each other and their difference is close to 0.

Remember that a mixed number is a whole number and a fraction combined. $4\frac{2}{5}$ is a mixed number.

Look at the number line below. Is $3\frac{5}{8}$ in. closest to 3 in., $3\frac{1}{2}$ in., or 4 in.?

$3\frac{5}{8}$ in. is closest to $3\frac{1}{2}$ in.

To estimate sums and differences of mixed numbers, compare each mixed number to the nearest whole number or the nearest $\frac{1}{2}$.

EXAMPLE 3

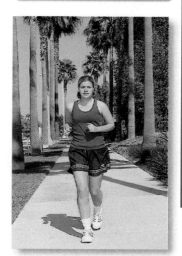

Bianca's goal is to jog 15 miles a week. So far this week she has jogged $2\frac{5}{6}$ mi and $3\frac{1}{10}$ mi. About how many more miles must she jog to meet her goal?

Estimate $15 - \left(2\frac{5}{6} + 3\frac{1}{10}\right)$.

$15 - \left(2\frac{5}{6} + 3\frac{1}{10}\right)$ *$2\frac{5}{6}$ is close to 3, and $3\frac{1}{10}$ is close to 3.*

        $\downarrow$ $\downarrow$

$15 - (3 \ + 3)$ *Add.*

$15 - 6 = 9$ *Subtract.*

So, Bianca needs to jog about 9 more miles to meet her goal.

You can find a range to estimate a sum or difference.

EXAMPLE 4

Estimate $5\frac{3}{4} - 4\frac{7}{8}$.

Since $5\frac{3}{4}$ is halfway between $5\frac{1}{2}$ and 6, find two estimates.

$5\frac{3}{4} - 4\frac{7}{8}$ *$5\frac{3}{4}$ is close to 6, and $4\frac{7}{8}$ is close to 5.* $5\frac{3}{4} - 4\frac{7}{8}$ *$5\frac{3}{4}$ is close to $5\frac{1}{2}$, and $4\frac{7}{8}$ is close to 5.*

$\downarrow$ $\downarrow$ *AND* $\downarrow$ $\downarrow$

$6 - 5 = 1$ *Subtract.* $5\frac{1}{2} - 5 = \frac{1}{2}$ *Subtract.*

The range is $\frac{1}{2}$ to 1. A good estimate of $5\frac{3}{4} - 4\frac{7}{8}$ would be $\frac{3}{4}$, halfway between $\frac{1}{2}$ and 1.

CHECK FOR UNDERSTANDING

Think and Discuss ▶ **Look back at the lesson to answer each question.**

1. **Explain** how to round fractions less than 1.

2. **Compare** the numerator and denominator of several fractions greater than $\frac{1}{2}$. What do you notice?

Guided Practice ▶ **Use the number line to tell whether the fraction is closest to 0, $\frac{1}{2}$, or 1. Write *close to 0*, *close to $\frac{1}{2}$*, or *close to 1*.**

3. $\frac{11}{12}$ 4. $\frac{1}{12}$ 5. $\frac{1}{4}$ 6. $\frac{2}{3}$

Estimate the sum or difference.

7. $\frac{3}{5} + \frac{6}{7}$ 8. $\frac{10}{12} - \frac{4}{5}$ 9. $\frac{7}{8} - \frac{4}{9}$ 10. $\frac{5}{16} + \frac{4}{9}$

11. $4\frac{3}{8} + 2\frac{3}{4}$ 12. $3\frac{2}{16} - \frac{8}{15}$ 13. $4\frac{11}{12} - 3\frac{1}{5}$ 14. $6\frac{7}{8} + 5\frac{5}{9}$

PRACTICE AND PROBLEM SOLVING

Independent Practice ▶ **Use the number line to tell whether the fraction is closest to 0, $\frac{1}{2}$, or 1. Write *close to 0*, *close to $\frac{1}{2}$*, or *close to 1*.**

15. $\frac{5}{6}$ 16. $\frac{5}{12}$ 17. $\frac{1}{6}$ 18. $\frac{3}{4}$

Estimate the sum or difference.

19. $\frac{12}{13} + \frac{15}{17}$ 20. $2\frac{1}{4} + 1\frac{1}{8}$ 21. $\frac{15}{16} - \frac{6}{10}$ 22. $7\frac{9}{16} - 4\frac{2}{9}$

23. $12\frac{1}{12} + 4\frac{4}{5}$ 24. $\frac{18}{20} - \frac{3}{50}$ 25. $4\frac{3}{8} - 1\frac{1}{3}$ 26. $\frac{3}{5} + \frac{6}{7}$

27. $\frac{5}{9} - \frac{1}{2}$ 28. $\frac{1}{12} + 1\frac{3}{19}$ 29. $9\frac{7}{12} - 5\frac{1}{5}$ 30. $\frac{13}{15} + \frac{8}{9} + \frac{1}{7}$

Use a range to estimate each sum or difference.

31. $6\frac{1}{4} + 2\frac{5}{6}$ 32. $5\frac{7}{10} - 2\frac{1}{4}$ 33. $8\frac{3}{4} - 4\frac{1}{8}$

34. $3\frac{1}{4} - 1\frac{1}{8}$ 35. $1\frac{3}{4} + 3\frac{4}{5}$ 36. $4\frac{1}{9} + 5\frac{3}{4}$

37. Estimate the sum. $3\frac{1}{5} + 2\frac{1}{3} + 1\frac{11}{12}$

38. Estimate the difference between 8 and $2\frac{7}{18}$.

39. About how much more is $8\frac{7}{10}$ feet than $2\frac{3}{8}$ feet?

40. Estimate the sum of $2\frac{1}{8}$, $7\frac{7}{9}$, $1\frac{7}{8}$, and $8\frac{1}{4}$.

41. Estimate the sum of 8.5, $5\frac{1}{4}$, $4\frac{2}{3}$, and 6.75.

42. Estimate the difference between $15\frac{3}{5}$ and $4\frac{2}{9}$.

43. Estimate the sum of $18\frac{3}{10}$, $5\frac{7}{8}$, $2\frac{1}{4}$, and $3\frac{4}{5}$.

44. About how much more is $24\frac{1}{3}$ yd than $11\frac{5}{6}$ yd?

Problem Solving ▶
Applications

Use Data The table shows data on the longest ski trails at five different mountains in the United States. For 45–46, use the table.

45. About how much longer is the trail at Killington than the trail at Mammoth?

46. Find the median of the trail lengths.

Mountain	Longest Ski Trail (mi)
Beaver Creek, CO	$2\frac{3}{4}$
Killington, VT	$10\frac{1}{5}$
Mammoth, CA	$2\frac{1}{2}$
Taos, NM	$5\frac{1}{4}$
Whiteface, NY	3

47. Leo needs $\frac{3}{4}$ yd of blue fabric, $\frac{7}{10}$ yd of purple fabric, and $\frac{1}{5}$ yd of white fabric for a sewing project. About how much fabric does he need for the sewing project?

48. ✏️ **Write a problem** about everyday life at home that can be solved by estimating fractions. Exchange with a classmate and solve.

49. Merika mowed her yard every ninth day, beginning on March 6 and continuing through April. On what other days did she mow her yard in March? in April?

MIXED REVIEW AND TEST PREP

Write the fraction or decimal as a percent. (p. 169)

50. $\frac{3}{4}$ **51.** 0.45

52. A survey question reads "Is blue your favorite color?" Is the question biased or unbiased? (p. 98)

53. TEST PREP The band members are setting up chairs for a concert in the school auditorium. They can put 25 chairs in each row. How many rows will they need to seat 350 people? (p. 22)

A 10   **B** 14   **C** 18   **D** 25

54. TEST PREP Find the GCF for 80, 96, and 112. (p.150)

F 6   **G** 8   **H** 14   **J** 16

Extra Practice page H40, Set A

Model Addition and Subtraction

Explore how to use fraction bars to add and subtract fractions with unlike denominators.

You need fraction bars.

Vocabulary

unlike fractions

QUICK REVIEW

Find the LCM for each set of numbers.

1. 2, 8   **2.** 6, 9

3. 4, 15   **4.** 4, 10

5. 2, 3, 10

Fractions with the same denominator, such as $\frac{5}{9}$ and $\frac{4}{9}$, are called like fractions. Fractions with different denominators are called **unlike fractions**. You can use fraction bars to rename the denominators before adding.

Remember that the LCM is the least of the common multiples of two or more numbers.
2: 2, 4, 6, 8, 10, 12, ...
3: 3, 6, 9, 12, 15, ...
The LCM of 2 and 3 is 6. The LCM can be used to write common denominators of two or more fractions.

Activity 1

Find $\frac{1}{6} + \frac{1}{4}$.

- Use fraction bars to show both fractions.

- Which fraction bars fit exactly across $\frac{1}{6}$ and $\frac{1}{4}$? Think about the LCM of 6 and 4.

- What is $\frac{1}{6} + \frac{1}{4}$?

$\frac{1}{6} = \frac{2}{12}$ $\frac{1}{4} = \frac{3}{12}$

Think and Discuss

- Look at the model for $\frac{1}{6} + \frac{1}{4}$. What do you know about $\frac{1}{6}$ and $\frac{2}{12}$? about $\frac{1}{4}$ and $\frac{3}{12}$?

- How are the denominators of $\frac{1}{6}$, $\frac{1}{4}$, and $\frac{1}{12}$ related? (HINT: Think about common multiples.)

Practice

Use fraction bars to find the sum. Draw a diagram of your model.

1. $\frac{1}{4} + \frac{1}{2}$ **2.** $\frac{1}{2} + \frac{1}{3}$ **3.** $\frac{1}{2} + \frac{2}{5}$ **4.** $\frac{2}{3} + \frac{1}{6}$

5. $\frac{1}{3} + \frac{1}{4}$ **6.** $\frac{3}{8} + \frac{1}{4}$ **7.** $\frac{1}{6} + \frac{1}{2}$ **8.** $\frac{3}{8} + \frac{1}{2}$

9. $\frac{1}{5} + \frac{1}{2}$ **10.** $\frac{3}{4} + \frac{1}{6}$ **11.** $\frac{1}{3} + \frac{1}{6}$ **12.** $\frac{5}{8} + \frac{1}{4}$

 CALIFORNIA STANDARDS NS 2.1 Solve problems involving addition, subtraction, multiplication, and division of positive fractions and explain why a particular operation was used for a given situation. *also,* O−π NS 2.0, NS 2.4, MR 2.4, MR 2.5

Fraction bars also can be used to subtract unlike fractions.

Activity 2

Find $\frac{1}{2} - \frac{1}{5}$.

- Use fraction bars to show $\frac{1}{2}$ and $\frac{1}{5}$.

- Which fraction bars fit exactly across $\frac{1}{2}$ and $\frac{1}{5}$? Think about the LCM.

- Compare $\frac{5}{10}$ and $\frac{2}{10}$. How much more is $\frac{5}{10}$ than $\frac{2}{10}$?

- What is $\frac{5}{10} - \frac{2}{10}$? What is $\frac{1}{2} - \frac{1}{5}$?

TECHNOLOGY LINK

More Practice: Use E-Lab, *Addition and Subtraction of Unlike Fractions.*
www.harcourtschool.com/elab2002

Think and Discuss

- How are the denominators of $\frac{1}{2}$, $\frac{1}{5}$, and $\frac{1}{10}$ related?

- Look at the model of $\frac{3}{4} - \frac{1}{3}$. Which fraction bars will fit exactly across $\frac{3}{4}$ and $\frac{1}{3}$? Explain.

- Which fraction bars will fit exactly across $\frac{1}{2} - \frac{1}{4}$?

Practice

Use fraction bars to subtract. Draw a diagram of your model.

1. $\frac{3}{4} - \frac{1}{3}$
2. $\frac{2}{5} - \frac{1}{10}$
3. $\frac{1}{3} - \frac{1}{4}$
4. $\frac{1}{2} - \frac{1}{3}$

5. $\frac{1}{2} - \frac{2}{5}$
6. $\frac{1}{2} - \frac{5}{12}$
7. $\frac{1}{4} - \frac{1}{6}$
8. $\frac{1}{3} - \frac{1}{6}$

9. $\frac{1}{4} - \frac{1}{8}$
10. $\frac{1}{2} - \frac{1}{4}$
11. $\frac{5}{6} - \frac{1}{3}$
12. $\frac{7}{8} - \frac{3}{4}$

MIXED REVIEW AND TEST PREP

Compare. Write <, >, or = for each ●. (p. 169)

13. $\frac{3}{8}$ ● 0.375
14. 0.15 ● $\frac{1}{4}$
15. 0.6 ● $\frac{11}{20}$

16. Multiply. 9.25×3.2 (p. 70)

17. **TEST PREP** Which is $9 \times 9 \times 9 \times 9 \times 9$ written with exponents? (p.40)

   A 9^9 **B** 9^5 **C** 9×5 **D** 9×9

Add and Subtract Fractions

Learn how to add and subtract fractions.

You can use a diagram to add and subtract fractions. To help you, think about the LCM of the denominators and about equivalent fractions.

EXAMPLE 1

Kayla is making two recipes. One recipe calls for $\frac{1}{4}$ c raisins, and the other recipe calls for $\frac{1}{3}$ c raisins. How many total cups of raisins does Kayla need?

Complete the diagram to find the sum of $\frac{1}{4}$ and $\frac{1}{3}$.

The LCM of 4 and 3 is 12.
Draw twelfths under $\frac{1}{4}$ and $\frac{1}{3}$.
THINK: $\frac{1}{4} = \frac{3}{12}$ and $\frac{1}{3} = \frac{4}{12}$.

So, Kayla needs $\frac{7}{12}$ c of raisins.

Math Idea ▶ To add fractions without using diagrams, you can write equivalent fractions by using the **least common denominator**, or **LCD**. The LCD is the LCM of the denominators.

EXAMPLE 2

Find $\frac{1}{2} + \frac{3}{5}$.

Estimate. Each fraction is close to $\frac{1}{2}$, so the sum is about 1.

$$\frac{1}{2} = \frac{1 \times 5}{2 \times 5} = \frac{5}{10}$$
$$+\frac{3}{5} = +\frac{3 \times 2}{5 \times 2} = +\frac{6}{10}$$

The LCM of 2 and 5 is 10, so the LCD of $\frac{1}{2}$ and $\frac{3}{5}$ is 10. Multiply to write equivalent fractions using the LCD.

$$\frac{1}{2} = \frac{5}{10}$$
$$+\frac{3}{5} = +\frac{6}{10}$$

Add the numerators.
Write the sum over the denominator.

$$\frac{11}{10}, \text{ or } 1\frac{1}{10}$$

Write the answer as a fraction or as a mixed number.

Compare the answer to your estimate. Since $1\frac{1}{10}$ is close to the estimate of 1, the answer is reasonable. So, $\frac{1}{2} + \frac{3}{5} = 1\frac{1}{10}$.

CALIFORNIA STANDARDS O—π NS 2.0 Students calculate and solve problems involving addition, subtraction, multiplication, and division. O—π NS 2.4 Determine the least common multiple and the greatest common divisor of whole numbers; use them to solve problems with fractions. MR 2.1 Use estimation to verify the reasonableness of calculated results. *also,* NS 2.1

You can use a similar method to subtract unlike fractions.

TECHNOLOGY LINK

More Practice: Use **Mighty Math Number Heroes**, *Fraction Fireworks*, Level W.

EXAMPLE 3

Kayla is preparing a pasta dish for a small dinner party. Kayla has $\frac{1}{2}$ c of grated mozzarella cheese. The recipe calls for $\frac{2}{3}$ c of grated mozzarella cheese. How much more mozzarella cheese does Kayla need to grate?

Find $\frac{2}{3} - \frac{1}{2}$.

Estimate. $\frac{2}{3}$ is a little more than $\frac{1}{2}$, so the difference is close to 0.

$$\frac{2}{3} = \frac{2 \times 2}{3 \times 2} = \frac{4}{6}$$ *The LCD of $\frac{2}{3}$ and $\frac{1}{2}$ is 6.*

$$-\frac{1}{2} = -\frac{1 \times 3}{2 \times 3} = -\frac{3}{6}$$ *Multiply to find the equivalent fractions using the LCD.*

$$\frac{2}{3} = \frac{4}{6}$$
$$-\frac{1}{2} = -\frac{3}{6}$$
$$\frac{1}{6}$$

Subtract the numerators. Write the difference over the denominator.

Compare the answer to your estimate. Since $\frac{1}{6}$ is close to the estimate of 0, the answer is reasonable.

So, Kayla needs to grate $\frac{1}{6}$ c more cheese.

EXAMPLE 4

A.
$$\frac{5}{6} = \frac{5 \times 3}{6 \times 3} = \frac{15}{18}$$

$$-\frac{7}{9} = -\frac{7 \times 2}{9 \times 2} = -\frac{14}{18}$$
$$\frac{1}{18}$$

B.
$$\frac{5}{12} = \frac{5}{12}$$

$$-\frac{1}{4} = -\frac{1 \times 3}{4 \times 3} = -\frac{3}{12}$$
$$\frac{2}{12} = \frac{1}{6}$$

CHECK FOR UNDERSTANDING

Think and Discuss   Look back at the lesson to answer each question.

1. **Tell** how much more cheese Kayla would need to grate if she had $\frac{1}{4}$ c of grated cheese.

2. *REASONING* **Tell** when the LCD of two fractions is equal to the product of the denominators.

Guided Practice   Use the LCD to rewrite the problem by using equivalent fractions.

3. $\frac{7}{10} + \frac{1}{5}$ 4. $\frac{1}{3} + \frac{1}{8}$ 5. $\frac{4}{5} - \frac{1}{3}$

183

Write the sum or difference in simplest form. Estimate to check.

6. $\frac{1}{5} + \frac{3}{5}$ **7.** $\frac{7}{9} - \frac{4}{9}$ **8.** $\frac{7}{9} - \frac{1}{6}$ **9.** $\frac{2}{3} + \frac{3}{4}$

10. $\frac{3}{4} - \frac{3}{8}$ **11.** $\frac{2}{5} - \frac{1}{3}$ **12.** $\frac{2}{5} + \frac{2}{4}$ **13.** $\frac{4}{9} + \frac{1}{3}$

PRACTICE AND PROBLEM SOLVING

Independent ▶
Practice

Use the LCD to rewrite the problem by using equivalent fractions.

14. $\frac{9}{10} - \frac{1}{5}$ **15.** $\frac{6}{7} - \frac{3}{4}$ **16.** $\frac{1}{4} + \frac{5}{8}$

Write the sum or difference in simplest form. Estimate to check.

17. $\frac{1}{6} + \frac{2}{3}$ **18.** $\frac{4}{7} - \frac{1}{7}$ **19.** $\frac{1}{2} + \frac{3}{10}$ **20.** $\frac{1}{3} - \frac{1}{4}$

21. $\frac{5}{7} - \frac{1}{2}$ **22.** $\frac{1}{3} + \frac{2}{3}$ **23.** $\frac{6}{10} - \frac{4}{10}$ **24.** $\frac{3}{8} + \frac{1}{3}$

25. $\frac{3}{4} + \frac{5}{8}$ **26.** $\frac{7}{12} + \frac{2}{3}$ **27.** $1 - \frac{3}{8}$ **28.** $\frac{1}{2} - \frac{2}{5}$

29. $\frac{4}{5} - \frac{1}{3}$ **30.** $\frac{1}{4} + \frac{2}{3}$ **31.** $\frac{5}{9} - \frac{1}{3}$ **32.** $\frac{3}{8} + \frac{3}{20}$

33. Find the sum of $\frac{1}{8}$, $\frac{3}{8}$, and $\frac{2}{8}$. **34.** Find $\frac{1}{2} + \frac{2}{3} + \frac{1}{6}$.

35. Find $\frac{3}{4} + 0.5 + 0.75$. **36.** Find $0.6 - \frac{3}{8}$.

37. How much longer than $\frac{1}{4}$ mile is $\frac{2}{3}$ mile?

Solve each equation mentally. Write the answer in simplest form.

38. $p + \frac{1}{4} = \frac{3}{4}$ **39.** $r = \frac{5}{12} + \frac{7}{12}$ **40.** $\frac{4}{5} - q = \frac{2}{5}$

41. $c = \frac{7}{10} - \frac{1}{10}$ **42.** $\frac{3}{7} + s = \frac{5}{7}$ **43.** $m - \frac{1}{6} = \frac{5}{6}$

Problem Solving ▶
Applications

Use Data For 44–46, use the recipe at the right.

44. Yuji has $\frac{7}{8}$ c of orange juice. How much orange juice does he have left to drink after he makes the fruit cups?

45. How many total teaspoons of vanilla and orange extract does Yuji need to make the fruit cups?

46. Yuji used $\frac{1}{4}$ tsp of orange extract. By how much did he exceed the amount of orange extract in the recipe?

Fruit Cups

2 cups of orange sections

1 cup blueberries

$\frac{1}{4}$ cup orange juice

1 tbsp sugar

1 tsp lemon juice

$\frac{1}{2}$ tsp vanilla extract

$\frac{1}{8}$ tsp orange extract

47. Each week, Reina spends $\frac{2}{3}$ of her allowance on school lunches and saves $\frac{1}{5}$ of it. What fraction of her allowance is left? Which operation(s) did you use? Why?

48. ❓ **What's the Question?** In Mrs. Lucero's class, $\frac{1}{10}$ of the students are wearing blue shirts and $\frac{3}{5}$ of the students are wearing white shirts. The answer is $\frac{3}{10}$ of the class.

49. *REASONING* Joaquin is thinking of two numbers. Each number is between 21 and 30. The GCF of the numbers is 4. What are the numbers?

50. One cup of whole milk contains 166 calories and one cup of skim milk contains 88 calories. How many more calories are there in 4 cups of whole milk than in 4 cups of skim milk?

MIXED REVIEW AND TEST PREP

For 51-52, write in simplest form. (p.160)

51. $\frac{36}{81}$ **52.** $\frac{95}{200}$ **53.** Subtract. $9,285 - 3,153$ (p. 20)

54. **TEST PREP** A teacher selects 50 students by picking names out of a box without looking. What type of sample is this? (p. 94)

A biased **B** convenience **C** random **D** systematic

55. **TEST PREP** Eric skated 500 meters in 37.14 seconds. Andre skated the same distance in 37.139 seconds, and Al in 37.12 seconds. What is the correct order of their times in seconds, from fastest to slowest times? (p. 52)

F 37.14, 37.139, 37.12 **H** 37.12, 37.139, 37.14

G 37.12, 37.14, 37.139 **J** 37.139, 37.12, 37.14

Thinker's CORNER

MATH FUN • FRACTION GAME
Materials: 9 index cards per group
Have students write these numbers on the cards, one to a card:

$$\frac{2}{3} \quad \frac{1}{4} \quad \frac{5}{8} \quad \frac{1}{2} \quad \frac{1}{3} \quad \frac{1}{5} \quad \frac{3}{8} \quad \frac{11}{12} \quad \frac{7}{10}$$

A student shuffles the cards. Each student in the group picks two cards and subtracts the lesser number from the greater number. Students compare answers. The one whose answer is the greatest number scores a point. Reshuffle the cards after each round. Three points are needed to win the game.

Add and Subtract Mixed Numbers

Learn how to add and subtract mixed numbers.

A flock of birds is migrating from Canada to Florida for the winter. One day they fly $2\frac{3}{4}$ hours before stopping to rest. Then they fly $1\frac{1}{6}$ more hours. For how long do the birds fly?

One Way   You can draw a diagram.

EXAMPLE 1

Remember that to write a fraction as a mixed number, you divide the numerator by the denominator. The quotient is the mixed number.

$$\frac{13}{3} \rightarrow \begin{array}{r} 4\frac{1}{3} \\ 3\overline{)13} \\ -12 \\ \hline 1 \end{array}$$

Show $2\frac{3}{4} + 1\frac{1}{6}$.

$2 + 1 = 3$ $\frac{9}{12} + \frac{2}{12} = \frac{11}{12}$

Combine whole numbers. Combine fractions. Draw equivalent fractions with the LCD, 12.

Add fractions. Add whole numbers.

So, the birds fly for $3\frac{11}{12}$ hours.

Another Way   You can use the LCD to write equivalent fractions.

EXAMPLE 2

Find $4\frac{2}{3} + 5\frac{4}{5}$.

$$\begin{array}{r} 4\frac{2}{3} = \quad 4\frac{10}{15} \\ +5\frac{4}{5} = +5\frac{12}{15} \\ \hline 9\frac{22}{15} = 9 + 1\frac{7}{15} = 10\frac{7}{15} \end{array}$$

Write equivalent fractions, using the LCD, 15.
Add fractions. Add whole numbers.

Rename the fraction as a mixed number. Rewrite the sum.

So, $4\frac{2}{3} + 5\frac{4}{5} = 10\frac{7}{15}$.

 CALIFORNIA STANDARDS O—∩NS 2.0 Students calculate and solve problems involving addition, subtraction, multiplication, and division. O—∩NS 2.4 Determine the least common multiple and the greatest common divisor of whole numbers; use them to solve problems with fractions. MR 2.1 Use estimation to verify the reasonableness of calculated results. *also,* O—∩NS 1.0, NS 2.1, MR 1.0, MR 1.1, MR 2.4, MR 2.7

Subtract Mixed Numbers

One Way You can use diagrams to subtract mixed numbers.

EXAMPLE 3

Find $1\frac{1}{2} - 1\frac{1}{10}$.

Draw $1\frac{1}{2}$.
Find the LCD for $\frac{1}{2}$ and $\frac{1}{10}$.
Change the half to tenths.

Subtract $1\frac{1}{10}$ from $1\frac{5}{10}$.

So, $1\frac{1}{2} - 1\frac{1}{10} = \frac{4}{10}$, or $\frac{2}{5}$.

- Draw a diagram to find $2\frac{1}{2} - 1\frac{1}{3}$.

Another Way When you subtract mixed numbers with unlike fractions, you can use the LCD to write equivalent fractions.

EXAMPLE 4

The rusty-spotted cat, the smallest meat-eating feline, lives in southern India and Sri Lanka. An adult has a head-and-body length of $13\frac{2}{5}$ in. to $18\frac{9}{10}$ in. Find the difference between the greatest and least lengths.

Find $18\frac{9}{10} - 13\frac{2}{5}$.

Estimate. $18\frac{9}{10}$ is close to 19 and $13\frac{2}{5}$ is close to 13. So, the difference is about $19 - 13$, or 6.

$$18\frac{9}{10} = 18\frac{9}{10}$$ *Write equivalent fractions, using the LCD, 10.*

$$-13\frac{2}{5} = -13\frac{4}{10}$$ *Subtract the fractions.*

$$\phantom{-13\frac{2}{5} = -13}5\frac{5}{10} = 5\frac{1}{2}$$ *Subtract the whole numbers.*

The exact answer is reasonable because it is close to the estimate of 6. So, the difference between the greatest and least lengths is $5\frac{1}{2}$ in.

CHECK FOR UNDERSTANDING

Think and Discuss ▸ Look back at the lesson to answer each question.

1. **Explain** why you must find equivalent fractions to add $1\frac{1}{5} + 1\frac{1}{2}$.

2. **Tell** how you know that $6\frac{1}{2} - 4\frac{1}{4}$ is more than 2.

Guided Practice ▸ Draw a diagram to find each sum or difference. Write the answer in simplest form.

3. $2\frac{3}{5} + 1\frac{1}{5}$ 4. $1\frac{1}{4} + 2\frac{2}{3}$ 5. $2\frac{4}{5} - 1\frac{1}{2}$

Write the sum or difference in simplest form.
Estimate to check.

6. $1\frac{1}{8} + 1\frac{5}{8}$ **7.** $2\frac{1}{4} + 4\frac{1}{3}$ **8.** $5\frac{3}{8} - 1\frac{1}{4}$

9. $4\frac{1}{3} - 3\frac{1}{6}$ **10.** $3\frac{3}{4} + 4\frac{5}{12}$ **11.** $6\frac{5}{6} - 5\frac{7}{9}$

PRACTICE AND PROBLEM SOLVING

Independent ▶ Practice

Draw a diagram to find each sum or difference. Write the answer in simplest form.

12. $1\frac{5}{12} + 1\frac{1}{4}$ **13.** $1\frac{1}{3} + 1\frac{1}{6}$ **14.** $4\frac{1}{2} - 2\frac{2}{5}$

Write the sum or difference in simplest form.
Estimate to check.

15. $4\frac{1}{2} + 3\frac{4}{5}$ **16.** $4\frac{1}{3} - 2\frac{1}{4}$ **17.** $5\frac{5}{6} + 4\frac{2}{9}$

18. $3\frac{1}{4} - 1\frac{1}{6}$ **19.** $7\frac{1}{2} - 3\frac{2}{5}$ **20.** $3\frac{2}{7} + 8\frac{1}{3}$

21. $7\frac{3}{4} + 3\frac{2}{5}$ **22.** $5\frac{5}{6} - 2\frac{7}{9}$ **23.** $4\frac{5}{7} + 3\frac{1}{2}$

TECHNOLOGY LINK
More Practice: Use
**Mighty Math
Calculating Crew,**
Nautical Number Line,
Level N.

24. How much greater is $5\frac{3}{4}$ than 3?

25. What is the sum of $25\frac{3}{8}$ and $2\frac{3}{4}$?

26. What is the sum of $4\frac{5}{8}$ and 7.8?

Find the missing number and identify which property of addition you used.

27. $3\frac{7}{8} + \blacksquare = 2\frac{1}{4} + 3\frac{7}{8}$ **28.** $3\frac{3}{4} + 0 = \blacksquare$ **29.** $\left(\frac{2}{3} + 1\frac{5}{6}\right) + \frac{1}{6} = \frac{2}{3} + \left(\blacksquare + \frac{1}{6}\right)$

Problem Solving ▶ Applications

Use Data The graph shows the head-and-body lengths of five small mammals. For 30–31, use the graph.

30. How much longer is the harvest mouse than the Kitti's hognosed bat?

31. The masked shrew is $1\frac{2}{3}$ in. long. Is it longer or shorter than the little brown bat? How much? Which operation did you use? Why?

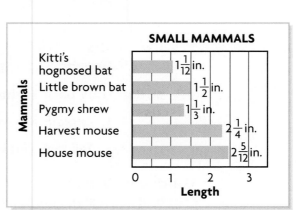

SMALL MAMMALS

Mammals	
Kitti's hognosed bat	$1\frac{1}{12}$ in.
Little brown bat	$1\frac{1}{2}$ in.
Pygmy shrew	$1\frac{1}{3}$ in.
Harvest mouse	$2\frac{1}{4}$ in.
House mouse	$2\frac{5}{12}$ in.

Length — 0   1   2   3

32. On its way to the shore, a sea turtle traveled $4\frac{1}{4}$ hr the first day. The second day, the turtle traveled $3\frac{1}{2}$ hr. How many hours did the sea turtle travel in the two days?

33. Mrs. Myers used $1\frac{1}{2}$ c of flour to make muffins, $4\frac{1}{4}$ c to make bread, and $\frac{3}{4}$ c to make gravy. If she has $3\frac{1}{4}$ c left, how much flour did Mrs. Myers have before she started the meal?

34. ⊙ **What's the Error?** Izumi added $3\frac{1}{4}$ and $2\frac{2}{3}$ and got $5\frac{3}{12}$. Explain the error. What is the correct sum?

35. Betty, Jim, Manuel, and Rosa won the first four prizes in a design contest. Jim won second prize. Manuel did not win third prize. Rosa won fourth prize. What prize did Betty win?

36. Alexis needs a new blender. She found the same blender on sale at five different stores. The prices are $22.95, $21.85, $22.05, $20.95, and $21.99. Order the prices from least to greatest.

MIXED REVIEW AND TEST PREP

37. Find the sum of $\frac{2}{3}$ and $\frac{2}{5}$. (p. 182)

38. Subtract. $425.2 - 51.05$ (p. 66)

39. Order $\frac{1}{2}$, $\frac{4}{5}$, and $\frac{2}{3}$ from greatest to least. (p. 166)

40. TEST PREP Find the value of $(6 + 4)^2 \div 5$. (p. 44)

    A 4   **B** 4.4   **C** 9.2   **D** 20

41. TEST PREP Which is the solution of $x + 3 = 10$? (p. 30)

    A $x = 3$ **B** $x = 7$ **C** $x = 10$ **D** $x = 13$

LiNKUP to Reading

Strategy • Choose Relevant Information
Sometimes a word problem contains more information than you need. You must decide which information is relevant, or needed to solve the problem.

The koala of eastern Australia feeds mostly on eucalyptus leaves. It nibbles on about 6 of the 500 species of eucalyptus per day and selects certain trees and leaves over others to find the $1\frac{1}{4}$ pounds of food that it needs. Suppose the koala finds only $1\frac{1}{8}$ pounds of food by the end of the day. How many more pounds of eucalyptus leaves does the koala need?

  1. What does the problem ask you to find?

  2. Identify relevant information.

  3. What information is not relevant?

  4. Solve the problem.

Rename to Subtract

MATH LAB

Explore how to use fraction bars to subtract mixed numbers.

You need fraction bars.

Sometimes you need to rename mixed numbers before you can subtract.

Activity

A. Find $2\frac{1}{5} - 1\frac{4}{5}$.

- Use fraction bars to model $2\frac{1}{5}$.

 ← $2\frac{1}{5}$

- Here is another way to model $2\frac{1}{5}$.

 ← $1\frac{6}{5}$

- From which model can you subtract $1\frac{4}{5}$?
- Subtract $1\frac{4}{5}$ from $1\frac{6}{5}$. What is $2\frac{1}{5} - 1\frac{4}{5}$?

B. Find $2\frac{1}{6} - 1\frac{5}{12}$.

- Use fraction bars to model $2\frac{1}{6}$.

 ← $2\frac{1}{6}$

- Since you are subtracting twelfths, think of the LCD for $\frac{1}{6}$ and $\frac{5}{12}$. Change the sixths to twelfths.

 ← $2\frac{2}{12}$

- Can you subtract $1\frac{5}{12}$ from either of these models?
- Here is another way to model $2\frac{2}{12}$.

← $1\frac{14}{12}$

- Subtract $1\frac{5}{12}$ from $1\frac{14}{12}$. What is $2\frac{1}{6} - 1\frac{5}{12}$?

CALIFORNIA STANDARDS O—n NS 2.0 Students calculate and solve problems involving addition, subtraction, multiplication, and division. O—n NS 2.4 Determine the least common multiple and the greatest common divisor of whole numbers; use them to solve problems with fractions. *also,* NS 2.1, MR 2.4, MR 2.5

C. Find $2\frac{1}{4} - 1\frac{3}{8}$.

- Use fraction bars to model $2\frac{1}{4}$.

$\leftarrow 2\frac{1}{4}$

- Since you are subtracting eighths, think of the LCD for $\frac{1}{4}$ and $\frac{3}{8}$. Change the fourth to eighths.

| 1 | 1 | $\frac{1}{8}$ | $\frac{1}{8}$ |

$\leftarrow 2\frac{2}{8}$

- Can you subtract $1\frac{3}{8}$ from either of these models?

- Here is another way to model $2\frac{2}{8}$.

| 1 | $\frac{1}{8}$ | $\frac{1}{8}$ | $\frac{1}{8}$ | $\frac{1}{8}$ | $\frac{1}{8}$ | $\frac{1}{8}$ | $\frac{1}{8}$ | $\frac{1}{8}$ | $\frac{1}{8}$ | $\frac{1}{8}$ |

$\leftarrow 1\frac{10}{8}$

- Subtract $1\frac{3}{8}$ from $1\frac{10}{8}$. What is $2\frac{1}{4} - 1\frac{3}{8}$?

TECHNOLOGY LINK

More Practice: Use E-Lab, *Subtraction of Mixed Numbers.*
www.harcourtschool.com/elab2002

Think and Discuss

- Think about $2\frac{5}{6} - 1\frac{1}{6}$. Do you need to rename before you subtract? Explain.

- Think about $5\frac{2}{5} - 3\frac{4}{5}$. Do you need to rename before you subtract? Explain.

Practice

Use fraction bars to subtract. Draw a diagram of your model.

1. $3\frac{1}{3} - 1\frac{2}{3}$
2. $3\frac{3}{8} - \frac{3}{4}$
3. $3\frac{1}{9} - 1\frac{2}{3}$
4. $2\frac{3}{8} - 1\frac{1}{2}$

5. $1\frac{1}{2} - \frac{4}{5}$
6. $3\frac{1}{5} - 1\frac{3}{10}$
7. $2\frac{2}{3} - 1\frac{3}{4}$
8. $3\frac{1}{12} - 2\frac{5}{6}$

MIXED REVIEW AND TEST PREP

9. Draw a diagram to find the sum of $2\frac{1}{2}$ and $1\frac{1}{4}$. Write the answer in simplest form. (p. 186)

10. Write $\frac{2}{5}$ as a decimal and as a percent. (p. 169)

11. Evaluate $a \times b$ for $a = 4.2$ and $b = 5.1$. (p. 82)

12. Write the prime factorization of 245 in exponent form. (p. 148)

13. **TEST PREP** Which is the quotient $2.79 \div 0.045$? (p. 76)

   A 0.12555   **B** 2.705   **C** 2.835   **D** 62

Subtract Mixed Numbers

Learn how to subtract mixed numbers involving renaming.

A 1993 blizzard called The Storm of the Century brought record cold, wind, and snow to many cities in the southern and eastern United States.

How much more snow did Mount LeConte receive than Birmingham received?

Find $4\frac{1}{3} - 1\frac{5}{12}$.

STORM OF THE CENTURY	
City	Snow (ft)
Birmingham, AL	$1\frac{5}{12}$
Asheville, NC	$1\frac{7}{12}$
Pittsburgh, PA	$2\frac{1}{2}$
Syracuse, NY	$3\frac{7}{12}$
Mount LeConte, TN	$4\frac{1}{3}$

Estimate. $4\frac{1}{3}$ is close to $4\frac{1}{2}$ and $1\frac{5}{12}$ is close to $1\frac{1}{2}$. So, the difference is about $4\frac{1}{2} - 1\frac{1}{2}$, or 3.

Subtract.

$$\begin{array}{r} 4\frac{1}{3} = 4\frac{4}{12} \\ -1\frac{5}{12} = -1\frac{5}{12} \end{array}$$

The LCD of $\frac{1}{3}$ and $\frac{5}{12}$ is 12.
Write equivalent fractions, using the LCD, 12.

$$\begin{array}{r} 4\frac{1}{3} = 4\frac{4}{12} = 3\frac{16}{12} \\ -1\frac{5}{12} = -1\frac{5}{12} = -1\frac{5}{12} \\ \hline 2\frac{11}{12} \end{array}$$

Since $\frac{5}{12}$ is greater than $\frac{4}{12}$, rename $4\frac{4}{12}$.
$4\frac{4}{12} = 3 + \frac{12}{12} + \frac{4}{12} = 3\frac{16}{12}$.
Subtract the fractions.
Subtract the whole numbers.

The answer is reasonable because it is close to the estimate of 3 ft. So, Mount LeConte received $2\frac{11}{12}$ ft more snow than Birmingham.

EXAMPLE

Jake is running in a 5-km race. So far he has run $1\frac{4}{5}$ km. How far does he have to go?

Find the difference. $5 - 1\frac{4}{5}$

$$\begin{array}{r} 5 = 4\frac{5}{5} \\ -1\frac{4}{5} = -1\frac{4}{5} \\ \hline 3\frac{1}{5} \end{array}$$

Since you are subtracting fifths, rename 5 as $4\frac{5}{5}$.

Subtract the fractions.
Subtract the whole numbers.

So, Jake has $3\frac{1}{5}$ km to go.

CALIFORNIA STANDARDS ○━ⁿNS 2.0 Students calculate and solve problems involving addition, subtraction, multiplication, and division. ○━ⁿNS 2.4 Determine the least common multiple and the greatest common divisor of whole numbers; use them to solve problems with fractions. **MR 2.1** Use estimation to verify the reasonableness of calculated results. *also,* **NS 1.0, NS 2.1, MR 2.7**

Think and Discuss ▸ Look back at the lesson to answer each question.

1. **Explain** how to rename $3\frac{1}{3}$ so you could subtract $1\frac{2}{3}$.

2. **What if** you wanted to find $4\frac{5}{12} - 2\frac{3}{8}$? What equivalent fractions would you write using the LCD?

Guided Practice ▸ Write the difference in simplest form. Estimate to check.

3. $4\frac{1}{3} - 2\frac{1}{4}$ 4. $6 - 2\frac{2}{3}$ 5. $7\frac{3}{10} - 3\frac{2}{5}$

6. $6\frac{1}{5} - 3\frac{7}{10}$ 7. $8\frac{5}{6} - 4\frac{8}{9}$ 8. $16\frac{3}{8} - 7\frac{1}{2}$

PRACTICE AND PROBLEM SOLVING

Independent Practice ▸ Write the difference in simplest form. Estimate to check.

9. $3\frac{1}{6} - 1\frac{1}{4}$ 10. $5\frac{1}{2} - 3\frac{7}{10}$ 11. $12\frac{1}{9} - 7\frac{1}{3}$

12. $4\frac{1}{4} - 2\frac{2}{5}$ 13. $8\frac{1}{4} - 5\frac{2}{3}$ 14. $7\frac{5}{9} - 2\frac{5}{6}$

15. $11\frac{1}{4} - 9\frac{7}{8}$ 16. $5.25 - 2\frac{3}{8}$ 17. $6.2 - 3\frac{1}{2}$

Evaluate each expression for $j = 5\frac{1}{2}$, $k = 4\frac{3}{5}$, and $m = 2\frac{7}{10}$.

18. $j - k$ 19. $j - m$ 20. $k - m$

Problem Solving Applications ▸

21. Whit usually drives $4\frac{7}{8}$ mi on the expressway to work. Sometimes traffic is bad due to weather conditions and he takes another route which is $5\frac{3}{4}$ mi long. How much shorter is his usual route?

22. ✎ **Write About It** Why do you write equivalent fractions before you rename? Can you rename before you write equivalent fractions?

23. **Number Sense** Prime numbers that differ by 2, such as 3 and 5, or 59 and 61, are called twin primes. Write two other pairs of twin primes between 1 and 50.

MIXED REVIEW AND TEST PREP

24. Find the sum of $1\frac{1}{3}$ and $2\frac{1}{6}$. (p. 186)

25. Tell which are equivalent numbers.
$\frac{11}{3}$, $7\frac{1}{3}$, $6\frac{1}{3}$, $\frac{22}{3}$ (p. 160)

26. Find the median for the data. 30, 36, 39, 38, 36, 33 (p. 106)

27. Evaluate $b \div d$, for $b = 2{,}260$ and $d = 41$.
(p. 28)

28. **TEST PREP** Alexander buys 3 boxes of computer paper for $10.79 per box, including tax. How much change will he get from $50? (p. 70)

A $39.21   C $17.63

B $32.37   D $11.00

Extra Practice page H40, Set D

PROBLEM SOLVING STRATEGY
Draw a Diagram

Analyze
Choose
Solve
Check

Learn how to use the strategy *draw a diagram* to solve problems.

QUICK REVIEW

1. $\frac{3}{4} - \frac{1}{4}$ 2. $\frac{4}{5} - \frac{1}{5}$ 3. $\frac{2}{3} - \frac{2}{3}$ 4. $\frac{1}{5} + \frac{4}{5}$ 5. $\frac{9}{10} - \frac{6}{10}$

A taxicab travels $1\frac{1}{3}$ mi west, 4 mi north, $8\frac{1}{4}$ mi east, $1\frac{1}{3}$ mi south, and then 10 mi west. How far has the taxicab traveled when it crosses its own path?

Analyze   What are you asked to find?

What information is given?

Is there numerical information you will not use? If so, what?

Choose   What strategy will you use?

You can draw a diagram that shows the taxicab's route.

Solve   How will you solve the problem?

Draw a diagram and label the distances and locations.

Add all the distances from A to E. Then add the distance from E to the point where the taxicab crosses its own path. This will be the same as the distance from C to D.

$$1\frac{1}{3} + 4 + 8\frac{1}{4} + 1\frac{1}{3} + 8\frac{1}{4}$$

Write equivalent fractions by using the LCD, 12.

$$1\frac{4}{12} + 4 + 8\frac{3}{12} + 1\frac{4}{12} + 8\frac{3}{12} = 22 + \frac{14}{12} = 22 + 1\frac{2}{12} = 23\frac{1}{6}$$

So, the taxicab has traveled $23\frac{1}{6}$ mi when it crosses its own path.

Check   How can you check to see if your answer is reasonable?

CALIFORNIA STANDARDS O⊸nNS 2.0 Students calculate and solve problems involving addition, subtraction, multiplication, and division. *also,* NS 1.0, NS 2.4, MR 2.0, MR 2.4, MR 2.7, MR 3.2

Solve by drawing a diagram.

1. A tour bus travels $7\frac{1}{2}$ mi south, $3\frac{1}{3}$ mi east, $4\frac{1}{3}$ mi north, and $11\frac{1}{2}$ mi west. How far has the tour bus traveled when it crosses its own path?

2. Tanisha needs a fence that is 33 ft long to separate her two gardens. If she puts one post in the ground every $5\frac{1}{2}$ ft, how many posts will Tanisha need for the fence?

For 3–4, use the information below.

Carla drives south $2\frac{1}{2}$ mi from her home. Next, she drives east $\frac{1}{3}$ mi. Then, she drives south 3 mi.

3. How many total miles does Carla drive from her home?

 A 5 mi   **C** $5\frac{5}{6}$ mi

 B $5\frac{1}{5}$ mi   **D** 6 mi

4. In which directions must Carla drive to return home?

 F east and north   **H** west and north

 G east and south   **J** west and south

MIXED STRATEGY PRACTICE

5. A display in a store has 24 cans in the bottom row, 21 in the second row, and 18 in the third row. If the pattern continues, how many cans are in the fifth row?

6. Tickets to see an Irish dance group cost $52 and $28. Mario bought 7 tickets for his family and paid a total of $316. How many tickets at each price did he buy?

7. Ms. Lopez travels north from home to pick up Maria at school. Then she travels $3\frac{1}{2}$ mi east to pick up Marcos and $4\frac{1}{4}$ mi south to pick up Carter. If Ms. Lopez has driven a total of $12\frac{1}{2}$ mi, what is the distance from home to Maria's school?

8. Tim has a job in the city. He has to commute $6\frac{1}{2}$ km to work. The bus he takes to work travels about $3\frac{1}{4}$ km in 10 minutes. About how much time does Tim spend on the bus going to and from work?

9. In a recent contest, Gary scored more points than Catherine, who scored more points than Clara. Christopher scored more points than Clara but fewer than Gary. Who had the most points?

10. Use the table below. What is Joshua's total bill if he rented 3 videos and kept them for 6 days?

11. ✎ **Write a problem** that can be solved by using the strategy *draw a diagram*. Explain the steps you would use to solve the problem, and draw the diagram.

Video Rental Prices	
1 movie for 1 day	$2.99
Each additional day	$0.99
5 additional days	$3.79

Estimate the sum or difference. (pp. 176–179)

1. $\frac{7}{12} + \frac{1}{4}$

2. $\frac{4}{5} - \frac{2}{7}$

3. $\frac{4}{5} + \frac{3}{8}$

4. $\frac{3}{4} - \frac{1}{3}$

5. $4\frac{2}{9} - \frac{1}{7}$

6. $\frac{9}{20} + 1\frac{4}{5}$

7. $6\frac{1}{4} + 3\frac{2}{9}$

8. $9\frac{4}{5} - 2\frac{7}{9}$

Write the sum or difference in simplest form.
Estimate to check. (pp. 182–185)

9. $\frac{1}{2} + \frac{1}{3}$

10. $\frac{3}{4} + \frac{1}{6}$

11. $\frac{2}{5} + \frac{2}{4}$

12. $\frac{2}{3} + \frac{3}{4}$

13. $\frac{5}{6} + \frac{2}{3}$

14. $\frac{1}{3} + \frac{5}{6}$

15. $\frac{3}{8} + \frac{3}{4}$

16. $\frac{5}{8} + \frac{1}{6}$

17. $\frac{3}{4} - \frac{1}{3}$

18. $\frac{7}{8} - \frac{1}{4}$

19. $\frac{5}{6} - \frac{2}{9}$

20. $\frac{8}{9} - \frac{1}{6}$

21. $\frac{7}{8} - \frac{5}{6}$

22. $\frac{7}{12} - \frac{5}{12}$

23. $\frac{3}{5} - \frac{1}{4}$

24. $\frac{4}{5} - \frac{1}{10}$

Draw a diagram to find the sum or difference. Write the answer in simplest form. (pp. 186–189)

25. $2\frac{3}{8} + 1\frac{1}{4}$

26. $6\frac{2}{3} - 3\frac{1}{4}$

27. $5\frac{2}{5} - 3\frac{3}{10}$

28. $2\frac{1}{6} + 1\frac{1}{3}$

Write the sum or difference in simplest form.
Estimate to check. (pp. 186–189, 192–193)

29. $1\frac{1}{6} + 3\frac{2}{3}$

30. $2\frac{3}{4} + 3\frac{1}{8}$

31. $1\frac{1}{2} + 2\frac{1}{4}$

32. $1\frac{1}{5} + 1\frac{3}{10}$

33. $7\frac{3}{4} - 5\frac{1}{3}$

34. $8\frac{1}{3} - 3\frac{1}{8}$

35. $3\frac{1}{4} - 2\frac{1}{2}$

36. $4\frac{1}{2} - 1\frac{2}{3}$

37. $7\frac{1}{5} - 5\frac{4}{9}$

Solve each problem by drawing a diagram. (pp. 194–195)

38. A minibus leaves the garage and travels $9\frac{5}{6}$ mi north to pick up Tanya. Then it travels $3\frac{1}{6}$ mi west to pick up Luis, $4\frac{1}{4}$ mi south to pick up Alissa, and $4\frac{5}{12}$ mi east to the school. How far does the minibus travel before crossing its own path?

39. Satoko has a board that is 9 ft long. She needs to cut the board into $2\frac{1}{4}$-ft sections. How many cuts will she have to make?

40. Del drives $3\frac{1}{4}$ mi north from his home. Next, he drives west $\frac{3}{4}$ mi. Then, he drives north $\frac{1}{2}$ mi. In which directions must Del drive to return home?

Decide on a plan.

See item **2.**

Show the hours practiced as an addition sentence. Find the *least common denominator* before you add.

Also see problem **4**, p. H64.

Choose the best answer.

1. Bryce had $\frac{13}{15}$ gallon of paint. He used $\frac{1}{3}$ gallon for a project. How much paint did he have left?

   A $\frac{8}{15}$ gal   **C** $\frac{1}{3}$ gal

   B $\frac{1}{2}$ gal   **D** $\frac{1}{5}$ gal

2. Tirzah practiced the piano for $1\frac{2}{3}$ hours on Monday, $\frac{3}{4}$ hour on Wednesday, and $1\frac{1}{2}$ hours on Friday. Which is the total amount of time she spent practicing the piano?

   F $3\frac{2}{5}$ hr   **H** $3\frac{11}{12}$ hr

   G $3\frac{3}{4}$ hr   **J** 4 hr

3. Last week Jessie's swimming practice lasted for $\frac{3}{5}$ hour on Monday, $\frac{5}{6}$ hour on Wednesday, and $\frac{9}{10}$ hour on Friday. How many hours did she have swimming practice during the week?

   A $1\frac{1}{3}$ hr   **C** 2 hr

   B $1\frac{2}{3}$ hr   **D** $2\frac{1}{3}$ hr

4. What is the value of $n + \frac{1}{2}$ for $n = \frac{5}{6}$?

   F $\frac{1}{3}$ **H** $1\frac{1}{6}$

   G $\frac{11}{12}$ **J** $1\frac{1}{3}$

5. In which pair are both numbers equivalent to $\frac{2}{5}$?

   A 0.2; 20%

   B 25%; $\frac{4}{10}$

   C 40%; 0.4

   D 0.6; $\frac{6}{15}$

6. $14\frac{2}{3} - 9\frac{5}{12}$

   F $4\frac{1}{3}$ **H** $5\frac{1}{4}$

   G $4\frac{1}{2}$ **J** Not here

7. The mean of 5 numbers is 25.6. What is the sum of the numbers?

   A 128   **C** 26.1

   B 30.6   **D** Not here

8. Bart passed a sign that read, "City Limit $5\frac{1}{2}$ miles." If he drives $2\frac{1}{4}$ miles farther, how far will he be from the city limit?

   F $2\frac{1}{4}$ mi   **H** 3 mi

   G $2\frac{1}{2}$ mi   **J** $3\frac{1}{4}$ mi

9. Which is a reasonable estimate of the sum of the fractions $\frac{1}{12}$, $\frac{4}{9}$, $\frac{5}{8}$, and $\frac{11}{12}$?

   A 1   **C** 4

   B 2   **D** $4\frac{1}{2}$

10. Mel hiked $\frac{9}{16}$ mile on Saturday and $\frac{7}{8}$ mile on Sunday. Which is a good estimate for how far Mel hiked in all?

    F about 2 mi

    G about $1\frac{1}{2}$ mi

    H about 1 mi

    J about $\frac{1}{2}$ mi

11. $1\frac{7}{8} - \frac{9}{10}$

    A $1\frac{1}{2}$ **C** $1\frac{1}{40}$

    B $1\frac{1}{10}$ **D** $\frac{39}{40}$

12. $115{,}371.9 + 22{,}671.25$

    F 138,043.015   **H** 138,043.259

    G 138,043.15   **J** Not here

10 Multiply and Divide Fractions and Mixed Numbers

The 86th Tour de France began on July 3, 1999, and ended on July 25. The cyclists rode for 21 days and rested for 2 days. The first day of the race is called the Prologue. On that day, riders traveled 6.8 km. Each of the days after the Prologue is called a stage. Each stage covers a different distance, from about 50 km to nearly 230 km. The total distance covered in the race was about 3,690 km. Estimate the number of kilometers the cyclists had ridden after they had completed $\frac{1}{3}$ of the race.

TOUR DE FRANCE DISTANCES

Race Stage	Kilometers Covered
Prologue–Stage 4	779.8
Stages 5-8	685.5
Stages 9-12	834.0
Stages 13-16	800.5
Stages 17-20	587.5

0 100 200 300 400 500 600 700 800 900 1000

Kilometers Covered

Check What You Know

Use this page to help you review and remember important skills needed for Chapter 10.

✓ Vocabulary

Choose the best term from the box.

fraction
mixed number
equation

1. An algebraic or numerical sentence that shows two quantities are equal is a(n) __?__ .

2. A number that is made up of a whole number and a fraction is a(n) __?__ .

Round Fractions (See p. H10.)

Round each fraction to 0, $\frac{1}{2}$, or 1.

3. $\frac{2}{9}$ **4.** $\frac{7}{8}$ **5.** $\frac{7}{15}$ **6.** $\frac{5}{8}$ **7.** $\frac{1}{4}$

8. $\frac{2}{15}$ **9.** $\frac{3}{8}$ **10.** $\frac{5}{6}$ **11.** $\frac{1}{3}$ **12.** $\frac{10}{11}$

13. $\frac{11}{20}$ **14.** $\frac{7}{12}$ **15.** $\frac{1}{6}$ **16.** $\frac{4}{7}$ **17.** $\frac{2}{6}$

Mental Math and Equations (See p. H11.)

Use mental math to solve.

18. $9.3 + x = 12.5$ **19.** $4c = 128$ **20.** $x - 160 = 520$ **21.** $5.11 = 5.28 - x$

22. $0.12 = \frac{x}{0.6}$ **23.** $35 - t = 21$ **24.** $6y = 0.24$ **25.** $r + 3.7 = 6.3$

✓ Fractions and Mixed Numbers (See p. H11.)

Write each fraction as a mixed number.

26. $\frac{18}{5}$ **27.** $\frac{7}{6}$

28. $\frac{16}{15}$ **29.** $\frac{4}{3}$

Write each mixed number as a fraction.

30. $1\frac{5}{8}$ **31.** $7\frac{1}{2}$

32. $3\frac{2}{3}$ **33.** $2\frac{4}{5}$

LOOK AHEAD

In Chapter 10 you will

- estimate products and quotients
- multiply and divide fractions and mixed numbers
- use fractions in expressions and equations

Estimate Products and Quotients

Learn how to estimate products and quotients of fractions and mixed numbers.

In a landfill, bulldozers spread and compact the garbage into 10-foot layers. Every layer is covered with clean soil.

The landfills in more than half of the states in the United States will soon be full. It is estimated that each person in the United States produces $4\frac{2}{5}$ pounds of garbage a day. If the population of the United States was $249\frac{7}{10}$ million, about how many pounds of garbage would be produced every day?

Remember that when rounding fractions, round to 0, $\frac{1}{2}$, or 1. When rounding mixed numbers, round to the nearest whole number.

One way to estimate the answer is to round the mixed numbers to the nearest whole number.

Estimate. $4\frac{2}{5} \times 249\frac{7}{10}$

$4\frac{2}{5} \times 249\frac{7}{10}$ *Round to the nearest whole number.*

$\downarrow$ $\downarrow$ THINK: $\frac{2}{5}$ *rounds to 0, and* $\frac{7}{10}$ *rounds to 1.*

$4 \times 250 = 1{,}000$ *Multiply.*

So, about 1,000 million pounds would be produced each day.

You can also estimate by averaging two estimates.

EXAMPLE

Estimate. $8 \div \frac{3}{4}$

Since $\frac{3}{4}$ is halfway between $\frac{1}{2}$ and 1, find the two estimates and then find their average.

Round up.   *Round down.*

$8 \div \frac{3}{4} \rightarrow 8 \div 1 = 8$ $8 \div \frac{3}{4} \rightarrow 8 \div \frac{1}{2} = 16$

$8 + 16 = 24;\ 24 \div 2 = 12$ So, $8 \div \frac{3}{4}$ is about 12.

You can use compatible numbers to estimate a product or quotient.

$23\frac{3}{4} \div 4\frac{1}{2} \rightarrow 25 \div 5 = 5$

So, $23\frac{3}{4} \div 4\frac{1}{2}$ is about 5.

CALIFORNIA STANDARDS O⎯n NS 20 Students calculate and solve problems involving addition, subtraction, multiplication, and division. **NS 2.2** Explain the meaning of multiplication and division of positive fractions and perform the calculations. *also* O⎯n NS 1.0, MR 2.5

Think and Discuss ▶ Look back at the lesson to answer each question. ▶

1. **What if** each person in the United States produced $2\frac{1}{5}$ pounds of garbage? About how many pounds of garbage would be produced?

2. **Tell** what compatible numbers you could use to find $81\frac{3}{5} \div 12\frac{7}{8}$.

Guided Practice ▶ Estimate each product or quotient.

3. $\frac{7}{8} \times \frac{7}{16}$

4. $10\frac{8}{11} \div 2\frac{1}{5}$

5. $78\frac{3}{7} \div 4\frac{1}{6}$

6. $\frac{3}{5} \times 38$

7. $1\frac{3}{4} \times 35$

8. $21\frac{3}{8} \div 17\frac{1}{3}$

9. $58\frac{3}{4} \times 1\frac{5}{6}$

10. $98\frac{7}{8} \div 23\frac{1}{5}$

Independent Practice ▶ Estimate each product or quotient.

11. $\frac{7}{9} \times \frac{1}{3}$

12. $10\frac{8}{9} \times \frac{5}{6}$

13. $\frac{5}{6} \div \frac{11}{12}$

14. $67\frac{9}{12} \div 2\frac{7}{10}$

15. $24\frac{9}{10} \div 6\frac{2}{3}$

16. $36\frac{5}{8} \div 13\frac{3}{5}$

17. $67\frac{2}{3} \div 23\frac{1}{8}$

18. $97\frac{2}{9} \div 52\frac{5}{8}$

19. $3\frac{11}{12} \times 4\frac{6}{7}$

20. $12\frac{5}{24} \div \frac{8}{12}$

21. $\frac{5}{9} \times \frac{7}{12}$

22. $\frac{2}{5} \div \frac{10}{21}$

Problem Solving Applications ▶ Estimate to compare. Write < or > for each ●.

23. $4\frac{1}{6} \times 3\frac{2}{3}$ ● $7\frac{5}{8} \div 2\frac{1}{3}$

24. $7\frac{2}{3} \div 5$ ● $2\frac{4}{8} \times 3\frac{1}{8}$

25. Cal runs $5\frac{3}{10}$ miles in $33\frac{4}{5}$ minutes. About how many minutes does it take for Cal to run one mile?

26. **Write About It** Explain how you would estimate a quotient of mixed numbers.

27. Doris picked up used newspapers around her neighborhood for six days. She picked up 0.5 kg the first day, 2 kg the second day, and 3.5 kg the third day. If this pattern continued, how much newspaper did she pick up on the sixth day?

MIXED REVIEW AND TEST PREP

28. Write $\frac{7}{20}$ as a percent. (p. 169)

29. $5\frac{7}{9} - 2\frac{1}{3}$ (p. 186)

30. Find the mean, median, and mode for the data.
27, 48, 83, 76, 48, 27 (p. 110)

31. Evaluate the expression $m - 4^2$ for $m = 3^3$. (p. 28)

32. **TEST PREP** Which is the LCM of 18 and 24? (p.150)

    A 2   **B** 8   **C** 72   **D** 128

Multiply Fractions

Learn how to multiply fractions.

Carolyn asked $\frac{3}{4}$ of her classmates what time they leave for school in the morning. Of those she asked, $\frac{1}{2}$ leave at 7:00 A.M. What fractional part of the class told her that they leave for school at 7:00 A.M.?

One way to find the fractional part of a fraction is to make a model.

Find $\frac{1}{2}$ of $\frac{3}{4}$, or $\frac{1}{2} \times \frac{3}{4}$.

Fold a piece of paper into 4 equal parts.

Shade 3 parts to show $\frac{3}{4}$.

Fold the paper in half. Shade $\frac{1}{2}$ of the paper.

Of the $2 \times 4 = 8$ parts, $1 \times 3 = 3$ are shaded, so $\frac{3}{8}$ of the paper is shaded twice. These parts represent $\frac{1}{2} \times \frac{3}{4}$.

$$\frac{1}{2} \times \frac{3}{4} = \frac{3}{8}$$

So, $\frac{3}{8}$ of the students Carolyn asked leave at 7:00 A.M.

• Compare the numerator and denominator of the product with the numerators and denominators of the factors. What relationship do you see?

You can see this relationship in the solution to the problem.

$$\frac{\text{numerator} \times \text{numerator}}{\text{denominator} \times \text{denominator}} = \frac{\text{numerator}}{\text{denominator}}$$

   ↑   ↑   ↑

   factor   factor   product

• Make a model to find $\frac{1}{3} \times \frac{3}{4}$.

CALIFORNIA STANDARDS O—n NS 2.0 Students calculate and solve problems involving addition, subtraction, multiplication, and division. **NS 2.2** Explain the meaning of multiplication and division of positive fractions and perform the calculations. **MR 2.1** Use estimation to verify the reasonableness of calculated results. *also* O—n NS 1.0, NS 2.1, O—n NS 2.4, MR 1.3, MR 2.5

You can use this relationship to multiply fractions without making a model.

EXAMPLE 1

Remember that to write a fraction in simplest form, divide the numerator and the denominator by the greatest common factor (GCF).

Find $\frac{1}{3} \times \frac{3}{7}$. Write the product in simplest form.

$$\frac{1}{3} \times \frac{3}{7} = \frac{1 \times 3}{3 \times 7} = \frac{3}{21}$$

Multiply the numerators.
Multiply the denominators.

$$= \frac{3 \div 3}{21 \div 3}$$

Divide the numerator and the denominator by the GCF, 3.

$$= \frac{1}{7}$$

Write the product in simplest form.

So, $\frac{1}{3} \times \frac{3}{7} = \frac{1}{7}$.

• Explain why the product, $\frac{1}{7}$, is less than the factor $\frac{3}{7}$.

You can also multiply a whole number and a fraction without making a model.

EXAMPLE 2

Ms. Jones's car is being repaired after being in an accident. Her daughter Cindy will have to walk a total of $\frac{9}{10}$ mi to and from school every day for 11 days. How far will Cindy walk in all?

Find $11 \times \frac{9}{10}$.

Estimate. $11 \times 1 = 11$

$$11 \times \frac{9}{10} = \frac{11}{1} \times \frac{9}{10}$$

Write the whole number as a fraction.

$$= \frac{11 \times 9}{1 \times 10}$$

Multiply the numerators.
Multiply the denominators.

$$= \frac{99}{10}, \text{ or } 9\frac{9}{10}$$

Write the answer as a fraction or as a mixed number in simplest form.

Compare the product to your estimate. $9\frac{9}{10}$ is close to the estimate of 11. The product is reasonable.

So, Cindy will walk $9\frac{9}{10}$ mi.

• **What if** Cindy walked to school for 21 days? How far would Cindy walk in all?

When a numerator and a denominator have a common factor, you can simplify before you multiply.

EXAMPLE 3

Cheryl has $\frac{3}{4}$ of a box of snacks left from a school party. She gives $\frac{2}{5}$ of the snacks to the people in the school office. What part of the box of snacks does she give the people in the office?

Find $\frac{2}{5} \times \frac{3}{4}$.

Estimate. $\frac{1}{2} \times 1 = \frac{1}{2}$

$\frac{2}{5} \times \frac{3}{4}$ ← The GCF of 2 and 4 is 2.

Look for a numerator and denominator with common factors. Find the GCF.

$\frac{\overset{1}{2}}{5} \times \frac{3}{\underset{2}{4}}$ ← $2 \div 2 = 1$
← $4 \div 2 = 2$

Divide the numerator and denominator by the GCF, 2.

$\frac{\overset{1}{2}}{5} \times \frac{3}{\underset{2}{4}} = \frac{1 \times 3}{5 \times 2} = \frac{3}{10}$

Multiply.

So, Cheryl gives away $\frac{3}{10}$ of the box of snacks.

• What is $\frac{1}{8} \times \frac{6}{7}$? Simplify the fractions before you multiply.

EXAMPLE 4

Find $\frac{8}{9} \times \frac{3}{4}$. Use the GCF to simplify the fractions before you multiply.

$\frac{8}{9} \times \frac{3}{4}$ ← The GCF of 8 and 4 is 4.
← The GCF of 3 and 9 is 3.

$\frac{\overset{2}{8}}{\underset{3}{9}} \times \frac{\overset{1}{3}}{\underset{1}{4}} = \frac{2 \times 1}{3 \times 1} = \frac{2}{3}$

Divide the numerators and denominators by the GCFs, 3 and 4. Multiply.

So, $\frac{8}{9} \times \frac{3}{4} = \frac{2}{3}$.

CHECK FOR UNDERSTANDING

Think and Discuss ▶ **Look back at the lesson to answer each question.**

1. **Explain** how to make a model to show $\frac{1}{3} \times \frac{3}{8}$.

2. **Tell** how you can rewrite a whole number before you multiply it by a fraction.

Guided Practice ▶ **Make a model to find the product.**

3. $\frac{3}{4} \times \frac{1}{2}$ **4.** $\frac{1}{3} \times \frac{5}{8}$ **5.** $\frac{2}{5} \times \frac{1}{2}$ **6.** $\frac{1}{3} \times \frac{1}{2}$

Multiply. Write the answer in simplest form.

7. $\frac{3}{4} \times \frac{2}{5}$ **8.** $\frac{2}{5} \times \frac{7}{8}$ **9.** $2 \times \frac{6}{7}$ **10.** $\frac{2}{3} \times 16$

11. $\frac{2}{3} \times 4$ **12.** $9 \times \frac{2}{3}$ **13.** $\frac{5}{6} \times \frac{2}{3}$ **14.** $\frac{3}{5} \times \frac{5}{6}$

PRACTICE AND PROBLEM SOLVING

Independent ▶
Practice

Make a model to find the product.

15. $\frac{3}{4} \times \frac{1}{4}$ **16.** $\frac{3}{4} \times \frac{2}{3}$ **17.** $\frac{1}{8} \times \frac{1}{2}$ **18.** $3 \times \frac{1}{4}$

Multiply. Write the answer in simplest form.

19. $\frac{1}{3} \times \frac{2}{3}$ **20.** $\frac{3}{4} \times \frac{1}{3}$ **21.** $\frac{1}{5} \times \frac{2}{3}$ **22.** $\frac{1}{4} \times \frac{2}{7}$

23. $\frac{4}{5} \times \frac{7}{8}$ **24.** $\frac{2}{9} \times \frac{3}{4}$ **25.** $\frac{1}{8} \times \frac{4}{5}$ **26.** $\frac{5}{9} \times \frac{3}{10}$

27. $\frac{4}{9} \times \frac{3}{5}$ **28.** $\frac{2}{3} \times 21$ **29.** $24 \times \frac{1}{12}$ **30.** $\frac{1}{8} \times 16$

Compare. Write $<$, $>$, or $=$ for ●.

31. $\frac{2}{9} \times \frac{3}{10}$ ● $\frac{2}{9}$ **32.** $\frac{5}{6} \times 5$ ● $\frac{5}{6}$ **33.** $8 \times \frac{1}{9}$ ● 8

Problem Solving ▶
Applications

34. Sandra takes $\frac{1}{2}$ hr to walk to school. She spends $\frac{1}{2}$ of that time walking down her street. What part of an hour does Sandra spend walking down her street? How many minutes is this?

35. There are 144 registered voters in Booker County. In the last election, $\frac{1}{4}$ of them did not vote. How many voters did vote?

36. ⓘ **What's the Question?** Natalie runs $\frac{2}{3}$ the distance her brother runs in one week. Her brother runs 15 miles one week. The answer is 10 miles.

37. **REASONING** Scott chose a number, added 2, multiplied the sum by 4, and divided the product by 8. The final number was 4. What number had Scott chosen?

MIXED REVIEW AND TEST PREP

38. $3\frac{2}{3} \times 2\frac{1}{7}$ (p. 206)   **39.** Solve. $5b = 60.45$ (p. 82)   **40.** 267.45×2.8 (p. 70)

41. **TEST PREP** The manager of a grocery store asks customers in the store on a given day to fill out a card naming their favorite cookie. Which type of sample is this? (p. 98)

     A convenience   **B** biased   **C** random   **D** systematic

42. **TEST PREP** Which shows the GCF for 130 and 75? (p. 150)

     F 3   **G** 4   **H** 5   **J** 6

Extra Practice page H41, Set B

Multiply Mixed Numbers

Learn how to multiply mixed numbers.

Write the missing numerator.

1. $1\frac{2}{3} = \frac{\blacksquare}{3}$ 2. $6\frac{3}{5} = \frac{\blacksquare}{5}$

3. $7\frac{3}{7} = \frac{\blacksquare}{7}$ 4. $9\frac{3}{8} = \frac{\blacksquare}{8}$

5. $3\frac{2}{5} = \frac{\blacksquare}{5}$

Remember that you can write a mixed number as a fraction.

$2\frac{3}{4} = \frac{(2 \times 4) + 3}{4}$

$= \frac{11}{4}$

Ann and Sheri are training for a bicycle race. On one day, Ann rides $3\frac{1}{5}$ mi. Sheri rides $2\frac{1}{2}$ times as far as Ann. How many miles does Sheri ride?

Find $2\frac{1}{2} \times 3\frac{1}{5}$.   Estimate. $3 \times 3 = 9$

$2\frac{1}{2} \times 3\frac{1}{5} = \frac{5}{2} \times \frac{16}{5}$ *Write the mixed numbers as fractions.*

$= \frac{\overset{1}{\cancel{5}}}{\underset{1}{\cancel{2}}} \times \frac{\overset{8}{\cancel{16}}}{\underset{1}{\cancel{5}}}$ *Simplify the fractions. Multiply.*

$= \frac{8}{1}$, or 8   *Write the answer in simplest form or as a whole or mixed number.*

So, Sheri rides 8 mi. The answer is reasonable since it is close to the estimate of 9 mi.

You can use the Distributive Property to multiply a whole number by a mixed number.

EXAMPLE

Multiply. $5 \times 2\frac{3}{8}$

$5 \times 2\frac{3}{8} = 5 \times \left(2 + \frac{3}{8}\right)$

$= (5 \times 2) + \left(5 \times \frac{3}{8}\right)$ *Use the Distributive Property.*

$= (5 \times 2) + \left(\frac{5}{1} \times \frac{3}{8}\right)$ *Write the whole number as a fraction. Find 5×2 and $\frac{5}{1} \times \frac{3}{8}$.*

$= 10 + \frac{15}{8}$

$= 10 + 1\frac{7}{8} = 11\frac{7}{8}$ *Write the fraction as a mixed number. Find the sum.*

So, $5 \times 2\frac{3}{8} = 11\frac{7}{8}$.

CALIFORNIA STANDARDS O━┓NS 2.0 Students calculate and solve problems involving addition, subtraction, multiplication, and division. **MR 2.1** Use estimation to verify the reasonableness of calculated results. **NS 2.2** Explain the meaning of multiplication and division of positive fractions and perform the calculations (e.g., $\frac{5}{8} \div \frac{5}{16} = \frac{5}{8} \times \frac{16}{15} = \frac{2}{3}$). *also* O━┓NS 1.0, O━┓NS 2.1, NS 2.4, MR 1.3, MR 2.5

Think and Discuss ▸ Look back at the lesson to answer each question.

1. Show how to use the Distributive Property to find $3 \times 4\frac{2}{3}$.

2. Discuss whether the product of two mixed numbers is greater than or less than the factors. Give two examples.

Guided Practice ▸ Multiply. Write your answer in simplest form.

3. $\frac{3}{4} \times 1\frac{1}{2}$ **4.** $\frac{1}{2} \times 2\frac{1}{3}$ **5.** $1\frac{1}{2} \times 1\frac{1}{2}$ **6.** $1\frac{2}{5} \times 2\frac{1}{4}$

Use the Distributive Property to multiply.

7. $6\frac{1}{8} \times 3$ **8.** $3 \times 9\frac{4}{5}$ **9.** $1\frac{1}{8} \times 2$ **10.** $6 \times 4\frac{1}{4}$

PRACTICE AND PROBLEM SOLVING

Independent Practice ▸ Multiply. Write your answer in simplest form.

11. $4\frac{2}{3} \times 1\frac{3}{4}$ **12.** $1\frac{3}{8} \times 4\frac{2}{3}$ **13.** $5\frac{1}{2} \times 6$ **14.** $2 \times 3\frac{1}{7}$

15. $4\frac{1}{6} \times 3\frac{3}{5}$ **16.** $1\frac{3}{4} \times 3$ **17.** $10\frac{1}{5} \times 8\frac{1}{3}$ **18.** $5 \times 1\frac{5}{6}$

Use the Distributive Property to multiply.

19. $3 \times 2\frac{2}{5}$ **20.** $4 \times 8\frac{5}{6}$ **21.** $3\frac{3}{4} \times 6$ **22.** $1\frac{1}{2} \times 12$

Compare. Write $<$, $>$, or $=$ for each ●.

23. $3\frac{1}{3} \times 2\frac{1}{7}$ ● $3\frac{1}{4} \times 5$ **24.** $7 \times 7\frac{3}{7}$ ● $6\frac{3}{4} \times 4\frac{4}{5}$

Problem Solving Applications ▸ **25.** Mr. Jackson rides his bicycle $1\frac{2}{3}$ mi every day. His wife rides $1\frac{1}{4}$ times as far as he does. How many miles does Mrs. Jackson ride her bicycle?

26. John works part time for $6.50 an hour. He works $3\frac{1}{2}$ hr each on Monday, Tuesday, and Thursday afternoons. How much does he earn those three days?

27. **Write About It** Without multiplying, tell whether the product of $\frac{2}{3} \times \frac{3}{4}$ is a fraction, a whole number, or a mixed number.

MIXED REVIEW AND TEST PREP

28. $\frac{4}{9} \times \frac{2}{3}$ (p. 202)   **29.** $8\frac{3}{4} - 2\frac{2}{5}$ (p. 186)   **30.** Simplify. $\frac{27}{45}$ (p. 160)   **31.** $414,089 - 62,036$ (p. 20)

32. TEST PREP Which is the mean of the data?
90, 94, 65, 90, 84, 94, 85 (p. 110)
    A 29   **B** 86   **C** 90   **D** 94

Division of Fractions

MATH LAB

Explore how to model division of fractions.

You need fraction circles.

Vocabulary

reciprocal

Using models will help you understand division of fractions.

Activity 1

A. Use fraction circles to find $4 \div \frac{1}{3}$, or the number of thirds in 4 wholes.

* Trace 4 whole circles on your paper.

* Model $4 \div \frac{1}{3}$ by tracing $\frac{1}{3}$-circle pieces on the 4 circles.

One whole equals three thirds.

* How many thirds are in 4 wholes? What is $4 \div \frac{1}{3}$?

B. Use fraction circles to find $\frac{1}{3} \div \frac{1}{6}$, or the number of sixths in $\frac{1}{3}$.

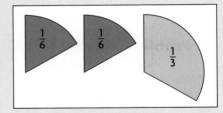

* Place as many $\frac{1}{6}$ pieces as you can on the $\frac{1}{3}$ piece.

* How many sixths are in $\frac{1}{3}$? What is $\frac{1}{3} \div \frac{1}{6}$?

Think and Discuss

* If $4 \div \frac{1}{3} = 12$, explain what $8 \div \frac{1}{3}$ must be.

* If $2 \div \frac{1}{6} = 12$, explain what $8 \div \frac{1}{6}$ must be.

Practice

Use fraction circles to model each problem. Draw a diagram of your model.

1. $3 \div \frac{1}{3}$ 2. $4 \div \frac{1}{4}$ 3. $\frac{1}{2} \div \frac{1}{8}$ 4. $\frac{3}{4} \div \frac{1}{8}$

CALIFORNIA STANDARDS O—n NS 2.0 Students calculate and solve problems involving addition, subtraction, multiplication, and division. **NS 2.2** Explain the meaning of multiplication and division of positive fractions and perform the calculations (e.g., $\frac{5}{8} \div \frac{15}{16} = \frac{5}{8} \times \frac{16}{15} = \frac{2}{3}$). *also* O—n NS 2.1, NS 2.4

Two numbers are **reciprocals** if their product is 1.

$$\frac{1}{2} \times 2 = 1 \qquad\qquad \frac{3}{4} \times \frac{4}{3} = 1 \qquad\qquad 6 \times \frac{1}{6} = 1$$

$$\uparrow\;\;\uparrow \qquad\qquad\qquad \uparrow\;\;\uparrow \qquad\qquad\qquad \uparrow\;\;\uparrow$$

reciprocals reciprocals reciprocals

By using inverse operations, you can write related number sentences.

$$1 \div \frac{1}{2} = 2 \qquad\qquad 1 \div \frac{3}{4} = \frac{4}{3} \qquad\qquad 1 \div 6 = \frac{1}{6}$$

$$1 \div 2 = \frac{1}{2} \qquad\qquad 1 \div \frac{4}{3} = \frac{3}{4} \qquad\qquad 1 \div \frac{1}{6} = 6$$

You can use reciprocals and inverse operations when you divide.

Activity 2

• Study these problems.

Find $6 \div \frac{1}{2}$.

$1 \div \frac{1}{2} = 2$ *Think of the reciprocal of $\frac{1}{2}$.*

Since $1 \div \frac{1}{2} = 1 \times 2$, $6 \div \frac{1}{2} = 6 \times 2$.

So, $6 \div \frac{1}{2} = 6 \times 2 = 12$.

Find $\frac{3}{4} \div \frac{1}{3}$.

$1 \div \frac{1}{3} = 3$ *Think of the reciprocal of $\frac{1}{3}$.*

Since $1 \div \frac{1}{3} = 1 \times 3$, $\frac{3}{4} \div \frac{1}{3} = \frac{3}{4} \times 3$.

So, $\frac{3}{4} \div \frac{1}{3} = \frac{3}{4} \times 3 = \frac{9}{4}$, or $2\frac{1}{4}$.

Think and Discuss

• When you divide 1 by a number, what is the quotient?

• If $1 \div \frac{1}{3} = 3$, then what is $2 \div \frac{1}{3}$? What is $5 \div \frac{1}{3}$?

• If $1 \div \frac{2}{5} = \frac{5}{2}$, then what is $2 \div \frac{2}{5}$? What is $\frac{1}{2} \div \frac{2}{5}$?

Practice

Find the value of n.

1. If $1 \div \frac{1}{2} = 2$, then $2 \div \frac{1}{2} = n$.

2. If $1 \div \frac{4}{5} = \frac{5}{4}$, then $4 \div \frac{4}{5} = n$.

3. If $1 \div \frac{2}{3} = \frac{3}{2}$, then $\frac{1}{2} \div \frac{2}{3} = n$.

4. If $1 \div \frac{3}{5} = \frac{5}{3}$, then $\frac{3}{4} \div \frac{3}{5} = n$.

Find the quotient.

5. $6 \div \frac{3}{4}$

6. $4 \div \frac{1}{2}$

7. $\frac{1}{2} \div \frac{1}{3}$

8. $\frac{2}{3} \div \frac{1}{8}$

MIXED REVIEW AND TEST PREP

9. $2\frac{4}{5} \times 3\frac{3}{8}$ (p. 206)

10. Complete. $\frac{3}{8} = \frac{\blacksquare}{24}$ (p. 160)

11. Compare 606.64 and 606.074. Use $<$, $>$, or $=$. (p. 52)

12. $46.08 - 19.204$ (p. 66)

13. **TEST PREP** Which shows $\frac{24}{60}$ in simplest form? (p. 160)

A $\frac{1}{3}$ B $\frac{2}{5}$ C $\frac{8}{20}$ D $\frac{6}{15}$

Divide Fractions and Mixed Numbers

Learn how to divide with fractions and mixed numbers.

QUICK REVIEW

Write each mixed number as a fraction.

1. $2\frac{1}{3}$ 2. $1\frac{3}{8}$ 3. $4\frac{1}{2}$

4. $3\frac{9}{10}$ 5. $6\frac{2}{3}$

Stacey is bringing juice to be served after the school council meeting. Stacey said she would bring enough juice for everyone at the meeting to have a glass. Each glass holds $\frac{1}{5}$ liter. How many glasses can be served if Stacey brings 8 liters of juice?

Find $8 \div \frac{1}{5}$.

Math Idea ▶ When you divide by a fraction, you can use multiplication to find the quotient.

Rewrite the division problem as a multiplication problem by using the reciprocal.

$8 \div \frac{1}{5} = \frac{8}{1} \div \frac{1}{5}$ *Write the whole number as a fraction.*

$= \frac{8}{1} \times \frac{5}{1}$ *Use the reciprocal of the divisor to write a multiplication problem.*

$= \frac{8}{1} \times \frac{5}{1} = \frac{40}{1}$, or 40   *Multiply.*

So, 8 liters can fill 40 glasses.

You can also use the reciprocal of the divisor to divide fractions and mixed numbers.

EXAMPLE 1

Find $\frac{2}{3} \div \frac{4}{7}$.

$\frac{2}{3} \div \frac{4}{7} = \frac{2}{3} \times \frac{7}{4}$ *Use the reciprocal of the divisor to write a multiplication problem.*

$= \frac{\overset{1}{2}}{3} \times \frac{7}{\underset{2}{4}}$ *Divide the numerator and denominator by the GCF, 2.*

$= \frac{7}{6}$, or $1\frac{1}{6}$ *Multiply.*

So, $\frac{2}{3} \div \frac{4}{7} = 1\frac{1}{6}$.

CALIFORNIA STANDARDS O━NS 2.0 Students calculate and solve problems involving addition, subtraction, multiplication, and division. **NS 2.2** Explain the meaning of multiplication and division of positive fractions and perform the calculations. *also* O━NS 1.0, NS 2.1, O━NS 2.4, MR 2.1

EXAMPLE 2

Each member of the school council will write his or her name on a strip of paper at the meeting. Jeremy has pieces of paper $5\frac{1}{4}$ in. long. Each strip of paper should be $1\frac{3}{4}$ in. How many strips can Jeremy cut from the length of one piece of paper?

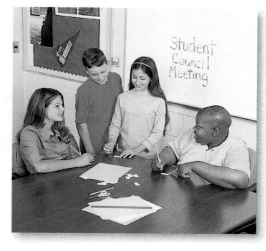

Find $5\frac{1}{4} \div 1\frac{3}{4}$.

Estimate. $5 \div 2 = 2\frac{1}{2}$

$5\frac{1}{4} \div 1\frac{3}{4} = \frac{21}{4} \div \frac{7}{4}$ *Write the mixed numbers as fractions.*

$= \frac{21}{4} \times \frac{4}{7}$ *Use the reciprocal of the divisor to write a multiplication problem.*

$= \frac{\overset{3}{\cancel{21}}}{\underset{1}{\cancel{4}}} \times \frac{\overset{1}{\cancel{4}}}{\underset{1}{\cancel{7}}}$ *Simplify and multiply.*

$= \frac{3}{1}$, or 3

Compare the product to your estimate. 3 is close to the estimate of $2\frac{1}{2}$. The product is reasonable.

So, Jeremy can cut 3 strips from each piece of paper.

• What is $2\frac{3}{4} \div 1\frac{2}{5}$?

Sometimes you can use mental math to divide whole numbers and fractions.

EXAMPLE 3

Use mental math to solve.

A $9 \div \frac{1}{2}$ **THINK:** $9 \times 2 = 18$.   *Dividing by $\frac{1}{2}$ is the same as multiplying by 2.*
There are 18 halves in 9.

So, $9 \div \frac{1}{2} = 18$.

B $13 \div \frac{1}{3}$ **THINK:** $13 \times 3 = 39$.   *Dividing by $\frac{1}{3}$ is the same as multiplying by 3.*
There are 39 thirds in 13.

So, $13 \div \frac{1}{3} = 39$.

C $20 \div \frac{2}{5}$ **THINK:** $20 \times 5 = 100$.
                    $100 \div 2 = 50$.   *Dividing by $\frac{2}{5}$ is the same as multiplying by $\frac{5}{2}$.*

So, $20 \div \frac{2}{5} = 50$.

• Use mental math to find $15 \div \frac{1}{6}$.

Think and ▶ Discuss

Look back at the lesson to answer each question.

1. **Tell** what the reciprocal of a number is. Give an example.

2. **Give** an example of a fraction or mixed-number division problem where the quotient is greater than the dividend.

Guided ▶ Practice

Write the reciprocal of the number.

3. $\frac{2}{3}$ 4. $\frac{3}{4}$ 5. 7   6. $2\frac{3}{8}$ 7. $4\frac{1}{3}$

Find the quotient. Write the answer in simplest form.

8. $\frac{1}{3} \div \frac{1}{2}$ 9. $\frac{1}{5} \div \frac{1}{4}$ 10. $\frac{1}{4} \div \frac{1}{2}$ 11. $\frac{1}{8} \div 3$

12. $4 \div \frac{2}{3}$ 13. $3\frac{3}{5} \div 1\frac{1}{5}$ 14. $2\frac{3}{5} \div 4$ 15. $3\frac{1}{4} \div 2\frac{2}{3}$

PRACTICE AND PROBLEM SOLVING

Independent ▶ Practice

Write the reciprocal of the number.

16. $\frac{5}{8}$ 17. 10   18. $\frac{1}{6}$ 19. $\frac{2}{9}$ 20. $3\frac{1}{2}$

21. $\frac{15}{7}$ 22. $\frac{1}{5}$ 23. 9   24. $1\frac{5}{6}$ 25. $2\frac{3}{5}$

Find the quotient. Write the answer in simplest form.

26. $\frac{3}{8} \div \frac{1}{2}$ 27. $\frac{2}{3} \div \frac{4}{7}$ 28. $\frac{7}{8} \div \frac{1}{3}$ 29. $8 \div \frac{6}{7}$

30. $12 \div \frac{3}{5}$ 31. $\frac{3}{4} \div \frac{1}{3}$ 32. $\frac{4}{9} \div \frac{3}{5}$ 33. $4 \div \frac{4}{5}$

34. $3\frac{2}{5} \div 1\frac{1}{5}$ 35. $3\frac{4}{5} \div \frac{3}{4}$ 36. $4\frac{1}{2} \div \frac{1}{4}$ 37. $4\frac{1}{5} \div 2\frac{3}{5}$

Use mental math to find each quotient.

38. $10 \div \frac{1}{2}$ 39. $12 \div \frac{2}{3}$ 40. $6 \div \frac{1}{4}$ 41. $8 \div \frac{1}{4}$

Algebra Evaluate the expression.

42. $4 \div a$ for $a = \frac{2}{3}$ 43. $b \div 2\frac{1}{3}$ for $b = 5\frac{1}{2}$

44. $1\frac{4}{5} \div a$ for $a = 2\frac{3}{5}$ 45. $c \div 7\frac{4}{5}$ for $c = 2\frac{1}{6}$

46. $b \div 7$ for $b = \frac{1}{5}$ 47. $3\frac{1}{5} \div b$ for $b = 4$

48. The members of the school council were told they could divide the 15 acres of land next to the school into $1\frac{1}{2}$-acre sections to be used for new building projects for the school. How many $1\frac{1}{2}$-acre sections will there be?

49. How many $\frac{1}{4}$-lb turkey burgers can Katie make with 12 lb of ground turkey?

50. In a $\frac{1}{4}$-mi relay, each runner on a team runs $\frac{1}{16}$ mi. How many runners are in the relay?

51. Gerald has $8\frac{3}{4}$ yd of fabric. This is 7 times the amount he needs to make one costume for the school play. How much fabric does he need for each costume?

52. ❓ **What's the Error?** Jamie worked the problem below. What mistake did Jamie make? What is the correct answer in simplest form?

$$3\frac{1}{3} \div 2\frac{2}{9} = \frac{10}{3} \div \frac{20}{9} = \frac{3}{10} \times \frac{\cancel{20}}{9} = \frac{2}{3}$$

53. Jacki wants to buy a blouse for $12.95, a T-shirt for $15.95, a book for $10.50, and two pens for $3.75 each. What is the total cost?

MIXED REVIEW AND TEST PREP

54. $3\frac{1}{3} \times 2\frac{2}{5}$ (p. 206)

55. $10\frac{1}{2} + 2\frac{7}{8}$ (p. 186)

56. Evaluate $5.2x$ for $x = 3.41$. (p. 82)

57. TEST PREP Which is the sum for $234,607 + 84,395$? (p. 20)

A 218,002   **B** 218,992   **C** 318,902   **D** 319,002

58. TEST PREP Which shows $4\frac{3}{8}$ as a fraction? (p. 172)

F $\frac{11}{8}$ **G** $\frac{15}{8}$ **H** $\frac{35}{8}$ **J** $\frac{39}{8}$

LiNKUP to Careers

Carpenter A carpenter builds and repairs structures. A carpenter can frame a house, put on siding and insulation, put up the rafters and roof, hang doors and windows, and lay floors. Carpenters usually work with wood. They need to measure accurately and work with fractions and mixed numbers when they buy and cut wood for projects.

• Mark is hired to do interior trim work. The molding that he will use comes in 8-ft lengths. He needs strips of wood that are $\frac{5}{6}$ ft long. How many $\frac{5}{6}$ ft pieces can he cut from each 8-ft length?

PROBLEM SOLVING SKILL
Choose the Operation

Analyze
Choose
Solve
Check

Learn how to choose the operation needed to solve a problem.

QUICK REVIEW

QUICK REVIEW

1. $\frac{5}{6}$ plus $\frac{3}{8}$ 2. $\frac{2}{3}$ divided by $\frac{1}{2}$

3. $\frac{5}{6}$ minus $\frac{1}{2}$ 4. $\frac{5}{7} \times \frac{2}{3}$ 5. $\frac{1}{2} \div \frac{3}{4}$

Use the chart to help you decide how the numbers in a problem are related.

Then choose the operation needed to solve each problem.

ADD	• Joining groups of different sizes
SUBTRACT	• Taking away or comparing groups
MULTIPLY	• Joining equal-sized groups
DIVIDE	• Separating into equal-sized groups • Finding out how many in each group

Read each problem and decide how you would solve it.

A In one week, José drove $3\frac{1}{4}$ mi to play practice, $3\frac{1}{4}$ mi to the hardware store to buy supplies, and $7\frac{4}{5}$ mi to Jim's house to work on the costumes for the play. How many miles did he drive in all?	**B** Carla bought 6 yd of terry cloth to make collars for some of the costumes for the play. Each collar takes $\frac{2}{3}$ yd of fabric. How many collars can Carla make?
C Leanne makes props for the school play. She uses $4\frac{1}{8}$ lb of clay to make each vase and $2\frac{3}{4}$ lb to make each bowl. How much more clay does Leanne use to make one vase than one bowl?	**D** Sean bought $12\frac{2}{3}$ yd of pine lumber for $6 per yd to build some scenery for the school play. How much money did he spend?

• What operation would you use to solve each problem?

• Which problems could you solve by using a combination of operations?

• How did the chart above help you decide which operation to use for each problem?

CALIFORNIA STANDARDS O—∎ NS 2.0 Students calculate and solve problems involving addition, subtraction, multiplication, and division. **MR 3.2** Note the method of deriving the solution and demonstrate a conceptual understanding of the derivation by solving similar problems. *also* NS 2.1, NS 2.2

Solve. Name the operations used.

1. It takes Karen $5\frac{1}{2}$ minutes to walk to the community center. It takes Jackie $7\frac{1}{3}$ minutes. How much longer does it take Jackie?

2. Rob makes candles that weigh $1\frac{3}{16}$ lb each. How much do 24 of them weigh?

3. Katie decorates travel bags. Each bag requires $1\frac{1}{3}$ yd of trim around the top and 1 yd of trim on the handle.

   a. Which operations could you use to find how much trim is needed for a number of bags?

      A subtraction and division

      B addition and multiplication

      C division and multiplication

      D addition and division

   b. How much trim would it take to make 4 bags?

      F $8\frac{1}{3}$ yd

      G 8 yd

      H $9\frac{1}{3}$ yd

      J 9 yd

4. Marcie has 18 oz of dough left in a container. She needs $3\frac{3}{5}$ oz of dough to make one ornament.

   a. Write the expression you would use to find the number of ornaments she can make. Solve the problem.

   b. How many ounces of dough are left in the container if she makes 4 ornaments?

Use Data   For 5–8, use the map.

5. Darrin hiked the shortest route from the trailhead to Hart Mountain in $7\frac{3}{4}$ hours. What is the average number of miles he hiked in 1 hour?

6. How much shorter was the trail that Darrin took than the trail through Warm Springs?

7. Elk Meadow is halfway along the trail from the trailhead to Hart Mountain. How far is it from the trailhead to Elk Meadow?

8. Sharlene left the trailhead at 8:30 A.M. for a daylong hike. She returned to the trailhead $7\frac{1}{2}$ hours after she left. What time did she return?

9. ✎ **Write About It**   Explain how you decide what operation to use when solving a problem.

ALGEBRA
Fraction Expressions and Equations

Learn how to evaluate expressions and solve equations with fractions.

Venus's-flytrap is found in the coastal regions of North and South Carolina.

Ann will be mailing some brochures about the different types of plants that eat insects. The price to mail them depends on the total weight. Each brochure weighs $\frac{3}{4}$ ounce. How much do the brochures weigh in all?

You can write and evaluate an expression. Choose a variable to represent the number of brochures.

Let $b =$ the number of brochures.

$b \times \dfrac{3}{4}$ or $\dfrac{3}{4}b$ *Write the expression.*

The number of brochures Ann mails changes every week. What was the total weight of the 23 brochures she mailed last week?

$b \times \dfrac{3}{4}$ or $\dfrac{3}{4}b$ *Write the expression.*

$23 \times \dfrac{3}{4}$ *Replace b with 23.*

$\dfrac{69}{4}$, or $17\frac{1}{4}$ *Multiply.*

So, the total weight is $17\frac{1}{4}$ oz.

You have solved equations with whole numbers and decimals by using mental math. You can use mental math to solve some equations with fractions.

EXAMPLE

Solve the equation $n \div \frac{1}{4} = 8$ by using mental math.

$n \div \dfrac{1}{4} = 8$ *Remember, when dividing by $\frac{1}{4}$, you multiply by the reciprocal, 4.*

$n = 2$ *What number times 4 equals 8?*

$2 \div \dfrac{1}{4} = 8$ *Check your answer. Replace n with 2.*

$\dfrac{2}{1} \times \dfrac{4}{1} = \dfrac{8}{1}$, or 8 So, $n = 2$.

• Use mental math to solve $x + \dfrac{2}{3} = \dfrac{5}{6}$.

CALIFORNIA STANDARDS ⊶ **AF 1.1** Write and solve one-step linear equations in one variable. **AF 1.0** Students write verbal expressions and sentences as algebraic expressions and equations; they evaluate algebraic expressions, solve simple linear equations, and graph and interpret their results. *also* **NS 2.1, NS 2.2**

CHECK FOR UNDERSTANDING

Think and ▶
Discuss

Look back at the lesson to answer each question.

1. **What if** Ann mails 32 brochures? Show how you would find the total weight of the brochures.

2. **Explain** how you would solve the equation $d \div \frac{1}{3} = 9$.

Guided ▶
Practice

Evaluate the expression.

3. $x - 1\frac{1}{5}$ for $x = 3\frac{2}{5}$ 4. $\frac{1}{4}y$ for $y = \frac{1}{2}$ 5. $t + 2\frac{1}{4}$ for $t = 3\frac{3}{8}$

Use mental math to solve the equation.

6. $x - 4\frac{1}{4} = 2\frac{1}{2}$ 7. $\frac{2}{3}y = \frac{1}{4}$ 8. $2\frac{5}{8} = t + 1\frac{1}{4}$

PRACTICE AND PROBLEM SOLVING

Independent ▶
Practice

Evaluate the expression.

9. $c + 2\frac{2}{3}$ for $c = 5\frac{1}{6}$ 10. $4\frac{5}{8}\,m$ for $m = 3\frac{1}{3}$ 11. $b \div 2\frac{1}{2}$ for $b = 3\frac{1}{8}$

12. $4\frac{5}{8} + a$ for $a = 4\frac{4}{8}$ 13. $b \times \frac{1}{5}$ for $b = \frac{1}{4}$ 14. $\frac{5}{8} \div c$ for $c = 5$

15. $\frac{21}{24} - a$ for $a = \frac{1}{2}$ 16. $c \div 1\frac{1}{4}$ for $c = 1\frac{2}{3}$ 17. $\frac{1}{3} + \frac{1}{2} + d$ for $d = \frac{1}{6}$

Use mental math to solve the equation.

18. $x - 3\frac{1}{2} = 2\frac{1}{4}$ 19. $\frac{1}{5}x = \frac{2}{5}$ 20. $14\frac{2}{3} + x = 28$

21. $8x = 2$ 22. $\frac{2}{3}t = 2$ 23. $3\frac{1}{3} + x = 4\frac{2}{3}$

Problem Solving ▶
Applications

24. Ann's brochures tell about the eating habits of pitcher plants, sundew plants, and Venus's-flytraps. Suppose one type of plant traps about $3\frac{1}{2}$ oz of insects each day. Write and solve an equation to find how many days the plant takes to trap a total of $17\frac{1}{2}$ oz of insects.

Sundew Plant

25. **?** **What's the Error?** Robin evaluated $\frac{3}{4}h$ for $h = \frac{3}{8}$ in this way: $\frac{3}{4} \times \frac{3}{8} = \frac{6}{8} \times \frac{3}{8} = \frac{9}{8}$. Explain her mistake.

MIXED REVIEW AND TEST PREP

26. $4\frac{5}{8} \div 2\frac{1}{6}$ (p. 210)

27. Order from greatest to least. $\frac{3}{4}, \frac{3}{8}, \frac{3}{5}$ (p. 166)

28. 25.95×13.3 (p. 70)

29. Solve. $1.5 = \frac{a}{3}$ (p. 82)

30. **TEST PREP** Which decimal is 10 times as great as 34.62? (p. 70)

   A 0.34   **B** 3.4   **C** 3.46   **D** 346.2

Extra Practice page H41, Set E

1. VOCABULARY Two numbers are ___?___ if their product is 1. (pp. 208–209)

Estimate each product or quotient. (pp. 200–201)

2. $\frac{2}{9} \times \frac{1}{6}$
3. $\frac{7}{8} \div \frac{11}{12}$
4. $\frac{7}{15} \div \frac{8}{9}$

5. $2\frac{4}{5} \times 3\frac{3}{10}$
6. $4\frac{2}{15} \times 5\frac{4}{7}$
7. $31\frac{3}{8} \div 4\frac{1}{2}$

Multiply. Write the answer in simplest form. (pp. 202–207)

8. $\frac{1}{6} \times \frac{3}{5}$
9. $\frac{2}{3} \times \frac{4}{7}$
10. $16 \times \frac{5}{12}$
11. $\frac{3}{8} \times 10$

12. $1\frac{1}{2} \times \frac{3}{4}$
13. $4\frac{1}{2} \times 2\frac{1}{3}$
14. $1\frac{1}{2} \times \frac{2}{3}$
15. $2\frac{1}{3} \times 3\frac{1}{7}$

Find the quotient. Write it in simplest form. (pp. 210–213)

16. $\frac{6}{7} \div \frac{3}{5}$
17. $\frac{3}{4} \div \frac{1}{3}$
18. $3\frac{1}{3} \div 2\frac{4}{5}$
19. $9\frac{1}{2} \div 1\frac{3}{8}$

20. $8 \div \frac{6}{7}$
21. $\frac{4}{5} \div 4$
22. $2\frac{3}{5} \div 4\frac{1}{5}$
23. $\frac{5}{8} \div 10$

Evaluate the expression. (pp. 216–217)

24. $x - 3\frac{1}{3}$ for $x = 6\frac{1}{5}$
25. $\frac{3}{4} \div r$ for $r = 4$
26. $25\frac{1}{8} + m$ for $m = 6\frac{2}{3}$

27. $x + 3\frac{1}{10}$ for $x = 1\frac{1}{2}$
28. $\frac{7}{8} a$ for $a = \frac{2}{5}$
29. $\frac{3}{4} b$ for $b = 1\frac{1}{6}$

Use mental math to solve the equation. (pp. 216–217)

30. $x - 6\frac{1}{9} = 3\frac{2}{3}$
31. $x \div \frac{1}{3} = 12$
32. $12\frac{3}{4} + x = 15\frac{7}{8}$

33. $1\frac{1}{8} b = 4\frac{1}{20}$
34. $c - \frac{1}{5} = \frac{1}{2}$
35. $\frac{3}{7} \div a = \frac{1}{28}$

Solve.

36. Mike rides his bicycle $6\frac{1}{2}$ min to school. Cami rides her bicycle $1\frac{1}{2}$ times as long. How long does it take Cami to ride to school? (pp. 206–207)

37. Eric practiced $\frac{3}{4}$ hr on Monday and $\frac{2}{5}$ hr on Saturday. How much longer did Eric practice on Monday? (pp. 214–215)

38. A race course is $\frac{3}{4}$ mi long. Racers want to run three equal sprints. How far apart should the markers be for the sprints? (pp. 210–213)

39. Sara has $\frac{5}{6}$ yd of ribbon and Mark has $\frac{3}{4}$ yd of ribbon. How much ribbon do Sara and Mark have altogether? (pp. 214–215)

40. The sum of Kiesha's and Brent's heights is $127\frac{1}{4}$ in. Kiesha's height is $64\frac{1}{2}$ in. Write an equation to find Brent's height. (pp. 216–217)

Understand the problem.
See item **3**.

You know that this week's playing time is $\frac{2}{3}$ of last week's playing time. Use a *variable* to write any relationship like this one as an equation.
Also see problem **1**, p. H62.

Choose the best answer.

1. Carla rode her bicycle $1\frac{2}{5}$ miles on Monday, $2\frac{1}{4}$ miles on Wednesday, and $\frac{4}{5}$ mile on Friday. Which is a reasonable estimate of the total distance Carla rode her bicycle?

   A 3 mi   **C** $5\frac{1}{2}$ mi

   B $4\frac{1}{2}$ mi   **D** $6\frac{1}{2}$ mi

2. Dan has $6\frac{2}{3}$ yards of cloth. He uses $\frac{5}{6}$ yard in each banner he makes. How many banners can Dan make from the cloth?

   F 4   **H** 11

   G 8   **J** Not here

3. Vince is trying to decrease the amount of time he spends playing video games. This week he spent $5\frac{2}{3}$ hours playing video games. That was $\frac{2}{3}$ of the amount of time he played video games last week. How long did Vince play video games last week?

   A $4\frac{1}{2}$ hr   **C** 7 hr

   B $6\frac{1}{3}$ hr   **D** $8\frac{1}{2}$ hr

4. Eva ordered $2\frac{1}{2}$ pounds of potato salad, $2\frac{3}{4}$ pounds of fruit salad, and $1\frac{5}{8}$ pounds of cole slaw. Which operation would be best to use to find out how much more fruit salad she ordered than coleslaw?

   F addition

   G multiplication

   H subtraction

   J division

5. Which expression can be used to find $\frac{3}{5} \div \frac{1}{3}$?

   A $\frac{3}{5} + \frac{1}{3}$

   B $\frac{3}{5} - \frac{1}{3}$

   C $\frac{3}{5} \times \frac{3}{1}$

   D $\frac{5}{3} \times \frac{1}{3}$

6. There are 397 students in the sixth grade at East Side School. During the first week of school, about $\frac{1}{5}$ of the sixth-grade students tried out for the band. About how many students tried out?

   F 70   **H** 90

   G 80   **J** 100

7. Which is a reasonable estimate for $21\frac{4}{5} \div 3$?

   A 3   **C** 5

   B 4   **D** 7

8. Solve.
$$b - 4\frac{1}{2} = 2\frac{2}{3}$$

   F $b = 1\frac{5}{6}$ **H** $b = 6\frac{5}{6}$

   G $b = 6\frac{3}{5}$ **J** $b = 7\frac{1}{6}$

9. Cal has 4 bags of apples with masses of 2.5 kilograms, 2.05 kilograms, 3.12 kilograms, and 2.18 kilograms. Which shows the masses in order from lightest to heaviest?

   A 2.05, 2.18, 2.5, 3.12

   B 2.5, 2.18, 3.12, 2.05

   C 3.12, 2.5, 2.18, 2.05

   D 2.18, 2.5, 2.05, 3.12

10. The sum of the prime factors of 12 is 7. What is the sum of the prime factors of 72?

   F 12   **H** 42

   G 27   **J** Not here

MATH DETECTIVE

On a Roll

REASONING Roll a number cube ten times. Record the numbers that you roll. Use your knowledge of fractions to attempt to solve the problems below. You may use each of your ten numbers exactly once. Once you use a number, cross it out so that you don't use it again. Depending on which numbers you roll, you may or may not be able to solve all four mysteries.

Fraction Mystery 1

Write a pair of equivalent fractions.

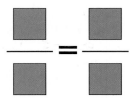

Fraction Mystery 2

Write a fraction with a value greater than 1.

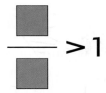

Fraction Mystery 3

Write a fraction with a value greater than $\frac{1}{2}$ and less than 1.

Fraction Mystery 4

Write a fraction that is not in simplest form.

Think It Over!

- **Write About It** Salvador rolled all 2's and 6's. Explain why he couldn't solve all four mysteries.

- **Stretch Your Thinking** Write a set of ten numbers with which you could solve only two of the mysteries.

Mixed Numbers and Time

Learn how to write times as mixed numbers, and how to add and subtract times.

The ticket agent at the airport told Mario that his flight from Portland, OR, to St. Paul, MN, would take "about $4\frac{1}{2}$ hours." The agent expressed the time as a mixed number.

To write a time as a mixed number, use the fact that there are 60 minutes in 1 hour.

EXAMPLE 1

Mario's flight from Portland to St. Paul took 4 hours 24 minutes. Write the time as a mixed number.

$24 \text{ min} = 24 \times \frac{1}{60} \text{ hr} = \frac{24}{60} \text{ hr}$ *Think: 60 min = 1 hr. So, 1 min = $\frac{1}{60}$ hr.*

$= \frac{2}{5} \text{ hr}$ *Write the fraction in simplest form.*

$4 \text{ hr } 24 \text{ min} = 4\frac{2}{5} \text{ hr}$ *Write the time.*

So, Mario's flight took $4\frac{2}{5}$ hr.

To add or subtract times, use the same methods you use to add or subtract mixed numbers.

EXAMPLE 2

Solve.

A.
$$\begin{array}{r} 2 \text{ hr } 18 \text{ min} \\ + 3 \text{ hr } 52 \text{ min} \\ \hline 5 \text{ hr } 70 \text{ min} \end{array}$$
$= 5 \text{ hr} + (60 + 10) \text{ min}$
$= 5 \text{ hr} + 1 \text{ hr} + \frac{10}{60} \text{ hr}$
$= 6 \text{ hr} + \frac{1}{6} \text{ hr}$
$= 6\frac{1}{6} \text{ hr}$

B.
$$\begin{array}{r} 8 \text{ hr } 14 \text{ min} \rightarrow 7 \text{ hr } (60 + 14) \text{ min} \\ -3 \text{ hr } 20 \text{ min} \rightarrow -3 \text{ hr } 20 \text{ min} \\ \hline \end{array}$$
$$\begin{array}{r} 7 \text{ hr } 74 \text{ min} \\ -3 \text{ hr } 20 \text{ min} \\ \hline 4 \text{ hr } 54 \text{ min} \end{array}$$
$= 4 \text{ hr} + \frac{54}{60} \text{ hr}$
$= 4\frac{9}{10} \text{ hr}$

TALK ABOUT IT

• Tell how you would write the time 7 hours 35 minutes as a mixed number.

TRY IT

Write as a mixed number.

1. 1 hr 30 min   **2.** 4 hr 55 min   **3.** 1 hr 19 min

4. 28 min   **5.** 45 min   **6.** 3 hr 42 min

Solve. Write as a mixed number.

7. 5 hr 12 min + 8 hr 18 min   **8.** 6 hr 10 min − 4 hr 20 min

VOCABULARY

1. The largest of the common factors of two numbers is the ___?___ . (p. 151)

2. To add unlike fractions, you can write equivalent fractions by using the ___?___ . (p. 182)

EXAMPLES

EXERCISES

Chapter 7

- **Write the prime factorization of a number.** (pp. 148–149)

  Find the prime factorization of 150.

  $2 \times 3 \times 5^2$

Write the prime factorization of each number in exponent form.

3. 52   **4.** 125   **5.** 180

6. 27   **7.** 100   **8.** 72

- **Find the GCF and LCM of two or more numbers.** (pp. 150–153)

  Find the LCM of 9 and 15.

  9: 9, 18, 27, 36, 45   *Find multiples of 9*
  15: 15, 30, 45   *and 15.*
  LCM = 45   *Find the LCM.*

Find the GCF for each set of numbers.

9. 15, 30   **10.** 9, 12   **11.** 8, 16, 12

Find the LCM for each set of numbers.

12. 6, 10   **13.** 6, 8   **14.** 5, 9, 12

Chapter 8

- **Write fractions in simplest form.** (pp. 160–163)

  Write $\frac{18}{30}$ in simplest form.

  $\frac{18}{30} = \frac{18 \div 6}{30 \div 6} = \frac{3}{5}$ *Divide by the GCF.*

Write the fraction in simplest form.

15. $\frac{8}{12}$ **16.** $\frac{15}{20}$ **17.** $\frac{40}{50}$

18. $\frac{36}{48}$ **19.** $\frac{18}{21}$ **20.** $\frac{90}{40}$

- **Write mixed numbers as fractions and fractions as mixed numbers.** (pp. 164–165)

  Write $3\frac{5}{6}$ as a fraction.

  $3\frac{5}{6} = \frac{(3 \times 6)}{6} + \frac{5}{6} = \frac{23}{6}$

Write the mixed number as a fraction.

21. $4\frac{2}{3}$ **22.** $2\frac{8}{9}$ **23.** $8\frac{1}{2}$

Write the fraction as a mixed number.

24. $\frac{17}{2}$ **25.** $\frac{40}{7}$ **26.** $\frac{33}{4}$

- **Convert among fractions, decimals, and percents.** (pp. 169–171)

  Write $\frac{1}{8}$ as a percent.

  $\frac{1}{8} = 1 \div 8 = 0.125$ *Change $\frac{1}{8}$ to a decimal.*

  $0.125 = 12.5\%$
  So, $\frac{1}{8} = 12.5\%$.

Write the decimal as a fraction.

27. 0.75   **28.** 0.6   **29.** 0.43

Write the fraction as a percent.

30. $\frac{5}{8}$ **31.** $\frac{2}{5}$ **32.** $\frac{5}{2}$

Chapter 9

- **Add and subtract fractions and mixed numbers.** (pp. 186–193)

  Subtract. $4\frac{1}{4} - 2\frac{2}{3}$

  $$
  \begin{aligned}
  4\frac{1}{4} &= 4\frac{3}{12} = 3\frac{15}{12} \\
  -2\frac{2}{3} &= -2\frac{8}{12} = -2\frac{8}{12} \\
  \hline
  & \qquad\qquad\quad\; 1\frac{7}{12}
  \end{aligned}
  $$

  Write equivalent fractions. Rename as needed.

  Subtract fractions. Subtract whole numbers.

Add or subtract. Write the answer in simplest form.

33. $\frac{3}{4} + \frac{1}{3}$ 34. $\frac{3}{5} + \frac{1}{2}$

35. $\frac{9}{10} - \frac{3}{5}$ 36. $\frac{5}{6} - \frac{1}{4}$

37. $2\frac{3}{8} + 3\frac{3}{4}$ 38. $4\frac{3}{5} + 2\frac{1}{3}$

39. $4\frac{1}{3} - 1\frac{5}{6}$ 40. $5\frac{1}{8} - 3\frac{2}{3}$

Chapter 10

- **Multiply and divide fractions and mixed numbers.** (pp. 202–213)

  Divide. $3\frac{3}{4} \div 4\frac{1}{2}$

  $$3\frac{3}{4} \div 4\frac{1}{2} = \frac{15}{4} \div \frac{9}{2}$$

  Write mixed numbers as fractions.

  $$= \frac{15}{4} \times \frac{2}{9}$$

  Multiply by the reciprocal.

  $$= \frac{5}{6}$$

Multiply or divide. Write the answer in simplest form.

41. $\frac{1}{3} \times \frac{2}{5}$ 42. $\frac{5}{8} \times \frac{7}{10}$

43. $\frac{5}{9} \div \frac{1}{3}$ 44. $9 \div \frac{4}{5}$

45. $3\frac{3}{4} \times 3\frac{1}{3}$ 46. $3\frac{1}{8} \times 1\frac{1}{5}$

47. $1\frac{2}{3} \div 3\frac{5}{6}$ 48. $1\frac{7}{9} \div 2\frac{2}{5}$

- **Evaluate an algebraic expression using fractions.** (pp. 216–217)

  Evaluate $f + 5\frac{1}{8}$ for $f = 2\frac{1}{3}$.

  $f + 5\frac{1}{8}$

  $2\frac{1}{3} + 5\frac{1}{8}$ *Replace f with $2\frac{1}{3}$.*

  $2\frac{8}{24} + 5\frac{3}{24} = 7\frac{11}{24}$ *Add.*

Evaluate each expression.

49. $d + 2\frac{2}{3}$ for $d = 2\frac{5}{6}$

50. $k \times \frac{3}{4}$ for $k = 1\frac{3}{5}$

51. $m - 3\frac{4}{9}$ for $m = 6\frac{5}{18}$

52. $w \div 1\frac{1}{3}$ for $w = 2\frac{1}{2}$

PROBLEM SOLVING APPLICATIONS

53. Katelynn has two sheets of paper $8\frac{1}{2}$ in. by 11 in. She cuts out a 3-in. square from the center top edge of one of the sheets. Which has the greater perimeter, the sheet with the square cut out or the uncut sheet? How much greater?

    (pp. 194–195)

54. Juan goes to the fruit market every 6 days. Anita goes to the market every 4 days. They meet at the market on August 31. When will they meet at the market again? (pp. 154–155)

Trees

California is home to some of the world's most famous trees. Here you will find the largest tree, the oldest tree, and the second-tallest tree. California also has the tallest individuals of many species of trees.

The General Sherman giant sequoia tree, in Sequoia National Park, is one of the largest trees in the world. Its weight is estimated to be equal to the combined weights of 740 elephants.

1. The world's tallest bluegum eucalyptus tree is located in Sonoma County, California. It is 165 ft tall. The General Sherman giant sequoia tree is 275 ft tall. Express the eucalyptus tree's height as a fraction of the General Sherman tree's height, in simplest form.

2. The diameter of the General Sherman tree is shown in the diagram. Write the diameter as a fraction.

$26\frac{12}{25}$ ft

Use Data   For 3, use the table.

3. The world's tallest grand fir, ponderosa pine, and sugar pine are found in California. The table shows how the heights of the three trees compare with that of the General Sherman tree. Order the trees from tallest to least tall.

HEIGHTS COMPARED TO THE GENERAL SHERMAN TREE	
Tallest grand fir	$\frac{19}{20}$
Tallest ponderosa pine	$\frac{41}{50}$
Tallest sugar pine	$\frac{21}{25}$

4. California's tallest tree, and the world's second tallest, is a coast redwood found in Redwoods State Park. It is 313 ft tall. Classify the number 313 as prime or composite.

5. The world's oldest tree is a 4,700-year-old bristlecone pine growing in California's White Mountains, near the town of Bishop. Write the prime factorization of 4,700.

Use Data   For 6, use the drawing.

6. The world's tallest incense cedar and the world's tallest sugar pine grow in California. The height of each tree is a multiple of the height of a redwood seedling. What is the tallest the redwood seedling could be?

232 ft
152 ft
redwood seedling   incense cedar   sugar pine

Joshua Tree National Park

The Joshua tree is one of the most unusual and easily recognized trees in the world. The Joshua tree is a member of the lily family. The best examples of this rare tree are found in Joshua Tree National Park, in Southern California.

The unusual spiny-leaved Joshua tree can grow up to 40 feet tall and live up to 1,000 years.

Use Data For 1–9, use the table.

1. Which trail is longer, Indian Cove or Oasis of Mara? How much longer?

2. One of the trails listed is half the length of the Ryan Mountain Trail. Which trail is it?

3. What is the combined length of the two longest trails listed?

4. How much longer than the Skull Rock Trail is the Mastodon Peak Trail?

5. Malcolm hiked the Boy Scout Trail at an average rate of $2\frac{2}{3}$ mph. How long did the hike take?

Mr. Ristorcelli has hiked the Lost Palms Oasis Trail every summer for the past 16 years.

6. How far has he hiked altogether?

7. How many times would he have to hike the 49 Palms Oasis Trail in order to cover the same total distance?

8. Kendra hiked four trails for a total of $11\frac{3}{4}$ mi. Three of the trails she hiked are Mastodon Peak, High View, and Ryan Mountain. What is the name of the fourth trail Kendra hiked?

9. The Lost Horse Mine trail is $8\frac{1}{2}$ times as long as the trail Mesha and her brother hiked. Which trail did Mesha and her brother hike?

| TRAILS IN JOSHUA TREE NATL. PARK ||
Trail	Approx. Length (mi)
Boy Scout	16
49 Palms Oasis	$3\frac{3}{5}$
High View	$1\frac{3}{10}$
Indian Cove	$\frac{3}{5}$
Lost Horse Mine	$4\frac{1}{4}$
Lost Palms Oasis	$7\frac{7}{8}$
Mastodon Peak	3
Oasis of Mara	$\frac{1}{2}$
Ryan Mountain	$3\frac{1}{5}$
Skull Rock	$1\frac{3}{5}$

11 Number Relationships

The elevation of Mt. McKinley is 20,320 feet. It is the highest mountain in North America and has one of Earth's steepest vertical rises. With the high altitude, unpredictable weather, and steep, icy slopes, this mountain is a challenge to climb. Climbers are flown into base camp, where they start their climb at 7,200 ft. If a climb takes 20 days, what is the average altitude gained per day?

Mt. McKINLEY		
Camp	Altitude	Possible Time
Base Camp	7,200 ft	0 day
Camp 2	9,500 ft	1st day
Camp 3	11,000 ft	3rd day
Camp 4	14,200 ft	10th day
Camp 5	16,200 ft	12th day
High Camp	17,200 ft	15th day
Summit	20,320 ft	20th day

Check What You Know

Use this page to help you review and remember important skills needed for Chapter 11.

✓ Vocabulary

Choose the best term from the box.

> positive numbers
> negative numbers
> whole numbers

1. Numbers to the left of zero on the number line are ___?___ .

2. Numbers greater than zero are ___?___ .

✓ Locate Points on a Number Line (See p. H13.)

Copy each number line. Graph the numbers on the number line.

3. 4   4. 0   5. ⁻5   6. ⁻1

7. 0.5   8. $\frac{-1}{2}$ 9. ⁻0.75   10. $1\frac{1}{4}$

✓ Sets of Numbers (See p. H12.)

Give four examples from each set of numbers.

11. whole numbers   12. counting numbers   13. odd numbers

✓ Compare Fractions (See p. H12.)

Compare. Write <, >, or = for each ●.

14. $9\frac{1}{2}$ ● $9\frac{3}{4}$ 15. $\frac{2}{3}$ ● $\frac{3}{2}$ 16. $\frac{3}{4}$ ● $\frac{3}{5}$

17. $\frac{16}{2}$ ● $\frac{24}{3}$ 18. $2\frac{3}{5}$ ● $2\frac{1}{4}$ 19. $\frac{12}{2}$ ● $2\frac{1}{12}$

✓ Temperature (See p. H12.)

Tell the temperature shown by each letter on the thermometer.

20. A   21. B   22. C

23. D   24. E   25. F

> ## LOOK AHEAD
>
> **In Chapter 11 you will**
> - identify integers and rational numbers
> - compare and order rational numbers
> - use number lines

Understand Integers

Learn how to identify integers and find absolute value.

Death Valley is the country's hottest and driest spot.

Vocabulary

integers
opposites
positive integers
negative integers
absolute value

Mount McKinley is 20,320 ft above sea level. Death Valley is 282 ft below sea level. Sea level equals 0 ft. You can use the integers $^+$20,320 and $^-$282 to represent these elevations.

Integers include all whole numbers and their **opposites**. Each integer has an opposite that is the same distance from 0 but on the opposite side of 0. The opposite of positive 8 ($^+$8) is negative 8 ($^-$8). The opposite of 0 is 0.

Integers greater than 0 are **positive integers**. Integers less than 0 are **negative integers**. The integer 0 is neither positive nor negative.

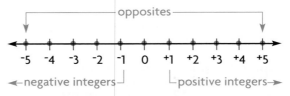

Negative integers are written with a negative sign, $^-$.

Positive integers can be written with or without a positive sign, $^+$.

EXAMPLE 1

Name an integer to represent the situation.
A. a gain of 12 yd   **B.** 30° below zero   **C.** a deposit of $100
      $^+$12 $^-$30° $^+$100

The **absolute value** of an integer is its distance from 0. Look at $^+$3 and $^-$3. They are both 3 units from 0.

Write: $|^-3| = 3$ **Read:** The absolute value of negative three is three.
Write: $|^+3| = 3$ **Read:** The absolute value of positive three is three.

EXAMPLE 2

Use the number line to find each absolute value.
A. $|^-4|$ **B.** $|^+2|$ **C.** $|^-1|$ **D.** $|^+3|$
      4   2   1   3

CALIFORNIA STANDARDS MR 2.4 Use a variety of methods, such as words, numbers, symbols, charts, graphs, tables, diagrams, and models, to explain mathematical reasoning. *also,* MR 2.2, MR 3.0, MR 3.2, MR 3.3

Think and ▶
Discuss

Look back at the lesson to answer each question.

1. **Give an example** of a scale in real life in which zero is used together with other integers.

2. **Tell** what the absolute value of an integer is.

Guided ▶
Practice

Write an integer to represent each situation.

3. 350 ft below sea level

4. an increase of 78 points

5. 14 degrees below zero

Write the opposite integer.

6. $^-289$ 7. $^-25$ 8. $^+315$ 9. $^+742$ 10. $^+993$

PRACTICE AND PROBLEM SOLVING

Independent ▶
Practice

Write an integer to represent each situation.

11. a $5.00 decline in value   12. an increased attendance of 477

13. a gain of 12,000 ft in altitude   14. a decrease of 50 points

Write the opposite integer.

15. $^-2$ 16. $^-14$ 17. $^-31$ 18. $^+88$ 19. $^+207$

Find the absolute value.

20. $|^+390|$ 21. $|^-28|$ 22. $|^-727|$ 23. $|^+660|$ 24. $|^+795|$

Problem Solving ▶
Applications

25. The elevation of the Dead Sea is about 1,310 ft below sea level. Write the elevation using an integer.

26. **REASONING** What values can n have if $|n| = 5$?

27. Chris has a stack of coins containing 5 quarters, 5 dimes, and 15 nickels. What fraction of the number of coins are quarters?

28. **Write a problem** about two campers who camped at altitudes of 9,470 ft and 7,200 ft. Use positive and negative integers to describe the change in elevation.

The Dead Sea

MIXED REVIEW AND TEST PREP

29. Multiply. $\frac{4}{5} \times \frac{3}{4}$ (p. 202)

30. Lisa works for $6.50 an hour, 3 hours a day, 4 days a week. How many weeks will it take her to earn more than $300? (p. 76)

31. Find the value of the expression $9 \times (12 - 4) \div 2^3 + 4$ (p. 44)

32. Write the prime factorization of 72. (p.148)

33. **TEST PREP** Six soccer teams compete for the regional play-offs. Each team plays each of the other teams only once. How many games do they play? (p. 154)

A 6   **B** 12   **C** 15   **D** 30

Rational Numbers

Learn how to classify rational numbers and find another rational number between two rational numbers.

Vocabulary

ratio
rational number
Venn diagram

QUICK REVIEW

Write as a decimal and as a fraction.

1. eight tenths   **2.** fifty-four hundredths   **3.** three tenths

4. nineteen hundredths   **5.** forty thousandths

A **ratio** is a comparison of two numbers, a and b, written as a fraction $\frac{a}{b}$. A **rational number** is any number that can be written as a ratio $\frac{a}{b}$, where a and b are integers and $b \neq 0$. The numbers below are all rational numbers since they can be expressed as a ratio $\frac{a}{b}$.

$$3\frac{2}{5} \qquad 0.6 \qquad 42 \qquad {}^-2.5$$

Write each rational number as a ratio $\frac{a}{b}$.

EXAMPLE 1

A. $3\frac{2}{5}$ **B.** 0.6 **C.** 42 **D.** $^-2.5$

$3\frac{2}{5} = \frac{17}{5}$ $0.6 = \frac{6}{10}$ $42 = \frac{42}{1}$ $^-2.5 = \frac{^-5}{2}$

The **Venn diagram** shows how the sets of rational numbers, integers, and whole numbers are related.

```
┌─────────────────────────────┐
│      Rational Numbers        │
│  ┌───────────────────────┐   │
│  │       Integers        │   │
│  │  ┌─────────────────┐  │   │
│  │  │  Whole Numbers  │  │   │
│  │  └─────────────────┘  │   │
│  └───────────────────────┘   │
└─────────────────────────────┘
```

EXAMPLE 2

Use the Venn diagram to determine in which set or sets each number belongs.

A. 80   The number 80 belongs in the sets of whole numbers, integers, and rational numbers.

B. $^-2$ The number $^-2$ belongs in the sets of integers and rational numbers but not in the set of whole numbers.

C. $6\frac{1}{2}$ The number $6\frac{1}{2}$ belongs in the set of rational numbers but not in the set of integers or the set of whole numbers.

D. 7.09   The number 7.09 belongs in the set of rational numbers but not in the set of integers or the set of whole numbers.

• Name two integers that are not also whole numbers.

CALIFORNIA STANDARDS ⊙━ᴨ NS 1.0 Students compare and order positive and negative fractions, decimals, and mixed numbers. Students solve problems involving fractions, ratios, proportions, and percentages. ⊙━ᴨ NS 1.1 Compare and order positive and negative fractions, decimals, and mixed numbers and place them on a number line. *also,* **MR 1.0, MR 2.0, MR 3.0, MR 3.2, MR 3.3**

Christopher is training to run in a 5-km road race. Yesterday he ran $4\frac{1}{4}$ km, and he plans to run $4\frac{1}{2}$ km tomorrow. What distance could he run today if he wants to run between $4\frac{1}{4}$ km and $4\frac{1}{2}$ km?

Think of the distances of Christopher's training runs as rational numbers.

One Way You can use a number line to find numbers between two rational numbers.

EXAMPLE 3

Find a distance between $4\frac{1}{4}$ km and $4\frac{1}{2}$ km, using the number line.

Notice that there is a mark between $4\frac{1}{4}$ and $4\frac{1}{2}$. That could be a distance Christopher could run.

So, Christopher could run $4\frac{3}{8}$ km today.

Another Way You can use a common denominator to find a number between two rational numbers.

EXAMPLE 4

Find a number between $4\frac{1}{4}$ and $4\frac{1}{2}$.

$4\frac{1}{4} = 4\frac{2}{8}$ $4\frac{1}{2} = 4\frac{4}{8}$ *Use a common denominator to write equivalent fractions.*

$4\frac{3}{8}$ is between $4\frac{2}{8}$ and $4\frac{4}{8}$.   *Find a rational number between the two numbers.*

So, $4\frac{3}{8}$ is between $4\frac{1}{4}$ and $4\frac{1}{2}$.

- What are some common denominators you could use to find other rational numbers between $4\frac{1}{4}$ and $4\frac{1}{2}$?

You can also find a number between two rational numbers in decimal form.

EXAMPLE 5

Find a rational number between ⁻8.4 and ⁻8.5.

⁻8.4 = ⁻8.40   *Add a zero to each decimal.*

⁻8.5 = ⁻8.50

Use a number line to find a number between the two decimals.

So, ⁻8.43, ⁻8.45, and ⁻8.48 are some of the numbers between ⁻8.4 and ⁻8.5.

Remember that you can add any number of zeros to the right of a decimal without changing its value.

CHECK FOR UNDERSTANDING

Think and ▶ Look back at the lesson to answer each question.
Discuss

1. **Tell** why every integer is a rational number. Give an example to support your answer.

2. **Tell** what numbers would be between $4\frac{1}{4}$ and $4\frac{1}{2}$ if you divided a number line into sixteenths.

Guided ▶ Write each rational number in the form $\frac{a}{b}$.
Practice

3. $^{-}0.37$ 4. $2\frac{4}{5}$ 5. 0.889 6. 7.31 7. $^{-}7\frac{1}{3}$

Use the number line to find a rational number between the two given numbers.

8. 1 and $1\frac{1}{2}$ 9. $^{-}\frac{3}{4}$ and $^{-}\frac{1}{4}$ 10. $\frac{1}{2}$ and 1 11. $^{-}\frac{1}{2}$ and $\frac{1}{2}$

PRACTICE AND PROBLEM SOLVING

Independent ▶ Write each rational number in the form $\frac{a}{b}$.
Practice

12. $9\frac{2}{3}$ 13. $^{-}0.71$ 14. 80.4 15. $^{-}2\frac{5}{8}$ 16. 3.18

Use the number line to find a rational number between the two given numbers.

17. $^{-}0.2$ and $^{-}0.4$ 18. 0 and 0.2 19. $^{-}0.6$ and $^{-}0.8$

Find a rational number between the two given numbers.

20. $\frac{3}{4}$ and $\frac{1}{2}$ 21. $^{-}7$ and $^{-}\frac{15}{2}$ 22. 104.1 and $103\frac{7}{8}$

23. 16.1 and 16.01 24. $3\frac{5}{8}$ and $\frac{27}{8}$ 25. $^{-}\frac{3}{4}$ and $^{-}\frac{3}{8}$

Tell if the first rational number is between the second and third rational numbers. Write *yes* or *no*.

26. $\frac{1}{3}; \frac{1}{2}$ and $\frac{3}{4}$ 27. $0.97; 0.85$ and 0.99 28. $3.29; 3.20$ and 3.25

29. $3.07; 3.1$ and 3.01 30. $^{-}8; \frac{^{-}29}{4}$ and $\frac{^{-}33}{4}$ 31. $^{-}\frac{1}{8}; \frac{^{-}1}{16}$ and $^{-}\frac{1}{4}$

Problem Solving ▶ 32. Marie had already completed $\frac{1}{2}$ of the running race but had not
Applications   yet reached the point $\frac{3}{4}$ of the way through the race. Could she have completed $\frac{5}{8}$ of the race? Explain.

33. Susan follows the directions of a treasure map and walks 45 steps north, then $\frac{24}{3}$ steps east, $\frac{75}{5}$ steps south, and 22 steps west. How many steps does Susan walk?

34. Is it easier to find a rational number between $\frac{1}{2}$ and $\frac{3}{4}$ or between 0.50 and 0.75? Explain your reasoning.

35. **? What's the Error?** Jeff says that every whole number is an integer and that every integer is a whole number. Explain his error.

MIXED REVIEW AND TEST PREP

36. Find the absolute value. $|^-88|$ (p. 228)

37. Write the decimal and fraction equivalents of 34%. (p. 169)

38. $\frac{2}{5} + \frac{4}{10} + \frac{4}{5}$ (p. 182)

39. **TEST PREP** Which is the range of the number set 28, 8, 13, 20, and 24? (p. 106)

    A 4   **B** 8   **C** 15   **D** 20

40. **TEST PREP** Which is the least common multiple of 3, 5, 9, and 15? (p. 150)

    F 15   **G** 30   **H** 45   **J** 90

 to Reading

Strategy • Use Graphic Aids Graphic aids such as Venn diagrams, charts, and tables provide specific or important information in a visual form rather than in text. Sometimes, the information needed to solve a problem may be provided only in a graphic aid.

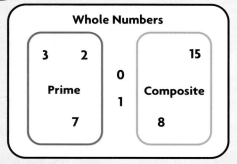

Look at the Venn diagram to the right. It shows the relationships among whole, prime, and composite numbers.

• Is a composite number always, sometimes, or never a whole number?

• Is a whole number always, sometimes, or never a prime number?

Use the Venn diagram to solve the following problems.

1. How would you describe the numbers 7 and 15? Are they whole numbers? Are they prime or composite?

2. Is 1 a whole number? Is it prime or composite? Explain.

Extra Practice page H42, Set B

Compare and Order Rational Numbers

Learn how to compare and order rational numbers.

QUICK REVIEW

Order the numbers from greatest to least.

1. 12, 14, 10

2. ⁻10, ⁻21, ⁻20

3. 3.10, 3.05, 3.15

4. $\frac{1}{4}, \frac{1}{2}, \frac{3}{4}$

5. $\frac{3}{5}, \frac{3}{6}, \frac{3}{4}$

Temperature commonly is measured on a scale in units of degrees. The scale contains both negative and positive numbers, like a number line. For example, the temperature in Death Valley has reached a high of 132°F, and the record low in Alaska is ⁻80°F.

You can use a number line to compare integers. On a number line, each number is greater than any number to its left and less than any number to its right. The number line below shows that ⁻80° < 132° and 132° > ⁻80°.

⁻150° ⁻120° ⁻90° ⁻60° ⁻30° 0° 30° 60° 90° 120° 150°

EXAMPLE 1

Compare the integers. Use < and >. Think about their positions on a number line.

A. 2 and ⁻3
⁻3 is to the left of 2 on the number line.
So, ⁻3 < 2, or 2 > ⁻3.

B. ⁻2 and ⁻4
⁻2 is to the right of ⁻4 on the number line.
So, ⁻2 > ⁻4, or ⁻4 < ⁻2.

It is easier to compare and order rational numbers when they are all expressed as decimals or as fractions with a common denominator.

EXAMPLE 2

Order $8\frac{1}{4}$, ⁻$8\frac{1}{2}$, and 8.3 from least to greatest.

Since ⁻$8\frac{1}{2}$ is the only negative number, it is the least number.

Compare $8\frac{1}{4}$ and 8.3.

$8\frac{1}{4} = 8.25$ and $8.3 = 8.30$ — *Write the numbers as decimals with the same number of decimal places.*

$8.25 < 8.30$, or $8\frac{1}{4} < 8.3$ — *Compare by looking at the place values.*

⁻$8\frac{1}{2} < 8\frac{1}{4} < 8.3$ — *Order the three numbers.*

So, from least to greatest the numbers are ⁻$8\frac{1}{2}$, $8\frac{1}{4}$, and 8.3.

CALIFORNIA STANDARDS ○━ᴨ **NS 1.0** Students compare and order positive and negative fractions, decimals, and mixed numbers. Students solve problems involving fractions, ratios, proportions, and percentages. *also,* **MR 1.0, MR 2.0, MR 2.4, MR 3.0**

Think and ▶ Discuss

Look back at the lesson to answer each question.

1. Describe how you would compare 2.62 and $2\frac{3}{5}$.

2. Give an example of a rational number that is greater than $^-2.5$ and one that is less than $^-2.5$.

Guided ▶ Practice

Compare. Write $<$, $>$, or $=$ for each ●.

3. $^-1.5$ ● $^-0.5$ **4.** 0.5 ● $^-1.0$ **5.** $1\frac{1}{4}$ ● 1.5 **6.** $^-2$ ● $^-1\frac{1}{2}$

Independent ▶ Practice

Compare. Write $<$, $>$, or $=$ for each ●.

7. $\frac{1}{2}$ ● $\frac{3}{4}$ **8.** $\frac{1}{4}$ ● $^-\frac{1}{2}$ **9.** 0.5 ● $\frac{3}{8}$ **10.** $^-\frac{1}{4}$ ● 0.25

11. 1.25 ● 1.75 **12.** $^-\frac{1}{4}$ ● $^-\frac{1}{3}$ **13.** 2 ● $^-3$ **14.** $\frac{4}{5}$ ● 0.9

15. $3.2 + 4.4$ ● $4\frac{3}{4} + 2\frac{3}{4}$ **16.** $2\frac{3}{4} \times 4$ ● $3\frac{1}{4} + 8.5$

Order the rational numbers from least to greatest.

17. $0.6, 0.4, 0.46$ **18.** $^-\frac{1}{8}, ^-\frac{1}{2}, ^-\frac{3}{8}, ^-\frac{3}{4}$ **19.** $^-0.2, \frac{1}{4}, 0, ^-\frac{1}{4}$

Problem Solving ▶ Applications

20. Lynda's times for running a mile are $5\frac{1}{2}$ min, 5.48 min, 5.51 min, and $5\frac{2}{5}$ min. What is the longest she has taken to run a mile?

21. The mean temperatures for three days were $^-3°$, $^-5°$, and $1°$. Order the temperatures from highest to lowest.

22. ✎ **Write About It** Explain how you would order three numbers that include a positive fraction and decimal and a negative integer.

23. Write an integer to represent 450 ft below sea level. (p. 228)

24. Round 2.0955 to the nearest thousandth. (p. 58)

25. Write $3\frac{3}{7}$ as a fraction. (p. 164)

26. $\frac{7}{8} - \frac{1}{2}$ (p. 182)

27. TEST PREP The swim team practiced for 8 hours the first week, $1\frac{1}{4}$ times as long the second week, 12 hours the third week, and $\frac{24}{4}$ hours the fourth week. How many hours total did the swim team practice? (p. 206)

A 26 **B** 28 **C** 36 **D** 38

PROBLEM SOLVING STRATEGY
Use Logical Reasoning

- Analyze
- Choose
- Solve
- Check

Learn how to solve problems by using logical reasoning.

Stephen, Leticia, Shelley, and Ed emptied their banks. They found $7.75, $4.35, $5.00, and $10.00, but not necessarily in that order. Leticia has twice as much money as Shelley. Stephen has an amount between Shelley's and Leticia's. Who has $4.35?

Analyze   What are you asked to find?

What information are you given?

Choose   What strategy will you use?

You can *use logical reasoning*.

Solve   How will you solve the problem?

Take the clues one at a time. Use a table to help.

Only one box in each row and column can have a "yes."

	$4.35	$5.00	$7.50	$10.00
Stephen	no	no	yes	no
Leticia	no	no	no	yes
Shelley	no	yes	no	no
Ed	yes	no	no	no

Leticia has twice as much money as Shelley. So, Leticia must have $10.00 and Shelley $5.00. Fill in "yes" in those boxes, and fill in "no" in the rest of the boxes in those rows and columns.

Stephen has an amount between Shelley's and Leticia's. $7.50 is between $5.00 and $10.00, so Stephen must have $7.50. Fill in the rest of the boxes with "yes" and "no."

So, Ed has $4.35.

Check   How can you check your answer?

What if the amounts were $7.50, $4.35, $5.00, and $12.50? How could you change the clues in the problem?

CALIFORNIA STANDARDS **MR 2.0** Students use strategies, skills, and concepts in finding solutions. **MR 3.2** Note the method of deriving the solution and demonstrate a conceptual understanding of the derivation by solving similar problems. *also,* **MR 1.0, MR 3.0, MR 3.3**

PROBLEM SOLVING PRACTICE

Solve the problems by using logical reasoning.

1. Arthur, Victoria, and Jeffrey are in sixth, seventh, and eighth grades, although not necessarily in that order. Victoria is not in eighth grade. The sixth grader is in chorus with Arthur and in band with Victoria. Which student is in each grade?

2. Use the following information to tell which numbers in the box at the right are *A, B, C, D,* and *E.*

$^-3.5$	$2\frac{1}{2}$
0.43	
4.3	$^-0.43$

   - *A* is greater than *D* and less than *C*.
   - *A* and *D* are opposites.
   - *E* is the greatest number.

For 3–4, use this information.

Amir, Katherine, Patrick, and Lee are comparing their stamp collections. Lee has twice as many stamps as Amir. Patrick has 5 fewer stamps than Katherine, who has 5 fewer stamps than Lee. Their collections consist of 15, 20, 25, and 30 stamps.

3. If Amir has 15 stamps, how many stamps does Patrick have?

   A 30   **C** 20
   B 25   **D** 15

4. The four have a total of 90 stamps. How many stamps does Katherine have?

   F 30   **H** 20
   G 25   **J** 15

MIXED STRATEGY PRACTICE

5. Kathy, Arlene, and Helene are in line for concert tickets, though not necessarily in that order. Helene is behind Kathy. Kathy is not first in line. Tell the order of the three in line.

6. Mel sold 25 beanbag toys at a fair. He sold some for $8 and some for $5. He made $170. How many toys did he sell for $8?

7. A guidebook costs $3.75 in Canadian dollars in a Vancouver, B.C., bookstore. It also has a price of $3.50 in United States dollars. If the exchange rate is $1.12 Canadian dollars for each U.S. dollar, would you rather pay in Canadian or U.S. dollars?

8. Shelters are located at regular intervals along a wilderness trail. The distance from the first shelter to the fourth shelter is 6 mi. What is the distance from the seventh shelter to the thirteenth shelter?

9. Sandy has to be at the airport for a 5:20 P.M. flight. She wants to arrive 1 hr 15 min early. If it takes 50 min to drive to the airport, when should she leave for the airport?

10. **? What's the Question?** Anne bought 28 fish. She bought three times as many goldfish as angelfish. The answer is 7.

PROBLEM SOLVING STRATEGIES

Draw a Diagram or Picture
Make a Model
Predict and Test
Work Backward
Make an Organized List
Find a Pattern
Make a Table or Graph
Solve a Simpler Problem
Write an Equation
▶ **Use Logical Reasoning**

237

1. **VOCABULARY** Positive whole numbers, their opposites, and 0 make up the set of ___?___ . (p. 228)

2. **VOCABULARY** A number that can be written as a ratio $\frac{a}{b}$, where a and b are integers and $b \neq 0$, is a(n) ___?___ . (p. 230)

Write an integer to represent each situation. (pp. 228–229)

3. an increase of 15 points

4. 6 degrees below zero

5. a loss of 20 pounds

Write the opposite integer. (pp. 228–229)

6. $^-32$

7. 12

8. $^-15$

9. $^-289$

10. 0

Find the absolute value. (pp. 228–229)

11. $|^-12|$

12. $|^-4|$

13. $|^+17|$

14. $|^+8|$

15. $|^-347|$

Write each rational number in the form $\frac{a}{b}$. (pp. 230–233)

16. 2

17. $^-0.89$

18. $3\frac{2}{3}$

19. 5.4

20. 14

21. 0.334

Find a rational number between the two given numbers. (pp. 230–233)

22. $\frac{1}{4}$ and $\frac{2}{3}$

23. 1.3 and 1.32

24. $\frac{2}{5}$ and $\frac{3}{4}$

25. $^-4.3$ and $^-4.4$

26. 3.4 and 3.52

27. $2\frac{1}{5}$ and $2\frac{1}{2}$

28. $3\frac{5}{8}$ and $3\frac{9}{10}$

29. 0.9 and 0.94

Compare. Write $<$, $>$, or $=$ for each ●. (pp. 234–235)

30. $7\frac{5}{8}$ ● $7\frac{10}{16}$ =

31. 1.01 ● 1.10

32. $^-3.4$ ● $^-4.3$

33. $\frac{^-25}{3}$ ● $\frac{^-17}{2}$

34. $4\frac{3}{4}$ ● 4.77

Order the rational numbers from greatest to least. (pp. 234–235)

35. 3.7, 3.2, $3\frac{5}{8}$

36. $^-8$, $^-3$, $\frac{^-77}{11}$

37. $^-1.2$, 1.2, 0.12

38. $\frac{5}{7}$, $\frac{9}{14}$, $\frac{11}{8}$

39. Elizabeth, Doria, Emily, and Claudia each earned money doing odd jobs. They earned $4.50, $6.50, $8.00, and $9.00. Doria earned twice as much as Elizabeth. Emily earned $1.50 more than Claudia. Claudia earned $2.00 more than Elizabeth. How much did each person earn? (pp. 236–237)

40. Louisa, Chris, and Vicki each have one pet: a dog, a cat, and an iguana. Louisa does not have a cat and Chris does not have a dog. Louisa's pet is a reptile. What pet does each girl have? (pp. 236–237)

Look for important words.

See item **8.**

Important words are *least to greatest*. Write the numbers as decimals and order them from least to greatest.

Also see problem **2**, p. H62.

Choose the best answer.

1. Which of the following can be used to represent the depth of a cave that goes 37 feet below ground level?

   A $^-37$ **C** $^+3.7$

   B $^-3.7$ **D** $^+37$

2. Which points on the number line show opposite integers?

   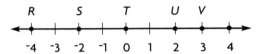

   F *R* and *S*   **H** *U* and *V*

   G *S* and *U*   **J** *T* and *R*

3. Four baseball players are standing in a line. Owen is standing between Paul and Jim. Casey is in front of Paul. Jim is last. How are they arranged in the line?

   A Paul, Casey, Owen, Jim

   B Casey, Paul, Jim, Owen

   C Owen, Casey, Jim, Paul

   D Casey, Paul, Owen, Jim

4. With respect to sea level, the average depth of the Atlantic Ocean is $^-11,730$ feet, the Pacific Ocean is $^-12,925$ feet, and the Gulf of Mexico is $^-5,297$ feet. Which shows these values in order from least to greatest?

   F $^-11,730, ^-12,925, ^-5,297$

   G $^-11,730, ^-5,297, ^-12,925$

   H $^-12,925, ^-11,730, ^-5,297$

   J $^-12,925, ^-5,297, ^-11,730$

5. Last week, the dance band practiced for $5\frac{1}{3}$ hours. The band usually practices for $8\frac{3}{4}$ hours per week. How many fewer hours did the band practice last week?

   A $2\frac{3}{4}$ hr   **C** $3\frac{5}{12}$ hr

   B $3\frac{1}{3}$ hr   **D** $4\frac{1}{5}$ hr

6. Earl read that $\frac{3}{8}$ of his town's budget is spent on road repairs. Which is that amount expressed as a decimal?

   F 0.375   **H** 0.67

   G 0.4   **J** 0.875

7. Which symbol makes this number sentence true?

   $$^-3 \; \bullet \; 0$$

   A $<$ **C** $=$

   B $>$ **D** $\geq$

8. Which shows the numbers in order from least to greatest?

   $$1.7, \frac{1}{2}, ^-0.3, \frac{^-3}{5}$$

   F $\frac{1}{2}, 1.7, \frac{^-3}{5}, ^-0.3$

   G $\frac{^-3}{5}, ^-0.3, 1.7, \frac{1}{2}$

   H $^-0.3, \frac{1}{2}, \frac{^-3}{5}, 1.7$

   J Not here

9. Which letter represents the opposite of 3 on the number line?

   A *A*   **B** *B*   **C** *C*   **D** *D*

10. Vicki has 8 mysteries, 6 biographies, and 10 adventure books. What part of her book collection is biographies?

   F $\frac{1}{8}$ **H** $\frac{1}{4}$

   G $\frac{1}{6}$ **J** $\frac{1}{3}$

CHAPTER 12

Add and Subtract with Integers

NASA's Pathfinder and Sojourner space probes were built to withstand extreme temperatures as they traveled from Earth to Mars. The south pole on Mars is made not of frozen water but of frozen carbon dioxide, which has a temperature of ⁻193°F. The coldest day recorded on Earth was ⁻129°F, in Antarctica. Which temperature is lower? By about how much?

Temperature	Earth	Mars
Lowest Recorded	⁻129°F	⁻220°F
Highest Recorded	136°F	68°F
Average	57°F	⁻81°F

Check What You Know

Use this page to help you review and remember important skills needed for Chapter 12.

 Vocabulary

Choose the best term from the box.

1. The set of whole numbers and their opposites is the set of __?__ .

> positive
> numbers
> integers

 Understand Integers (See p. H12.)

Write a positive or negative integer to represent each situation.

2. 14° below zero

3. 62 degrees above zero

4. 10 ft above sea level

5. 13 m below sea level

6. bottom of a well, 50 ft below the surface

7. a gain of 12 yards

8. a bank deposit of $280

9. 3 ft below ground level

 Number Lines (See p. H13.)

Name the integer that corresponds to the point.

10. A   11. B   12. C   13. D

Use the number line to find the ending point for each move.

	Starting Point	Move	Ending Point
14.	⁻8	4 to the right	
15.	⁻7	6 to the left	
16.	⁺6	9 to the left	
17.	⁻1	5 to the right	
18.	⁻12	3 to the left	
19.	⁺5	5 to the left	
20.	⁺3	7 to the left	

LOOK AHEAD

In Chapter 12 you will
- add integers
- subtract integers

ALGEBRA
Model Addition of Integers

MATH LAB

Explore how to use two-color counters to add integers.

You need two-color counters.

QUICK REVIEW

1. 40 + 50   **2.** 120 − 32

3. 19 + 21   **4.** 118 − 70

5. 120 + 130

Vocabulary

additive inverse

Lin and Nina are playing a board game. To keep track of points, they are using yellow counters to represent positive points, or points gained, and red counters to represent negative points, or points lost.

Activity 1

• Lin earned 6 points during the first round. She earned 3 points during the second round. Use yellow counters to model Lin's total score.

First-round points   Second-round points

$$6 + 3 = 9 \; \leftarrow \; \text{total score}$$

• Nina lost 2 points in the first round. Then she lost 5 points in the second round. Use red counters to model Nina's total score.

First-round points   Second-round points

$$^-2 + {}^-5 = {}^-7 \; \leftarrow \; \text{total score}$$

Think and Discuss

• How is adding the scores like adding whole numbers? How is it different?

• Would changing the order when adding Lin's or Nina's points change their scores? Why or why not?

• How would you model 2 + ⁻7? ⁻2 + ⁻7?

Practice

Use counters to find the sum.

1. 4 + 9   **2.** ⁻3 + ⁻7   **3.** ⁻6 + ⁻4   **4.** 5 + 8

CALIFORNIA STANDARDS ○━ⁿNS 2.3 Solve addition, subtraction, multiplication, and division problems, including those arising in concrete situations, that use positive and negative integers and combinations of these operations. *also,* ○━ⁿNS 2.0, MR 2.4, MR 2.5

The **additive inverse** of an integer is its opposite. 1 and ⁻1 are the additive inverses of each other. When you add an integer and its additive inverse, the sum is always 0. You can model this using counters.

- Model the sum of 1 and its additive inverse, ⁻1.

 = 0

- Model the sum of 5 and its additive inverse, ⁻5.

 = 0

- During a game, Carmen gained 8 points and then lost 5 points. To find her total score, Carmen paired points gained with points lost. Use yellow and red counters to model Carmen's total score. Remember that pairs of red and yellow counters equal 0.

8 + ⁻5 = 3

- Robert gained 3 points and then lost 7 points. Use yellow and red counters to model Robert's total score.

3 + ⁻7 = ⁻4

Think and Discuss

- Why is Carmen's score positive?

- Why is Robert's score negative?

Practice

Use counters to find each sum.

1. 4 + ⁻6   **2.** ⁻2 + 6   **3.** 7 + ⁻7   **4.** ⁻3 + 8

5. 5 + 2   **6.** 3 + 1   **7.** ⁻4 + ⁻5   **8.** ⁻3 + ⁻8

MIXED REVIEW AND TEST PREP

Order the rational numbers from least to greatest. (p. 234)

9. ⁻6.4, ⁻6.2, ⁻6.8

10. $\frac{1}{3}, \frac{2}{5}, \frac{1}{4}, \frac{3}{5}$

11. $-3\frac{4}{7}, -3.6, -3\frac{1}{2}$

12. Evaluate the expression $c + \frac{1}{2}$ for $c = \frac{1}{3}$. (p. 216)

13. TEST PREP Which shows the difference $6\frac{5}{6} - 5\frac{3}{4}$? (p. 186)

A $\frac{1}{9}$ **B** $1\frac{1}{12}$ **C** $1\frac{1}{2}$ **D** $1\frac{3}{4}$

ALGEBRA
Add Integers

Learn how to use a number line to add integers.

Jeb and Raul made up a game using a number line. Play starts at 0. A spinner is used to show positive moves and negative moves.

Jeb's first spin was ⁻3, and his second spin was ⁻2. What is Jeb's position on the number line?

⁻3 + ⁻2 = ⁻5 ← Jeb is at ⁻5.

Remember that you can write a positive number without the + sign.

⁺7 = 7

Raul's first spin was 4, and his second spin was ⁻9. Where is Raul on the number line?

4 + ⁻9 = ⁻5 ← Raul is at ⁻5.

You can use a number line to find the sum of two integers.

EXAMPLE 1

Use a number line to find the sum 4 + ⁻6.

Draw a number line.

Start at 0. Move 4 units to the right to show 4.

From 4, move 6 units to the left to show ⁻6. This takes you to ⁻2.

So, 4 + ⁻6 = ⁻2.

• When integers are added on a number line, when do the arrows point in the same direction and when do the arrows point in different directions?

CALIFORNIA STANDARDS O⁻ⁿNS 2.3 Solve addition, subtraction, multiplication, and division problems, including those arising in concrete situations, that use positive and negative integers and combinations of these operations. *also,* O⁻ⁿNS 2.0, MR 2.2, MR 2.4, MR 3.2

Remember that the absolute value of an integer is its distance from 0 on the number line.

Adding with the Same Sign
When adding integers with the same sign, add the absolute values of the integers. Use the sign of the addends for the result.

EXAMPLE 2

Find the sum $^-7 + {}^-2$.

$^-7 + {}^-2$

$|^-7| + |^-2| = 7 + 2$ *Add the absolute values of the integers.*

$\quad\quad\quad = 9$

So, $^-7 + {}^-2 = {}^-9$.   *Use the sign of the original addends.*

Adding with Different Signs
When adding integers with different signs, subtract the lesser absolute value from the greater absolute value. Use the sign of the addend with the greater absolute value for the result.

EXAMPLE 3

A. Find the sum $^-8 + 3$.

$^-8 + 3$

Subtract the lesser absolute value from the greater absolute value.

$|^-8| - |3| = 8 - 3$

$\quad\quad\quad = 5$

Use the sign of the addend with the greater absolute value.

$|^-8| > |3|$ *The sum is negative.*

So, $^-8 + 3 = {}^-5$.

B. Find the sum $^-5 + 9$.

$^-5 + 9$

Subtract the lesser absolute value from the greater absolute value.

$|9| - |^-5| = 9 - 5$

$\quad\quad\quad = 4$

Use the sign of the addend with the greater absolute value.

$|9| > |^-5|$ *The sum is positive.*

So, $^-5 + 9 = 4$.

• Find the sum $9 + {}^-12$.

EXAMPLE 4

On the first play of a football game, the Cobras gained 21 yards. On the second play, they lost 9 yards. Find the total number of yards gained or lost by the Cobras on the first two plays.

$21 + {}^-9$ *Use 21 for yards gained and $^-9$ for yards lost.*

$|21| - |^-9| = 21 - 9 = 12$ *Subtract the lesser absolute value from the greater absolute value.*

$|21| > |^-9| \rightarrow 21 + {}^-9 = 12$ *Use the sign of the addend with the greater absolute value.*

So, the Cobras gained a total of 12 yards on the first two plays.

CHECK FOR UNDERSTANDING

Think and Discuss ▶ Look back at the lesson to answer each question.

1. **Explain** how you determine the sign of the sum of two integers with the same sign.

2. **Explain** how you determine if the sum of two integers with different signs is positive or negative.

3. **Tell** how you know the Cobras had a gain of 12 yards instead of a loss of 12 yards in Example 4.

Guided Practice ▶ Write the addition problem modeled on the number line.

4.

5.

Find the sum.

6. $^-9 + 6$ **7.** $^-3 + ^-4$ **8.** $^-8 + 2$ **9.** $5 + ^-7$

10. $^-3 + 7$ **11.** $^-8 + ^-2$ **12.** $11 + ^-5$ **13.** $6 + ^-6$

PRACTICE AND PROBLEM SOLVING

Independent Practice ▶ Write the addition problem modeled on the number line.

14.

15.

Find the sum.

16. $^-5 + 8$ **17.** $2 + ^-3$ **18.** $7 + 2$ **19.** $^-1 + 4$

20. $8 + 7$ **21.** $^-12 + ^-8$ **22.** $^-15 + ^-10$ **23.** $^-17 + 25$

24. $^-2 + 5$ **25.** $^-12 + ^-16$ **26.** $^-17 + 5$ **27.** $25 + ^-37$

28. $24 + 12$ **29.** $30 + ^-41$ **30.** $|16| + |^-9|$ **31.** $|^-64| + |36|$

Algebra Use mental math to find the value of x.

32. $^-5 + x = ^-7$ **33.** $x + ^-6 = ^-13$

34. $x + ^-10 = ^-4$ **35.** $^-8 + x = ^-3$

Problem Solving ▶
Applications

36. In the morning the temperature was ⁻6°F. By noon it had risen 11°F. What was the temperature at noon?

37. In the evening the temperature was ⁻9°F. By midnight it had dropped 3°F. What was the temperature at midnight?

38. Tina played a game in which she owned stock worth $33. During the game, the stock increased $11 in value and then decreased $15 in value. Write an addition sentence to find the new value of the stock.

39. On the first three plays of a football game, the Wildcats gained 15 yards, lost 9 yards, and lost 8 yards. Find the total number of yards gained or lost by the Wildcats on the first three plays.

40. **What's the Error?** Ken says that ⁻6 + 2 = 8. What is his error? What is the correct sum?

41. **Geometry** A rectangle with an area of 180 cm² has a length of 15 cm. Find the width.

MIXED REVIEW AND TEST PREP

42. Is ⁻1.5 less than, greater than, or equal to ⁻$1\frac{3}{6}$? (p. 234)

43. Is $8\frac{3}{5}$ less than, greater than, or equal to 8.5? (p. 234)

44. Is ⁻3.4 less than, greater than, or equal to ⁻$3\frac{1}{5}$? (p. 234)

45. **TEST PREP** Find the quotient $4\frac{2}{3} \div 3\frac{1}{2}$. (p. 210)

A $\frac{3}{4}$ **B** $1\frac{1}{4}$ **C** $1\frac{1}{3}$ **D** $1\frac{1}{2}$

46. **TEST PREP** Find the sum $5\frac{3}{4} + 6\frac{1}{3}$. (p. 186)

F $11\frac{1}{12}$ **G** $11\frac{4}{7}$ **H** $12\frac{1}{12}$ **J** $12\frac{11}{12}$

Math Fun • Opposites Distract Remember that every number has an opposite that is the same distance from zero but is on the opposite side on the number line. Use a number line to solve these riddles.

1. I am the opposite of a number between $2\frac{2}{3}$ and $3\frac{5}{6}$.

2. I am the first integer that is less than the opposite of a number between $2\frac{1}{2}$ and $2\frac{3}{4}$.

3. I am the opposite of a number between ⁻4.3 and ⁻$4\frac{3}{8}$.

4. We are between ⁻2.4 and ⁻2.6. If you add our opposites, you get 5.

5. I am the opposite of the integer between the sum ⁻5 + 2 and the sum ⁻15 + 10.

6. I am the second integer that is less than the sum ⁻22 + ⁻17.

7. I am the opposite of the even integer between the sum 35 + ⁻14 and the sum 12 + 6.

8. I am the integer that is 4 times the sum 42 + ⁻35.

ALGEBRA
Model Subtraction of Integers

MATH LAB

Explore how to use two-color counters to subtract integers.

You need two-color counters.

QUICK REVIEW

1. 17 − 12   **2.** 42 + 20
3. 224 − 19   **4.** 132 − 18
5. 89 + 19

You can use red and yellow counters to subtract integers. Subtracting integers is similar to subtracting whole numbers.

Activity 1

• Find ⁻9 − ⁻4. First, make a row of 9 red counters.

• Then, take away 4 of them.

     → ⁻9 − ⁻4

• How many counters are left? What is ⁻9 − ⁻4?

Using red and yellow counters, model ⁻7 − 5.

• First, make a row of 7 red counters.

• Recall that a red counter paired with a yellow counter equals 0. Adding a red counter paired with a yellow counter does not change the value of ⁻7. Show another way to model ⁻7 that includes 5 yellow counters.

• Use your model to find ⁻7 − 5. Take away 5 yellow counters.

• What does your model show now? What is ⁻7 − 5?

Now model 7 − ⁻2.

• Model 7. Put down pairs of yellow and red counters until you can take away ⁻2. What is 7 − ⁻2?

CALIFORNIA STANDARDS NS 2.3 Solve addition, subtraction, multiplication, and division problems, including those arising in concrete situations, that use positive and negative integers and combinations of these operations. *also,* ⊶NS 2.0, MR 2.4, MR 2.5

Activity 2

Addition and subtraction of integers are related.

- Copy the model for $^-7$ below.

- Use the model to find $^-7 - {}^-3$. What is $^-7 - {}^-3$?

- Model $^-7$ again. Then add three yellow counters to find $^-7 + 3$. What is $^-7 + 3$?

Think and Discuss

- The models above show that $^-7 - {}^-3 = {}^-7 + 3$. How are $^-3$ and 3 related? How are subtraction and addition related?

- How are the two models different?

  You can write a subtraction problem as an addition problem by adding the opposite of the number you are subtracting.

$$6 - {}^-2 = 6 + 2 \leftarrow \text{Add the opposite of } ^-2.$$

- How can you write $^-6 - 2$ as an addition problem?

Practice

Use counters to find the difference.

1. $^-7 - 4$ 2. $^-8 - {}^-5$ 3. $^-13 - 9$ 4. $^-9 - 4$

Complete the addition problem.

5. $^-6 - {}^-2 = {}^-6 + \blacksquare$ 6. $^-9 - {}^-3 = {}^-9 + \blacksquare$ 7. $^-7 - 5 = {}^-7 + \blacksquare$ 8. $^-19 - 12 = {}^-19 + \blacksquare$

MIXED REVIEW AND TEST PREP

Find the sum. (p. 244)

9. $^-3 + 9$ 10. $2 + {}^-7$ 11. $^-1 + {}^-7$

12. Write an integer to represent 217 m below sea level. (p. 228)

13. **TEST PREP**   Which is the value of $b + 1\frac{3}{4}$ for $b = 5\frac{2}{3}$? (p. 216)

   A $6\frac{5}{7}$ B $6\frac{5}{12}$ C $7\frac{1}{12}$ D $7\frac{5}{12}$

ALGEBRA
Subtract Integers

Learn how to use a number line to subtract integers.

Pathfinder endured temperatures as low as $^-89°F$.

During the summer of 1997, NASA landed the Mars Pathfinder on the planet Mars. On July 9, Pathfinder reported a temperature of $^-1°F$. On July 10, Pathfinder reported a temperature of $8°F$. Find the range of temperatures reported by Pathfinder from July 9 to July 10.

To find the range of temperatures, you need to find the difference of 8 and $^-1$, or $8 - {}^-1$. You can find the difference of two integers by adding the opposite of the integer you are subtracting. You can then use the rules for addition of integers.

The opposite of $^-1$ is 1. So, $8 - {}^-1$ becomes $8 + 1$.

$8 - {}^-1 = 8 + 1 = 9$

So, the range of temperatures was $9°F$.

EXAMPLE

During an experiment, a scientist recorded a high temperature of $^-9°C$ and a low temperature of $^-22°C$. What was the range of temperatures during the experiment?

$^-9 - {}^-22 = {}^-9 + 22$ *Write the subtraction problem as an addition problem. Use the rules for addition of integers.*

$^-9 + 22$

$|22| - |^-9| = 22 - 9$ *Subtract the lesser absolute value from the greater absolute value.*

$ = 13$

$|22| > |^-9| \ \rightarrow \ {}^-9 - {}^-22 = 13$ *Use the sign of the addend with the greater absolute value.*

So, the range of temperatures during the experiment was $13°C$.

• During the afternoon, the temperature fell from $7°F$ to $^-5°F$. What was the range of temperatures?

CALIFORNIA STANDARDS NS 2.3 Solve addition, subtraction, multiplication, and division problems, including those arising in concrete situations, that use positive and negative integers and combinations of these operations. *also,* ⊶NS 2.0, MR 2.2, MR 2.5

Think and Discuss ▶ Look back at the lesson to answer the question.

1. **Tell** how you would write the subtraction problem as an addition problem if the temperature at the end of the experiment in the example was $^-40°C$.

Guided Practice ▶ Rewrite the subtraction problem as an addition problem.

2. $7 - 10$ 3. $3 - {}^-6$ 4. $^-1 - {}^-8$ 5. $^-4 - 6$

Find the difference.

6. $4 - 8$ 7. $^-7 - {}^-2$ 8. $4 - {}^-5$ 9. $1 - 8$

Independent Practice ▶ Rewrite the subtraction problem as an addition problem.

10. $12 - 15$ 11. $8 - {}^-11$ 12. $^-6 - {}^-13$ 13. $^-9 - 11$

Find the difference.

14. $6 - 11$ 15. $^-9 - {}^-5$ 16. $2 - {}^-1$ 17. $3 - 5$

18. $7 - 11$ 19. $^-5 - {}^-5$ 20. $8 - {}^-3$ 21. $4 - 9$

22. $31 - 37$ 23. $35 - 39$ 24. $\left|{}^-43\right| - \left|12\right|$ 25. $\left|{}^-27\right| - \left|{}^-32\right|$

Evaluate.

26. $^-3 - {}^-5 + {}^-8$ 27. $6 - {}^-4 + {}^-5$ 28. $8 - {}^-6 - 10$

Problem Solving Applications

29. In the morning, the temperature in Anchorage, Alaska, is $^-12°F$. Later that evening, the temperature is $^-19°F$. What is the range of temperatures?

30. The water level of a river was 3 ft above normal. After an unusually dry season, the water level is 6 ft below normal. Find the range of the water levels of the river.

31. ❓ **What's the Question?**   The temperature at noon was 15°F. By midnight the temperature was $^-3°F$. The answer is 18°F.

MIXED REVIEW AND TEST PREP

32. Find the sum $^-4 + 9$. (p. 244)

33. Write the opposite integer for 213. (p. 228)

34. Find the quotient $5\frac{3}{4} \div 2\frac{1}{4}$. (p. 210)

35. Find the prime factorization of 84. (p. 148)

36. **TEST PREP**   Solve $x + 9 = 21$ using mental math. (p. 30)

   A $x = 9$ **B** $x = 11$ **C** $x = 12$ **D** $x = 30$

1. VOCABULARY The opposite of an integer is its ___?___ . (p. 243)

Write the addition problem modeled on each number line. (pp. 244–247)

2.

3.

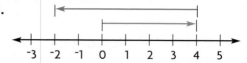

Find the sum. (pp. 244–247)

4. $7 + {}^-6$	**5.** ${}^-5 + {}^-3$	**6.** ${}^-7 + 4$	**7.** $4 + {}^-1$
8. ${}^-3 + {}^-4$	**9.** ${}^-2 + 6$	**10.** ${}^-8 + 3$	**11.** ${}^-7 + {}^-9$
12. ${}^-9 + 4$	**13.** $9 + {}^-1$	**14.** ${}^-3 + 9$	**15.** ${}^-5 + {}^-6$
16. ${}^-37 + 24$	**17.** $17 + {}^-19$	**18.** ${}^-17 + {}^-41$	**19.** $21 + 17$

Use mental math to find the value of x. (pp. 244–247)

20. $x + 4 = {}^-4$ **21.** $x + {}^-6 = {}^-1$ **22.** $9 + x = 0$ **23.** ${}^-4 + x = {}^-3$

Find the difference. (pp. 250–251)

24. $7 - 11$	**25.** ${}^-2 - 8$	**26.** ${}^-3 - {}^-5$	**27.** $4 - {}^-4$
28. ${}^-1 - {}^-8$	**29.** ${}^-2 - 2$	**30.** $3 - 8$	**31.** $6 - 12$
32. ${}^-1 - 4$	**33.** ${}^-6 - {}^-8$	**34.** $51 - {}^-23$	**35.** ${}^-41 - 18$

Solve. (pp. 242–251)

36. A submarine started one leg of its voyage at ${}^-300$ ft. At the end of that leg of the voyage, it was at ${}^-1,250$ ft. What was the difference between the two depths?

37. Neal was measuring the motion of an object. He observed that it moved 12 cm forward and 17 cm backward. What was its distance from the starting point?

38. Tasha needed to find the average temperature during three days in the winter. Her first step was to add the temperatures together. If the temperatures were ${}^-6°$, ${}^-8°$, and $11°$, what was the sum?

39. At midnight the temperature was ${}^-15°$F. The temperature continued to fall until 7:00 A.M., when it was ${}^-22°$F. How many degrees had the temperature fallen during the seven hours?

40. Two explorers start a hike in a valley 12 feet below sea level. When they finish their hike they are at 22 feet above sea level. What is the difference between the two elevations?

Get the information you need.

See item **4.**

Think about how you can write the subtraction of integers as addition. Then find the expression that has the same solution as the given subtraction problem.

Also see problem **3**, p. H63.

Choose the best answer.

1. $^-7 + 5$

A $^-12$ **C** 2

B $^-2$ **D** Not here

2. On Sunday, the low temperature was $^-4°$F. The low temperature on Monday was $13°$ higher. What was the low temperature on Monday?

F $^-13°$F

G $^-9°$F

H $4°$F

J $9°$F

3. What is the value of $29 + k$ for $k = {}^-26$?

A $^-55$ **C** 3

B $^-3$ **D** 55

4. Which problem has the same solution as $^-3 - {}^-5$?

F $^-3 - 5$ **H** $^-3 + 5$

G $^-3 + {}^-5$ **J** $3 + 5$

5. $5 + {}^-8$

A $^-3$ **C** 13

B 3 **D** Not here

6. $^-5 - {}^-3$

F $^-8$ **H** 2

G $^-2$ **J** Not here

7. The graph shows the number of sixth-grade students at a middle school who are in each club listed. How many more students are in the Science Club than in the Chess Club?

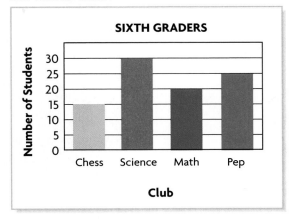

A 2 **C** 20

B 15 **D** 30

8. How is $2 \times 2 \times 2 \times 2$ written in exponent form with a base of 2?

F 2^4 **H** 4^2

G 2^5 **J** 2^{10}

9. Use mental math to find the value of x.

$$x + {}^-9 = 5$$

A $x = {}^-14$ **C** $x = 4$

B $x = {}^-4$ **D** $x = 14$

10. Ellen got a score of 50 on her history exam. The mean score on that exam was 45. Which claim is valid?

F Ellen's score is above the average score.

G Ellen got the top score.

H Ellen's score is higher than most of the other students' scores.

J Ellen's score is below the average score.

11. $0.053 - 0.019$

A 3.4 **C** 0.034

B 0.34 **D** Not here

CHAPTER 13 Multiply and Divide with Integers

One of the most unusual hotels in the world is located above the Arctic Circle in Sweden. The Ice Hotel is made entirely of ice, including each room, bed, fireplace, chair, and eating utensil. Each winter, adventurous travelers pay from $165 to $250 per night to stay at the hotel. On some nights, it may be 25°F in the hotel when a traveler goes to bed and the temperature will drop an average of 3°F each hour. What will be the temperature when the traveler gets up 8 hours later?

AVERAGE TEMPERATURE IN SWEDEN
(Above the Arctic Circle)

Temperature (°F) *

Month

Key:
- - - high
——— low

* To convert °F to °C use the formula
°C = (°F - 32) x 0.555

Check What You Know

Use this page to help you review and remember important skills needed for Chapter 13.

 Multiplication and Division Facts (See p. H13.)

Find the product or the quotient.

1. 4×6 **2.** 9×7 **3.** $80 \div 8$

4. $24 \div 3$ **5.** 4×1 **6.** 8×7

7. 12×6 **8.** $108 \div 9$ **9.** $45 \div 9$

10. 3×11 **11.** $120 \div 12$ **12.** $81 \div 9$

Solve for _n_.

13. $2 \times n = 14$ **14.** $n \times 5 = 15$ **15.** $n \div 6 = 9$

16. $88 \div n = 11$ **17.** $42 = n \times 6$ **18.** $11 = n \div 12$

Add and Subtract Integers (See p. H14.)

Find the sum or difference.

19. $8 - 15$ **20.** $^-9 + {}^-5$ **21.** $^-15 - {}^-8$

22. $^-9 + 21$ **23.** $^-5 - 12$ **24.** $16 - {}^-13$

25. $^-31 + {}^-37$ **26.** $82 + {}^-59$ **27.** $^-15 - 32$

28. $^-50 - {}^-80$ **29.** $12 + {}^-17$ **30.** $^-43 - {}^-43$

Patterns (See p. H14.)

Describe a possible pattern and find the next three numbers in your pattern.

31. $1, 2, 3, 4, 5, \ldots$ **32.** $16, 14, 12, 10, 8, \ldots$

33. $256, 128, 64, 32, 16, \ldots$ **34.** $1, 4, 7, 10, 13, \ldots$

35. $3, 6, 12, 24, 48, \ldots$ **36.** $50{,}000, 10{,}000, 2{,}000, 400, \ldots$

37. $120, 112, 104, 96, 88, \ldots$ **38.** $^-6, 12, {}^-24, 48, {}^-96, \ldots$

Draw a picture of the next 3 figures in the pattern. Describe the pattern.

39.

40.

> **LOOK AHEAD**

In Chapter 13 you will
- multiply integers
- divide integers
- perform more than one operation with integers

ALGEBRA
Model Multiplication of Integers

MATH LAB

Explore how to use two-color counters and number lines to multiply integers.

You need two-color counters.

Remember that 4 × 5 can be written as 4 · 5.

You can use yellow and red counters to model multiplication of integers. A yellow counter represents ⁺1 and a red counter represents ⁻1.

Activity 1

• Use yellow counters to model the product 2 · 3.

     ← 2 groups of 3

2 · 3 = 6

• Use red counters to model the product 2 · ⁻3.

     ← 2 groups of ⁻3

2 · ⁻3 = ⁻6

• Use red counters to model the product ⁻2 · 3. Using the Commutative Property, you can write ⁻2 · 3 as 3 · ⁻2.

       ← 3 groups of ⁻2

⁻2 · 3 = 3 · ⁻2 = ⁻6

Think and Discuss

• How could you model the product 3 · ⁻4?

• How could you model the product ⁻3 · 4?

• What do you notice about the product of two positive integers? of a positive integer and a negative integer?

Practice

Use counters to find the product.

1. 2 · 4   **2.** 3 · ⁻5   **3.** ⁻5 · 2   **4.** ⁻3 · 6

CALIFORNIA STANDARDS O—∎ NS 2.3 Solve addition, subtraction, multiplication, and division problems, including those arising in concrete situations, that use positive and negative integers and combinations of these operations. *also* O—∎ NS 2.0, MR 2.4

Activity 2

You can also use a number line to multiply integers.

- Use a number line to find the product 3 · 4. Start at 0.

The number line shows that 3 · 4 = 4 + 4 + 4 = 12.

- Use a number line to find the product 5 · ⁻2. Start at 0.

The number line shows that 5 · ⁻2 = ⁻2 + ⁻2 + ⁻2 + ⁻2 + ⁻2 = ⁻10.

- Use a number line to find the product ⁻7 · 2. Using the Commutative Property, write ⁻7 · 2 as 2 · ⁻7. Then represent 2 · ⁻7 on the number line.

The number line shows that 2 · ⁻7 = ⁻7 + ⁻7 = ⁻14. So, by the Commutative Property, ⁻7 · 2 = ⁻14.

Think and Discuss

- How could you use a number line to find the product 2 · 8?

- How could you use a number line to find the product 3 · ⁻6?

Practice

Use a number line to find each product.

1. 4 · 5   **2.** 3 · ⁻7   **3.** 6 · ⁻3   **4.** ⁻5 · 6

MIXED REVIEW AND TEST PREP

Find the difference. (p. 250)

5. ⁻3 − ⁻7   **6.** 5 − 9   **7.** 12 − ⁻9

8. Order 0.2, $\frac{1}{8}$, and $\frac{1}{4}$ from least to greatest. (p. 234)

9. TEST PREP Which is the prime factorization of 28? (p. 148)

     A $4 \cdot 7$ **B** $2^2 \cdot 7$ **C** $2 \cdot 7^2$ **D** $2^2 \cdot 7^2$

ALGEBRA
Multiply Integers

*L*earn how to multiply integers.

A submarine is diving and its depth is changing by ⁻30 m every minute. If the submarine started at the surface of the ocean, how far below the surface is it after 4 minutes?

Use a number line to find the product 4 · ⁻30.

-140 -120 -100 -80 -60 -40 -20 0 20 40

The number line shows that the depth of the submarine changed ⁻120 m. So, the submarine is 120 m below the surface.

You can use patterns to find rules for multiplying integers.

EXAMPLE 1

Complete the pattern.

4 · 3 = 12	4 · 3 = 12
4 · 2 = 8	4 · 2 = 8
4 · 1 = 4	4 · 1 = 4
4 · 0 = 0	4 · 0 = 0
4 · ⁻1 = ■	4 · ⁻1 = ⁻4
4 · ⁻2 = ■	4 · ⁻2 = ⁻8
4 · ⁻3 = ■	4 · ⁻3 = ⁻12

Study the pattern. As the second factor decreases by 1, the product decreases by 4. Assume the pattern continues.

So, the missing products are ⁻4, ⁻8, and ⁻12.

• What is the sign of the product of a positive and negative integer?

EXAMPLE 2

Complete the pattern.

⁻4 · 3 = ⁻12	⁻4 · 3 = ⁻12
⁻4 · 2 = ⁻8	⁻4 · 2 = ⁻8
⁻4 · 1 = ⁻4	⁻4 · 1 = ⁻4
⁻4 · 0 = 0	⁻4 · 0 = 0
⁻4 · ⁻1 = ■	⁻4 · ⁻1 = 4
⁻4 · ⁻2 = ■	⁻4 · ⁻2 = 8
⁻4 · ⁻3 = ■	⁻4 · ⁻3 = 12

Study the pattern. As the second factor decreases by 1, the product increases by 4. Assume the pattern continues.

So, the missing products are 4, 8, and 12.

• What is the sign of the product of two positive integers? two negative integers?

Examples 1 and 2 lead to the rules below.

> The product of two integers with like signs is positive.
> The product of two integers with unlike signs is negative.

CALIFORNIA STANDARDS ○━ⁿ NS 2.3 Solve addition, subtraction, multiplication, and division problems, including those arising in concrete situations, that use positive and negative integers and combinations of these operations. *also* ○━ⁿ NS 2.0, AF 1.3, MR 1.1, MR 2.2, MR 2.4, MR 3.3

Think and ▶
Discuss

Look back at the lesson to answer each question.

1. **Tell** if the product of a negative number and zero is positive or negative. Explain.

2. **Name** two possible choices for two numbers which are the same and whose product is 16.

Guided ▶
Practice

Find the product.

3. $^-9 \cdot 6$　　　4. $^-3 \cdot {}^-4$　　　5. $^-8 \cdot 2$　　　6. $8 \cdot {}^-7$

7. $^-3 \cdot 7$　　　8. $^-8 \cdot {}^-2$　　　9. $18 \cdot {}^-5$　　　10. $^-22 \cdot 4$

PRACTICE AND PROBLEM SOLVING

Independent ▶
Practice

Find the product.

11. $^-5 \cdot 8$　　　12. $2 \cdot {}^-3$　　　13. $7 \cdot 2$　　　14. $^-9 \cdot 4$

15. $12 \cdot 7$　　　16. $^-12 \cdot {}^-8$　　　17. $^-11 \cdot {}^-10$　　　18. $^-8 \cdot 3$

19. $24 \cdot 12$　　　20. $30 \cdot {}^-12$　　　21. $(^-8 - 2) \cdot (^-52 + 7)$

 Algebra Use mental math to find the value of *y*.

22. $y \cdot {}^-2 = 20$　　　23. $4 \cdot y = {}^-12$　　　24. $^-7 \cdot y = {}^-56$

25. $y \cdot {}^-6 = {}^-18$　　　26. $y \cdot {}^-10 = 40$　　　27. $^-8 \cdot y = 32$

Problem Solving ▶
Applications

28. Drought conditions have caused the water level in a lake to change by $^-3$ in. every month for the past 4 months. Write the total change in water level over the past 4 months as a negative number.

29. Use the Associative Property to help you find the product of $^-3 \cdot {}^-2 \cdot {}^-5$.

30. A water tank has a leak. The amount of water in the tank changes by $^-6$ gal a day. What is the total change after 7 days?

31. **(?) What's the Error?** Beth says that $^-5 \cdot {}^-2 = {}^-10$. What is her error? What is the correct product?

MIXED REVIEW AND TEST PREP

Find the difference. (p. 250)

32. $^-25 - {}^-42$　　　　33. $30 - 42$　　　　34. $56 - {}^-18$

35. Find the LCM of 26 and 28. (p. 150)

36. **TEST PREP** Evaluate $2k$ for $k = 9.6$. (p. 70)

    A 19.2　　　**B** 11.6　　　**C** 7.6　　　**D** 4.8

Extra Practice page H44, Set A

ALGEBRA
Divide Integers

Learn how to divide integers.

QUICK REVIEW

1. 420 ÷ 6   **2.** 225 ÷ 25   **3.** 50 ÷ 2   **4.** 3,600 ÷ 4   **5.** 540 ÷ 60

Multiplication and division are inverse operations. To solve a division problem, think about the related multiplication problem.

42 ÷ 7 = ■   **THINK:** 7 · 6 = 42, so 42 ÷ 7 = 6.

You can use related multiplication problems to determine the sign of the quotient when dividing integers.

8 · 3 = 24, so 24 ÷ 8 = 3.   ⁻8 · ⁻3 = 24, so 24 ÷ ⁻8 = ⁻3.

⁻8 · 3 = ⁻24, so ⁻24 ÷ ⁻8 = 3.   8 · ⁻3 = ⁻24, so ⁻24 ÷ 8 = ⁻3.

In the problems above, look at the sign of the quotients of two positive integers and the sign of the quotients of two negative integers. Then look at the sign of the quotients of a positive and negative integer. You can see that the rules below apply when dividing integers.

> The quotient of two integers with like signs is positive.
> The quotient of two integers with different signs is negative.

TECHNOLOGY LINK

More Practice: Use **Mighty Math Astro Algebra,** Red, Level M.

EXAMPLE 1

Find the quotient.

A. ⁻84 ÷ ⁻7

Divide as with whole numbers. The quotient is positive since the integers have like signs.

⁻84 ÷ ⁻7 = 12

B. ⁻55 ÷ 11

Divide as with whole numbers. The quotient is negative since the integers have unlike signs.

⁻55 ÷ 11 = ⁻5

EXAMPLE 2

The low temperatures for five days in Fairbanks, Alaska, were ⁻3°F, ⁻8°F, 2°F, 3°F, and ⁻4°F. Find the average low temperature for the five days.

Find the sum. Divide the sum by 5.
⁻3 + ⁻8 + 2 + 3 + ⁻4 = ⁻10

⁻10 ÷ 5 = ⁻2

So, the average low temperature was ⁻2°F.

CALIFORNIA STANDARDS O¬ **NS 2.3** Solve addition, subtraction, multiplication, and division problems, including those arising in concrete situations, that use positive and negative integers and combinations of these operations. *also* O¬ **NS 2.0, AF 1.3, SDAP 1.1, MR 1.1, MR 2.5, MR 3.3**

Think and Discuss ▶ **Look back at the lesson to answer the question.**

1. **Tell** how the sign rules for multiplying two integers compare with the sign rules for dividing two integers.

Guided Practice ▶ **Find the quotient.**

2. $^-9 \div 3$ 3. $^-24 \div ^-4$ 4. $^-28 \div 7$ 5. $56 \div ^-7$

6. $^-21 \div 7$ 7. $^-16 \div ^-2$ 8. $120 \div ^-5$ 9. $^-132 \div 6$

PRACTICE AND PROBLEM SOLVING

Independent Practice ▶ **Find the quotient.**

10. $^-40 \div 8$ 11. $6 \div ^-3$ 12. $14 \div 2$ 13. $^-36 \div 4$

14. $84 \div 7$ 15. $^-72 \div ^-8$ 16. $^-63 \div ^-7$ 17. $^-70 \div 7$

18. $432 \div 12$ 19. $255 \div ^-15$ 20. $(^-40 - 4) \div (^-86 + 97)$

 Algebra **Use mental math to find the value of x.**

21. $x \div ^-6 = ^-3$ 22. $x \div ^-7 = 5$ 23. $^-18 \div x = 2$

Problem Solving Applications ▶

24. The low temperatures for a winter festival were $^-8°F$, $^-6°F$, $^-9°F$, and $^-1°F$. Find the average low temperature for the four days.

25. During the first 6 months of business, a sporting goods store showed its losses as $^-\$3,054$. How would it show the average monthly loss?

26. The depth of a submarine changed $^-225$ ft over a period of 9 minutes. If the submarine descended at a constant rate, what was the change of depth per minute?

27. The water level in a leaking swimming pool is changing $^-4$ in. per week. How many weeks will it take for the water level to change $^-20$ in.?

28. **? What's the Question?** In a division problem, the numbers are $^-250$ and 10. The answer is $^-25$.

MIXED REVIEW AND TEST PREP

29. Find the product $16 \cdot ^-6$. (p. 258)

30. Find the sum $^-32 + ^-42$. (p. 244)

31. Find the product $5\frac{2}{3} \cdot 6\frac{3}{4}$. (p. 206)

32. Find the quotient $6\frac{3}{5} \div 2\frac{3}{4}$. (p. 210)

33. **TEST PREP** Write the fraction $\frac{5}{8}$ as a percent. (p. 169)

    A 0.625%   **B** 6.25%   **C** 62.5%   **D** 625%

Combine Operations with Integers

Learn how to evaluate an expression with more than one operation.

QUICK REVIEW

1. $(4 + 8) \cdot 6$ **2.** $100 \div (30 - 5)$ **3.** $50 - 3^2$
4. $(5 + 4) \cdot (3 + 2)$ **5.** $(25 - 5) \div (10 - 6)$

Sometimes you have to perform more than one operation when working with expressions involving integers. Remember the order of operations when performing more than one operation.

EXAMPLE 1

Remember that when performing more than one operation in an expression, you must use the order of operations.

1. Operate inside parentheses.
2. Clear exponents.
3. Multiply and divide from left to right.
4. Add and subtract from left to right.

Find the value of $^-3 \cdot (12 + 3) \div 3^2$.

$^-3 \cdot (12 + 3) \div 3^2$	*Operate inside parentheses.*
$^-3 \cdot 15 \div 3^2$	*Clear exponents.*
$^-3 \cdot 15 \div 9$	*Multiply.*
$^-45 \div 9$	*Divide.*
$^-5$	

So, the value of the expression is $^-5$.

• How do you know the final answer is negative?

When performing combined operations with integers, you can use properties to simplify the problem.

EXAMPLE 2

Find the value of the expression.

Commutative Property	Associative Property
A. $92 + 365 + {}^-2$	**B.** $(52 + 14) + {}^-14$
$92 + {}^-2 + 365$	$52 + (14 + {}^-14)$
$90 + 365$	$52 + 0$
455	52

So, the value is 455.   So, the value is 52.

• Which property would you use to evaluate $47 + 131 + {}^-17$? Explain.

CHECK FOR UNDERSTANDING

Think and ▶ Discuss

Look back at the lesson to answer each question.

1. **Find** the value of the expression $(^-3 + 7) \cdot 3^2 \div {}^-18$.

2. **Explain** how the Commutative and Associative Properties simplify the problems in Example 2.

CALIFORNIA STANDARDS O─ NS 2.3 Solve addition, subtraction, multiplication, and division problems, including those arising in concrete situations, that use positive and negative integers and combinations of these operations. *also* O─ NS 2.0, AF 1.3, AF 1.4, MR 1.3, MR 2.2

Guided ▶
Practice

Evaluate the expression.

3. $42 \div 21 - 4 \cdot 9$

4. $^-15 + (^-45 \div ^-9)^2 - 29$

Use a property to simplify the expression. Then evaluate the expression and identify the property you used.

5. $47 + 127 + 3$ 　　　**6.** $(223 + 24) + 6$ 　　　**7.** $(576 + ^-53) + 53$

PRACTICE AND PROBLEM SOLVING

Independent ▶
Practice

Evaluate the expression.

8. $^-4 + 7 \cdot ^-4 \div 14 - 12$ 　　　**9.** $(3 - ^-15) \div 6 \cdot (22 - 32)$

10. $2 - 14 + ^-8 \cdot 2^4$ 　　　　　**11.** $(^-15 + 7) \cdot (^-3 - 12)$

12. $4^3 \div (2 - 18) \cdot (^-8 - 3)$ 　　　**13.** $^-(3^2) \div \left|^-3\right| - ^-10$

Use a property to simplify the expression. Then evaluate the expression and identify the property you used.

14. $(921 + 27) + ^-27$ 　**15.** $85 + 321 + 15$ 　　　**16.** $^-7 + 29 + 17$

Problem Solving ▶
Applications

Use Data　For 17–19, use the line graph.

17. By how much did the price of Integer stock increase from January to February?

18. By how much did the price of Integer stock decrease from February to March?

19. What was the change in the price of Integer stock from January to May? Express your answer as an integer.

20. 📓 **Write About It** Write a problem that must be solved by performing more than one operation with integers.

21. Number Sense How many prime numbers less than 1,000 are even?

MIXED REVIEW AND TEST PREP

Find the quotient. (p. 260)

22. $^-56 \div 8$ 　　　　**23.** $^-77 \div ^-11$ 　　　　　**24.** $236 \div ^-4$

25. Find the sum $^-79 + 42$. (p. 244)

26. TEST PREP Find the prime factorization of 54. (p. 148)

　A $2 \cdot 3$ 　　　**B** $2^3 \cdot 3$ 　　　**C** $2 \cdot 3^3$ 　　　**D** $2^3 \cdot 3^3$

Find the product. (pp. 258–259)

1. $4 \times {}^-8$

2. ${}^-9 \times {}^-5$

3. $7 \times {}^-2$

4. ${}^-11 \times {}^-7$

5. ${}^-5 \times 4$

6. ${}^-8 \times {}^-10$

7. ${}^-12 \times 5$

8. ${}^-19 \times {}^-5$

9. ${}^-51 \times 14$

10. ${}^-18 \times {}^-7$

11. $50 \times {}^-7$

12. $83 \times {}^-21$

Find the quotient. (pp 260–261)

13. $27 \div {}^-3$

14. ${}^-18 \div 2$

15. ${}^-36 \div {}^-6$

16. ${}^-49 \div 7$

17. $144 \div {}^-12$

18. ${}^-132 \div {}^-12$

19. $63 \div {}^-9$

20. ${}^-220 \div {}^-10$

21. $180 \div {}^-9$

22. ${}^-216 \div 18$

23. $324 \div {}^-36$

24. ${}^-1{,}458 \div {}^-54$

Evaluate the expression. (pp. 262–263)

25. ${}^-4 + 5 \times {}^-4 \div 2 - 12$

26. $(2^2 - 4^2)^2 \div {}^-24 - {}^-10$

27. $(8 - {}^-4) \div 6 \times (12 - 16)$

28. $(6^2 - 3^2)^2 \div {}^-3 - {}^-45$

29. $2 - 11 + {}^-4 \times 3^3$

30. $2 + ({}^-11 + {}^-4) \times 5^3$

31. $(5 - 13) \times ({}^-8 \times 3)$

32. ${}^-32 \times 11 + {}^-4 \div 2^2$

33. $({}^-15 \times 7) \times ({}^-12 \div {}^-3)$

Use a property to simplify the expression. Then evaluate the expression and identify the property you used. (pp. 262–263)

34. $(113 + 58) + {}^-28$

35. ${}^-85 + 113 + {}^-15$

36. ${}^-117 + 49 + 17$

37. $(125 + {}^-13) + {}^-112$

Solve.

38. During the last glacial age, sea level in one body of water changed by an average of ${}^-3$ feet every 200 years. The glacial age lasted 10,000 years. What was the total change in sea level? (pp. 258–259)

39. During the first nine months of the year, a worldwide club showed its total loss of members as ${}^-2{,}718$. What integer shows the average loss of members per month? (pp. 260–261)

40. One way to study climate is to compare the current conditions to conditions in previous years. One group studying climate found that rainfall during five months in an area showed the following changes compared with the previous year: ${}^-45$ in., ${}^-12$ in., 7 in., 23 in., ${}^-3$ in. What was the average change in rainfall for these five months? (pp. 260–261)

Eliminate choices.

See item **8.**

You can eliminate any choices that do not have the right sign.

Also see problem **5**, p. H64.

Choose the best answer.

1. $(^-5)^2 + (^-17 \cdot 20)$

  A $^-315$ **C** 315

  B 3 **D** 340

2. Which expression has a value of 10 for $m = ^-3$?

  F $9 \div m + 7$

  G $m - (9 + 2)$

  H $9m + 2$

  J $9 - (m + 2)$

3. $^-54 \div ^-9$

  A $^-486$

  B $^-6$

  C 6

  D Not here

4. $8 \times ^-2 + 17$

  F $^-544$ **H** 1

  G $^-15$ **J** 33

5. $108 + (4 \times 3^2) - 76 \div 4$

  A 233

  B 125

  C 24.5

  D Not here

6. $(4^2) - (^-7 \times 2)$

  F 30 **H** $^-30$

  G 2 **J** Not here

7. Data that are divided into parts totaling 100% are best displayed in which kind of graph?

  A Line graph

  B Bar graph

  C Circle graph

  D Histogram

8. Use mental math to find the value of x.

$$x \div ^-6 = 90$$

  F $x = ^-540$ **H** $x = 15$

  G $x = ^-15$ **J** $x = 96$

9. Jason's test scores are listed below. Which is the mean score?

TEST SCORES					
75	85	100	91	93	87
82	76	78	100	79	98

  A 25 **C** 82

  B 75 **D** 87

10. $\frac{7}{9} - \frac{4}{6}$

  F $\frac{1}{18}$ **H** $\frac{9}{18}$

  G $\frac{1}{9}$ **J** $\frac{3}{3}$

11. $9\frac{1}{6} \div 1\frac{7}{8}$

  A $4\frac{8}{9}$ **C** $9\frac{7}{48}$

  B 5 **D** $31\frac{3}{7}$

12. $|^-2,187|$

  F $^-2,187$

  G $^-|2,187|$

  H $2,187$

  J Not here

MATH DETECTIVE

Play by the Rules

REASONING Each function machine is missing something. For Machines 1, 2, and 3, an input or an output value is missing in each step. For Machine 4, the rule is missing. Use your knowledge of integer operations to find what is missing for each machine.

Function Machine 1

Rule: Add ⁻6.

INPUT	OUTPUT
3	■
■	0
9	■
⁻3	■
■	⁻12

Function Machine 2

Rule: Subtract 4.

INPUT	OUTPUT
5	■
■	7
1	■
⁻2	■
■	0

Function Machine 3

Rule: Multiply by ⁻3. Then add 1.

INPUT	OUTPUT
3	■
1	■
0	■
⁻2	■
⁻10	■

Function Machine 4

Rule: ?

INPUT	OUTPUT
⁻8	4
⁻4	2
6	⁻3
10	⁻5
0	0

Think It Over!

- **Write About It** Explain how you found a rule for Function Machine 4.

- **Stretch Your Thinking** Find a rule for this function machine.

INPUT	OUTPUT
0	3
1	5
2	7
3	9

Negative Exponents

Learn how to work with negative exponents and how to write small numbers by using scientific notation.

Activity

Copy and complete the pattern. Use a calculator as needed.

$10^3 = 1,000$

$10^2 = 100$

$10^1 = 10$

$10^0 = $ ■

$10^{-1} = $ ■

$10^{-2} = $ ■

$10^{-3} = $ ■

Notice that as the value of the exponent decreases, each number is 0.1, or $\frac{1}{10}$, as great as the previous number.

- For powers of 10, how is the negative exponent related to the number of decimal places?

Negative exponents are used to write very small numbers in scientific notation.

$$0.004 = 4 \times 0.001 = 4 \times 10^{-3}$$

Replace 0.001 with 10^{-3}.

Look at the relationship between 0.004 and 4×10^{-3}. To write 0.004 as 4, move the decimal point 3 places to the right, multiplying by 1,000. Use $^-3$ as the exponent of 10 to show the corresponding division by 1,000.

EXAMPLE

Write 0.0000245 in scientific notation.

0.0000245
5 places

Count the number of places the decimal point must be moved to the right to form a number that is at least 1 but less than 10.

2.45×10^{-5}

Write the number. Since the decimal point moved 5 places to the right, the exponent of 10 is 10^{-5}.

TALK ABOUT IT

- When writing a very small number in scientific notation, how can you tell what should be the exponent?

TRY IT

Write using scientific notation.

1. 0.002
2. 0.00034
3. 0.06
4. 0.005365
5. 0.0084
6. 0.0000794
7. 0.0000008
8. 0.000202

VOCABULARY

1. The distance an integer is from zero is its __?__. (p. 228)

2. All whole numbers and their opposites are the set of __?__. (p. 228)

3. A comparison of two numbers, a and b, written as a fraction $\frac{a}{b}$, is a __?__. (p. 230)

EXAMPLES

EXERCISES

Chapter 11

- **Write the absolute value of an integer.**

  (pp. 228–229)

  $|{-5}| = 5$ *Write the distance -5 is from 0 on the number line.*

Write the absolute value.

4. $|{-8}|$ 5. $|{+6}|$ 6. $|{-28}|$

7. $|{-73}|$ 8. $|{+49}|$ 9. $|0|$

- **Classify numbers as whole numbers, integers, and rational numbers.**

  (pp. 230–233)

  -5 is an integer and a rational number.

  $\frac{3}{5}$ is a rational number.

Name the sets to which each number belongs.

10. $3\frac{5}{9}$ 11. 0.35

12. -42 13. $\frac{6}{3}$

- **Find a rational number between two rational numbers.** (pp. 230–233)

  Find a rational number between -8.5 and -8.45.

  $-8.50 \quad -8.45$ *Write each decimal using hundredths.*

  -8.46 is between -8.5 and -8.45. Some other numbers are -8.47, -8.48, and -8.482.

Find a rational number between the two given numbers.

14. $\frac{1}{8}$ and $\frac{1}{2}$ 15. -2.3 and -2.4

16. 18.01 and 18.02   17. $\frac{-1}{8}$ and $\frac{1}{10}$

18. $\frac{1}{10}$ and 0.2   19. $-1\frac{1}{2}$ and $-1\frac{3}{4}$

- **Compare and order rational numbers.**

  (pp. 234–235)

  Compare. Write $<$, $>$, or $=$ for the ●.

  -5.8 ● $-5\frac{3}{4}$

  $-5\frac{3}{4} = -5.75$ *Write the fraction as a decimal.*

  $-5.80 < -5.75$ *Compare the decimals.*

  So, $-5.8 < -5\frac{3}{4}$.

Compare. Write $<$, $>$ or $=$ for each ●.

20. $\frac{2}{5}$ ● 0.38   21. $3\frac{3}{4}$ ● 3.89

22. -0.25 ● $\frac{-1}{4}$ 23. $4\frac{3}{10}$ ● 4.03

Order from least to greatest.

24. 1.55, $\frac{2}{3}$, -3, 4.2, -1.8

25. $\frac{-3}{4}$, 0, 6.5, -8, -3.6

Chapter 12

- **Add and subtract integers.** (pp. 242–251)

  Subtract. $6 - {}^-4$

  $6 - {}^-4 = 6 + 4$ *Write as an addition*
  $\quad\quad = 10$ *sentence.*

  Add. $15 + {}^-3$

  $15 + {}^-3 = 12$ *Subtract the absolute values.*

Find the sum or difference.

26. $^-3 + 9$
27. $^-8 + {}^-9$
28. $12 + {}^-5$
29. $^-4 - 10$
30. $^-9 - {}^-13$
31. $15 - {}^-6$

Chapter 13

- **Multiply and divide integers.** (pp. 258–261)

  $^-4 \times 7 = {}^-28$ *The product or quotient of a positive and a negative integer is negative.*

  $^-72 \div {}^-9 = 8$ *The product or quotient of two positive or two negative integers is positive.*

Find the product or quotient.

32. $^-16 \cdot 3$
33. $^-12 \cdot {}^-12$
34. $15 \cdot {}^-6$
35. $^-14 \cdot {}^-9$
36. $^-54 \div 9$
37. $124 \div {}^-4$
38. $^-75 \div {}^-15$
39. $^-119 \div 7$

- **Combine operations with integers.** (pp. 262–263)

  Evaluate the expression.

  $^-3 + 9 \times {}^-4 \div 3 - 8$ *Do all multiplication and division from left to right.*

  $^-3 + {}^-12 - 8$ *Do all addition and*
  $^-3 + {}^-12 + {}^-8$ *subtraction from*
  $^-23$ *left to right.*

Evaluate the expression.

40. $5 - 12 \div {}^-2 + 4 \times {}^-5$
41. $(16 - {}^-8) \div (3 \times {}^-2)$
42. $2^3 + (4 - {}^-6) \div {}^-2$
43. $(^-3 + 9)^2 \div (^-4 \times 3)$

PROBLEM SOLVING APPLICATIONS

44. Four students spent money on books. The amounts they spent were $2.50, $3.25, $5.00, and $5.75. Ari spent $3.25 more than Nina. Nina spent half as much as Jamal. Kenny spent $1.75 less than Jamal. How much did each person spend? (pp. 236–237)

45. Natalie, Aaron, Toral, and Casey are each on a different sports team. Toral is not on the baseball team. Casey's team plays on the ice. Aaron's team practices in the water. They play on soccer, hockey, swimming, and baseball teams. Who plays on each team? (pp. 236–237)

46. Rachel is the owner of a restaurant. Her profits and losses for the past four weeks are $^-$380, $420, $^-$145 and $620. How much was her profit or loss for the past four weeks? (pp.244–247)

47. Frances claimed that the average of the low temperatures last week was greater than $^-4°C$. Patty says that the average was less than $^-4°C$. The low temperatures last week were $^-6°C$, $^-8°C$, $^-3°C$, $^-4°C$, $0°C$, $^-9°C$, and $^-5°C$. Who is correct? Explain. (pp.260–261)

California Connections

Temperatures

California is a state of temperature extremes. Parts of the high and cold Sierra Nevada receive 50 *feet* or more of snow during the winter. Yet Death Valley, California, had the highest temperature ever recorded in the United States, 134°F.

In some parts of California, you can snow ski in the morning and play tennis in the afternoon.

1. It was 64°F when Cindy's family left Santa Barbara. The temperature was 30°F when they got to Pine Mountain Summit in Los Padres National Forest. Find the temperature difference between Pine Mountain and Santa Barbara.

2. Truckee, California, sometimes has the lowest temperature outside of Alaska. One morning it was 53°F in San Francisco and ⁻15°F in Truckee. Find the temperature difference between San Francisco and Truckee.

3. Cindy's family camped near Pine Mountain. The next day, they went on to Bakersfield, where the temperature was 86°F. Find the temperature difference between Bakersfield and Pine Mountain.

4. On a February morning, these temperatures were recorded in three California mountain towns: Tahoe City, ⁻6°F; Mammoth Lakes, 0°F; Kirkwood, ⁻11°F. Order the temperatures from coldest to warmest.

5. Here's a useful rule for estimating temperature change: temperature decreases about 5°F for every 1,000 ft that you go up in elevation. The temperature in Oakdale, California (elevation 155 ft), is 23°F. Estimate the temperature at Lake Tahoe (elevation 6,229 ft).

Because of the "wind-chill factor," air temperatures feel colder when the wind is blowing.

Use Data For 6, use the table for 10 mph winds.

CURRENT TEMPERATURES	35°F	30°F	25°F	20°F
WIND-CHILL TEMPERATURES	22°F	16°F	10°F	4°F

6. The air temperature in Quincy, California, is 5°F. A 10 mph wind is blowing. Use the patterns in the table to estimate the wind-chill temperature.

Elevations

California is a state of elevation extremes, too. In California is Mount Whitney, which, at 14,494 ft, is one of the highest peaks in the country. Here, too, is Death Valley, at 282 ft below sea level (⁻282), the country's lowest point.

Mount Whitney, California's highest point, is only about 80 miles from Death Valley, its lowest point.

Use Data **For 1–2, use the diagram at the right.**

1. What is the elevation difference between the lowest point in Death Valley and the top of Mount Whitney?

2. Hikers walked from the bottom of Death Valley to the top of Mount Whitney. By about how many feet did their elevation increase, on average, every mile?

Use Data **For 3–4, use the diagram at the right.**

3. What is the elevation difference between point *P* and the lowest point in Death Valley?

4. About how many times the elevation of point *P* is the elevation of the deepest point in the valley?

Use Data **For 5–6, use the table.**

5. Write the absolute value of Death Valley's lowest elevation and the absolute value of Long Beach's elevation. Which point is farther from sea level? Explain.

6. **REASONING** Recall that air temperature decreases about 5°F for every 1,000 ft that you go up in elevation. If the temperature at Death Valley's lowest point is 90°F, which of the cities is most likely to have a temperature of about 75°F? Explain.

ELEVATIONS OF CALIFORNIA CITIES	
Mammoth Lakes	7,860 ft
Lone Pine	3,733 ft
Victorville	2,714 ft
Ridgecrest	2,289 ft
Bakersfield	408 ft
Long Beach	170 ft

CHAPTER 14

Expressions

Kruger National Park in South Africa was established in 1898. Many different types of vegetation and wildlife are contained within its 7,523 square miles. The wildlife includes more than 800 species of mammals, birds, reptiles, and amphibians. There are more than 500 species of birds alone. Suppose there is an average of *n* buffalo per square mile. Write an expression with *n* to show how many buffalo live in the park.

ESTIMATED NUMBERS OF ANIMALS IN KRUGER NATIONAL PARK

Type of Animal	Number of Animals
Lions	1,500
Elephants	10,000
Buffalo	15,000
Crocodiles	3,000
Antelope	

Check What You Know

Use this page to help you review and remember important skills needed for Chapter 14.

✓ Vocabulary

Choose the best term from the box.

numerical expression
algebraic expression
exponent
operation

1. The number that shows how many times the base is used as a factor is a(n) __?__ .

2. A mathematical phrase that includes only numbers and operation symbols is a(n) __?__ .

3. An expression that includes a variable is called a(n) __?__ .

✓ Exponents (See p. H15.)

Find the value.

4. 4^3 5. 2^5 6. 3^4

7. 5^3 8. 3^3 9. 2^4

10. 8^2 11. 9^2 12. 2^8

13. 7^2 14. 12^2 15. 6^3

✓ Order of Operations (See p. H15.)

Evaluate each expression.

16. $6 + 3 \cdot 5$ 17. $3^2 + 4^2$ 18. $(5 + 2) \cdot 6$

19. $(8 + 4) - (3 + 2)$ 20. $24 \div (8 - 5)$ 21. $\frac{(15 - 7)}{8} \cdot 2$

22. $\frac{(5 + 9)}{7} + 4^2$ 23. $5^2 - (4 - 2)$ 24. $3^2 \div (5 - 2)$

25. $24 - 8 \div 2$ 26. $2^3 \cdot (15 - 8)$ 27. $8^2 \div (9 + 7)$

28. $3^2 + 4 - (2 + 1)$ 29. $20 - 5 \cdot 3 + 16$ 30. $3^3 \div (13 - 4) + 8$

✓ Factors (See p. H8.)

Find all the factors of the number.

31. 16   32. 24

33. 33   34. 28

35. 17   36. 64

37. 42   38. 81

39. 40   40. 39

> **LOOK AHEAD**
>
> **In Chapter 14 you will**
> - write algebraic expressions
> - evaluate algebraic expressions
> - evaluate expressions with squares and square roots

Write Expressions

Learn how to write algebraic expressions.

Remember that $a \times b$ can be written as $a \cdot b$ or ab.

John mows lawns to earn spending money. He charges a flat rate of $15.75 for each lawn plus $2.00 for every hour it takes him to mow, trim, and edge the lawn. To find the total amount of money he earns, John uses the algebraic expression below.

$15.75 for each lawn	plus	$2.00 for every hr
↓	↓	↓
15.75	+	$2 \cdot h$
		↑
		number of hours

As shown above, word expressions can be written as algebraic expressions.

EXAMPLE 1

Write an algebraic expression for the word expression.

18 more than twice s	$2s + 18$
q divided by 8, added to the product of t and r	$tr + \dfrac{q}{8}$
5.6 less than $\frac{4}{5}m$	$\frac{4}{5}m - 5.6$
the sum of $\frac{1}{2}a$, $\frac{1}{6}b$, and $\frac{1}{4}c$	$\frac{1}{2}a + \frac{1}{6}b + \frac{1}{4}c$

You can also write word expressions for numerical and algebraic expressions.

EXAMPLE 2

Write a word expression for the numerical or algebraic expression.

a. $15 + 4.5$ — the sum of 15 and 4.5

b. abc — the product of a, b, and c

c. $\frac{2}{3}m + 8$ — 8 more than $\frac{2}{3}m$

d. $xy - \frac{z}{5}$ — the difference of the product of x and y and the quotient of z divided by 5

CALIFORNIA STANDARDS AF 1.0 Students write verbal expressions and sentences as algebraic expressions and equations; they evaluate algebraic expressions, solve simple linear equations, and graph and interpret their results. **AF 1.2** Write and evaluate an algebraic expression for a given situation, using up to three variables. *also* **MR 1.1, MR 2.5, MR 3.2**

Think and ▸ Discuss

Look back at the lesson to answer the question.

1. Give another word expression for Example 2c.

Guided ▸ Practice

Write an algebraic expression for the word expression.

2. 7 more than $\frac{3}{8}y$　　　　　　　**3.** y divided by 1.5

Write a word expression for each.

4. $19 + 8.7$　　　　　**5.** $19x - 54$　　　　　**6.** $p \cdot q \cdot r$

PRACTICE AND PROBLEM SOLVING

Independent ▸ Practice

Write an algebraic expression for the word expression.

7. 2.9 more than 3 times c　　　　**8.** d divided by 8, less $\frac{1}{3}$

9. the product of h, $4j$, and k　　　**10.** the quotient of a and 3.5, less 2

11. the difference between $6.3p$ and $5m$

12. $\frac{1}{2}$ multiplied by the sum of $2m$ and $4n$

Problem Solving ▸ Applications

Write a word expression for each.

13. $42 + 2.5$　　　**14.** $8c - \frac{5}{6}$　　　**15.** $4.5 \cdot a \cdot b$

16. $yz + \frac{1}{2}x$　　　**17.** $a + b + c$　　　**18.** $98 \div 4.1y$

19. What if John charged a flat rate of $14.25 per lawn and $3.50 for each hour he mows, trims, and edges? What expression could he use to find the total amount he charges?

20. Let x represent José's height, in inches. His nephew is $\frac{1}{4}$ as tall as José. Write an algebraic expression for Jose's nephew's height.

21. Jill's niece is 2 inches taller than twice her nephew's height. What algebraic expression could you write for her niece's height?

22. A taxi charges $2.75 for the first mile and $2.25 for each additional mile. How much would a 15-mi trip cost?

23. ✎ **Write a problem** in which you have to write an algebraic expression using the variable x.

MIXED REVIEW AND TEST PREP

Evaluate the expression. (p. 262)

24. $^-15 + 10 \div 2 \cdot {}^-6$　　　**25.** $(^-2 - {}^-5)^2 \cdot {}^-4$　　　**26.** $(14 - 47) \div (2 - 5)$

27. Find a pair of numbers that have a LCM of 24 and a GCF of 2. (p. 150)

28. TEST PREP Use mental math to tell which is the solution for $x + 2.5 = 10.5$. (p. 82)

A $x = 13$　　**B** $x = 8.5$　　**C** $x = 8$　　**D** $x = 2.5$

Evaluate Expressions

Learn how to evaluate algebraic expressions.

Vocabulary

terms

like terms

Ricardo runs the concession stand at school. To find out how much money the stand makes or loses, he uses the algebraic expression $s - c - a$ where s is the amount of sales, c is the cost of food products, and a is advertising costs.

During the month of November, Ricardo spent $327 on food products and $50 on advertising for the concession stand. The amount of sales was $349. How much money did the concession stand make or lose during November?

EXAMPLE 1

Use the algebraic expression $s - c - a$.

$s - c - a$	*Replace s with 349, c with 327, and a with 50.*
$349 - 327 - 50$	*Subtract.*
$22 - 50$	*Subtract.*
$^-28$	*A negative result represents a loss.*

So, the concession stand lost $28 during November.

Algebraic expressions can be evaluated for different values.

EXAMPLE 2

Evaluate $^-4x + 7$ for $x = ^-2, ^-1, 0,$ and 1.

Remember that the order of operations is:
1. Operate inside parentheses.
2. Clear exponents.
3. Multiply and divide from left to right.
4. Add and subtract from left to right.

$x = ^-2$		$x = ^-1$	
$^-4x + 7$	*Replace x with $^-2$.*	$^-4x + 7$	*Replace x with $^-1$.*
$^-4 \cdot ^-2 + 7$	*Multiply.*	$^-4 \cdot ^-1 + 7$	*Multiply.*
$8 + 7$	*Add.*	$4 + 7$	*Add.*
15		11	
So, for $x = ^-2$, $^-4x + 7 = 15$.		So, for $x = ^-1$, $^-4x + 7 = 11$.	

$x = 0$		$x = 1$	
$^-4x + 7$	*Replace x with 0.*	$^-4x + 7$	*Replace x with 1.*
$^-4 \cdot 0 + 7$	*Multiply.*	$^-4 \cdot 1 + 7$	*Multiply.*
$0 + 7$	*Add.*	$^-4 + 7$	*Add.*
7		3	
So, for $x = 0$, $^-4x + 7 = 7$.		So, for $x = 1$, $^-4x + 7 = 3$.	

CALIFORNIA STANDARDS AF 1.2 Write and evaluate an algebraic expression for a given situation, using up to three variables. **AF 1.3** Apply algebraic order of operations and the commutative, associative, and distributive properties to evaluate expressions; and justify each step in the process. *also* **AF 1.0, NS 2.3, MR 2.2, MR 2.5, MR 3.2**

Some algebraic expressions are more complicated. To evaluate them, replace each variable with its given value. Then follow the order of operations.

EXAMPLE 3

Evaluate $2(a + b)^2 \div c$ for $a = {}^-5$, $b = 14$, and $c = {}^-3$.

$2(a + b)^2 \div c$	*Replace a, b, and c with their given values.*
$2({}^-5 + 14)^2 \div {}^-3$	*Operate inside parentheses.*
$2 \cdot 9^2 \div {}^-3$	*Clear the exponent.*
$2 \cdot 81 \div {}^-3$	*Multiply.*
$162 \div {}^-3$	*Divide.*
${}^-54$	

So, for $a = {}^-5$, $b = 14$, and $c = {}^-3$, the expression $2(a + b)^2 \div c = {}^-54$.

TECHNOLOGY LINK

More Practice: To learn more about *Evaluating Expressions,* watch the **Harcourt Math Newsroom Video** *Bangladesh Bank.*

The parts of an algebraic expression that are separated by an addition or subtraction sign are called **terms**. Before you evaluate some algebraic expressions, you can simplify them by combining like terms. **Like terms** have the same variable raised to the same power.

ALGEBRAIC EXPRESSION	LIKE TERMS
$6x + 5x + 17$	$6x$ and $5x$
$42 + 13y^2 - 10y^2$	$13y^2$ and $10y^2$

To make an algebraic expression simpler, combine like terms by adding or subtracting like terms.

ALGEBRAIC EXPRESSION	SIMPLIFIED
$6x + 5x + 17$	$11x + 17$
$42 + 13y^2 - 10y^2$	$42 + 3y^2$

EXAMPLE 4

Simplify $5x + 3x + 7$ by combining like terms. Then evaluate the expression for $x = {}^-2$.

$5x + 3x + 7 = 8x + 7$	*Simplify by combining 5x and 3x.*
$8 \cdot {}^-2 + 7$	*Replace x with ⁻2 and multiply.*
${}^-16 + 7$	*Add.*
${}^-9$	

So, $5x + 3x + 7 = {}^-9$ when $x = {}^-2$.

- Simplify $12a - 9a + 42$ by combining like terms. Then evaluate the expression for $a = {}^-4$.

You can use the Distributive Property to solve some problems mentally.

EXAMPLE 5

Evaluate pq for $p = {}^-8$ and $q = 57$.

pq

$\begin{aligned} {}^-8 \cdot 57 \\ {}^-8(50 + 7) \\ {}^-8 \cdot 50 + {}^-8 \cdot 7 \\ {}^-400 + {}^-56 \\ {}^-456 \end{aligned}$

Replace p with ${}^-8$ and q with 57.
Think of 57 as $50 + 7$. Then
use the Distributive Property
and mental math.

CHECK FOR UNDERSTANDING

Think and ▶ Discuss

Look back at the lesson to answer each question.

1. **Tell** why it can help to simplify an algebraic expression before evaluating it.

2. **Show** that evaluating the following expression before you simplify gives the same result as evaluating it after you simplify. $4x + 6x + 17$ for $x = {}^-2$

Guided ▶ Practice

Evaluate the expression for $x = {}^-2$, ${}^-1$, 0, and 1.

3. ${}^-6 + 5x$

4. $2x + 3$

5. $14 + 8x$

6. ${}^-6x - (4 - 9)$

7. $x^2 - 3$

8. $\dfrac{{}^-24}{3x - 2}$

Simplify the expression. Then evaluate the expression for the given value of the variable.

9. $4x + 3x - 8$ for $x = {}^-3$

10. $15y - 11y + 10$ for $y = {}^-3$

PRACTICE AND PROBLEM SOLVING

Independent ▶ Practice

Evaluate the expression for $x = 1, 2, 3,$ and 4.

11. ${}^-12 - 4x$

12. $0.25x + 4$

13. $21 - \dfrac{1}{2}x$

14. $x^2 + 4$

15. $\dfrac{{}^-12}{x} + 7$

16. $\dfrac{{}^-48}{2x + 4}$

Simplify the expression. Then evaluate the expression for the given value of the variable.

17. $3x + 9x - 41$ for $x = {}^-4$

18. $22y - 19y + 17$ for $y = {}^-9$

19. $356 + 13a - 7a$ for $a = {}^-20$

20. $10x + 6x - 59 + 8y$ for $x = {}^-12$ and $y = 8$

21. $2a^2 + bc$ for $a = 7$, $b = {}^-25$, and $c = 4$

22. $\dfrac{1}{2}x - 2y + z$ for $x = 20$, $y = 8$, and $z = {}^-10$

Name a property you could use to help evaluate the expression. Then evaluate the expression.

23. $x + y + z$ for $x = 17$, $y = 58$, and $z = {}^-8$

24. ab for $a = {}^-3$ and $b = 87$

Find a value of x for which the expressions are equal.

25. $5x$; $x + 8$ **26.** $x - 3$; $17 - x$ **27.** $2x - 1$; $3x$

Problem Solving ▶ Applications

28. To determine if there is enough money in her checking account to write the last check, Julie uses the expression $a + d - c$, where a is the starting amount, d is the total of the deposits made, and c is the total of her checks. Does Julie have enough money in her checking account if she started with $334?

Deposits	Checks
$72	$177
$18	$53
$28	$70
$14	$48
	$122

29. **?** **What's the Error?** Rick evaluates $x + y \cdot z$ for $x = 3$, $y = 4$, and $z = {}^-2$. He says the answer is $^-14$. What is his error? What is the correct answer?

MIXED REVIEW AND TEST PREP

30. Write an algebraic expression for 3 more than twice x. (p. 274)

31. $^-15 \div 3$ (p. 260)   **32.** $9 \cdot {}^-4$ (p. 258)

33. **TEST PREP** Find the sum. $^-26 + 9$ (p. 244)

    A $^-35$ **B** $^-17$ **C** 17   **D** 35

34. **TEST PREP** Find the difference. $37 - {}^-11$ (p. 250)

    F 48   **G** 26   **H** $^-26$ **J** $^-48$

LINKUP to Reading

Strategy • Use Graphic Aids Sometimes a graphic aid, such as a table, can help you to see the solution to a problem. Chad is ordering CDs from a catalog. Each CD costs $11 and there is a $7 charge for shipping no matter how many CDs are ordered. If Chad has $57 to spend, what is the maximum number of CDs he can order?

Use the expression $11 \cdot x + 7$ and an input/output table.

INPUT (CDs)	RULE 11 · x + 7	OUTPUT (COST)
1	11 · 1 + 7	$18
2	11 · 2 + 7	$29
3	11 · 3 + 7	$40
4	11 · 4 + 7	$51
5	11 · 5 + 7	$62

So, Chad can order a maximum of 4 CDs.

• Suppose Chad had $90, each CD cost $13, and the shipping cost $5. What is the maximum number of CDs he could order?

Extra Practice page H45, Set B

279

Squares and Square Roots

Explore how to use a model to find squares and square roots.

You need square tiles.

Vocabulary

square

square root

A **square** is the product of a number and itself. The square of 4 is 16 since $4 \cdot 4 = 16$. A square can be expressed with the exponent 2. You read 4^2 as "4 squared." Use square arrays to model square numbers.

Activity 1

• Make a square array with 4 square tiles on each side. How many tiles did you use?

• Make a square array with 5 square tiles on each side.

• Make a square array with 6 square tiles on each side.

4

4

Think and Discuss

• How many tiles are in the array with 5 tiles on each side? Complete: $5^2 = \blacksquare$.

• How many tiles are in the array with 6 tiles on each side? Complete: $6^2 = \blacksquare$.

• Without making an array, tell how many tiles are in these: a 7×7 array, an 8×8 array, a 9×9 array, and a 10×10 array.

• Suppose you have a square with n tiles on each side. How many tiles would be in the array?

Practice

Use square tiles to find each square.

1. 2^2 **2.** 3^2 **3.** 7^2

You can also use a calculator to find squares.

12 $\boxed{x^2}$ $\boxed{144}$ *Enter the value of the base, 12. Then press* $\boxed{x^2}$.

So, $12^2 = 144$.

4. Use a calculator to find 17^2, 25^2, and 51^2.

CALIFORNIA STANDARDS O—n **NS 2.0** Students calculate and solve problems involving addition, subtraction, multiplication, and division. *also* **MR 2.4**

When you find the two equal factors of a number, you are finding the **square root** of the number. The symbol for a positive square root is $\sqrt{}$.

Since $5^2 = 25$, $\sqrt{25} = 5$. It is also true that $(^-5)^2 = 25$, but we use $\sqrt{}$ for the positive square root. You read $\sqrt{25}$ as "the square root of 25."

Activity 2

You can think about the sides of a square array to help you find the square root of a number.

- Make a square array with 9 square tiles. How many tiles are on each side of your square array? What is $\sqrt{9}$?

- How many tiles are on each side of a square array with 16 tiles? What is $\sqrt{16}$?

TECHNOLOGY LINK

More Practice: Use E-Lab, *Squares and Square Roots.*
www.harcourtschool.com/ elab2002

Think and Discuss
- How many tiles do you think would be on each side of a square array with 225 square tiles? 400 squares tiles? Explain.

- Can you form a square with 10 square tiles? 12 square tiles? Explain.

You can use a calculator to find square roots.

49 (2nd) ($\sqrt{}$) [7]

So, $\sqrt{49}$ is 7.

Enter the number, 49.
Then press (2nd)
and ($\sqrt{}$) .

Practice

Find the square roots.

1. $\sqrt{64}$
2. $\sqrt{144}$
3. $\sqrt{196}$
4. $\sqrt{441}$
5. $\sqrt{484}$
6. $\sqrt{1{,}764}$
7. If $9 \cdot 9 = 81$, then $\sqrt{\blacksquare} = \blacksquare$.

MIXED REVIEW AND TEST PREP

8. Evaluate $3xy + z$ for $x = 2$, $y = ^-3$, and $z = ^-4$. (p. 276)

9. Simplify and evaluate $5a + 7a + 35$ for $a = ^-7$. (p. 276)

10. Write an algebraic expression for $\frac{1}{2}$ more than twice k. (p. 274)

11. Compare. Write $<$ or $>$ for ●. $4\frac{3}{4} \cdot 3$ ● $11\frac{1}{4} + 2.5$ (p. 234)

12. **TEST PREP** Which equals $2^4 \cdot (3 - 5) \div (^-2 - 6)$? (p. 262)

    A $^-16$　　**B** $^-4$　　**C** 2　　**D** 4

Expressions with Squares and Square Roots

Learn how to evaluate expressions with squares and square roots.

You can evaluate expressions containing squares and square roots by using the order of operations. Evaluate square roots at the same time you evaluate exponents.

EXAMPLE 1

Mayla Sue wants to put a fence around her square garden. Jay told her that the area of the garden is 64 ft². Find the perimeter of the garden.

Since the garden is square, the length of one side is $\sqrt{64}$.

$\sqrt{64} = 8$ *Evaluate* $\sqrt{64}$. $\sqrt{64} = 8$ *since* $8 \cdot 8 = 64$.

So, the length of one side is 8 ft. The formula for the perimeter of a square is $P = 4s$. Evaluate for $s = 8$.

$P = 4s = 4 \cdot 8$ *Replace s with 8. Then multiply.*

$P = 32$ So, the perimeter of the garden is 32 ft.

EXAMPLE 2

Evaluate $18 - (\sqrt{100} + 5^2)$.

$18 - (\sqrt{100} + 5^2)$ *Evaluate* $\sqrt{100}$ *and* 5^2.

$18 - (10 + 25)$ *Operate inside parentheses.*

$18 - 35$ *Subtract.*

$^-17$ So, $18 - (\sqrt{100} + 5^2) = {}^-17$.

Sometimes the value of a variable is a square root.

EXAMPLE 3

Evaluate $(x - 2)^2 + 8$ for $x = \sqrt{25}$.

$(x - 2)^2 + 8$ *Substitute* $\sqrt{25}$ *for x.*

$(\sqrt{25} - 2)^2 + 8$ *Evaluate* $\sqrt{25}$ *inside the parentheses.*

$(5 - 2)^2 + 8$ *Operate inside parentheses.*

$3^2 + 8$ *Clear the exponent.*

$9 + 8$ *Add.*

17 So, for $x = \sqrt{25}$, the expression $(x - 2)^2 + 8 = 17$.

CALIFORNIA STANDARDS O━ₙ NS 2.0 Students calculate and solve problems involving addition, subtraction, multiplication, and division. **AF 1.3** Apply algebraic order of operations and the commutative, associative, and distributive properties to evaluate expressions; and justify each step in the process. *also* **MR 2.5**

Think and ▶
Discuss

Look back at the lesson to answer the question.

1. **Explain** how to evaluate the expression $\sqrt{36} \cdot (7 - 5)$. Then evaluate.

Guided ▶
Practice

Evaluate the expression.

2. $\sqrt{25} - 8^2 + 42$

3. $10 + \sqrt{36} - 7^2$

4. $\sqrt{100} \cdot \sqrt{225}$

5. $45 \div \sqrt{81} \cdot {}^-7 + 59$

Independent ▶
Practice

Evaluate the expression.

6. $\sqrt{4} - 5^2 + 22$

7. $246 + \sqrt{16} - 10^2$

8. $^-25\,(\sqrt{49} - 10)$

9. $88 \div \sqrt{121} \cdot {}^-3$

Evaluate the expression for the given value of the variable.

10. $(y + 2) - 42$ for $y = \sqrt{1}$

11. $(5^2 + \sqrt{p}) \div {}^-4$ for $p = 225$

12. $4a + b - 125$ for $a = \sqrt{9}$ and $b = 9$

13. $(x - 5)^2 + 7y$ for $x = \sqrt{144}$ and $y = \sqrt{16}$

Compare for **a = 5, b = 20, and c = 16. Use <, >, or = .**

14. $a^2 + 10$ ⬤ $10a^2$

15. $\sqrt{ab}$ ⬤ ^-c

Problem Solving ▶
Applications

16. **REASONING** Insert one square root symbol so the value of the expression $25 + 64 - 16 \cdot 81$ is $^-55$.

17. A square flower garden and a sidewalk around the garden have an area of 400 ft². The sidewalk has a width of 3 ft. What is the length of one side of the flower garden?

18. **⟨?⟩ What's the Question?** The expression is $\sqrt{x} + {}^-2$. The answer is 7.

Evaluate the expression for **x = 1, 2, 3, and 4.** (p. 276)

19. $^-4x + 2$

20. $20 - 6x$

21. $\dfrac{12}{x} - 10$

22. Find $5\frac{2}{3} \div 2\frac{1}{3}$. (p. 210)

23. **TEST PREP** A bag held $6\frac{3}{4}$ c of flour. Bill took out $2\frac{1}{2}$ c. He had too much, so he put back $\frac{3}{4}$ c. How much flour was left in the bag? (p. 186)

A $3\frac{1}{2}$ c **B** 5 c **C** $8\frac{1}{2}$ c **D** 10 c

1. **VOCABULARY** When you find the two equal factors of a number, you are finding the __?__ of the number. (p. 281)

2. **VOCABULARY** The parts of an algebraic expression that are separated by an addition or subtraction sign are __?__. (p. 277)

Write an algebraic expression for the word expression.
(pp. 274–275)

3. the product of 2, $^-2$, and b

4. the quotient of s divided by 1.6

5. 15 more than twice p

6. $\frac{1}{2}$ less than the product of 6 and r

7. the sum of $4a$ and $5b$

8. the product of ^-2q and r

Evaluate the expression for $y = 5$. (pp. 276–279)

9. $^-4y + 12$

10. $y^3 + 12$

11. $4 + 2y$

12. $5 - 3y$

13. $\frac{10}{y} - 1$

14. $10 + 3y$

Evaluate the expression for $a = ^-4$, $b = 3$, and $c = ^-5$. (pp. 276–279)

15. $3b + (a - c)$

16. $12 - a + c$

17. $28 + (c + 2)^2$

18. $18c + \frac{8b}{a}$

19. $(3c - a)^2 - b$

20. $338 \div (a + {}^-2bc)^2$

Simplify the expression. Then evaluate the expression for the given value of the variable. (pp. 276–279)

21. $6x - 2x$, $x = ^-2$

22. $3x + 2x + 6$, $x = 1$

23. $10x - 2x$, $x = ^-10$

Evaluate the expression. (pp. 282–283)

24. $\sqrt{49} - 2 + 5$

25. $3(6^2 - 16)$

26. $4\sqrt{81} - 3 \times 6$

27. $\sqrt{25} \cdot \sqrt{16}$

28. $^-124\,(\sqrt{64} - 6)$

29. $10 + \sqrt{100} - 4^2$

Evaluate the expression for the given value of the variable. (pp. 282–283)

30. $4 - \sqrt{c}$ for $c = 49$

31. $^-(5^2) - d$ for $d = \sqrt{121}$

32. A cashier uses the expression $d + s - c$ to check the amount of money in her cash drawer, where d is the initial amount in the drawer, s is the amount of cash that customers pay, and c is the amount of change that customers receive. Suppose the initial amount is $125.00, customers pay $15.00, $18.25, and $24.00, and the amount of change returned is $1.50 and $1.95. How much money is in the cashier's drawer? (pp. 276–279)

33. Write an expression for the perimeter of the rectangle. Then evaluate the expression to find the perimeter when $a = 4$ cm and $b = 7$ cm. (pp. 276–279)

Choose the answer.

See item **6.**

Think about the symbols and what they tell you to do. Check your computation to make sure it is correct before you choose the answer.

Also see problem **6**, p. H64.

Choose the best answer.

1. Which of the following expressions is represented by the algebraic expression $m - 6$?

    A m decreased by 6

    B 6 more than m

    C the product of m and 6

    D m divided by 6

2. Mac is buying items for a cookout. Hot dogs are sold 8 in a pack. Buns are sold 6 in a pack. What is the least number of packs of each he should buy to have an equal number of hot dogs and buns?

    F 2 packs of hot dogs and 3 packs of buns

    G 3 packs of hot dogs and 3 packs of buns

    H 3 packs of hot dogs and 4 packs of buns

    J 4 packs of hot dogs and 3 packs of buns

3. The attendance at a concert on Saturday night was 42 more than the attendance on Friday night. If a represents the attendance on Friday night, which expression represents the attendance for Saturday night?

    A $42a$

    B $42 - a$

    C $a + 42$

    D $a - 42$

4. $600 \div (52 - 32) \times 3$

    F 10   **H** 90

    G 57   **J** Not here

5. The cost of a bag of apples was shared equally among 5 friends. If a represents the cost of the apples, which algebraic expression represents the amount each friend paid?

    A $\frac{5}{a}$ **C** $5 - a$

    B $a + 5$ **D** $\frac{a}{5}$

6. $6^2 + \sqrt{25}$

    F 11   **H** 41

    G 37   **J** Not here

7. What is the value of $c - 4$ for $c = {}^-5$?

    A $^-9$ **C** 2

    B $^-2$ **D** Not here

8. Martha kept track of her quiz scores. During one week she received these scores: 12, 17, 27, 23, 19. What is the difference between the mean and the median for the quiz scores?

    F 0.4   **H** 1.6

    G 0.6   **J** 19

9. $8 \times {}^-4 + 17$

    A $^-544$ **C** 49

    B $^-15$ **D** Not here

10. Which algebraic expression represents "29.1 times a number, n"?

    F $29.1 - n$

    G $29.1 + n$

    H $n \div 29.1$

    J $29.1 \times n$

CHAPTER 15

Addition and Subtraction Equations

Beginning on January 27, 1984, Klaus Friedrich single-handedly set up and toppled the greatest number of dominoes ever. Not all of the dominoes he set up toppled on the first push. The expression $281{,}581 + x$ can be used to model the toppled and untoppled dominoes. How can you use the equation $320{,}236 = 281{,}581 + x$ to find the number of untoppled dominoes?

KLAUS FRIEDRICH'S DOMINO TOPPLING		
	Set Up	Toppled
Amount of Time	18,600 min	13 min
Number of Dominoes	320,236	281,581

Check What You Know

Use this page to help you review and remember important skills needed for Chapter 15.

Vocabulary

Choose the best term from the box.

1. A statement showing that two quantities are equal is a(n) __?__ .

2. The property which states that $5 + x = x + 5$ is the __?__ Property of Addition.

> Associative
> Commutative
> equation
> expression

✓ Add and Subtract (See p. H16.)

Find the sum or difference.

3. $25 + 12$	**4.** $256 + 125$	**5.** $49 - 26$
6. $321 - 265$	**7.** $6.25 + 5.12$	**8.** $9.14 + 4.27$
9. $8.4 - 5.7$	**10.** $14.72 - 11.82$	**11.** $\frac{5}{6} + \frac{3}{4}$
12. $8\frac{2}{3} + 5\frac{1}{8}$	**13.** $\frac{5}{7} - \frac{3}{14}$	**14.** $6\frac{1}{8} - 2\frac{3}{5}$
15. $4\frac{1}{4} + 3\frac{2}{3}$	**16.** $\frac{4}{5} - \frac{1}{3}$	**17.** $4\frac{4}{7} - 2\frac{1}{4}$

Inverse Operations (See p. H16.)

Use the inverse operation to check the equation.

18. $14 + 5 = 19$	**19.** $24 + 19 = 43$	**20.** $125 + 219 = 344$
21. $19 - 3 = 16$	**22.** $37 - 14 = 23$	**23.** $242 - 196 = 46$
24. $8 \cdot 6 = 48$	**25.** $25 \cdot 5 = 125$	**26.** $14 \cdot 34 = 476$
27. $12 \div 6 = 2$	**28.** $72 \div 8 = 9$	**29.** $252 \div 12 = 21$
30. $52 \cdot 4 = 208$	**31.** $400 \div 16 = 25$	**32.** $19 \cdot 17 = 323$

✓ Words for Operations (See p. H17.)

Write the operation described by the phrase.

33. the sum of p and 12

34. the product of x and 3

35. the difference of 30 and k

36. the quotient of b and 4

37. 15 times y

38. m decreased by 4

39. n reduced by 72

40. 234 increased by z

> **LOOK AHEAD**
>
> In Chapter 15 you will
> - write equations
> - solve addition equations
> - solve subtraction equations

Connect Words and Equations

Learn how to translate words into numbers, variables, and operations.

Mt. Whitney is the highest mountain in California.

Remember that an equation is a statement showing that two quantities are equal.

You can translate word expressions into equations by translating words into numbers, variables, and operations.

The lowest point in the United States is Death Valley, California, with an altitude of $^-86$ m. The altitude of Mt. Whitney, also in California, is 4,504 m greater than the altitude of Death Valley.

EXAMPLE 1

Write an equation to find the altitude of Mt. Whitney.

Choose a variable. Let *a* be the altitude of Mt. Whitney. Then translate the words into an equation.

altitude of Mt. Whitney	is	4,504 m	greater than	altitude of Death Valley
↓	↓	↓	↓	↓
a	=	4,504	+	$^-86$

So, the equation is $a = 4{,}504 + {}^-86$.

• Write an equation to show that the altitude of Mt. Whitney is 1,777 m less than the altitude of Mt. McKinley, which is 6,195 m.

Equations can also include subtraction, multiplication, and division.

EXAMPLE 2

Write an equation for the following word sentence: The number of hours worked, less $3\frac{1}{2}$ hr, is $15\frac{1}{4}$ hr.

Choose a variable. Let *h* be the number of hours worked.

number of hours worked	less	$3\frac{1}{2}$ hr	is	$15\frac{1}{4}$ hr
↓	↓	↓	↓	↓
h	−	$3\frac{1}{2}$	=	$15\frac{1}{4}$

Translate the sentence into an equation.

So, the equation is $h - 3\frac{1}{2} = 15\frac{1}{4}$.

• Write an equation for the following word sentence: Four times the original price is $24.96.

CALIFORNIA STANDARDS **AF 1.0** Students write verbal expressions and sentences as algebraic expressions and equations; they evaluate algebraic expressions, solve simple linear equations, and graph and interpret their results. **AF 1.1** Write and solve one-step linear equations in one variable. *also* **MR 2.5**

Think and Discuss ▶ Look back at the lesson to answer each question.

1. Tell what the variable represents when you translate a word sentence into an algebraic equation.

2. Tell what kind of equation matches this sentence: The quotient of $265.75 and the number of payments is $53.15. Then write the equation.

Guided Practice ▶ Write an equation for the word sentence.

3. 7 more than a number is 20.

4. 9 fewer than a number is $17\frac{5}{8}$.

5. 5 times the number of marbles is 35.

6. The quotient of 567 points and the number of test scores is 81.

PRACTICE AND PROBLEM SOLVING

Independent Practice ▶ Write an equation for the word sentence.

7. 14 is 12 more than a number.

8. 8.09 less than a number equals 44.2.

9. A number divided by $2\frac{3}{4}$ is $\frac{5}{6}$.

10. $^{-}16$ times a number is $^{-}144$.

Problem Solving Applications

11. Science The diameter of Saturn is 72,000 miles. This is about 9 times as great as the diameter of Earth. Write an equation that describes this relationship.

12. Translate the following sentence into an equation: $^{-}2$ is 10 less than a number. Could 12 be the number? Explain.

13. **Write a problem** using facts found in your science book. Then translate the problem into an equation.

14. Number Sense Picnic forks come in packs of 20, and picnic knives come in packs of 32. What is the least number of packs of each that you could buy to have the same number of forks and knives?

MIXED REVIEW AND TEST PREP

Simplify the expression. Then evaluate the expression for the given value of the variable. (p. 276)

15. $5x + 3x - 15$ for $x = ^{-}2$

16. $10y - 8y + 22$ for $y = ^{-}6$

17. $25 + 22z - 13z$ for $z = ^{-}8$

18. Evaluate the expression $^{-}3 \cdot (5 + 2) - 17$. (p. 262)

19. TEST PREP Find $^{-}25 - ^{-}16$. (p. 250)

A $^{-}41$ **B** $^{-}9$ **C** 9   **D** 41

Model and Solve One-Step Equations

Explore how to model the solutions to one-step equations.

You need algebra tiles, or green paper rectangles and yellow squares.

Activity

• Model an equation by using algebra tiles. Use a green rectangle to represent the variable. Use a yellow square to represent 1. Model and solve the equation $x + 3 = 4$.

$$x \quad + \quad 3 \quad = \quad 4 \qquad \leftarrow x + 3 = 4$$

• To solve an equation, you must get the variable alone on one side. To do this, take away 3 ones from each side.

• Taking 3 ones from each side of the equation leaves the variable alone on one side. The equation is now solved.

$$x \quad = \quad 1 \qquad \leftarrow \textit{Green rectangle is alone.}$$

Math **I**dea ▶ The value of the variable that makes the equation true is called the solution of the equation.

Think and Discuss

• What is the solution of the equation $x + 3 = 4$?

• How can you tell this from the model?

• What operation did you model in the second step?

• **REASONING** What would you do differently to solve the equation $x - 3 = 4$?

TECHNOLOGY LINK

More Practice: Use E-Lab, Modeling Addition and Subtraction Equations.
www.harcourtschool.com/ elab2002

Practice

Solve each equation by using algebra tiles.

1. $x + 2 = 6$ **2.** $5 = x + 3$ **3.** $x + 1 = 8$ **4.** $x + 4 = 9$

5. $x + 1 = 4$ **6.** $x + 4 = 7$ **7.** $6 = x + 5$ **8.** $x + 2 = 7$

CALIFORNIA STANDARDS AF 1.0 Students write verbal expressions and sentences as algebraic expressions and equations; they evaluate algebraic expressions, solve simple linear equations, and graph and interpret their results. **○━┓ AF 1.1** Write and solve one-step linear equations in one variable. **○━┓ NS 2.0** Students calculate and solve problems involving addition, subtraction, multiplication, and division. *also* **MR 2.5**

Solve Addition Equations

Learn how to solve addition equations.

Vocabulary

Subtraction Property of Equality

Ralf Laue of Leipzig, Germany, set a world record in February of 1999 for the most dominoes stacked. He stacked 555 dominoes on a single supporting domino.

How many more dominoes would you need to stack to match the world record if you have already stacked 123? One way to solve the problem is by representing it with an equation.

dominoes left to stack		dominoes stacked		total
↓		↓		↓
d	$+$	123	$=$	555

Addition and subtraction are inverse operations. To solve the addition equation, you will use the inverse operation, subtraction, to get the variable alone on one side of the equation. The Subtraction Property of Equality justifies that step.

Subtraction Property of Equality	$5 = 5$
If you subtract the same number from both	$5 - 2 = 5 - 2$
sides of an equation, the two sides remain equal.	$3 = 3$

EXAMPLE 1

Solve the equation $d + 123 = 555$ to find the number of additional dominoes you need to stack. Check your solution.

$d + 123 = 555$	*Write the equation.*
$d + 123 - 123 = 555 - 123$	*Use the Subtraction Property of Equality.*
$d + 0 = 432$	*Use the Identity Property of Zero.*
$d = 432$	
$d + 123 = 555$	*Check your solution.*
$432 + 123 \overset{?}{=} 555$	*Replace d with 432.*
$555 = 555$ ✓	*The solution checks.*

Remember that the Identity Property of Zero states: For all numbers a, $a + 0 = a$.

So, you must stack 432 more dominoes to match the world record.

CALIFORNIA STANDARDS **AF 1.0** Students write verbal expressions and sentences as algebraic expressions and equations; they evaluate algebraic expressions, solve simple linear equations, and graph and interpret their results. ○━ **AF 1.1** Write and solve one-step linear equations in one variable. ○━ **NS 2.0** Students calculate and solve problems involving addition, subtraction, multiplication, and division. *also* NS 2.1, MR 2.2, MR 2.7, MR 3.1

291

You can sometimes use the Commutative Property of Addition to help solve equations. Remember that equations can have decimals, fractions, or negative integers.

EXAMPLE 2

Solve and check. $8.4 + x = 12.7$

$$8.4 + x = 12.7 \qquad \text{\textit{Use the Commutative Property}}$$
$$x + 8.4 = 12.7 \qquad \text{\textit{of Addition.}}$$
$$x + 8.4 - 8.4 = 12.7 - 8.4 \quad \text{\textit{Use the Subtraction Property of Equality.}}$$
$$x + 0 = 4.3 \qquad \text{\textit{Use the Identity Property of Zero.}}$$
$$x = 4.3$$

$$8.4 + x = 12.7 \qquad \text{\textit{Check your solution.}}$$
$$8.4 + 4.3 \stackrel{?}{=} 12.7 \qquad \text{\textit{Replace x with 4.3.}}$$
$$12.7 = 12.7 \checkmark \qquad \text{\textit{The solution checks.}}$$

So, $x = 4.3$.

• Solve. $1.4 + x = {}^-8$.

Sometimes the variable will be on the right side of the equation.

EXAMPLE 3

Solve and check. $14 = 8\frac{2}{3} + k$

$$14 = 8\frac{2}{3} + k \qquad \text{\textit{Use the Commutative Property of}}$$
$$14 = k + 8\frac{2}{3} \qquad \text{\textit{Addition.}}$$
$$14 - 8\frac{2}{3} = k + 8\frac{2}{3} - 8\frac{2}{3} \qquad \text{\textit{Use the Subtraction Property of Equality.}}$$
$$5\frac{1}{3} = k + 0 \qquad \text{\textit{Use the Identity Property of Zero.}}$$
$$5\frac{1}{3} = k$$

$$14 = 8\frac{2}{3} + k \qquad \text{\textit{Check your solution.}}$$
$$14 \stackrel{?}{=} 8\frac{2}{3} + 5\frac{1}{3} \qquad \text{\textit{Replace k with }} 5\frac{1}{3}.$$
$$14 = 14 \checkmark \qquad \text{\textit{The solution checks.}}$$

So, $k = 5\frac{1}{3}$.

• Solve. $6\frac{5}{6} = 3\frac{1}{4} + y$.

CHECK FOR UNDERSTANDING

Think and ▶
Discuss

Look back at the lesson to answer each question.

1. **Explain** why the choice of a variable does not affect the solution of the equation.

2. **Describe** how to solve an addition equation.

Solve and check.

3. $x + 10 = 17$ **4.** $3 + b = {}^-9$ **5.** $5.2 = c + 2.5$

6. $y + 1.2 = 18$ **7.** $21 = m + 5\frac{3}{4}$ **8.** $7\frac{1}{5} + d = 22$

PRACTICE AND PROBLEM SOLVING

Independent ▸
Practice

Solve and check.

9. $x + 4 = 12$ **10.** $29 = a + 17$ **11.** $9 + k = {}^-32$

12. $z + 6.3 = 11.7$ **13.** $4.2 = b + 1.3$ **14.** $14.9 = 12.5 + c$

15. $y + 9\frac{1}{4} = 18$ **16.** $16 = 3\frac{2}{5} + s$ **17.** ${}^-12 = t + 7$

Problem Solving ▸
Applications

For 18–19, choose a variable, write an equation, and solve it.

18. What if you stack 221 dominoes? How many more would you have to stack to match the world record?

19. Geometry The perimeter of the triangle at the right is 29 cm. Use an equation to find the unknown length.

20. ❓ **What's the Question?** Jay opened a savings account with a deposit of $124.50. Later he made another deposit and his account balance was $350.27. The answer is $225.77.

MIXED REVIEW AND TEST PREP

For 21–22, write an equation for the word sentence. (p. 288)

21. 11 is 8 more than a number.

22. 3.65 less than a number is 12.2.

23. Evaluate. $\sqrt{64} - 4 + 9$ (p. 282)

24. Evaluate. $25 + \sqrt{81} - 6^2$ (p. 282)

25. TEST PREP Evaluate $3(a - b)^2 \cdot c$ for $a = {}^-3$, $b = 1$, and $c = {}^-2$. (p. 276)

A 96   **B** 46   **C** ${}^-24$ **D** ${}^-96$

Thinker's CORNER

Equation Relation Solve the equations below. What is the relationship among the solutions of the blue equations? Which one of the green equations should be blue?

$x + 5 = 12$	$59 = a + 17$	$b + 19 = 22$
$10 = s + 2$	$b + 15 = 43$	$95 = a + 25$
$r + 6 = 29$	$78 = m + 1$	$c + 4 = 48$
$79 = p + 23$	$x + 12 = 103$	$105 = t + 42$

Solve Subtraction Equations

Learn how to solve subtraction equations.

Vocabulary

Addition Property of Equality

Tamika was practicing the trick of catching dimes stacked on her elbow. On her first try, she caught 13 dimes. How many dimes were originally stacked on her elbow if 11 fell to the floor?

One way to solve this problem is by writing and solving an equation. Let d be the number of dimes originally stacked on Tamika's elbow.

dimes on elbow		dimes caught		dimes on floor
↓		↓		↓
d	$-$	13	$=$	11

Addition and subtraction are inverse operations. To solve the subtraction equation, use the inverse operation, addition, to get the variable alone on one side of the equation. The Addition Property of Equality justifies that step.

Addition Property of Equality	$5 = 5$
If you add the same number to both sides of an equation, the two sides remain equal.	$5 + 2 = 5 + 2$
	$7 = 7$

EXAMPLE

Solve the equation $d - 13 = 11$ to find the number of dimes originally stacked on Tamika's elbow. Check your solution.

$$d - 13 = 11 \qquad \text{\textit{Write the equation.}}$$
$$d - 13 + 13 = 11 + 13 \qquad \text{\textit{Use the Addition Property of Equality.}}$$
$$d + 0 = 24 \qquad \text{\textit{Use the Identity Property of Zero.}}$$
$$d = 24$$

$$d - 13 = 11 \qquad \text{\textit{Check your solution.}}$$
$$24 - 13 \overset{?}{=} 11 \qquad \text{\textit{Replace d with 24.}}$$
$$11 = 11 \checkmark \qquad \text{\textit{The solution checks.}}$$

So, Tamika had 24 dimes stacked on her elbow.

• Solve. $k - 4.7 = 5.8$

CALIFORNIA STANDARDS AF 1.0 Students write verbal expressions and sentences as algebraic expressions and equations; they evaluate algebraic expressions, solve simple linear equations, and graph and interpret their results. O─n AF 1.1 Write and solve one-step linear equations in one variable. *also* O─n NS 2.0, NS 2.1, MR 1.3, MR 2.7, MR 3.1

Think and ▶
Discuss

Look back at the lesson to answer the question.

1. **Explain** how you know what number to add to both sides of a subtraction equation.

Guided ▶
Practice

Solve and check.

2. $x - 8 = 15$

3. $b - 23 = {}^-29$

4. $7.9 = c - 3.4$

5. $y - 1.2 = 22$

6. $7\frac{1}{8} = m - 4\frac{3}{4}$

7. $d - 8\frac{4}{5} = 25\frac{2}{3}$

PRACTICE AND PROBLEM SOLVING

Independent ▶
Practice

Solve and check.

8. $x - 6 = 10$

9. $29 = a - 4$

10. $k - 5 = {}^-2$

11. $z - 5.8 = 11.2$

12. $7.2 = b - 1.9$

13. $14.5 = c - 8.8$

14. $y - 9\frac{1}{4} = 18$

15. $17\frac{2}{3} = s - 5\frac{1}{4}$

16. ${}^-8 = t - 5$

17. $2.8 = m - 8.3$

18. $h - 11 = {}^-3$

19. $f - \frac{3}{8} = \frac{5}{12}$

Problem Solving ▶
Applications

For 20–21, choose a variable, write an equation, and solve it.

20. **What if** Tamika had caught 17 dimes but 2 dimes fell on the floor? How many dimes would she have had stacked on her elbow?

JUANITA'S BACK-TO-SCHOOL SHOPPING

$40 School Supplies

$90 New Bicycle

$110 New Clothes

21. **Use Data** Look at the graph. Juanita withdrew money from her savings account for back-to-school shopping. She has $527 left in her savings account. How much did she have in her account before she withdrew the money?

22. **❓ What's the Error?** Frank said the solution of the equation $x - 7 = 10$ is $x = 3$. Find his error and then solve the equation.

23. Change one operation sign so the value of the expression $61 - 12 - 13 - 14 - 5 - 7$ is doubled.

MIXED REVIEW AND TEST PREP

For 24–26, solve and check. (p. 291)

24. $x + 3 = 9$

25. $5.6 = b + 1.2$

26. $5\frac{1}{4} = 3 + p$

27. Evaluate ${}^-2x + 5$ for $x = {}^-2, {}^-1, 0,$ and 1. (p. 276)

28. **TEST PREP** The low temperatures for five days in Yorktown were ${}^-2°F, {}^-3°F, 4°F, {}^-1°F,$ and $2°F$. Find the average low temperature for the five days. (p. 260)

A ${}^-2°F$ **B** $0°F$ **C** $2°F$ **D** $3°F$

1. **VOCABULARY** If you subtract the same number from both sides of an equation, you are using the __?__ Property of Equality. (p. 291)

2. **VOCABULARY** Subtraction is the __?__ of addition. (p. 291)

3. **VOCABULARY** If you add the same number to both sides of an equation, you are using the __?__ Property of Equality. (p. 294)

Write an equation for each word sentence. (pp. 288–289)

4. 72 is 15 more than a number.

5. 8.6 decreased by a number is 9.

6. ⁻18 times a number is ⁻45.

7. A number divided by $\frac{3}{8}$ is 16.

8. The product of $1\frac{3}{4}$ and a number is $21\frac{3}{5}$.

9. A number increased by 17 is 24.

10. 72 less than a number is 125.

11. The quotient of a number and 8 is 17.

Solve and check. (pp. 291–295)

12. $x + 7 = 10$

13. $9 + y = {}^-19$

14. $x - 6 = 3$

15. $3.78 = c - 2.33$

16. $21 + n = 52$

17. $x + 3\frac{1}{3} = 4\frac{2}{3}$

18. $3.4 + x = 19.5$

19. $1.9 = k - 1.2$

20. $y - 3\frac{1}{9} = 11\frac{5}{9}$

21. $d - 23 = {}^-18$

22. $42\frac{1}{5} + v = 51\frac{2}{3}$

23. $x - 33.7 = 10.9$

24. $w + 9\frac{3}{8} = 11\frac{2}{3}$

25. $2.3 + x = 5.4$

26. $q - 9.1 = 1.37$

27. $3\frac{1}{4} = x - 2\frac{1}{3}$

28. $17.5 = 2.7 + x$

29. $\frac{3}{5} + y = \frac{9}{10}$

For 30–33, choose a variable, write an equation, and solve it.

30. After soccer practice Li saw that there were 14 bottles of water left. During the practice the team consumed 18 bottles. How many bottles had there been at the start of practice? (p. 294)

31. Mirabelle has collected $37.50 so far for her favorite charity. The most she had collected in the past was $55.75. How much more does she need to collect to equal her record? (p. 291)

32. A triangle has a perimeter of 42 in. The sum of the lengths of two sides of the triangle is 29 in. Find the length of the third side of the triangle. (p. 291)

33. Thelma removed 12 cans of tomato paste from the pantry in her restaurant. There were 17 cans left. How many cans were originally in the pantry? (p. 294)

Check your work.

See item **6.**

Use number sense to decide if c will be greater or less than the difference. Write the inverse of the equation to find the value of c.

Also see problem **7**, p. H65.

For 1–12, choose the best answer.

1. Solve for y.
$$y - 8 = 4$$
   A $y = 2$ **C** $y = 12$

   B $y = 4$ **D** $y = 32$

2. Solve for p.
$$6 + p = 18$$
   F $p = 3$ **H** $p = 24$

   G $p = 12$ **J** $p = 108$

3. A backpack and a spiral notebook cost $12.78 before tax. The notebook costs $1.49. If b represents the cost of the backpack, which equation could be used to find the cost of the backpack?

   A $\$12.78 - b = \$14.27$

   B $\$1.49 + b = \$12.78$

   C $b - \$1.49 = \$12.78$

   D $\$1.49 \times b = \$12.78$

4. Solve for j.
$$j + 13.4 = {}^-26.8$$
   F $j = {}^-40.2$ **H** $j = {}^-18.2$

   G $j = {}^-22.2$ **J** $j = {}^-13.4$

5. Solve for k.
$$6.2 + k = 20$$
   A $k = 13.8$ **C** $k = 18.8$

   B $k = 14.2$ **D** $k = 26.2$

6. Solve for c.
$$c - 18.95 = 72.5$$
   F $c = 53.55$ **H** $c = 90.145$

   G $c = 54.45$ **J** $c = 91.45$

7. What is the value of $a \div b$ for $a = 25.5$ and $b = 1.5$?

   A 17   **C** 170

   B 24   **D** Not here

8. Terri blends $\frac{3}{4}$ cup of water, $\frac{5}{8}$ cup of milk, and $\frac{1}{3}$ cup of ice. Which is a reasonable total for the amount she blended?

   F Less than 1 c

   G Between $1\frac{1}{2}$ and 2 c

   H More than 2 c

   J About $2\frac{1}{2}$ c

9. Which fraction is equivalent to 0.003?

   A $\frac{3}{10}$ **C** $\frac{3}{1,000}$

   B $\frac{3}{100}$ **D** $\frac{3}{10,000}$

10. Solve for n.
$$n - 8 = 37$$
   F $n = 25$ **H** $n = 37$

   G $n = 29$ **J** $n = 45$

11. What is the prime factorization of 210?

   A $2 \times 3 \times 9$

   B $2 \times 3 \times 5 \times 7$

   C $2^2 \times 3 \times 5$

   D $2 \times 3^2 \times 5$

12. What is the GCF of 39 and 91?

   F 3   **H** 13

   G 7   **J** Not here

16 Multiplication and Division Equations

Travelers may rent cars, use public transportation, or take taxis in the cities they visit. In San Francisco, for example, it costs $2.50 when you get in a cab. Added to that is a mileage charge of $1.80 per mile. You can use the equation $m = \$1.80x$ to find the mileage charge, m, for any number of miles, x. How would you find the mileage charge for a 2-mile trip? How would you find the number of miles traveled if the mileage charge is $4.50?

TAXI CHARGES		
City	Starting Fare	Per-Mile Rate
Chicago, IL	$1.60	$1.40
Dallas, TX	$2.70	$1.20
New York City, NY	$2.00	$1.50
San Francisco, CA	$2.50	$1.80
Seattle, WA	$1.80	$1.80

Check What You Know

Use this page to help you review and remember important skills needed for Chapter 16.

✓ Vocabulary

Choose the best term from the box.

> Fahrenheit
> Celsius
> equation
> expression

1. The temperature scale on which water freezes at 0° and boils at 100° is the __?__ scale.

2. A statement showing that two quantities are equal is a(n) __?__.

✓ Words and Equations (See p. H17.)

Write an equation for the word expression.

3. 18 is 6 more than a number.

4. 12.5 less than a number equals 23.7.

5. A number divided by 2 is $\frac{2}{3}$.

6. 6 times a number is $^-90$.

7. Twice a number is 47.

8. The quotient of a number and 3 is $^-24$.

✓ Evaluate Expressions (See p. H18.)

Evaluate the expressions for the given values of the variables.

9. $2x$ for $x = 7$

10. $3a + 7$ for $a = {^-5}$

11. $5t - 13$ for $t = {^-12}$

12. pq for $p = 5$ and $q = 14$

13. $4mn$ for $m = {^-5}$ and $n = 7$

14. ^-8ab for $a = {^-5}$ and $b = 3$

15. $\frac{1}{2}x$ for $x = 18$

16. $\frac{2}{3}k$ for $k = 24$

17. $\frac{z}{4}$ for $z = 100$

18. $\frac{5}{9}f$ for $f = 27$

19. $\frac{9}{5}c$ for $c = 55$

20. $\frac{9}{5}c + 32$ for $c = 18$

21. $3.9b$ for $b = 7$

22. $^-7c + 3$ for $c = {^-5}$

23. $\frac{4}{5} + \frac{1}{2}c$ for $c = 6$

✓ Mental Math and Equations (See p. H11.)

Solve the equation by using mental math.

24. $5x = 20$

25. $3y = 15$

26. $42 = 6z$

27. $72 = 8a$

28. $\frac{b}{3} = 4$

29. $\frac{c}{5} = 6$

30. $4 = \frac{p}{4}$

31. $7 = \frac{q}{8}$

32. $\frac{4.5}{x} = 0.5$

33. $0.36 = 9y$

> ### LOOK AHEAD
>
> **In Chapter 16 you will**
> - solve multiplication and division equations
> - use formulas
> - explore two-step equations

Model Multiplication Equations

Explore how to model the solutions of multiplication equations.

You need algebra tiles or paper rectangles and squares.

You can use algebra tiles to model solving multiplication equations.

Activity

• A green rectangle represents a variable, and a yellow square represents the value 1. Here is a model of the multiplication equation $3y = 12$. To represent $3y$, use 3 green rectangles.

$$3y \qquad = \qquad 12$$

• Divide each side of the model into 3 equal groups.

• Look at one group on each side. What is in each pair? What equation does this model?

Think and Discuss

• What is the solution of the equation $3y = 12$?

• What operation did you model in the second step? How does this operation relate to multiplication?

• **REASONING** How do you think you would solve an algebraic equation involving division?

Practice

Use algebra tiles to model and solve each equation. Record each step.

1. $2c = 8$ **2.** $3a = 9$ **3.** $4b = 8$ **4.** $15 = 5y$

CALIFORNIA STANDARDS **AF 1.0** Students write verbal expressions and sentences as algebraic expressions and equations; they evaluate algebraic expressions, solve simple linear equations, and graph and interpret their results. **O━ AF 1.1** Write and solve one-step linear equations in one variable. *also* MR 2.4, MR 2.5, MR 3.2

Solve Multiplication and Division Equations

Learn how to solve multiplication and division equations.

Vocabulary

Division Property of Equality

Multiplication Property of Equality

Over a period of 7 days, a koala can sleep as many as 154 hours. Excluding animals that hibernate, the koala averages the most hours of sleep per day of any animal. Write and solve an equation to find the average number of hours per day a koala sleeps.

Choose a variable to represent the average number of hours per day that a koala sleeps. Then write an equation.

number of days	·	average daily hr of sleep	=	total hr of sleep
↓		↓		↓
7	·	h	=	154

Remember division can be written as a fraction.

$2 \div 3 = \frac{2}{3}$

$x \div 3 = \frac{x}{3}$

Multiplication and division are inverse operations. To solve the multiplication equation, use the inverse operation, division, so that the variable is alone on one side of the equation. The Division Property of Equality explains that step.

Division Property of Equality	$10 = 10$
If you divide both sides of an equation by the same nonzero number, the two sides remain equal.	$\frac{10}{2} = \frac{10}{2}$ $5 = 5$

EXAMPLE 1

Solve and check the equation $7h = 154$.

$7h = 154$ *Write the equation.*

$\frac{7h}{7} = \frac{154}{7}$ *Use the Division Property of Equality.*

$1 \times h = 22$ *$7 \div 7 = 1; 154 \div 7 = 22$*

$h = 22$ *Use the Identity Property of One.*

$7h = 154$ *Check your solution.*

$7 \times 22 \stackrel{?}{=} 154$ *Replace h with 22.*

$154 = 154$ ✓   *The solution checks.*

So, a koala sleeps an average of 22 hr a day.

CALIFORNIA STANDARDS **AF 1.0** Students write verbal expressions and sentences as algebraic expressions and equations; they evaluate algebraic expressions, solve simple linear equations, and graph and interpret their results. **○━┓ AF 1.1** Write and solve one-step linear equations in one variable. *also* MR 2.5, MR 3.2

301

Sometimes equations involve fractions.

EXAMPLE 2

Solve and check. $16 = \frac{2}{3}x$

$16 = \frac{2}{3}x$ *Write the equation.*

$\frac{16}{\frac{2}{3}} = \frac{\frac{2}{3}x}{\frac{2}{3}}$ *Use the Division Property of Equality.*

$16 \div \frac{2}{3} = x$

$16 \cdot \frac{3}{2} = x$ *Multiply by the reciprocal.*

$24 = x$

$16 = \frac{2}{3}x$ *Check your solution.*

$16 \stackrel{?}{=} \frac{2}{3} \cdot 24$ *Replace x with 24.*

$16 = 16 \checkmark$ *The solution checks.*

So, $x = 24$.

- Solve and check. $\frac{1}{4}x = 22$

TECHNOLOGY LINK

More Practice: Use
***Mighty Math Astro
Algebra,*** Green, Level P.

To solve a division equation, use the inverse operation, multiplication, and the Multiplication Property of Equality.

Multiplication Property of Equality	$10 = 10$
If you multiply both sides of an equation by the same number, the two sides remain equal.	$5 \times 10 = 5 \times 10$ $50 = 50$

EXAMPLE 3

Solve and check. $\frac{a}{5} = {}^{-}14$

$\frac{a}{5} = {}^{-}14$ *Write the equation.*

$5 \cdot \frac{a}{5} = 5 \cdot {}^{-}14$ *Use the Multiplication Property of Equality.*

$\frac{5}{1} \cdot \frac{a}{5} = {}^{-}70$

$\frac{5a}{5} = {}^{-}70$

$a = {}^{-}70$ *$5 \div 5 = 1$ and $1a = a$*

$\frac{a}{5} = {}^{-}14$ *Check your solution.*

$\frac{{}^{-}70}{5} \stackrel{?}{=} {}^{-}14$ *Replace a with $^{-}70$.*

${}^{-}14 = {}^{-}14 \checkmark$ *The solution checks.*

So, $a = {}^{-}70$.

- Solve and check. ${}^{-}9 = \frac{c}{12}$

Think and Discuss ▶ Look back at the lesson to answer the question.

1. **Tell** how you solve multiplication and division equations and how you solve addition and subtraction equations.

Guided Practice ▶ Solve and check.

2. $\frac{x}{3} = 8$　　　3. $4x = {}^-28$　　　4. $45 = \frac{1}{2}x$　　　5. $1.8 = \frac{y}{4}$

PRACTICE AND PROBLEM SOLVING

Independent Practice ▶ Solve and check.

6. $3x = 6$　　　7. $8k = 48$　　　8. $\frac{y}{5} = 7$　　　9. $\frac{a}{6} = 3$

10. $4 = \frac{m}{3}$　　　11. ${}^-48 = {}^-3p$　　　12. ${}^-15 = \frac{s}{5}$　　　13. $60 = {}^-12n$

14. $\frac{1}{4}d = 60$　　　15. $4.12 = \frac{a}{4}$　　　16. $24 = \frac{4}{5}b$　　　17. $2.1 = \frac{w}{1.3}$

Solve and check.

18. $\frac{8}{15} = \frac{c}{45}$　　　19. $\frac{a}{4} = 2\frac{1}{2}$　　　20. $\frac{3}{5}m = 2\frac{2}{5}$

Problem Solving Applications ▶

21. Mike divided his marbles equally among 3 friends. Each friend got 14 marbles. Write and solve an equation to find how many marbles Mike had originally.

22. In the equation $\frac{8.2}{n} = 4$, the variable is in the denominator. Explain how you would solve the equation. Then solve.

23. Write and solve a multiplication equation to find the width of a rectangle with an area of 0.45 cm^2 and a length of 0.9 cm.

24. **? What's the Error?** Mike's steps to solve the equation $2x = 12$ are shown at the right. Find his error and give the correct solution.

$$2x = 12$$
$$2x - 2 = 12 - 2$$
$$x = 10$$

MIXED REVIEW AND TEST PREP

Solve and check. (p. 294)

25. $x - 5 = 17$　　　26. $6 = y - 10$　　　27. ${}^-15 = z - 3$

28. Evaluate the expression $\sqrt{16} - 3^2 + 12$. (p. 282)

29. **TEST PREP** The low temperatures in Anchorage for five days were ${}^-2°F$, ${}^-6°F$, $1°F$, ${}^-9°F$, and ${}^-4°F$. Find the average low temperature for the five days. (p. 260)

A ${}^-20°F$　　　**B** ${}^-4°F$　　　**C** $4°F$　　　**D** $20°F$

Extra Practice page H47, Set A

303

Use Formulas

Learn how to use formulas to solve problems.

The Golden Gate Bridge in San Francisco, California, is one of the world's largest suspension bridges. The bridge is 8,981 feet long. Elena walked across the bridge at an average speed of 264 ft per min. How long did it take her to walk across the bridge?

If you know two of the three parts of the formula *distance = rate · time*, or $d = rt$, you can solve for the third part.

EXAMPLE 1

Find how long it took Elena to walk across the Golden Gate Bridge.

$d = rt$	*Write the formula.*
$8{,}981 = 264t$	*Replace d with 8,981 and r with 264.*
$\dfrac{8{,}981}{264} = \dfrac{264t}{264}$	*Solve the equation.*
$34\dfrac{5}{264} = t$	

So, it took $34\frac{5}{264}$ min, or about 34 min, for Elena to walk across the Golden Gate Bridge.

• Suppose Elena walked at an average speed of 250 ft per min. How long would it take her to cross the bridge?

You can solve for the distance if you know the rate and the time.

EXAMPLE 2

How far would Elena travel if she drove her car an average speed of 57 mi per hr for $3\frac{1}{2}$ hr?

$d = rt$	*Write the formula.*
$d = 57 \cdot 3\frac{1}{2}$	*Replace r with 57 and t with $3\frac{1}{2}$.*
$d = 199\frac{1}{2}$	*Multiply.*

So, Elena would travel $199\frac{1}{2}$ mi.

You can solve for the rate if you know the distance and the time.

EXAMPLE 3

Suppose Elena took 5 hours to drive 306.5 mi. What was Elena's average speed?

$d = rt$	*Write the formula.*
$306.5 = r \cdot 5$	*Replace d with 306.5 and t with 5.*
$\dfrac{306.5}{5} = \dfrac{5r}{5}$	*Solve the equation.*
$61.3 = r$	

So, Elena's average speed was 61.3 mi per hr.

Mark Twain once said, "The coldest winter I ever spent was a summer in San Francisco." The monthly normal temperature for July in San Francisco is about 17°C. You may be more familiar with the Fahrenheit temperature scale. To convert from degrees Celsius to degrees Fahrenheit, use the formula below.

$$F = \left(\frac{9}{5} \cdot C\right) + 32$$

EXAMPLE 4

Find the monthly normal temperature for July in San Francisco in degrees Fahrenheit (°F). Write your answer as a decimal.

$F = \left(\frac{9}{5} \cdot C\right) + 32$	*Write the formula.*
$F = \left(\frac{9}{5} \cdot 17\right) + 32$	*Replace C with 17. Operate inside parentheses.*
$F = 30\frac{3}{5} + 32$	*Add.*
$F = 62\frac{3}{5} = 62.6$	

So, the monthly normal temperature for July is about 62.6°F.

To convert from degrees Fahrenheit to degrees Celsius, use the formula below.

$$C = \frac{5}{9} \cdot (F - 32)$$

EXAMPLE 5

The monthly normal temperature in San Francisco for January is 49°F. Find the temperature in degrees Celsius (°C). Write your answer as a decimal and round to the nearest tenth of a degree.

$C = \frac{5}{9} \cdot (F - 32)$	*Write the formula.*
$C = \frac{5}{9} \cdot (49 - 32)$	*Replace F with 49. Operate inside parentheses.*
$C = \frac{5}{9} \cdot 17$	*Multiply.*
$C = 9\frac{4}{9} \approx 9.4$	

So, the monthly normal temperature for January is about 9.4°C.

Think and Discuss

Look back at the lesson to answer the questions.

1. **Explain** how you know the unit of time in Example 1 is minutes.

2. **Explain** how you know the unit of length in Example 2 is miles.

3. *REASONING* **Tell** what a comfortable room temperature would be in degrees Celsius.

Guided Practice

Use the formula $d = rt$ to complete.

4. $d = $ ■ ft
   $r = 22$ ft per min
   $t = 4$ min

5. $d = 100$ mi
   $r = $ ■ mi per hr
   $t = 5$ hr

6. $d = 300$ mi
   $r = 50$ mi per hr
   $t = $ ■ hr

Convert the temperature to degrees Fahrenheit. Write your answer as a decimal.

7. 10°C   **8.** 20°C   **9.** 30°C   **10.** 40°C   **11.** 5°C

Convert the temperature to degrees Celsius. Write your answer as a decimal and round to the nearest tenth of a degree.

12. 85°F   **13.** 57°F   **14.** 212°F   **15.** 100°F   **16.** 32°F

Independent Practice

Use the formula $d = rt$ to complete.

17. $d = $ ■ ft
    $r = 8$ ft per sec
    $t = 4.6$ sec

18. $d = 44$ km
    $r = $ ■ km per hr
    $t = 2.2$ hr

19. $d = 162.5$ mi
    $r = 65$ mi per hr
    $t = $ ■ hr

Convert the temperature to degrees Fahrenheit. Write your answer as a decimal.

20. 10°C   **21.** 37°C   **22.** 100°C   **23.** 95°C   **24.** 32°C

Convert the temperature to degrees Celsius. Write your answer as a decimal and round to the nearest tenth of a degree.

25. 32°F   **26.** 104°F   **27.** 47°F   **28.** 72°F   **29.** 94°F

Solve.

30. Rita drove for $3\frac{1}{2}$ hours at a rate of 40 mi per hr. How far did Rita travel?

31. A train traveled 350 mi in 4 hours. Find the train's average rate of speed.

32. A hot air balloon rose 4,200 ft in 14 min. Find the balloon's average rate of ascent.

Problem Solving ▶
Applications

33. Science In order to remain in orbit, the space shuttle must reach speeds of about 17,500 mi per hr. Suppose the space shuttle travels 52,650 mi in 3 hr. Is the space shuttle traveling fast enough to remain in orbit? Explain.

Use Data **For 34–36, use the data in the table below.**

34. Space Does Mercury or Venus have the higher extreme temperature? Explain.

35. Does Earth or Mars have the higher extreme temperature? Explain.

Extreme Planet Temperatures	
Mercury	430°C
Earth	60°C
Venus	867°F
Mars	81°F

36. Graphs Make a bar graph comparing all of the extreme temperatures shown in the table. Use °C.

37. **Write a problem** where you must use $d = rt$ to solve it.

MIXED REVIEW AND TEST PREP

Solve and check. (p. 301)

38. $3x = 15$ **39.** $^-42 = 7y$ **40.** $\frac{z}{8} = 3$

41. TEST PREP Which equation shows 18 is 6 more than a number? (p. 288)

   A $18 + 6 = x$ **B** $18 + x = 6$ **C** $18 = x + 6$ **D** $18 = x - 6$

42. TEST PREP Simplify $2x + 4x - 12$ by combining like terms. Then evaluate the expression for $x = {}^-4$. (p. 276)

   F $^-36$ **G** $^-10$ **H** 10 **J** 36

Thinker's CORNER

Mental Math Sometimes you can use mental math to solve equations.

$3x = 6$ **THINK:** $3 \cdot 2 = 6$, so $x = 2$.

You can use similar reasoning to solve a more complex problem.

$3x - 1 = 5$ **THINK:** $6 - 1 = 5$, so $3x = 6$ and $x = 2$.

Use mental math to solve each equation.

1. $2x - 4 = 2$ **2.** $3y - 2 = 4$ **3.** $2b - 3 = 1$

4. $4c - 1 = 3$ **5.** $2n - 3 = 5$ **6.** $3m - 1 = 8$

7. $3e - 4 = 5$ **8.** $5x - 2 = 8$ **9.** $4p - 3 = 13$

Two-Step Equations

MATH LAB

Explore how to use models to solve two-step equations.

You need algebra tiles or paper rectangles and squares.

Sometimes equations involve more than one step. In this lab you will use models to solve two-step equations.

Activity

- Use algebra tiles or paper rectangles and squares to model $2x + 2 = 6$. Use a green rectangle to represent the variable x, and use a yellow square to represent 1.

$$2x \qquad + 2 \quad = \qquad 6$$

- To solve an equation, arrange the model so that a rectangle is alone on one side. First, remove two squares from each side.

- The model now shows $2x = 4$.

- Separate the model into two equal groups on each side.

- What does each pair of groups model?

- What is the solution of the equation?

CALIFORNIA STANDARDS ○━┓ **AF 1.1** Write and solve one-step linear equations in one variable.
MR 2.4 Use a variety of methods, such as words, numbers, symbols, charts, graphs, tables, diagrams, and models, to explain mathematical reasoning. *also* **MR 2.5**

Think and Discuss

- How can you use the original model to check your solution?

- What did you do to the model so that only rectangles were alone on one side? What operation does this model?

- What operation did you model when you separated the model into two equal groups?

- Look at this model. What equation does it represent? Describe how to solve it.

Practice

1. Copy the model below. What equation does it represent? Use the model to solve the equation.

TECHNOLOGY LINK

More Practice: Use E-Lab, *Exploring Two-Step Equations.*

www.harcourtschool.com/ elab2002

Use a model to solve each equation.

2. $3x + 1 = 10$ **3.** $4y + 2 = 10$ **4.** $2b + 5 = 9$

5. Explain how you would solve the equation $2x - 1 = 5$. Solve the equation.

6. Can you solve $3x + 5 = 2$ by using a model? Explain.

MIXED REVIEW AND TEST PREP

Use the formula $d = rt$ to complete. (p. 304)

7. $d = 165$ mi
$r = 55$ mi per hr
$t = \blacksquare$ hr

8. $d = 679.5$ ft
$r = \blacksquare$ ft per min
$t = 4.5$ min

9. $d = \blacksquare$ km
$r = 88$ km per hr
$t = 4.25$ hr

10. Solve. $x - 7 = {}^-7$ (p. 294)

11. **TEST PREP** Evaluate $(x + 5)^2 - 9 \cdot 12$ for $x = \sqrt{64}$. (p. 282)

     A 1,920   **B** 960   **C** 61   **D** ${}^-19$

PROBLEM SOLVING STRATEGY
Work Backward

Analyze
Choose
Solve
Check

Learn how to use the strategy *work backward* to solve problems.

Felipe's parents paid $155 to rent a game room for his birthday party. The rates for renting the game room were $75 for the first hour and then $20 for each additional half hour. For how long did Felipe's parents rent the game room?

Analyze   What are you asked to find?

What information is given?

Is there information you will not use? If so, what?

Choose   What strategy will you use?

You can use inverse operations and the *work backward* strategy to solve the problem.

Solve   How will you solve the problem?

The cost of renting the game room was calculated in this way:

number of additional half hours	$\times$	cost for additional half hours	$+$	cost for first hour	$=$	total cost of renting game room
▨	$\times$	$20	$+$	$75	$=$	$155

You can work backward by reversing the operations and the order.

   ($155   − $75)   ÷ $20   =   4

first hour + additional half hours = total time
1 hour + 4 half hours = 3 hours

So, Felipe's parents rented the game room for 3 hours.

Check   How can you check your answer?

What if Felipe's parents had paid $175 to rent the game room? For how long would they have rented the game room?

PROBLEM SOLVING STRATEGIES

Draw a Diagram or Picture

Make a Model

Predict and Test

▶ **Work Backward**

Make an Organized List

Find a Pattern

Make a Table or Graph

Solve a Simpler Problem

Write an Equation

Use Logical Reasoning

Solve the problem by working backward.

1. The local cab company charges $1.25 for the first mile traveled and then $0.35 for each additional mile. Natalie spent $7.20 on a ride in a cab. How many miles did Natalie travel?

2. Abby buys three CDs in a set for $29.98. She saved $6.44 by buying the set instead of buying the individual CDs. If each CD costs the same amount, how much does each of the three CDs cost when purchased separately?

MIXED STRATEGY PRACTICE

For 3–4, use this information.

There are 141 students on the field trip to the Space Museum. It takes 3 full buses to transport the students. The buses are all identical. There will be 51 more students on the next field trip.

3. Which expression could you use to find the number of buses needed for the next field trip if b is the number of students per bus?

A $141 - (b + 51)$ **C** $(141 + 51) \div b$
B $51b + 3$ **D** $141 + 51 - b$

4. Which of the equations below could you use to find b, the number of students per bus?

F $\frac{50}{b} = 3$ **H** $\frac{b}{50} = 141$
G $50 + b = 141$ **J** $3b = 141$

5. Maria has chorus practice at 5:30 P.M. She needs 25 min to walk there, $\frac{1}{4}$ hr to get ready, and $\frac{1}{2}$ hr to fix and eat a snack before she gets ready. When should she begin to fix the snack?

6. Jana has 89 CDs in her CD collection. She wants to take all of them with her on vacation. A CD carrying case holds 25 CDs. How many carrying cases does Jana need to carry all 89 CDs?

7. Elliot has nickels and dimes in his pocket. He has twice as many dimes as nickels. If the total value of the coins is $1.00, how many of each coin does he have?

8. Amanda is making a small quilt that is 36 in. by 54 in. She wants to add a border of trim. She has 10 feet of the trim. How much more trim does she need?

9. Half of the computer club members use $\frac{3}{8}$ of the computer stations in the technology lab. What is the least number of students that could be in the computer club?

10. **? What's the Question?** Sean has a balance of ⁻$183 in his checking account. Last week he wrote a check for $25 and one for $217. The answer is $59.

1. **VOCABULARY** The property which states that if you divide both sides of an equation by the same nonzero number, the two sides remain equal is the ___?___. (p. 301)

2. **VOCABULARY** The property which states that if you multiply both sides of an equation by the same number, the two sides remain equal is the ___?___. (p. 302)

Solve and check. (pp. 301–303)

3. $6x = 24$

4. $28 = 4a$

5. $\frac{y}{7} = 11$

6. $1.9 = \frac{a}{12}$

7. $1.1n = 55$

8. $\frac{x}{5} = 3$

9. $9y = {}^-72$

10. ${}^-6 = \frac{x}{3}$

11. $\frac{k}{5} = {}^-12$

12. $\frac{3}{4} = \frac{m}{12}$

13. $2.5x = 42.5$

14. $\frac{1}{3}b = 36$

Use the formula $d = rt$ to complete. (pp. 304–307)

15. $d = 1{,}485$ mi
    $r = 55$ mi per hr
    $t = \blacksquare$ hr

16. $d = 220$ km
    $r = \blacksquare$ km per hr
    $t = \frac{1}{2}$ hr

17. $d = \blacksquare$ ft
    $r = 10$ ft per sec
    $t = 36$ sec

18. $d = 3{,}570$ ft
    $r = 140$ ft per min
    $t = \blacksquare$ min

19. $d = 1{,}625$ mi
    $r = \blacksquare$ mi per hr
    $t = 25$ hr

20. $d = \blacksquare$ m
    $r = 32$ m per sec
    $t = 9.6$ sec

21. $d = 273.42$ km
    $r = 18.6$ km per hr
    $t = \blacksquare$ hr

22. $d = 157.5$ in.
    $r = \blacksquare$ in. per min
    $t = 42$ min

Convert the temperature to degrees Fahrenheit. Write your answer as a decimal. (pp. 304–307)

23. $10°C$

24. $25°C$

25. $2°C$

26. $41°C$

Convert the temperature to degrees Celsius. Write your answer as a decimal and round to the nearest tenth of a degree. (pp. 304–307)

27. $59°F$

28. $89°F$

29. $72°F$

30. $38°F$

Solve.

31. The maximum speed of the X-15A-2 jet is 4,534 mi per hr. How long would it take this jet to travel the 2,787 mi from Los Angeles to New York if it is traveling at its maximum speed? Round your answer to the nearest tenth. (pp. 304–307)

32. André paid $14.35 for a cab ride. The rate for riding in the cab was $1.75 for the first mile and then $0.35 for each additional mile. How many miles long was André's cab ride? (pp. 310–311)

33. Two days ago, Raul had a certain amount of money in his wallet. He then added $2 to his wallet. Yesterday, he was paid for his chores, and the amount of money in his wallet tripled. He now has $18 in his wallet. How much was originally in his wallet? (pp. 310–311)

Decide on a plan.

See item **3.**

Decide what should replace F in the formula. Solve for C and round to the nearest degree.

Also see problem **4**, p. H63.

Choose the best answer.

1. Solve for x.

$$3x = {}^-18$$

  A $x = {}^-54$ **C** $x = 6$

  B $x = {}^-6$ **D** $x = 54$

2. Solve for x.

$$\frac{x}{3} = {}^-9$$

  F $x = {}^-27$ **H** $x = 3$

  G $x = {}^-3$ **J** $x = 27$

3. Use the formula $C = \frac{5}{9} \times (F - 32)$ to convert 60°F to degrees Celsius. What is that amount rounded to the nearest degree?

  A 14°C

  B 16°C

  C 50°C

  D 51°C

4. Solve for a.

$$9a = {}^-108$$

  F $a = {}^-12$ **H** $a = 12$

  G $a = 11$ **J** $a = 972$

5. Solve for x.

$$4x = 32$$

  A $x = 128$ **C** $x = 28$

  B $x = 36$ **D** $x = 8$

6. It took Mike 3 hours to drive to his college at an average speed of 52 miles per hour. Which equation can be used to find the distance Mike drove?

  F $d = 52 + 3$

  G $d = 52 - 3$

  H $d = 52 \times 3$

  J $d = 52 \div 3$

7. When the distance is 50 feet and the rate is 10 feet per second, what is the time?

  A 5 sec

  B 50 sec

  C 500 sec

  D Not here

8. Ten bicyclists rode the following number of miles: 5, 4, 3, 6, 4, 7, 5, 3, 4, 6. What is the mode of the distances?

  F 3   **H** 5

  G 4   **J** 6

9. Rita had $7\frac{1}{2}$ yards of fabric. She used $3\frac{7}{8}$ yards for a cushion. How much fabric is left?

  A $9\frac{3}{8}$ yd   **C** $3\frac{5}{8}$ yd

  B $6\frac{3}{8}$ yd   **D** $2\frac{1}{8}$ yd

10. Tim is surveying members of a hockey club. Which of the following methods should he use to produce a systematic sample?

  F A survey of the top-scoring champions

  G A survey of 2 out of every 9 goalies

  H A survey of the 25 oldest club members

  J A random survey of 1 out of every 8 club members

MATH DETECTIVE

Puzzle Power

REASONING The number at each triangle vertex is the sum of the lengths of the two sides that form the angle. For example, in Mystery Triangle 1, $a + b = 9$, $a + c = 13$, and $b + c = 12$. Use your knowledge of triangles and the strategy *predict and test* to find a, b, and c.

Mystery Triangle 1

Mystery Triangle 2

Mystery Triangle 3

Mystery Triangle 4

Mystery Triangle 5

Mystery Triangle 6

Think It Over!

- **Write About It** Explain how you found the lengths of the sides of Mystery Triangle 1.
- **Stretch Your Thinking** Find a, b, and c shown on the triangle at the right. Then explain why it is actually impossible to draw a triangle with these measurements.

Reflexive, Symmetric, and Transitive Properties

Two numbers can relate to each other in many different ways. For example, one number may be *less than* the other. The most familiar relationship between numbers is *equality*. Reflexive, Symmetric, and Transitive Properties describe ways that numbers can relate to each other.

Reflexive Property of Equality

A number is always equal to itself.

$a = a$ $7 = 7$

Symmetric Property of Equality

Equal numbers remain equal when their order is changed.

If $a = b$, then $b = a$.

If $4 + 9 = 10 + 3$, then $10 + 3 = 4 + 9$.

Transitive Property of Equality

Numbers equal to the same number are equal to each other.

If $a = b$ and $b = c$, then $a = c$.

If $5 + 3 = 8$ and $8 = 10 - 2$, then $5 + 3 = 10 - 2$.

EXAMPLE

Decide whether the Reflexive, Symmetric, or Transitive Property is used.

a. If $2 + 5 = 6 + 1$, then $6 + 1 = 2 + 5$.   *Symmetric*
b. $12 + t = 12 + t$ *Reflexive*
c. If $4 + 6 = 10$ and $10 = 13 - 3$, then $4 + 6 = 13 - 3$. *Transitive*
d. $15 + 6 = 15 + 6$ *Reflexive*

TALK ABOUT IT

*   In your own words, describe Reflexive, Symmetric, and Transitive Properties.

TRY IT

Identify the property.

1. If $s = 3$ and $3 = d$, then $s = d$.   **2.** $4 + n = 4 + n$

3. If $4 + 6 = 3 + 7$, then $3 + 7 = 4 + 6$.

Decide whether each of the Reflexive, Symmetric, and Transitive Properties holds for the given relationship.

4. "is greater than"   **5.** "is a sister of"

VOCABULARY

1. The parts of an algebraic expression that are separated by an addition or subtraction sign are called __?__. (p. 277)

2. The terms $3x^2$ and $2x^2$ have the same variable raised to the same power. They are called __?__ terms. (p. 277)

3. The product of a number and itself is a __?__. (p. 280)

EXAMPLES

EXERCISES

Chapter 14

• **Write numerical and algebraic expressions.** (pp. 274–275)

Write an algebraic expression for the word expression.

3.8 less than a	$a - 3.8$
the sum of twice b and 5	$2b + 5$

Write an algebraic expression for the word expression.

4. c divided by 2.6

5. the product of h and $\frac{3}{4}$

6. the difference between b and 45

• **Evaluate numerical and algebraic expressions.** (pp. 276–279, 282–283)

Evaluate $12 + 8x$ for $x = {}^-4$.

$12 + 8 \times ({}^-4)$	*Replace x with $^-4$.*
$12 + {}^-32$	*Multiply.*
$^-20$	*Add.*

For 7–11, evaluate the expressions for $x = {}^-3$ and $y = 9$.

7. $^-6x - {}^-8$

8. $4x + 2x + 9$

9. $12x - 7x + 15$

10. $4x^2 + xy$

11. $\sqrt{144} - 4^2 + 35y$

12. Evaluate $a - c + b$ for $a = 2.5$, $b = 7.8$, and $c = {}^-12$.

Chapter 15

• **Write word sentences as equations.** (pp. 288–289)

Forty is fifteen more than a number.

$$\downarrow \quad \downarrow \quad \downarrow \quad \downarrow \quad \downarrow$$
$$40 \quad = \quad 15 \quad + \quad n$$

Equation: $40 = 15 + n$

Write an equation for the word sentence.

13. $^-9$ is six less than a number.

14. A number divided by 2.5 is fifteen.

15. A number times $\frac{2}{3}$ is eighteen.

• **Solve addition equations.** (pp. 291–293)

$x + 6.4 = 9.8$	*Use the*
$x + 6.4 - 6.4 = 9.8 - 6.4$	*Subtraction Property of*
$x = 3.4$	*Equality.*

Solve.

16. $6 + m = {}^-24$

17. $r + 3\frac{2}{3} = 4$

18. $p + 5.8 = 13$

19. $15.1 = w + {}^-3.89$

- **Solve subtraction equations.** (pp. 294–295)

$$x - 17 = {}^-8$$

$$x - 17 + 17 = {}^-8 + 17 \quad \textit{Use the Addition}$$
$$\textit{Property of Equality.}$$

$$x + 0 = 9$$

$$\textit{Use the Identity}$$
$$\textit{Property of Zero.}$$

$$x = 9$$

Solve.

20. $m - 8 = 1$

21. $r - 1\frac{1}{2} = 3\frac{1}{4}$

22. $0.6 = p - 1.4$

23. ${}^-5 = w - 9$

Chapter 16

- **Solve multiplication and division equations.** (pp. 301–303)

$$\frac{m}{6} = 4.8$$

$$6 \cdot \frac{m}{6} = 4.8 \cdot 6 \quad \textit{Use the Multiplication}$$
$$\textit{Property of Equality.}$$

$$m = 28.8$$

Solve.

24. $18 = \frac{3}{4}c$

25. ${}^-7.2t = 14.4$

26. ${}^-6z = {}^-96$

27. $\frac{n}{8} = 3.4$

28. $2x = {}^-36$

- **Use formulas.** (pp. 304–307)

Use the formula $d = rt$ to complete.

rate = 54 miles per hour
time = 5 hours   distance = ▥ mi

$$d = rt$$

$$d = 54 \times 5 \quad \textit{Replace r with 54 and t}$$
$$\textit{with 5.}$$

$$d = 270 \quad \textit{Multiply.}$$

The distance is 270 miles.

Use the formula $A = lw$ to complete.

29. length = 8.6 in., width = 3 in.,
    area = ▥

30. Area = 420 in.^{2}, length = 70 in.,
    width = ▥

Use the formula $P = 2l + 2w$ to complete.

31. length = $9\frac{1}{2}$ ft, width = $4\frac{1}{2}$ ft,
    perimeter = ▥ ft

32. perimeter = 34 cm, length = 12 cm,
    width = ▥ cm

PROBLEM SOLVING APPLICATIONS

Solve. (pp. 310–311)

33. Anna Maria has soccer practice at 4:30 P.M. It takes her 5 min to change her clothes, 10 min to have a snack, and 15 min to get to the field. She wants to spend some time with her grandmother before practice. If she arrives home from school at 3:15 P.M., how much time does Anna Maria have to spend with her grandmother?

34. Andrew is thinking of a number between 1 and 10. If he multiplies the number by 4 and then subtracts 5, the result is 31. What is the number?

Hearst Castle

Two of California's many state parks are devoted to castles. Hearst Castle, the larger of the two, is located near the town of San Simeon. It was built over a period of 28 years by newspaper publisher William Randolph Hearst.

The Hearst Castle was transformed from an informal campsite into a world-famous castle plus guest houses, pools, and 8 acres of gardens.

1. La Casa Grande, the main house at Hearst Castle, has 115 rooms. There are a total of 165 rooms in all of the buildings at Hearst Castle. Write an equation you can use to find the number of rooms in buildings other than La Casa Grande. Then solve the equation.

2. La Casa Grande has $11\frac{1}{2}$ times as many rooms as a guest cottage called Casa del Monte. Write an equation you can use to find the number of rooms in Casa del Monte. Then solve the equation.

Use Data   For 3–5, use the map.

It is 15 mi farther from Morro Bay to San Simeon than it is from San Luis Obispo to Morro Bay.

3. Use the variable m to write an expression for the distance from Morro Bay to San Simeon.

4. Write and simplify an expression for the distance from Los Angeles to Hearst Castle.

5. It is 13 mi from San Luis Obispo to Morro Bay. How far is it from Los Angeles to Hearst Castle?

Hearst Castle □ 5 miles
San Simeon
Morro Bay ● m
San Luis Obispo ●
254 miles
Los Angeles ■

Use Data   For 6–8, use the price list.

6. Let A represent the price of an adult's ticket to tour Hearst Castle, and let C represent the price of a child's ticket. Write an algebraic expression for the total cost of a tour for 2 adults and 3 children, ages 6 to 12.

7. Evaluate your expression for a daytime tour.

8. Evaluate your expression for an evening tour.

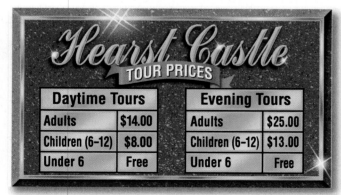

Hearst Castle TOUR PRICES

Daytime Tours		Evening Tours	
Adults	$14.00	Adults	$25.00
Children (6–12)	$8.00	Children (6–12)	$13.00
Under 6	Free	Under 6	Free

Vikingsholm

Vikingsholm is the smaller of the two castles in California's state parks. Vikingsholm is constructed to look like a Viking fortress of 1,000 years ago. Mrs. Lora Josephine Knight purchased the property at Emerald Bay and at Fannette Island, both at Lake Tahoe, and began its construction in 1928.

Vikingsholm is located on the western shore of Emerald Bay. Ideas for its construction came from buildings as far back as the 11th century.

Hearst Castle stands atop a mountain called the Enchanted Hill that is about 10 times as high as Fannette Island.

1. Let F represent the height of Fannette Island. Write an expression for the height of the Enchanted Hill.

2. Fannette Island is about 150 feet high. Use the expression you wrote to find the height of the Enchanted Hill.

La Casa Grande at Hearst Castle has 1 more than 3 times as many rooms as Vikingsholm has.

3. Let r represent the number of rooms in Vikingsholm. Write an equation relating the numbers of rooms in Vikingsholm and La Casa Grande.

4. Solve the equation you have written. How many rooms does Vikingsholm have?

5. There is a small stone building on Fannette Island called the Tea House. This building is in the shape of a square measuring $\sqrt{256}$ ft on a side. Write the dimensions of the building without the square root sign.

Lake Tahoe is located in the Sierra Nevada at an altitude of 6,229 feet. It is the largest mountain lake in North America and the third-deepest lake on the continent.

6. Carson City, Nevada, is located near Lake Tahoe at an elevation 76 ft higher than the bottom of the lake. Carson City's elevation is 4,660 ft. How deep is Lake Tahoe?

7. Lake Tahoe is 10 miles longer than it is wide. The sum of its length and its width is 34 miles. Find the length and the width of the lake.

Lake Tahoe elevation: 6,229 ft

Carson City elevation: 4,660 ft

76 ft

Bottom of Lake

319

Frank Lloyd Wright greatly influenced modern architecture. He designed many buildings around nature. His buildings often had large walls, terraces, and roof overhangs that reached out into the landscape. Wright built this house between 1936 and 1939. He named it Fallingwater. What types of angles do you see in the decks that hang over the waterfalls?

TYPES OF ANGLES

Name	Figure
acute	
right	
obtuse	
straight	

Check What You Know

Use this page to help you review and remember important skills needed for Chapter 17.

✓ Vocabulary

Choose the best term from the box.

1. Two rays with a common endpoint form a(n) __?__ .

2. An exact location is named by a(n) __?__ .

3. A straight path that extends without end in opposite directions is called a(n) __?__ .

4. A flat surface that extends without end in every direction is called a(n) __?__ .

> acute
> angle
> line
> plane
> point

✓ Classify Angles (See p. H18.)

Classify each angle by stating whether it is *acute, obtuse, right,* or *straight*.

5.

6.

7.

8.

9.

10.

11.

12.

✓ Name Angles (See p. H19.)

Name the angle formed by the blue rays.

13.

14.

15.

16.

17.

18.

19.

20.

LOOK AHEAD

In Chapter 17 you will

- solve problems involving figures in two and three dimensions
- classify parallel, intersecting, and perpendicular lines
- name, measure, and draw angles
- explore relationships among angles

Points, Lines, and Planes

Learn how to describe figures by using the terms of geometry.

Vocabulary

line segment
ray

QUICK REVIEW

QUICK REVIEW

1. $3x = 36$ **2.** $5y = 30$ **3.** $9b = {}^-36$ **4.** $56 = {}^-8d$ **5.** $24 = 10k$

Points, lines, and planes are the building blocks of geometry.

A *point* is an exact location. Think of a star in space.	•*P* Use the letter to point *P* name the point.
A *line* is a straight path that extends without end in opposite directions. Think of a narrow beam of light flashing across space in both directions.	line *AB*, $\overleftrightarrow{AB}$, line *BA*, $\overleftrightarrow{BA}$ Use two points on the line to name the line.
A *plane* is a flat surface that extends without end in every direction. Think of the surface of a large lake.	plane *LMN* Use three points in any order that are not on a line to name the plane.

Line segments and rays are parts of a line. Use points on the line to name line segments and rays.

A **line segment** has two endpoints.	line segment *XY*, $\overline{XY}$ line segment *YX*, $\overline{YX}$
A **ray** has one endpoint. From the endpoint, the ray extends forever in one direction only.	ray *JK*, $\overrightarrow{JK}$

You see models of points, lines, and planes every day. Specks of dust on a window model points. Flagpoles and pencils model line segments. Table tops and ceilings of rooms model planes.

Lines extend without end. So, real-world objects like goal "lines" and boundary "lines" are really just parts of lines. For the same reason, the sides of a skyscraper or of a gemstone are just portions of planes.

CALIFORNIA STANDARDS MG 2.0 Students identify and describe the properties of two-dimensional figures. *also* MR 2.5

CHECK FOR UNDERSTANDING

Think and Discuss ▶ Look back at the lesson to answer the question.

1. **Name** the geometric figure modeled by each of the following: a page of a book; a single dot, or "pixel," on a computer screen; the path of a passenger jet across the sky.

Guided Practice ▶ Name the geometric figure.

2.

3. •R

4.

PRACTICE AND PROBLEM SOLVING

Independent Practice ▶ Name the geometric figure.

5. C•

6.

7.

8. M N

9.

10.

For 11–14, use the figure at the right.

11. Name three different line segments.

12. Name four different rays.

13. Give six names for the line.   14. Give another name for ray *RQ*.

Problem Solving Applications ▶ Name the geometric figure formed by the intersection of each pair of figures.

15. two lines   16. two planes   17. line and plane

Give a real-world example of each term.

18. a ray   19. a point

20. a line   21. a line segment

22. 📖 **Write About It**   Can a line segment be part of a ray? Explain.

MIXED REVIEW AND TEST PREP

23. If $n = pq$, find n when $p = 24$ and $q = 4$. (p. 276)

24. Order from least to greatest: 0.4, 3, 1.19, 0.085 (p. 52)

25. Evaluate $\sqrt{36} - 2^3$ (p. 282)

26. Solve. $5.2 = k - 3.7$ (p. 294)

27. **TEST PREP**   Two apples weigh $\frac{3}{16}$ lb and $\frac{1}{4}$ lb. What is the difference in their weights? (p. 182)

   A $\frac{1}{6}$ lb   **B** $\frac{1}{16}$ lb   **C** $\frac{7}{16}$ lb   **D** $\frac{1}{4}$ lb

Extra Practice page H48, Set A)

Angles

Explore how to name, measure, and draw angles.

You need protractor, straightedge.

Vocabulary

vertex

An angle is formed by two rays with a common endpoint, called the **vertex**. You can name an angle by using the letter of its vertex alone. You can also use the letter of a point on each side, along with the letter of the vertex, to name the angle. The middle letter should always be for the vertex.

$\angle BAC$, $\angle CAB$, or $\angle A$

An angle is measured in degrees. One degree measures $\frac{1}{360}$ of a circle. The number of degrees determines the type of angle.

An acute angle measures less than 90°.	
A right angle measures 90°.	
An obtuse angle measures more than 90° but less than 180°.	
A straight angle measures 180°.	

Activity 1

You can measure an angle by using a protractor.

- On a sheet of paper, draw an angle that looks approximately like $\angle XYZ$.

- Place the center point of the protractor on the vertex of the angle.

- Place the base of the protractor along ray YZ.

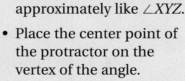

- Read the scale that starts at ray YZ. The measure of $\angle XYZ$ above is 75°. This can be written as $m\angle XYZ = 75°$.

CALIFORNIA STANDARDS MG 1.0 Students deepen their understanding of the measurement of plane and solid shapes and use this understanding to solve problems. *also* **MR 2.0**

Activity 2

You can also use a protractor to draw angles of a given measure.

- Draw a ray on a sheet of paper.

- Place the center point of the protractor on the endpoint of the ray. Align the protractor so that the ray passes through 0°.

- Draw an angle of 150° by making a mark on the paper above 150 on the scale you are using.

- Use your straightedge to draw a ray to complete the angle.

Activity 3

To estimate the size of an angle, you can compare it to an angle whose measure you know.

- Draw an angle on a sheet of paper.

- Compare its measure to the right angle shown here, and estimate how many degrees it contains.

- Measure the angle to check your estimate.

Think and Discuss

- What kind of angle did you draw in Activity 1?

- The protractor has two scales, a top scale and a bottom scale. How can you decide which scale to use?

Practice

Trace the angle. Then measure it and tell what kind of angle it is.

1.   2.   3.

TECHNOLOGY LINK

More Practice: Use **Mighty Math Cosmic Geometry,** *Amazing Angles,* Levels A and B.

Draw an angle with the given measure.

4. 48°   **5.** 180°   **6.** 154°   **7.** 90°   **8.** 45°

Estimate the measure. Then trace the angle and use a protractor to check your estimate.

9.   10.   11.

MIXED REVIEW AND TEST PREP

12. Name the geometric figure. (p. 322)

E   F

13. Solve. $96 = 12b$. (p. 301)

14. Find the product. 31.82×4 (p. 70)

15. Find $\frac{4}{5} \div 6$. (p. 210)

16. TEST PREP Find the mean, median, and mode: 50, 52, 56, 51, 50, 53. (p. 106)

A 51.5, 50, 52   **B** 50, 52, 51.5   **C** 52, 51.5, 50   **D** 51.5, 52, 50

325

Angle Relationships

Learn how to understand relationships of angles.

QUICK REVIEW

1. 90 − 15   **2.** 90 − 40   **3.** 180 − 50   **4.** 180 − 12

5. What number added to 25 gives 90?

Vocabulary

vertical angles

adjacent angles

complementary angles

supplementary angles

Angle relationships play an important role in many sports and games. Miniature-golf players must understand angles to know where to aim the ball. In the miniature-golf hole shown, m∠1 = m∠2, m∠3 = m∠4, and m∠5 = m∠6.

Certain angle pairs are given special names in geometry.

Remember that congruent figures have the same size and shape.

Vertical angles are formed opposite each other when two lines intersect. Vertical angles have the same measure, so they are always congruent.

∠*MRP* and ∠*NRQ* are vertical angles.

∠*MRN* and ∠*PRQ* are vertical angles.

Adjacent angles are side-by-side and have a common vertex and ray.

∠*MRN* and ∠*NRQ* are adjacent angles.

EXAMPLE 1

Identify the type of angle pair shown.

a.

b.

∠1 and ∠2 are opposite each other and are formed by two intersecting lines. They are vertical angles.

∠3 and ∠4 are side-by-side and have a common vertex and ray. They are adjacent angles.

• If ∠1 measures 40° in Example 1, what does ∠2 measure? If ∠4 measures 150°, what does ∠3 measure?

CALIFORNIA STANDARDS O—n **MG 2.2** Use the properties of complementary and supplementary angles and the sum of the angles of a triangle to solve problems involving an unknown angle. **MG 2.1** Identify angles as vertical, adjacent, complementary, or supplementary and provide descriptions of these terms. *also* **MR 2.5**

Complementary and Supplementary Angles

Two angles are complementary if the sum of their measures is 90°.
$65° + 25° = 90°$ $\angle XWY$ and $\angle RST$ are **complementary angles**.

Two angles are supplementary if the sum of their measures is 180°.
$65° + 115° = 180°$ $\angle RPS$ and $\angle ABC$ are **supplementary angles**.

In some cases you can use the definitions of complementary and supplementary angles to find the measures of angles.

EXAMPLE 2

The angles are complementary. Find the unknown measure.

$75° + \blacksquare = 90°$ *The sum of the measures is 90°.*

$90° - 75° = \blacksquare$ *Subtract to find the unknown measure.*

$90° - 75° = 15°$

So, the unknown angle measure is 15°.

EXAMPLE 3

The two adjacent angles are supplementary. Find the unknown measure.

$80° + \blacksquare = 180°$ *The sum of the measures is 180°.*

$180° - 80° = \blacksquare$ *Subtract to find the unknown measure.*

$180° - 80° = 100°$

So, the unknown angle measure is 100°.

When an object strikes a surface, it always rebounds at the same angle. In the figure, $\angle SWQ$ and $\angle TWR$ have the same measure.

EXAMPLE 4

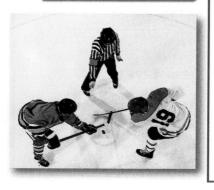

A hockey puck hits $\overline{BD}$, the wall of an ice rink, at an angle of 35°. $\overline{AC}$ is drawn so that $\angle BCA$ and $\angle DCA$ are right angles. Find the measures of $\angle 1$ and $\angle 3$.
The puck rebounds at the same angle, so the measure of $\angle 3$ is 35°.

$\angle 1$ and $\angle ECB$ form a right angle, so they are complementary.
$m\angle 1 + 35° = 90°$
$90° - 35° = 55°$, so the measure of $\angle 1$ is 55°.

327

CHECK FOR UNDERSTANDING

Think and Discuss ▶ **Look back at the lesson to answer each question.**

1. **Describe** vertical, adjacent, complementary, and supplementary angles.

2. **Explain** how you can find the measure of an angle complementary to a given angle; an angle supplementary to a given angle.

Guided Practice ▶ **For 3–6, use the figure at the right.**

3. Name two pairs of vertical angles.

4. Name two pairs of adjacent angles.

5. Name a pair of complementary angles.

6. Name two pairs of supplementary angles.

Find the unknown angle measure.

7.

8.

9.

PRACTICE AND PROBLEM SOLVING

Independent Practice ▶ **For 10–13, use the figure at the right.**

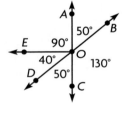

10. Name two angles adjacent to ∠COD.

11. Name the angle vertical to ∠AOB.

12. Name the angle complementary to ∠AOB.

13. Name two angles supplementary to ∠BOC.

Find the unknown angle measure. The angles are complementary.

14.

15.

16.

Find the unknown angle measure. The angles are supplementary.

17.

18.

19.

Tell if the angles are *vertical, adjacent, complementary, supplementary,* or *none of these.*

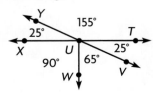

20. ∠TUV and ∠XUY

21. ∠XUY and ∠TUY

22. ∠VUW and ∠TUV

23. ∠XUW and ∠TUV

Problem Solving ▸ Applications

For 24–27, use the figure at the right. Trace the figure. Find the measure of the numbered angles.

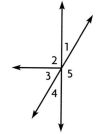

24. Name each pair of vertical angles. Tell how you know the angles are vertical.

25. Name each pair of adjacent angles. Tell how you know the angles are adjacent.

26. Name each pair of complementary angles. Tell how you know the angles are complementary.

27. Name each pair of supplementary angles. Tell how you know the angles are supplementary.

28. Which number should you aim at to hit the golf ball into the hole in one bounce? Use a protractor to decide. Explain your reasoning.

29. *REASONING* Two angles are each complementary to the same angle. What can you conclude about them.

30. (?) **What's the Question?** Eva drew angles for a project. She drew 6 angles every 2 minutes. The answer is 21 angles.

MIXED REVIEW AND TEST PREP

31. Name the geometric figure. (p. 322)

P   Q

32. Find the distance, d, for $r = 12.6$ ft per sec and $t = 22.5$ sec. (p. 304)

33. $^-9 - (^-3) = $ ▩ (p. 250)

34. TEST PREP Find $5,500 \div 25$. (p. 22)

    A 200   **B** 200 r5   **C** 220   **D** 2,200

35. TEST PREP Three sweaters cost $49, $29.95, and $36.75. Find the total cost. (p. 66)

    F $115.70 **G** $114.60 **H** $71.60 **J** $67.19

Thinker's CORNER

Get a complement If you know the complement or the supplement of an angle, you can write and solve an equation to find the missing angle measure.

The complement of an angle is 36°. What is the measure of the angle?

THINK: $x + 36 = 90$
$x + 36 - 36 = 90 - 36$
$x = 54$ So, the measure is 54°.

Write and solve an equation to find the measure of the angle.

1. The complement of the angle is 18°.   **2.** The supplement of the angle is 18°.

Extra Practice page H48, Set B

Classify Lines

Learn how to classify the different types of lines.

The chart shows some of the ways lines can relate to one another.

Parallel lines are lines in a plane that are always the same distance apart. They never intersect and have no common points.		Line AB is parallel to line ML. $\overleftrightarrow{AB} \parallel \overleftrightarrow{ML}$ (∥ means "is parallel to.")
Intersecting lines cross at exactly one point.		Line EF intersects line CD at point H.
Perpendicular lines intersect to form 90° angles, or right angles.		Line RS is perpendicular to line TU. $\overleftrightarrow{RS} \perp \overleftrightarrow{TU}$ (⊥ means "is perpendicular to.")

Line segments can also be parallel, intersecting, or perpendicular.

Vocabulary

parallel lines

perpendicular lines

EXAMPLE

The map shows a section of New York City. Describe each relationship as parallel, intersecting, or perpendicular.

a. streets to streets
Streets are parallel to each other.

b. avenues to avenues
Avenues are parallel to each other.

c. streets to avenues
Streets are perpendicular to avenues and intersect avenues.

d. Broadway to streets and avenues
Broadway intersects streets and 7th Avenue.

46th St.
45th St.
44th St.
43rd St.
42nd St.
41st St.
8th Ave.
7th Ave.
Broadway
6th Ave.

CHECK FOR UNDERSTANDING

Think and Discuss ▶ Look back at the lesson to answer the question.

1. Draw line 1 parallel to line 2. Draw line 3 perpendicular to line 1. Describe the relationship between line 2 and line 3.

CALIFORNIA STANDARDS MG 1.0 Students deepen their understanding of the measurement of plane and solid shapes and use this understanding to solve problems. *also* **MR 2.2, MR 2.5**

Classify the lines.

2.   3.   4.   5.

PRACTICE AND PROBLEM SOLVING

**Independent ▶
Practice**

Classify the lines.

6.   7.   8.   9.

The lines in the figure at the right intersect to form a cube.

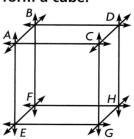

10. Name all the lines that are parallel to $\overleftrightarrow{AC}$.

11. Name all the lines that intersect $\overleftrightarrow{CD}$.

12. Name all the lines that are perpendicular to and intersect $\overleftrightarrow{DB}$.

13. Name all lines that are not parallel to and do not intersect $\overleftrightarrow{AE}$.

**Problem Solving ▶
Applications**

For 14–15 use this information. Main Street is a straight street. It intersects 3rd, 4th, and 5th Streets, which are parallel to each other. Pine and Oak Streets are perpendicular to 5th Street.

14. Draw a map showing the six streets.

15. Suppose that Pine and Oak Streets are perpendicular to Main Street rather than to 5th Street. Draw a map showing the streets.

16. **Write About It**   Can two lines be both parallel and intersecting? Explain.

17. **What's the Error?**   Aidan said that all perpendicular lines are intersecting lines and all intersecting lines are perpendicular. What is wrong with his statement?

MIXED REVIEW AND TEST PREP

18. Two angles are complementary. If one angle measures 72°, what does the other measure? (p. 326)

19. Suppose it took Cosetta 4.5 hours to drive 266.4 mi. What was Cosetta's average speed? (p. 304)

20. Find the sum. $3\frac{5}{8} + 5\frac{3}{8}$ (p. 186)

21. Which is greater, $\frac{5}{12}$ or $\frac{1}{3}$? (p. 166)

22. **TEST PREP**   Protractors sell for $0.09 each. How many can you buy for $9.00? (p. 76)

   A 0.01   **B** 10   **C** 100   **D** 1,000

1. **VOCABULARY** Part of a line that has two endpoints is a(n) ___?___ . (p. 322)

2. **VOCABULARY** ___?___ angles are opposite angles formed when two lines intersect. (p. 326)

3. **VOCABULARY** ___?___ angles are two angles whose sum is 180°. (p. 327)

4. **VOCABULARY** Lines that intersect to form right angles are called ___?___ . (p. 330)

Name the geometric figure. (pp. 322–323)

5. F •————————• G →   6. C •————————• D   7. ← A •————————• B →

Tell whether the angle is acute, right, obtuse, or straight. (pp. 324-325)

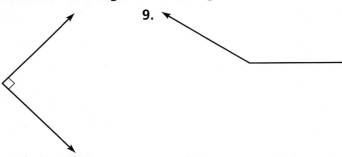

8.   9.   10.   11.

Use the figure to find the measure of each angle. (pp. 326–329)

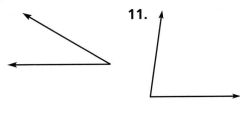

12. ∠ABF   13. ∠ABE

14. ∠CBD   15. ∠CBF

16. ∠FBE   17. ∠FBD

For 18–21, use the figure at the right. (pp. 330–331)

18. $\overleftrightarrow{HJ} \perp$ ___?___   19. $\overleftrightarrow{KH}$ intersects ___?___

20. $\overleftrightarrow{HI} \perp$ ___?___   21. $\overleftrightarrow{HJ} \parallel$ ___?___

22. Complete: If two angles are complementary, both angles must be ___?___ (acute, right, obtuse, straight). (pp. 326–329)

23. Complete: If two lines are perpendicular, then they are also ___?___ (parallel, intersecting) (pp. 330–331)

24. Two streets intersect and form a 38° angle. Find the unknown angle measure. (pp. 326–329)

25. Two sections of a picture frame join to form a right angle. Find the unknown angle measure.

(pp. 326–329)

Understand the problem.

See item **7**.

Be sure you know the meaning of each term in the answer choices. Look at the diagram to determine the relationship of ∠DCA and ∠ACB.
Also see problem **1**, p. H62.

Choose the best answer.

1. What geometric figure is modeled by a beam of light from the sun?

   A Point   **C** Ray

   B Circle   **D** Plane

2. Which term describes an angle that measures 60°?

   F Acute   **H** Obtuse

   G Right   **J** Straight

For 3–4, use the diagram below.

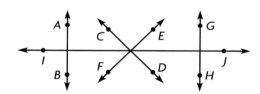

3. Which appears to be parallel to $\overleftrightarrow{AB}$?

   A $\overleftrightarrow{CD}$ **C** $\overleftrightarrow{GH}$

   B $\overleftrightarrow{EF}$ **D** $\overleftrightarrow{IJ}$

4. Which appears to be perpendicular to $\overleftrightarrow{GH}$?

   F $\overleftrightarrow{AB}$ **H** $\overleftrightarrow{EF}$

   G $\overleftrightarrow{CD}$ **J** $\overleftrightarrow{IJ}$

5. What is the measure of a straight angle?

   A 180°   **C** 90°

   B 100°   **D** 75°

For 6–7, use the diagram below.

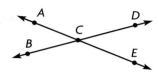

6. Which of these pairs of angles are vertical angles?

   F ∠ACB and ∠ACD

   G ∠ACB and ∠DCE

   H ∠DCE and ∠BCE

   J ∠ACD and ∠DCE

7. Which term describes the relationship of ∠DCA and ∠ACB?

   A Complementary angles

   B Supplementary angles

   C Right angles

   D Vertical angles

8. The orchestra rehearsed for $2\frac{1}{2}$ hours on Saturday and $3\frac{1}{2}$ hours on Sunday. Altogether, how many hours did the orchestra rehearse?

   F $6\frac{1}{2}$ **H** $5\frac{1}{2}$

   G 6   **J** 5

9. Paula made 190 tote bags. She sold 115 of them at the craft fair and made $2,070. Which equation can be used to find p, the price she sold each tote bag for?

   A $190 \times 115 = p$

   B $190 \times p = 115$

   C $190 \times p = \$2,070$

   D $115 \times p = \$2,070$

10. What is the value of $y - 54$ for $y = 17$?

    F 71   **H** ⁻71

    G 37   **J** Not here

CHAPTER 18 Plane Figures

Computer-aided design, know as CAD, uses computer software to create models that show the geometry and other characteristics of objects. The advantage of CAD is that a design, such as the pool design above, can be easily adjusted or reworked if needed. Notice the geometric figures used to design the pool and patio. What geometric figures do you see in the umbrella table?

Check What You Know

Use this page to help you review and remember important skills needed for Chapter 18.

☑ Vocabulary

Choose the best term from the box.

| circle |
| isosceles |
| polygon |
| scalene |

1. A triangle that has exactly two congruent sides is a(n) __?__ triangle.

2. A closed plane figure formed by three or more line segments is a(n) __?__ .

☑ Identify Polygons (See p. H19.)

Name each polygon.

3.

4.

5.

6.

7.

8.

9.

10.

☑ Angles in Polygons (See p. H18.)

Name each angle in the polygon. Tell whether it is *acute*, *obtuse*, *right*, or *straight*.

11.

12.

13.

14.

☑ Patterns (See p. H14.)

Describe a possible pattern. Then give the next three numbers in the pattern.

15. 4, 8, 12, 16

16. 2, 6, 18, 54

17. 37, 31, 25, 19

18. 243, 81, 27, 9

19. $1, \frac{1}{2}, \frac{1}{4}, \frac{1}{8}, \frac{1}{16}$

20. 9, 6, 3, 0

LOOK AHEAD

In Chapter 18 you will

- find measures of angles in triangles and solve problems involving quadrilaterals
- use the strategy *find a pattern* to solve problems
- draw two-dimensional figures
- identify parts of circles

Triangles

Learn how to classify triangles and solve problems involving the angle measures of triangles.

Vocabulary

acute triangle
obtuse triangle
right triangle

A triangle can be classified by the angles it contains. An **acute triangle** contains only acute angles. An **obtuse triangle** contains one obtuse angle. A **right triangle** contains one right angle.

acute triangle   obtuse triangle   right triangle

In this activity, you will explore the angle measures in a triangle.

Activity

MATH LAB

You need: protractor, straightedge, scissors

- Use a straightedge to draw a large triangle. Use scissors to cut it out.

- Tear off the three angles. Arrange them around a point on a line as shown.

- What is the sum of the measures of the three angles? Explain.

- Compare your results with those of your classmates. What guess can you make?

Math Idea ▶ The sum of the measures of the angles of a triangle is 180°. To decide whether a triangle is acute, obtuse, or right, you need to know the measures of its angles.

EXAMPLE 1

Three stars form a triangle known as "Triangulum," with two angles measuring 20° and 50°. Classify the triangle.

Remember that an *equilateral* triangle has three congruent sides, an *isosceles* triangle has two congruent sides, and a *scalene* triangle has no congruent sides.

To classify the triangle, find the measure of ∠B in the figure.

180° − (20° + 50°) = 180° − 70°
= 110°

Subtract the sum of the known angle measures from 180°.

So, the measure of ∠B is 110°. Because △ABC contains one obtuse angle, Triangulum forms an obtuse triangle.

CALIFORNIA STANDARDS ○─ㅠ **MG 2.2** Use the properties of complementary and supplementary angles and the sum of the angles of a triangle to solve problems involving an unknown angle. **MG 2.0** Students identify and describe the properties of two-dimensional figures. *also* **MR 1.2, MR 2.2**

Sometimes you can find the measure of an angle by using what you know about vertical angles.

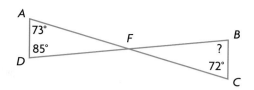

EXAMPLE 2

Find the measure of ∠FBC.

$180° - (73° + 85°) = 180° - 158°$ *Find m ∠AFD.*
$= 22°$

m ∠BFC = 22°   *Vertical angles are congruent.*

$180° - (22° + 72°) = 180° - 94°$ *Find m ∠FBC.*
$= 86°$

So, the measure of ∠FBC is 86°.

Another way to find the measure of a missing angle is to apply what you know about complementary and supplementary angles.

EXAMPLE 3

Remember that the sum of the measures of two *complementary* angles is 90°. The sum of the measures of two *supplementary* angles is 180°.

In the figure at the right, ∠NPQ is a right angle, ∠LNM and ∠MNP are supplementary angles, and ∠NPM and ∠QPM are complementary angles. Find the measure of ∠M.

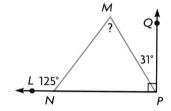

∠MNP = 180° − 125° = 55°   *The sum of the measures of two supplementary angles is 180°.*

∠MPN = 90° − 31° = 59°   *The sum of the measures of two complementary angles is 90°.*

55° + 59° = 114°   *Find the sum of the known angle measures in △NMP.*

180° − 114° = 66°   *Subtract the sum of the known angle measures for △NMP from 180°.*

So, ∠M measures 66°.

CHECK FOR UNDERSTANDING

Think and ▶ Discuss

Look back at the lesson to answer each question.

1. **Explain** why a triangle cannot have two obtuse angles.

2. **Draw** an example of each: scalene acute triangle; isosceles right triangle; scalene obtuse triangle.

Guided ▶ Practice

Find the measure of ∠B and classify △ABC.

3.

4.

5.

Independent ▶ Practice

Find the measure of ∠B and classify △ABC.

6.

7.

8.

9.

10.

11.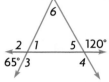

For 12–17, use the figure at the right. Find the measure of each angle.

TECHNOLOGY LINK

More Practice: Use **Mighty Math Cosmic Geometry**, *Amazing Angles*, Level C.

12. ∠*DAO*   **13.** ∠*ADO*

14. ∠*ABD*   **15.** ∠*AOD*

16. ∠*AOB*   **17.** ∠*DCA*

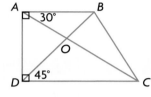

For 18–23, use the figure at the right. Find the measure of each angle.

18. ∠1   **19.** ∠2   **20.** ∠3

21. ∠4   **22.** ∠5   **23.** ∠6

For 24–27, use the figure at the right.

24. What is the relationship between ∠*RSW* and ∠*VST*?

25. What is the measure of ∠*W*?

26. Find the measure of ∠*RSW*.

27. Find the measure of ∠*V*.

Problem Solving ▶ Applications

28. In a video game, a moving ball always bounces off an edge at the same angle at which it hit the edge. Find the measure of the unknown angle.

a.

b.

c.

d.

29. The three main stars that are in the constellation known as Leo Minor form an isosceles triangle, with congruent base angles. The third angle in the triangle measures 144°. Find the measure of each base angle.

Leo Minor
144°
base angles

30. REASONING How many total degrees are in the angles of a quadrilateral? (**Hint:** Divide it into triangles.)

For 31–32, use the figure at the right.

A   exterior angle
105°
B 48°   D
         C

31. If you extend a side of a triangle, you form an *exterior angle*. In the figure, ∠ACD is an exterior angle of △ABC. Find the measure of ∠ACD.

32. **What's the Question?** The answer is ∠BCA and ∠ACD.

MIXED REVIEW AND TEST PREP

33. Classify the lines. (p. 330)

34. Which is greatest, 0.03, 0.009, or 0.107? (p. 52)

35. What is the median of the values 24, 17, 39, 19, and 21? (p. 106)

36. TEST PREP Which is the value of x in the expression $\frac{6}{x} = \frac{18}{12}$? (p. 216)

   A $x = 2$ **B** $x = 4$ **C** $x = 9$ **D** $x = 24$

37. TEST PREP Miguel wants to record a TV movie that begins at 9:40 P.M. and ends at 12:10 A.M. How long does the movie last? (p. 214)

   F 9 hr 30 min   **G** 3 hr 50 min   **H** 2 hr 30 min   **J** 2 hr 10 min

---

Thinker's CORNER

Figure It Out With a partner, draw a triangle, a quadrilateral, a pentagon, and a hexagon. Do not draw the polygons with congruent sides.

For each figure you draw, extend one side at each vertex, as shown for the quadrilateral at the right. The angles you form outside the polygon are exterior angles.

For each figure, cut out the exterior angles and arrange them around a common point.

1. Tell the sum of the measures of the exterior angles for the triangle, the quadrilateral, the pentagon, and the hexagon.

2. What can you guess about the sum of the measures of the exterior angles of a polygon?

Extra Practice page H49, Set A

PROBLEM SOLVING STRATEGY
Find a Pattern

Analyze
Choose
Solve
Check

Learn how to use the strategy *find a pattern* to solve problems.

Vocabulary

regular polygon

Mrs. Jones is building a deck in the shape of a regular octagon. A **regular polygon** is a polygon in which all sides are congruent and all angles are congruent. So, a regular octagon has 8 congruent sides and 8 congruent angles. What is the measure of each angle of the octagon?

Analyze

What are you asked to find?

What information is given?

Choose

What strategy will you use?

Look for a pattern in the sums of the measures of the angles of regular polygons that have fewer sides than an octagon.

Solve

How will you solve the problem?

Draw regular polygons that have fewer sides than an octagon. Divide each into triangles. Make a table to record your data and help you look for a pattern.

triangle

quadrilateral

pentagon

hexagon

POLYGON	SIDES	TRIANGLES	SUM OF ANGLE MEASURES
Triangle	3	1	180°
Quadrilateral	4	2	2 x 180° = 360°
Pentagon	5	3	3 x 180° = 540°
Hexagon	6	4	4 x 180° = 720°

The number of triangles is always 2 fewer than the number of sides. So, an octagon can be divided into $8 - 2 = 6$ triangles.

$$6 \times 180° = 1,080°$$

The sum of the angle measures of an octagon is 1,080°. To find the measure of each angle in a regular octagon, divide the sum by 8: $1,080° \div 8 = 135°$.

Check

How can you check your answer?

What if Shalana and her mother want to build a deck in the shape of a decagon? What would be the measure of each angle?

CALIFORNIA STANDARDS MR 1.3 Determine when and how to break a problem into simpler parts. MR 2.2 Apply strategies and results from simpler problems to more complex problems. MR 3.3 Develop generalizations of the results obtained and the strategies used and apply them in new problem situations. *also* MG 1.0, MR 1.0, MR 1.1, MR 2.0, MR 3.2

Draw a Diagram or Picture

Make a Model

Predict and Test

Work Backward

Make an Organized List

▶ **Find a Pattern**

Make a Table or Graph

Solve a Simpler Problem

Write an Equation

Use Logical Reasoning

Solve the problem by finding a possible pattern.

1. Macklin saves 1 dime on Monday, 2 dimes on Tuesday, 4 dimes on Wednesday, 7 dimes on Thursday, and 11 dimes on Friday. If he continues the pattern, how many dimes would you expect him to save on each of the next three days?

2. There were 10 players in the checkers tournament. If each player shook the hand of every other player 1 time, how many handshakes were there?

Steadman is making a design for a quilt using blue and white squares.

3. How many squares will there be in the eighth row?

   A 8   **B** 9   **C** 15   **D** 17

4. After he has 8 rows, how many squares will he have used in all?

   F 8   **G** 16   **H** 36   **J** 64

5. Silvia, David, and Rhoda have blue, green, and brown eyes, though not necessarily in that order. Silvia's eyes are the color of the sky. Rhoda does not have brown eyes. Give the color of each person's eyes.

6. This week Cheryl earned $23 more than last week. Last week she earned $225. She plans to spend $\frac{3}{8}$ of this week's earnings, put half in her checking account, and save the rest. How much will she save?

Use Data   For 7–8, use the table.

7. Estimate the amount the Alpha Corporation earns from publishing if its total revenues are $11.6 billion.

8. Draw a circle graph to show the sources of revenue for the Alpha Corporation. Then use the graph to find two revenue sources that together produce more than half of the company's revenues.

SOURCES OF REVENUE FOR ALPHA CORPORATION	
Publishing	35%
Movies and TV	28%
Broadcasting and Cable	20%
Other	17%

9. Julia noted that her car's mileage gauge read 12,445 miles at the start of a trip. At the end of the trip, the gauge read 12,829. If she used 16 gallons of gas on the trip, how many miles per gallon did she get?

10. **Write a problem** that can be solved by finding a pattern. Exchange with a classmate and solve. Explain your solution.

Quadrilaterals

Learn how to identify, classify, and compare quadrilaterals.

Find each angle measure.

1. ∠1   **2.** ∠2   **3.** ∠3

4. ∠4   **5.** ∠5

Remember that a quadrilateral is a polygon with four sides and four angles.

These 2 figures are quadrilaterals:

Five types of quadrilaterals are especially important. Their names and properties are listed below. Note: The same marking on two sides or more of a figure indicates that the sides are congruent.

QUADRILATERAL	FIGURE	PROPERTIES
parallelogram		opposite sides parallel and congruent
rectangle		parallelogram with four right angles
rhombus		parallelogram with four congruent sides
square		rectangle with four congruent sides
trapezoid		quadrilateral with exactly two parallel sides

Notice that rectangles and rhombuses are types of parallelograms.

Rectangle:

Parallelogram with **four right angles**

Rhombus:

Parallelogram with **four congruent sides**

A square is both a type of rectangle and a type of rhombus.

Square:

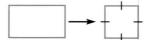

Rectangle with **four congruent sides**

Square:

Rhombus with **four right angles**

CALIFORNIA STANDARDS MG 2.0 Students identify and describe the properties of two-dimensional figures. **MR 2.0** Students use strategies, skills, and concepts in finding solutions. *also* **MR 2.4, MR 3.0, MR 3.2, MR 3.3**

The figure at the right has four sides, so it is a quadrilateral. A more *exact* name for the figure, however, is *rectangle*, because the name *rectangle* describes the figure's properties more completely than the name *quadrilateral* does.

Quadrilateral:
4 sides
Rectangle:
parallelogram with 4 right angles

EXAMPLE 1

Give the most exact name for the figure.

The figure is a rhombus.

The figure is a square.

The figure is a parallelogram.

• Why wouldn't it be more exact to name the figure above at the far right a rhombus or a rectangle?

Statements about geometric figures are often given in "If...then" form.

• *If* a polygon has five sides, *then* it is a pentagon.

• *If* a figure is a triangle, *then* the sum of the measures of its angles is 180°.

You can use what you know about quadrilaterals to write "If...then" statements about quadrilaterals.

EXAMPLE 2

Complete the statement, giving the most exact name for the figure: If a rhombus is also a rectangle, then the rhombus is a ___?___ .

THINK: A rhombus has 4 congruent sides and opposite sides parallel. If it is also a rectangle, then it has 4 right angles, which makes it a square.

rhombus   square

In the activity below, you will name quadrilaterals by using their properties.

Activity

A   B   C   D

1. List the properties of each figure above.

2. Give the most exact name for each figure.

Think and Discuss ▶ Look back at the lesson to answer each question.

1. **Compare and contrast** a rectangle and a rhombus.

2. **Explain** how to determine if a quadrilateral is a trapezoid.

Guided Practice ▶ Give the most exact name for the figure.

3.

4.

5.

Complete the statement, giving the most exact name for the figure.

6. If the opposite sides of a quadrilateral are parallel, then the quadrilateral is a ___?___.

7. If a quadrilateral has four right angles, then it is a ___?___.

8. If a rectangle has four congruent sides, then it is a ___?___.

Independent Practice ▶ Give the most exact name for the figure.

9.

10.

11.

12.

13.

14.

Complete the statement, giving the most exact name for the figure.

15. If a parallelogram has four congruent sides, then it is a ___?___.

16. If a polygon has five sides, then it is a ___?___.

17. If a quadrilateral has four congruent sides and four congruent angles, then the quadrilateral is a ___?___.

18. If exactly two sides of a quadrilateral are parallel, then it is a ___?___.

19. If a polygon has four sides, then it is a ___?___.

In 20–22, portions of quadrilaterals are hidden from view. Name some possible quadrilaterals each figure could be.

20. _____

21. _____

22. _____

23. A rectangular picture frame is 4 inches wider than it is tall. The total length of the four sides of the frame is 52 inches. What are the dimensions of the frame?

24. Kazumi cut two square frames from a 90-in. length of aluminum. If each side of one frame was 16 in. long, how long was each side of the second frame?

25. **?** **What's the Error?** Elton says that a quadrilateral with one pair of parallel sides and two right angles is a rectangle. What is wrong with his statement?

26. To raise money for his scout troop, Aldrick sold 16 cans of popcorn for $6.50 each and 18 calendars for $3.75 each. What were Aldrick's total sales?

MIXED REVIEW AND TEST PREP

27. Two angles in a triangle each measure 48°. Find the measure of the third angle. (p. 336)

28. $\angle M$ is complementary to a 75° angle. What is the measure of $\angle M$? (p. 326)

29. Write 8% as a decimal. (p. 169)

30. **TEST PREP** Which is the greatest common factor of 18 and 24? (p. 150)

   A 2   **B** 3   **C** 6   **D** 72

31. **TEST PREP** Which number is between 17.09 and 17.3? (p. 230)

   F 17.01   **G** 17.08   **H** 17.29   **J** 17.312

LiNKUP to Reading

Strategy • Make Generalizations Sometimes you need to *make generalizations* to solve a problem. When you generalize, you make a statement that is true about a whole group of similiar situations or objects.

Use the Venn diagram to make generalizations about quadrilaterals.

QUADRILATERALS

rhombuses

rectangles

squares

Parallelograms

Trapezoids

Complete. Write *always*, *sometimes*, or *never*.

1. A square is __?__ a parallelogram.

2. A trapezoid is __?__ a rectangle.

3. A rhombus is __?__ a trapezoid.

4. A rectangle is __?__ a square.

Draw Two-Dimensional Figures

Learn how to draw two-dimensional geometric figures.

Remember that an isosceles triangle has two congruent sides.

EXAMPLES

You can use the properties of geometric figures to draw the figures on dot paper.

A. Draw a right isosceles triangle.

Use square dot paper to draw figures with right angles.

THINK: *The triangle must have one right angle.*

The triangle also must have two congruent sides.

B. Draw a pentagon with two right angles.

THINK: *The figure must have two right angles.*

The figure also must have five sides.

C. Draw a quadrilateral with four congruent sides and no right angles.

Use isometric dot paper to draw figures with congruent sides and no right angles.

CHECK FOR UNDERSTANDING

Think and Discuss

Look back at the lesson to answer each question.

1. **Describe** how you could draw a trapezoid on dot paper.

2. ***REASONING***   Could you draw an isosceles triangle with two obtuse angles? Explain.

Guided Practice

Draw the figure. Use square dot paper or isometric dot paper.

3. a triangle with no congruent sides

4. an equilateral triangle

5. a hexagon with no congruent sides

CALIFORNIA STANDARDS MG 2.3 Draw quadrilaterals and triangles from given information about them (e.g., a quadrilateral having equal sides but no right angles, a right isosceles triangle). *also* MG 2.0, MR 1.0, MR 2.0, MR 3.2, MR 3.3

Independent ▶ Practice

Draw the figure. Use square dot paper or isometric dot paper.

6. an acute isosceles triangle

7. an obtuse scalene triangle

8. a quadrilateral with 4 right angles

9. a quadrilateral with no congruent sides or parallel sides

10. an octagon with 2 right angles

11. a pentagon with no congruent sides

Complete the sentence by using *must*, *can*, or *cannot*.

12. A rhombus __?__ have all sides congruent.

13. A right triangle __?__ have two right angles.

14. A trapezoid __?__ have all sides congruent.

15. A scalene triangle __?__ be obtuse.

Problem Solving ▶ Applications

16. Anita drew a quadrilateral. Then she drew a line segment connecting one pair of opposite corners. She saw that she had divided the quadrilateral into two right isosceles triangles. Classify the quadrilateral she began with.

17. On a separate sheet of paper, draw an equilateral triangle with sides 4 in. long. Cut out the triangle. Then cut from each corner of the triangle an equilateral triangle with sides 1 in. long. What is the perimeter of the figure that remains?

18. **?** **What's the Error?** Victor said he drew a triangle with one acute, one right, and one obtuse angle. Explain his mistake.

19. Ted drew three sides of a quadrilateral as shown at the right. If he wants a third angle to be 90°, what will be the measure of the fourth angle?

90° 125°

MIXED REVIEW AND TEST PREP

20. Give the most exact name for a parallelogram with four congruent sides. (p. 342)

21. Name the figure. (p. 322)

Y
X

22. Evaluate the expression. (p. 282)
$8^2 + \sqrt{100} - 7$

23. 3.6×0.25 (p. 70)

24. **TEST PREP** A box with a height of $1\frac{9}{16}$ ft is stacked on top of a box with a height of $2\frac{1}{2}$ ft. What is the combined height of the two boxes? (p. 186)

A $3\frac{1}{16}$ ft **B** $3\frac{5}{9}$ ft **C** $3\frac{13}{16}$ ft **D** $4\frac{1}{16}$ ft

Circles

Learn how to identify parts of circles.

radius = 6.25 cm
diameter = 12.5 cm

Vocabulary

radius
diameter
chord
arc
sector

A compact disc (CD) is shaped like a *circle*. A circle is the set of all points a given distance from its center. At the edge of the CD, every point on the circle is 6.25 cm from the center.

A line segment with one endpoint at the center of a circle and the other endpoint on the circle is a **radius** (plural, *radii*). A line segment that passes through the center of a circle and has both endpoints on the circle is a **diameter**. The length of a diameter is always twice the length of a radius.

CD radius = 6.25 cm   CD diameter = 2 × 6.25 = 12.5 cm

A circle is named by its center. Like a polygon, a circle is a plane figure. A circle is not made up of line segments, so it is not a polygon.

EXAMPLE 1

Name the center, the circle, a diameter, and three radii of the circle.

O is the center. The circle is circle *O*. $\overline{AB}$ is a diameter. $\overline{OA}$, $\overline{OB}$, and $\overline{OC}$ are radii.

Two points on a circle can indicate a chord, an arc, or a sector. A **chord** is a line segment with its endpoints on a circle. An **arc** is part of a circle. A **sector** is a region enclosed by two radii and the arc joining their endpoints.

chord *AB*   arc *AB*, or $\overarc{AB}$

sector *AB*

EXAMPLE 2

Music used to be recorded on "long-playing" records (LPs). The shaded sector shows how far an LP turns in one second. Name the radii and arc enclosing the sector.

The radii are $\overline{KR}$ and $\overline{KP}$. The arc is $\overarc{PR}$. Since there are two arcs between *P* and *R*, you can use point *Q* to indicate the longer one: $\overarc{PQR}$.

CALIFORNIA STANDARDS **MG 2.0** Students identify and describe the properties of two-dimensional figures. **MR 2.0** Students use strategies, skills, and concepts in finding solutions. *also* **MR 2.4, MR 2.5,**

Think and ▶
Discuss

Look back at the lesson to answer each question.

1. **Explain** how you can find the radius of a circle if you know the diameter.

2. **REASONING** What is the longest chord in a circle?

Guided ▶
Practice

For 3–6, use the circle at the right.
Name the given parts of the circle.

3. center
4. chords

5. radii
6. diameters

PRACTICE AND PROBLEM SOLVING

Independent ▶
Practice

For 7–12, use the circle at the right.
Name the given parts of the circle.

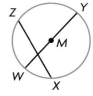

7. center
8. chords

9. radii
10. diameter

11. a sector
12. two arcs

Problem Solving ▶
Applications

13. **REASONING** What happens to the length of a radius as the size of a circle increases?

TECHNOLOGY LINK

More Practice: Use
**Mighty Math Cosmic
Geometry,** *Geo
Academy,* Level G.

Use Data For 14–15, use the table.

14. Find the radius of the LP, the 45, and the 78.

15. How much longer will it take an LP to revolve (spin) 1,000 times than it will take a CD?

16. 📓 **Write About It** The radius of a circle is 6 cm long. Chord *AB* is 8 cm long. Explain why $\overline{AB}$ cannot be a diameter.

TYPES OF RECORDING DISKS		
Disk	Diameter	Revolutions Per Minute
CD	12.5 cm	200
LP	12 in.	$33\frac{1}{3}$
45	$6\frac{7}{8}$ in.	45
78	10 in.	78

MIXED REVIEW AND TEST PREP

17. Give the most exact name for the figure. (p. 342)

18. Find the unknown angle measure. (p. 326)

19. Evaluate $18 - 9 \div 3 + 5$. (p. 44)

20. What is the LCM of 15 and 20? (p. 150)

21. **TEST PREP** Find $24 \div \frac{3}{4}$. (p. 210)

A 16 **B** 18 **C** 32 **D** 36

1. **VOCABULARY** A triangle containing one obtuse angle is a(n) ___?___ triangle. (p. 336)

2. **VOCABULARY** A triangle containing one right angle is a(n) ___?___ triangle. (p. 336)

3. **VOCABULARY** Any line segment with endpoints on a circle is called a(n) ___?___. (p. 348)

Find the measure of the angle and classify the triangle. (pp. 336–339)

4.

5.

6.

7.

Give the most exact name for the figure. (pp. 342–345)

8.

9.

10.

11.

Complete the statement, giving the most exact name for the figure. (pp. 342–345)

12. If the opposite sides of a quadrilateral are parallel, then it is a(n) ___?___.

13. If a polygon has three sides, then it is a(n) ___?___.

14. If a parallelogram has four right angles, then it is a(n) ___?___.

15. If a polygon has eight sides, then it is a(n) ___?___.

16. If exactly two sides of a quadrilateral are parallel, then it is a(n) ___?___.

Draw the figure. Use square dot paper or isometric dot paper. (pp. 346–347)

17. a right triangle with two congruent sides

18. a trapezoid with two right angles

19. a parallelogram with four congruent sides

Name the given parts of the circle. (pp. 348–349)

20. center

21. chords

22. radii

23. diameters

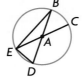

24. Jamie decorated her bike wheels by stringing ribbons between the ends of spokes. Classify △ABO. (O is the center.)

25. The figure shows a roof with vertical support $\overline{BD}$. Find the measure of the top roof angle, ∠ABC.

Get the information you need.

See item **3**.

Think about the attributes used to classify polygons. Recall which polygons have all sides the same length.

Also see problem **3**, p. H63.

Choose the best answer.

1. Which statement about the four angles in a rectangle is always true?

   A All are acute angles.

   B All are obtuse angles.

   C Some angles may be obtuse.

   D All are right angles.

2. What is the value of x in the figure below?

   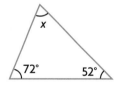

   F 124°   **H** 64°

   G 90°   **J** 56°

3. In which type of figure do all the sides always have the same length?

   A Rectangle   **C** Rhombus

   B Trapezoid   **D** Isosceles Triangle

4. What type of triangle has an angle that measures 90°?

   F Equilateral   **H** Obtuse

   G Right   **J** Acute

5. The diameter of a circular garden is 24 meters. What is the radius of the garden?

   A 3.14 m   **C** 48 m

   B 12 m   **D** Not here

6. Which term best describes this triangle?

   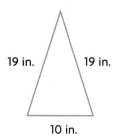

   F Isosceles triangle

   G Equilateral triangle

   H Scalene triangle

   J Right triangle

7. These integers follow a multiplying pattern: ⁻1, ⁻2, ⁻4, . . . Which number might be next?

   A ⁻8   **C** ⁻1

   B ⁻6   **D** ⁺8

8. Kate had $5\frac{1}{4}$ cups of flour. She used $1\frac{3}{8}$ cups in a recipe. How much flour is left?

   F $1\frac{3}{4}$ cups   **H** $3\frac{1}{8}$ cups

   G $2\frac{3}{8}$ cups   **J** $3\frac{7}{8}$ cups

9. Which term best describes this figure?

   A Line   **C** Ray

   B Point   **D** Line segment

10. Lenora drove 156.5 miles in the morning and 204.7 miles in the afternoon. What is the total distance she drove?

    F 361.2 mi

    G 389.2 mi

    H 461.2 mi

    J Not here

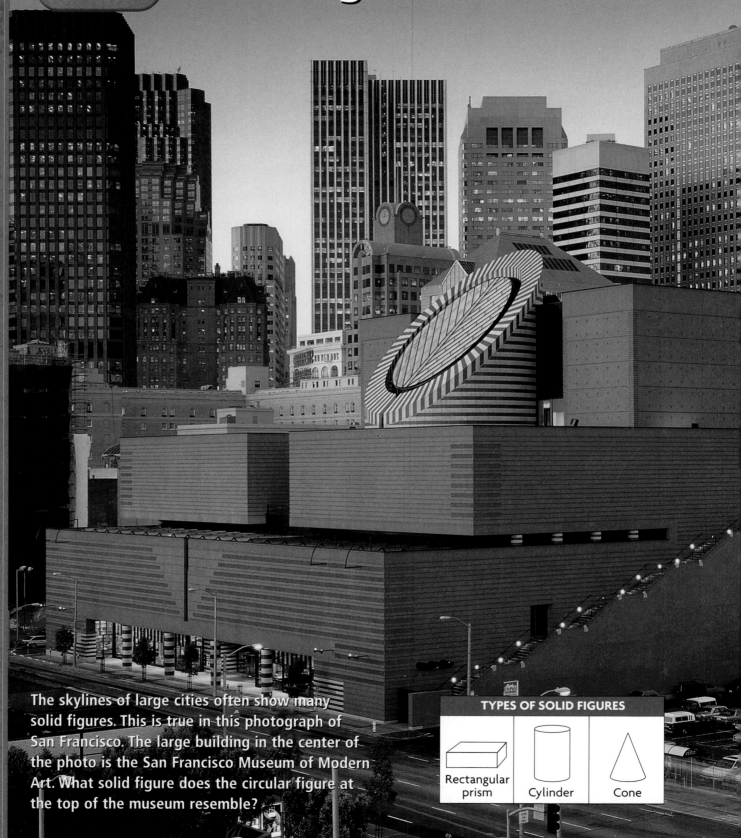

The skylines of large cities often show many solid figures. This is true in this photograph of San Francisco. The large building in the center of the photo is the San Francisco Museum of Modern Art. What solid figure does the circular figure at the top of the museum resemble?

TYPES OF SOLID FIGURES

Rectangular prism

Cylinder

Cone

Check What You Know

Use this page to help you review and remember important skills needed for Chapter 19.

✔ Identify Solid Figures (See p. H20.)

Name each solid figure.

1.

2.

3.

4.

5.

6.

7.

8.

✔ Faces, Vertices, and Edges (See p. H20.)

Copy and complete the table.

	Figure	Number of faces	Number of vertices	Number of edges
9.		▦	▦	▦
10.		▦	▦	▦
11.		▦	▦	▦
12.		▦	▦	▦
13.		▦	▦	▦

> ### LOOK AHEAD
>
> **In Chapter 19 you will**
> - identify solid figures
> - draw different views of solids
> - use the strategy *solve a simpler problem* to solve problems
> - build models of prisms and pyramids

Types of Solid Figures

Learn how to name solid figures.

You see many solid figures every day.

All solid figures are three-dimensional. In the following activity, you'll explore ways that solid figures can be alike and ways they can be different.

Vocabulary

polyhedron
lateral faces
bases
vertex

Activity

You need: isometric dot paper, colored pencils

• Copy the rectangular prism on isometric dot paper. Use three different colors: one for length, one for width, and one for height.

• Draw a rectangular prism with different dimensions.

• Copy the pyramid. Then draw a rectangular pyramid with different dimensions.

• How are your rectangular prisms alike? How are they different?

• How are your pyramids alike? How are they different?

Math **I**dea ▶ Many common geometric shapes have special names.

A **polyhedron** is a solid figure with flat faces that are polygons.

A prism is a polyhedron with two congruent, parallel bases. Its **lateral faces**, or faces that are not bases, are rectangles. A prism is named for the shape of its **bases**.

Prisms can have any polygon as a base.

Rectangular Prism

Triangular Prism

Hexagonal Prism

CALIFORNIA STANDARDS MR 2.4 Use a variety of methods, such as words, numbers, symbols, charts, graphs, tables, diagrams, and models, to explain mathematical reasoning.

EXAMPLE 1

Name the figure.

All the faces are flat and are polygons, so the figure is a polyhedron.

The lateral faces are rectangles, so the figure is a prism.

The bases are congruent pentagons.

So, the figure is a pentagonal prism.

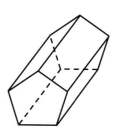

A cylinder is not a polyhedron. It has two flat, parallel, congruent circular bases and a curved lateral surface. Many cans are examples of cylinders.

curved lateral surface

flat circular bases

Cylinder

Remember that a *pyramid* is a solid figure whose one base is a polygon and whose faces are triangles that have a common vertex.

If you connect the single center point on the top base of a cylinder to all the points around the bottom base, you form a cone.

A cone is not a polyhedron. It has one flat circular base, a curved lateral surface, and a **vertex**. The vertex is the top (or bottom) point of this cone.

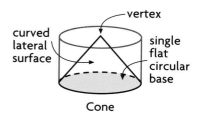

vertex

curved lateral surface

single flat circular base

Cone

A pyramid is related to a prism as a cone is related to a cylinder.

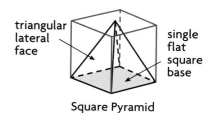

triangular lateral face

single flat square base

Square Pyramid

EXAMPLE 2

Name the figure.

All the faces are flat and are polygons, so the figure is a polyhedron.

The lateral faces are triangles with a common vertex, so it is a pyramid.

The base is a rectangle.
So, the figure is a rctangular pyramid.

CHECK FOR UNDERSTANDING

Think and ▶ Discuss

Look back at the lesson to answer each question.

1. **Tell** the name of each solid figure: a prism with octagonal bases; a pyramid with a pentagon as a base; a solid figure with flat faces that are polygons.

2. *REASONING* **Tell** how a cylinder and a prism are alike. How are they different? Tell how a cone and a cylinder are alike. How are they different?

Guided ▶
Practice

Name the figure. Is it a polyhedron?

3. 4. 5. 6.

PRACTICE AND PROBLEM SOLVING

Independent ▶
Practice

Name the figure. Is it a polyhedron?

7. 8. 9. 10.

Write *true* or *false* for each statement. Rewrite each false statement as a true statement.

11. A cone has no flat surfaces.

12. A cylinder is a polyhedron.

13. A cube is a rectangular prism.

14. The bases of a cylinder are congruent.

15. All pyramids have triangular bases.

16. A hexagonal prism has six faces.

Problem Solving ▶
Applications

17. Niles carved a wooden pyramid with seven faces. What is the shape of the base of the pyramid?

18. Gina made two square pyramids and glued the congruent bases together to make an ornament. How many faces does her ornament have?

19. **REASONING** The lateral faces of a pyramid are equilateral triangles. Is the base a regular polygon? Draw a diagram to explain.

Name the top base of the solid you would obtain by cutting off the top of each figure along the red line.

20. 21. 22.

MIXED REVIEW AND TEST PREP

23. Name the part of a circle that is enclosed by two radii and the arc joining their endpoints. (p. 348)

24. Let *a* represent Alan's age. Write an equation stating that Alan's age 5 years ago was 13. (p. 288)

25. Write $\frac{27}{7}$ as a mixed number. (p. 164)

26. $^-9 + 5$ (p. 244)

27. **TEST PREP** Maury worked 9.25 hours and earned $7.45 per hour. Which is the total amount he earned to the nearest cent? (p. 70)
  A $16.70 **B** $68.91 **C** $74.50 **D** $689.13

356 Chapter 19

Extra Practice page H50, Set A

Different Views of Solid Figures

Learn how to draw and identify different views of a solid.

QUICK REVIEW

Name the base of each solid figure.

1. rectangular prism   2. cone   3. cylinder

4. triangular prism   5. square pyramid

Car designers use computer-aided design (CAD) to view their designs from different angles. These differing views help designers to create cars that are safe, efficient, and attractive.

You can draw different views of a solid.

MATH LAB

Activity 1

You need: cylinder

• Look at the top of the cylinder. Draw the top view.

• Look at the front of the cylinder. Draw the front view.

• Look at a side of the cylinder. Draw the side view.

Math Idea ▶ You can use different views of a solid figure to identify the figure.

EXAMPLE 1

Name the solid figure that has the given views.

top view

The top view shows that the base is square and that the faces come together at a point.

front view   side view

The front and side views show that the solid has triangular sides.

So, this solid is a square pyramid.

CALIFORNIA STANDARDS MR 2.4 Use a variety of methods, such as words, numbers, symbols, charts, graphs, tables, diagrams, and models, to explain mathematical reasoning.

Activity 2

How would this solid look if you viewed it from the top, the front, and the side? You can build a model to find out.

You need: centimeter cubes, centimeter graph paper

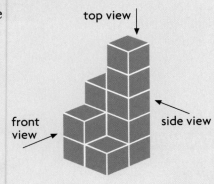

top view

front view

side view

• Use centimeter cubes to build the solid at the right.

• The top view of the solid is shown below:

Draw the front view and the side view on graph paper.

• How many cubes do you see in the top view? the front view? the side view?

• Which views show how tall the solid is?

EXAMPLE 2

Draw the front view of the above stack of cubes without building a model.

Think of the front view of each numbered cube. In your drawing, show only the numbered faces, placing them in correct relationship to one another.

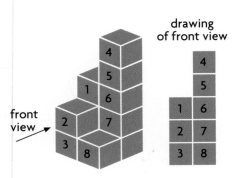

front view

drawing of front view

CHECK FOR UNDERSTANDING

Think and ▶ Discuss

Look back at the lesson to answer each question.

1. **Explain** why a side view of a solid figure might change if you change your viewpoint from the left side to the right side.

2. **Draw** a top view of the triangular prism.

Guided ▶ Practice

Name the solid figure that has the given views.

3.

top
front   bottom

4.
top
front   bottom

5.
top
front   side

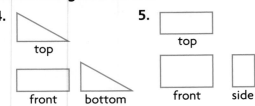

358 Chapter 19

Independent ▶ Practice

Name the solid figure that has the given views.

6.
top

front   side

7.
top

front   side

8.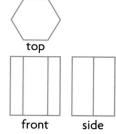
top

front   side

Draw the front, top, and bottom views of each solid.

9.

10.

11.

12.

Each solid is made with 10 cubes. On graph paper, draw a top view, a front view, and a side view for each solid.

13.

14.

15.

16.

17. Draw views of the hexagonal prism as seen from points A, back, and B, side.

A

B

Problem Solving ▶ Applications

18. Pick an object in your classroom. Draw the top, front, and side views.

19. Every side view of a triangular prism shows what shape?

20. ✎ **Write About It**   Name objects in the classroom or at home that have a top view of a rectangle.

21. Name the figure. (p. 354)

22. Name a parallelogram that has four congruent sides. (p. 342)

23. Evaluate $x^2 + 3x$ for $x = {}^-3$. (p. 276)

24. $\dfrac{3}{8} + \dfrac{3}{4}$ (p. 182)

25. **TEST PREP**   Find the sum. $22.7 + 27.9 + 24.3 + 25.1$ (p. 66)

    A 25   **B** 50   **C** 100   **D** 200

Models of Solid Figures

Explore how to build models of prisms and pyramids.

You need 4 in. × 6 in. index card, inch ruler, tape, scissors, protractor, patterns for cylinder and cone, grid paper.

Vocabulary

net

QUICK REVIEW

Tell the number of faces each solid figure has.

1. cube
2. triangular prism
3. square pyramid
4. pentagonal prism
5. triangular pyramid

You can build a solid figure by cutting its faces from paper, taping them together, and then folding them to form the solid.

3 in.

2 in.   1 in.

Activity 1

• Follow these steps to make a pattern for the prism.

Step 1: Draw the faces on the index card.

2 × 3	2 × 3
1 × 3	1 × 3
1 × 2	1 × 2	waste

Step 2: Cut out the six rectangles.

Step 3: Tape the pieces together to form the prism.

Step 4: Remove the tape from some of the edges so that the pattern lies flat.

A pattern of polygons that can be folded to form a prism is called a **net** for the prism.

Think and Discuss

• Can you make a different net for the rectangular prism? Draw a sketch to explain.

• How are the nets for a cube and a rectangular prism different?

Practice

On grid paper, draw four nets that can be folded to form a cube. Then draw four nets that cannot form a cube. One of each type is shown at the right.

Can   Cannot

CALIFORNIA STANDARDS MR 2.4 Use a variety of methods, such as words, numbers, symbols, charts, graphs, tables, diagrams, and models, to explain mathematical reasoning.

- How many triangles make up the pyramid?

- Draw an equilateral triangle on a sheet of paper. Cut out the triangle.

- Use the triangle as a pattern to trace three more triangles. Cut out these triangles.

- Tape the triangles together to form a triangular pyramid.

- Remove the tape from some of the edges to make a net of the triangular pyramid.

60°

60°   60°

Equilateral triangle

Think and Discuss

- What shapes will always appear in the net of a pyramid? of a prism?

- What solid figure could you make from the net shown at the right? How do you know?

- Explain why the nets shown below cannot be folded to make solid figures.

a.   b.   c.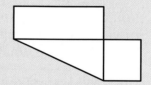

Practice

1. Draw a net for a pentagonal pyramid.

2. Draw a net for a pentagonal prism.

3. Use the cylinder pattern to build a cylinder.

4. Use the cone pattern to build a cone.

MIXED REVIEW AND TEST PREP

5. Draw a top view of a square pyramid. (p. 357)

6. A triangle contains two 48° angles. Find the measure of the third angle. (p. 336)

7. Write an integer to represent a loss of $432. (p. 228)

8. $2\frac{2}{3} \times 1\frac{7}{8}$ (p. 206)

9. TEST PREP Let t represent today's temperature. If tomorrow's temperature is 8 degrees less than twice today's temperature, which expression represents tomorrow's temperature? (p. 274)

A $2t - 8$ **B** $t - 16$ **C** $8 - 2t$ **D** $2t + 8$

PROBLEM SOLVING STRATEGY
Solve a Simpler Problem

Analyze
Choose
Solve
Check

Learn how to use the strategy *solve a simpler problem* to solve problems.

QUICK REVIEW

Find the next number in the pattern.

1. 4, 8, 16, 32   **2.** 20, 17, 14, 11   **3.** 3, 6, 9, 12

4. 80, 40, 20, 10   **5.** 1, 4, 9, 16, 25

Andrew is building models of prisms. He is using balls of clay for the vertices and straws for the edges. How many of each will he need to make a prism with 15-sided bases?

Analyze

What are you asked to find?

What information is given?

Is there information you will not use? If so, what?

Choose

What strategy will you use?

You can use the strategy *solve a simpler problem*. Find the number of vertices and edges for some prisms whose bases have fewer sides. Then use what you learn to solve the harder problem.

Solve

How will you solve the problem?

Find the number of vertices and edges there are on prisms whose bases have 3, 4, and 5 sides. Record your findings in a table.

Sides on base	3	4	5	
Vertices	3 + 3, or 6	4 + 4, or 8	5 + 5, or 10	*top + bottom*
Edges	3 + 3 + 3, or 9	4 + 4 + 4, or 12	5 + 5 + 5, or 15	*top + bottom + sides*

Now, apply the patterns shown in the table to a prism whose bases have 15 sides:

Vertices: 15 + 15 = 30

Edges: 15 + 15 + 15 = 45

So, Andrew will need 30 balls of clay for the vertices and 45 straws for the edges.

Check

What other strategy could you use to solve the problem?

What if the prism had bases with 25 sides each? How many vertices and how many edges would it have?

CALIFORNIA STANDARDS **MR 1.1** Analyze problems by identifying relationships, distinguishing relevant information from irrelevant, identifying missing information, sequencing and prioritizing information, and observing patterns. **MR 3.2** Note the method of deriving the solution and demonstrate a conceptual understanding of the derivation by solving similar problems. *also* **MR 2.2**

Solve the problem by first *solving a simpler problem*.

1. Margo wants to make a model of a prism whose bases have 10 sides each. She will use balls of clay for the vertices and straws for the edges. How many balls of clay will she need?

2. Amy wants to make a model of a pyramid whose base has 10 sides. She will use balls of clay for the vertices and toothpicks for the edges. How many balls of clay will she need? How many toothpicks will she need? How many faces will her pyramid have?

PROBLEM SOLVING STRATEGIES

Draw a Diagram or Picture

Make a Model

Predict and Test

Work Backward

Make an Organized List

Find a Pattern

Make a Table or Graph

▶ **Solve a Simpler Problem**

Write an Equation

Use Logical Reasoning

3. By drawing 1 vertical line and 1 horizontal line, you can divide a sheet of paper into 4 sections. Into how many sections can you divide a sheet of paper using 8 vertical and 8 horizontal lines?

1	2
3	4

A 10   **B** 16   **C** 64   **D** 81

4. Boxes are stacked with 1 box in the first row, 3 in the second, 5 in the third, and so on. How many boxes are needed altogether to make a 10-row stack?

F 12   **G** 19   **H** 81   **J** 100

MIXED STRATEGY PRACTICE

Use Data For 5–7, use the graph.

5. On which day was the difference between the high and low temperatures the greatest? What was the difference?

6. Between which two days was the temperature change the greatest? Did the high or low temperature change more?

7. ✎ **Write a problem** that can be solved by using the graph.

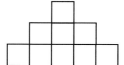

8. Corbin centered a table against a wall that was 13 ft wide. The table was $3\frac{1}{2}$ ft wide. How far was the left end of the table from the left end of the wall?

9. Helmut used 27 toothpicks as edges on a model of a prism. How many sides did the base of the prism have?

10. Keith, Fran, and Lanni divided a basket of tennis balls among themselves. Keith took half of the balls. Fran took half of those that remained. Lanni took the remaining 5 balls. How many balls were in the basket to begin with?

1. **VOCABULARY** A solid figure with flat faces that are polygons is called a(n) __?__. (p. 354)

2. **VOCABULARY** A pattern that can be folded to form a solid figure is called a(n) __?__. (p. 360)

3. **VOCABULARY** A solid figure with two flat, parallel, congruent circular bases and a curved lateral surface is a(n) __?__. (p. 355)

Name the figure. Is it a polyhedron? (pp. 354–356)

4.

5.

6.

7.

Name the solid figure that has the given views. (pp. 357–359)

8.

top

front   side

9.

top

front   side

10.

top

front   side

Draw the front, top, and side views of each solid. (pp. 357–359)

11.

12.

13.

14.

Will the net fold to form a cube? Write *yes* or *no*. (pp. 360–361)

15.

16.

17.

18.

Solve.

19. April made a model of a prism whose bases have 9 sides each. How many vertices and how many edges did her model have? (p. 362)

20. Mark made a model of a pyramid with 16 edges. How many sides did the base of his pyramid have? (p. 362)

Check your work.

See item **5.**

Draw a representation of a prism that has 8 vertices. Be sure to account for the edges in all three dimensions.

Also see problem **7**, p. H65.

Choose the best answer.

1. Which solid figure has 4 faces, 4 vertices, and 6 edges?

A Pentagonal prism

B Rectangular prism

C Triangular pyramid

D Triangular prism

2. What is a name for the solid figure shown below?

F Pyramid **H** Cube

G Cone **J** Cylinder

3. Which of these does a cone have?

A One flat, circular base

B A vertex

C A curved lateral surface

D All of the above

4. What model of a solid figure can be made from the net shown below?

F Rectangular prism

G Triangular prism

H Pentagonal prism

J Not here

5. Zack is building a model of a prism. He will use balls of clay for the vertices and straws for the edges. How many straws will he use if he uses 8 balls of clay for his model?

A 4 **C** 12

B 8 **D** 16

6. Which is a name for the solid figure shown below?

F Square pyramid

G Triangular prism

H Pentagonal prism

J Hexagonal prism

7. Which solid figure has triangles in all of its views?

A Triangular pyramid

B Cone

C Rectangular prism

D Triangular prism

8. Evaluate.

$$54 \div (3 \times 1)^2 + 17$$

F 6 **H** 35

G 23 **J** Not here

9. Solve for x.

$$x - 34 = 95$$

A $x = 69$ **C** $x = 129$

B $x = 121$ **D** Not here

10. What is the sum of the measures of two angles that are complementary?

F 45° **H** 180°

G 90° **J** 360°

MATH DETECTIVE

Truth or Consequences

REASONING Use your knowledge of quadrilaterals to decide whether each statement is true or false. If a statement is false, use words like *all*, *some*, or *no* to rewrite it to make it true.

Truth or Consequences 1
Every square is a parallelogram.

Truth or Consequences 2
Every rectangle is a square.

Truth or Consequences 3
Every square is a rhombus.

Truth or Consequences 4
Every rhombus is a quadrilateral.

Truth or Consequences 5
A quadrilateral with exactly one pair of parallel sides is a parallelogram.

Truth or Consequences 6
Every rectangle is a trapezoid.

Think It Over!

- **Write About It** Complete this sentence three different ways: "A square is __?__."

- **Stretch Your Thinking** I'm a quadrilateral that is symmetric about each of my diagonals. My diagonals are congruent. What am I?

Explore Perspective

Learn how to make one-point perspective drawings.

Perspective is a method used to make objects drawn on a flat surface appear three-dimensional.

Compare the two drawings on the right. The first drawing looks more flat. The second drawing appears to have depth because it is drawn in perspective.

Parallel lines   Perspective lines

EXAMPLE

Make a one-point perspective drawing of a rectangular prism.
- Draw the front face.
- Draw a point above the face and to one side. This is called a *vanishing point*.
- Connect the four vertices to the vanishing point.
- Draw the top rear edge of the prism. It should be parallel to the top front edge.
- Draw the remaining visible edge of the prism.
- Draw the "hidden" edges, using dashed lines. Erase the segments leading to the vanishing point.

TALK ABOUT IT

- Describe how to draw a triangular prism by using one-point perspective.

TRY IT

Draw each figure. Use one-point perspective.

1. a cube with a vanishing point to the left

2. a cube with a vanishing point to the right

3. a cereal box with a vanishing point in the center

4. The figure on the right shows a rectangular prism drawn using two-point perspective. Draw a rectangular prism by using two-point perspective. Place the two vanishing points in different positions from those shown here.

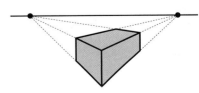

VOCABULARY

1. Two angles whose measures have a sum of 180° are __?__. (p. 327)

2. A line segment that has both endpoints on a circle and that passes through the center of the circle is a __?__. (p. 348)

3. A solid figure with flat faces that are polygons is called a __?__. (p. 354)

EXAMPLES   **EXERCISES**

Chapter 17

- **Identify and classify angles.** (pp. 324–325)

  Tell if the angle *is acute, right,* or *obtuse.*

  The angle is acute since its measure is less than 90°.

Tell if each angle is *acute, right,* or *obtuse*.

4.   5.

6.   7.

- **Identify angle relationships.** (pp. 326–329)

  Find the measure of ∠*BOF*.

  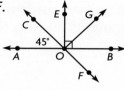

  ∠*BOF* and ∠*COA* are vertical angles and are congruent. So, the measure of ∠*BOF* is 45°.

Find the measure of each angle.

8. ∠*XTY*

9. ∠*ZTR*

10. ∠*YTZ*

11. ∠*RTV*

12. ∠*VTZ*

13. ∠*XTR*

Chapter 18

- **Classify triangles.** (pp. 336–339)

  Find the unknown angle measure and classify the triangle.

  $180° - (60° + 60°) =$
  $180° - 120° = 60°$

  The measure of the angle is 60°.

  The triangle is an acute triangle.

Find the unknown angle measure and classify the triangle.

14.   15.

16.   17.

- **Classify quadrilaterals.** (pp. 342–345)

  Give the most exact name for the figure.

  Opposite sides are congruent and parallel. There are 4 right angles. So, the figure is a square.

Give the most exact name for the figure.

18.

19.

20.

21.

- **Identify parts of a circle.** (pp. 348–349)

  $\overline{AB}$ is a diameter.

  $\overline{OC}$ is a radius.

  $\overline{DE}$ is a chord.

  $\overarc{CD}$ is an arc.

Name the given parts of the circle.

22. chord

23. diameter

24. radius

25. arc

Chapter 19

- **Name solid figures.** (pp. 354–356)

  Name the figure. Is it a polyhedron?

  square pyramid
  It is a polyhedron.

Name the figure. Is it a polyhedron? Write *yes* or *no*.

26.

27.

- **Identify nets for solid figures.** (pp. 360–361)

  Name the figure that can be formed from the net.

  triangular pyramid

  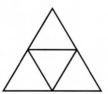

Will the arrangement of squares fold to form a cube? Write *yes* or *no*.

28.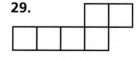

29.

PROBLEM SOLVING APPLICATIONS

30. Eight players were in a tennis match. Each player had to play every other player. How many matches did they play altogether? (pp. 340–341)

31. Juan made a model of an octagonal prism. How many vertices and how many edges did his model have?

    (pp. 362–363)

Devil's Postpile National Monument

Nearly 1 million years ago, lava flowed from a volcanic vent near today's town of Mammoth Lakes, California. Some of the lava cooled and cracked into tall columns. You can see the columns at Devil's Postpile National Monument.

The tallest columns at Devil's Postpile stand 60 feet high.

For 1–2, use the photo of the "floor," at the right.

1. The tops of the columns at Devil's Postpile form a smooth "floor" of many-sided shapes. Study the photograph of the floor. What shapes do you see?

2. If the bases of the columns form the same shapes as the tops, what solid figures do the columns form?

3. Complete a row in the table for each solid figure you named in Exercise 2.

Solid Figure	Number of Faces	Number of Edges	Number of Vertices
?	■	■	■
?	■	■	■

Use Data For 4, use the drawing below.

4. **What if** an ant crawled from *A* to *B* to *C* to *D* and back to *A* on the longest column at Devil's Postpile? The perimeter of the column's base is 9 ft. The ant moved at a rate of 3 ft per minute. How long did the journey take?

60 ft

9 ft

hexagonal prism

5. Explain how you could make a model of a Devil's Postpile column by using a common item found in every classroom.

Buildings

California is famous, too, for its human-made wonders. Because of their unique shapes, some of the state's buildings are known around the world.

The Transamerica Pyramid is the tallest building in San Francisco.

The twin ARCO Towers in Los Angeles are identical from the outside.

This unique building in San Francisco has a 40-sided roof.

Use Data **For 1–6, use the pictures of the buildings.**

Give the geometric term that best describes the two- or three-dimensional figure.

1. the Transamerica Pyramid

2. Arco Tower Number 1

3. each face of Arco Tower Number 2

4. the type of triangle forming each face of the Transamerica Pyramid

5. each base of Arco Tower Number 1

6. the base of the Transamerica Pyramid

Use Data **For 7, use the figure.**

7. The figure shows one section of the 40-sided roof. $\overline{BC}$ is one side of the polygon. A is the center of the circle and m$\angle B$ = m$\angle C$. Find the measure of each angle in $\triangle ABC$. (Hint: Recall that there are 360° in a circle.)

Floors A, B, and C, shown below, are the 6th, 27th, and 48th floors of one of the buildings pictured at the top of the page, though not necessarily in that order.

Use Data **For 8–10, use the floor diagrams.**

8. Which building do the floors belong to?

9. What is the floor number of Floor A? Floor B? Floor C?

10. Estimate the dimensions of the 17th floor.

20 Ratio and Proportion

Mini-Europe is an attraction in Belgium. The models in the display are made using the scale 1:25. The tall building is a replica of the Brussels Town Hall in Brussels, Belgium, and is about 4 m tall. The actual building is about 96 m tall. Another model in Mini-Europe is the Eiffel Tower. The actual Eiffel Tower is 320.75 m tall. Using the scale 1:25, about how tall is the model?

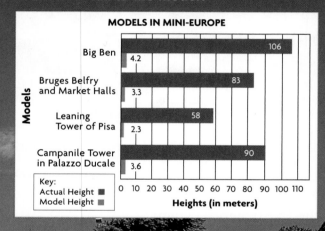

MODELS IN MINI-EUROPE

Models

Big Ben — 106 / 4.2
Bruges Belfry and Market Halls — 83 / 3.3
Leaning Tower of Pisa — 58 / 2.3
Campanile Tower in Palazzo Ducale — 90 / 3.6

Key:
Actual Height ■
Model Height ■

Heights (in meters)
0 10 20 30 40 50 60 70 80 90 100 110

Check What You Know

Use this page to help you review and remember important skills needed for Chapter 20.

Vocabulary

Choose the best term from the box.

> enlargement
> equivalent
> polygon
> polyhedron
> reduction

1. Fractions that name the same number or amount are ___?___ fractions.

2. When the size of a figure is increased, the new figure is a(n) ___?___ of the original figure.

3. A closed plane figure formed by line segments is a(n) ___?___ .

Write Equivalent Fractions (See p. H21.)

Complete each number sentence.

4. $\frac{4}{5} = \frac{\blacksquare}{20}$

5. $\frac{5}{9} = \frac{15}{\blacksquare}$

6. $\frac{\blacksquare}{3} = \frac{28}{21}$

7. $\frac{5}{6} = \frac{\blacksquare}{24}$

8. $\frac{\blacksquare}{9} = \frac{2}{3}$

9. $\frac{5}{\blacksquare} = \frac{25}{30}$

10. $\frac{3}{4} = \frac{12}{\blacksquare}$

11. $\frac{\blacksquare}{6} = \frac{45}{54}$

12. $\frac{1}{2} = \frac{15}{\blacksquare}$

13. $\frac{3}{\blacksquare} = \frac{21}{49}$

14. $\frac{\blacksquare}{25} = \frac{3}{5}$

15. $\frac{24}{64} = \frac{\blacksquare}{16}$

Solve Multiplication Equations (See p. H21.)

Solve.

16. $5x = 20$

17. $2p = 14$

18. $48 = 3k$

19. $45 = 9m$

20. $96 = 8b$

21. $65 = 2.5d$

22. $0.4n = 6$

23. $187 = 11y$

24. $\frac{1}{2}h = 21$

25. $7.2 = 0.03m$

26. $12 = \frac{2}{3}a$

27. $23h = 23$

Congruent and Similar Figures (See p. H22.)

Decide whether the figures in each pair appear to be *similar*, *congruent*, *both*, or *neither*.

28.

29.

30.

31.

32.

33.

LOOK AHEAD

In Chapter 20 you will

- write ratios and rates
- use the strategy *write an equation* to solve problems
- use ratios to identify similar figures
- use proportions to find unknown measures
- read and use scales

Ratios and Rates

Learn how to write ratios and rates and how to find unit rates.

Vocabulary

equivalent ratios
rate
unit rate

A model of the Liberty Bell on a scale of 1:45 is small enough to hold in your hand. A scale of 1:45 means that each measurement on the model is $\frac{1}{45}$ as long as the corresponding measurement on the real bell. Both 1:45 and $\frac{1}{45}$ are ratios.

A ratio is a comparison of two numbers, a and b, and can be written as a fraction $\frac{a}{b}$. You can write a ratio in three ways.

Write: 1 to 45 or 1:45 or $\dfrac{1}{45}$ $\begin{array}{l}\leftarrow \text{ first term}\\ \leftarrow \text{ second term}\end{array}$ **Read:** one to forty-five

You can write a ratio to compare two amounts—a part to a part, a part to the whole, or the whole to a part.

EXAMPLE 1

The world's largest bells are made from "bell metal." Each 4 pounds of bell metal contains 3 pounds of copper and 1 pound of tin. Write the following ratios.

a. pounds of copper to pounds of tin → $\frac{3}{1}$ *part to part*

b. pounds of copper to total pounds → $\frac{3}{4}$ *part to whole*

c. total pounds to pounds of tin → $\frac{4}{1}$ *whole to part*

Equivalent ratios are ratios that name the same comparisons.

EXAMPLE 2

Write three equivalent ratios to compare the number of stars with the number of stripes.

$\dfrac{\text{number of stars}}{\text{number of stripes}}$ → $\dfrac{6}{9}$

$\dfrac{6}{9}$ → $\dfrac{6 \div 3}{9 \div 3} = \dfrac{2}{3}$ *Divide both terms by a common factor.*

$\dfrac{6}{9}$ → $\dfrac{6 \times 2}{9 \times 2} = \dfrac{12}{18}$ *Multiply both terms by the same number.*

So, $\frac{6}{9}$, $\frac{2}{3}$, and $\frac{12}{18}$ are equivalent ratios.

CALIFORNIA STANDARDS O—■NS 1.2 Interpret and use ratios in different contexts (e.g., batting averages, miles per hour) to show the relative sizes of two quantities, using appropriate notations ($\frac{a}{b}$, a to b, $a:b$). O—■AF 2.2 Demonstrate an understanding that *rate* is a measure of one quantity per unit value of another quantity. *also* NS 1.0, AF 2.0, AF 2.3

Math Idea ▶ A **rate** is a ratio that compares two quantities having different units of measure.

Suppose 24 ounces of cereal costs $4.32.

rate: $\dfrac{\text{price}}{\text{number of ounces}} \rightarrow \dfrac{\$4.32}{24 \text{ oz}}$ $4.32 for 24 oz

A **unit rate** is a rate that has 1 unit as its second term.

unit rate: $\dfrac{\$4.32}{24} = \dfrac{\$4.32 \div 24}{24 \div 24} = \dfrac{\$0.18}{1}$ $0.18 for 1 oz

The unit rate of the cereal is $0.18 per ounce.

Average speed is average distance traveled in 1 unit of time, such as 1 hour or 1 second. So, average speed is a unit rate.

EXAMPLE 3

The Howards are driving their car from Richmond, VA, to Philadelphia, PA, a distance of 250 mi. They have traveled 200 mi in 4 hr. At this rate, how long will it take them to make the complete trip?

$\dfrac{\text{miles}}{\text{hours}} \rightarrow \dfrac{200}{4}$ *Write a ratio to compare miles to hours.*

$\dfrac{200}{4} = \dfrac{200 \div 4}{4 \div 4} = \dfrac{50}{1} \begin{array}{l} \leftarrow \text{miles} \\ \leftarrow \text{hours} \end{array}$ *Find the unit rate, or average speed.*

$\dfrac{50}{1} = \dfrac{50 \times 5}{1 \times 5} = \dfrac{250}{5} \begin{array}{l} \leftarrow \text{miles} \\ \leftarrow \text{hours} \end{array}$ **THINK:** *$50 \times 5 = 250$. Multiply each term by 5 to find the number of hours for 250 mi.*

So, it will take the Howards 5 hr to travel 250 mi.

CHECK FOR UNDERSTANDING

Think and ▶ Discuss

Look back at the lesson to answer each question.

1. **Explain** how to find equivalent ratios.

2. **Tell** which is the better buy, 14 oz of cereal at $2.52 or 20 oz of cereal at $3.20. Explain.

Guided ▶ Practice

Write two equivalent ratios.

3. $\dfrac{4}{8}$ 4. $\dfrac{6}{10}$ 5. $\dfrac{9}{12}$ 6. $\dfrac{10}{15}$

Write a ratio in fraction form. Then find the unit rate.

7. 150 points in 10 games   8. $64 in 16 hr

9. 90 words in 2 min   10. $2.56 for 4 pencils

11. $15 for 6 lb   12. 120 mi in 3 hr

Independent ▸ Practice

Write two equivalent ratios.

13. $\frac{2}{4}$ **14.** $\frac{4}{5}$ **15.** $\frac{3}{9}$ **16.** $\frac{15}{21}$

17. $\frac{6}{16}$ **18.** $\frac{21}{30}$ **19.** $\frac{5}{9}$ **20.** $\frac{14}{24}$

Write each ratio in fraction form. Then find the unit rate.

21. $15 for 5 tapes   **22.** 8 pages in 2.5 hr

23. 60 mi on 3 gal   **24.** $1.89 for 3 pens

25. $2.10 for 6 fish   **26.** 900 students for 30 teachers

For 27–28, use the figure at the right.

27. Find the ratio of red sections to blue sections. Then write three equivalent ratios.

28. Find the ratio of blue sections to all the sections. Then write three equivalent ratios.

Find the missing term, *m*, that makes the ratios equivalent.

29. 5 to 4, as *m* to 12   **30.** 20:*m*, as 8:10   **31.** $\frac{9}{12}$, as $\frac{45}{m}$

32. Write the following ratio: number of tricycles to number of tricycle wheels.

33. Write the following ratio: number of tires to number of cars.

Problem Solving ▸ Applications

34. Of the 10 members on a team, 4 are girls. What is the ratio of girls to all the members? of boys to girls?

Use Data   For 35–38, use the table.

35. *REASONING*   How can you decide if Brand B or Brand C is the better buy without dividing?

36. Which brand is the best buy?

37. Find the cost of 6 boxes of Brand D.

38. **What's the Question?**   The answer is $0.42 per box.

HOW JUICES COMPARE		
Brand	Cost	Number of Boxes
A	$1.17	3
B	$1.23	3
C	$2.52	6
D	$3.15	9

MIXED STRATEGY PRACTICE

39. Describe the pattern. Then find the next term. (p. 340) 1, 4, 9, 16, ■

40. If $k = 4p - 3t$, what is the value of k when $p = 2$ and $t = {}^-2$? (p. 276)

41. Give the prime factorization of 54. (p. 148)

42. Write 45.6% as a decimal. (p. 160)

43. TEST PREP   Solve $n + 3 = {}^-5$. (p. 291)

A $n = 8$ **B** $n = {}^-2$ **C** $n = {}^-6$ **D** $n = {}^-8$

Extra Practice page H51, Set A

ALGEBRA
Explore Proportions

Explore how to find equivalent ratios to form a proportion.

You need two-color counters.

Atoms are tiny building blocks of matter. Atoms join in equivalent ratios to form larger building blocks called molecules. For example, the ratio of oxygen atoms to hydrogen atoms in 1 molecule of water is $\frac{1}{2}$. The ratio of oxygen atoms to hydrogen atoms in 2 molecules of water is $\frac{2}{4}$, which is the same ratio as $\frac{1}{2}$.

These ratios can be used to write a proportion. A **proportion** is an equation that shows two equivalent ratios.

$$\frac{1}{2} = \frac{2}{4} \leftarrow \text{ proportion}$$

You can use counters to model proportions.

Activity

• Make the model shown here. Write the ratio of red counters to yellow counters.

• Separate the red counters into three equal groups. Do the same with the yellow counters. Write the ratio of red to yellow in each group.

• Separate the counters into six equal groups. Write the ratio of red to yellow in each group.

Think and Discuss

• One proportion shown by your model is $\frac{6}{12} = \frac{2}{4}$. What are two other proportions shown by your model?

• Is $\frac{4}{8} = \frac{3}{4}$ a proportion? Use your counters to explain your reasoning.

Practice

1. How many hydrogen atoms combine with 4 oxygen atoms to form water? How many combine with 6 oxygen atoms? Model the ratios with counters.

Use models to determine whether the ratios form a proportion.

2. $\frac{1}{3}$ and $\frac{4}{12}$ 3. $\frac{3}{4}$ and $\frac{6}{9}$ 4. $\frac{2}{4}$ and $\frac{3}{6}$

CALIFORNIA STANDARDS O⎯ℸ **NS 1.0** Students compare and order positive and negative fractions, decimals, and mixed numbers. Students solve problems involving fractions, ratios, proportions, and percentages. O⎯ℸ**NS 1.2** Interpret and use ratios in different contexts to show the relative sizes of two quantities, using appropriate notations. *also* O⎯ℸ**NS 1.3, MR 2.4**

PROBLEM SOLVING STRATEGY
Write an Equation

Analyze
Choose
Solve
Check

Learn how to use the strategy *write an equation* to solve problems.

QUICK REVIEW

1. 24 × 10   2. 30 × 7   3. 50 × 9
4. 400 ÷ 5   5. 300 ÷ 60

Remember that a *proportion* is an equation that shows two equivalent ratios.

The weight of an object depends on the mass and radius of the planet or moon where the object is located. A person who weighs 60 lb on Earth would weigh only 10 lb on the moon. The ratio $\frac{\text{weight on moon}}{\text{weight on Earth}}$ is the same for all objects. Danielle weighs 90 lb. How much would she weigh on the moon?

Analyze

What are you asked to find?

What information is given?

Is there information you will not use? If so, what?

Choose

What strategy will you use?

You can *write an equation* in the form of a proportion. Write two ratios that compare weight on Earth with weight on the moon. Use *n* to represent Danielle's weight on the moon.

Solve

Write and solve a proportion.

$$\frac{\text{weight on moon}}{\text{weight on Earth}} \rightarrow \frac{10}{60} = \frac{n}{90} \leftarrow \frac{\text{Danielle's weight on moon}}{\text{Danielle's weight on Earth}}$$

One Way Use the Multiplication Property of Equality.

$$\frac{10}{60} = \frac{n}{90}$$

Multiply both sides by 60 and 90.

$$\frac{10}{\cancel{60}_{1}} \times \frac{\cancel{60}^{1}}{1} \times \frac{90}{1} = \frac{n}{\cancel{90}_{1}} \times \frac{60}{1} \times \frac{\cancel{90}^{1}}{1}$$

Simplify.

$$900 = 60n$$

Solve the equation.

$$\frac{900}{60} = \frac{60n}{60}$$

$$n = 15$$

Another way Use cross products.

$$\frac{60}{10} \diagdown = \diagup \frac{90}{n}$$

Find the cross products.

$$60 \times n = 90 \times 10$$
$$60n = 900$$

Solve the equation.

$$\frac{60n}{60} = \frac{900}{60}$$

$$n = 15$$

So, Danielle would weigh 15 lb on the moon.

Check

How are the two ways of solving the proportion alike?

What if Danielle weighed 78 lb? What would she weigh on the moon?

CALIFORNIA STANDARDS O→NS 1.3 Use proportions to solve problems. Use cross-multiplication as a method for solving such problems, understanding it as the multiplication of both sides of an equation by a multiplicative inverse. MR 3.2 Note the method of deriving the solution and demonstrate a conceptual understanding of the derivation by solving similar problems. *also* O→NS 1.0, MR 1.1

PROBLEM SOLVING PRACTICE

PROBLEM SOLVING STRATEGIES

Draw a Diagram or Picture
Make a Model
Predict and Test
Work Backward
Make an Organized List
Find a Pattern
Make a Table or Graph
Solve a Simpler Problem
▶ **Write an Equation**
Use Logical Reasoning

Solve the problem by writing an equation.

1. Astronaut Neil Armstrong, the first person to walk on the moon, weighed 165 lb on Earth. How much did he weigh on the moon?

2. Abdul's car uses about 10 gal of gasoline on a 180-mi trip. Predict how much gasoline Abdul will need to travel 540 mi.

3. A map uses a scale of 2 in. for every 9 mi. If the map shows a distance of 18 in., what is the actual distance?

For 4–5, use this information: A machine fills 4 bottles every 6 seconds. The machine fills 12 bottles.

4. Which proportion could you use to find s, the amount of time it takes to fill 12 bottles?

   A $\frac{4}{6} = \frac{s}{12}$ C $\frac{4}{6} = \frac{12}{s}$

   B $\frac{6}{4} = \frac{12}{s}$ D $\frac{4}{s} = \frac{6}{12}$

5. How long does it take the machine to fill 12 bottles?

   F 8 sec   H 16 sec

   G 12 sec   J 18 sec

MIXED STRATEGY PRACTICE

6. At 5:00 P.M. the temperature was 72°F. It fell at a steady rate. At 8:00 P.M. it was 54°F. If the pattern continues, what will the temperature be at 9:00 P.M.?

7. Debra is making 20 invitations. Each invitation is 6 in. long and 5 in. wide. How many square inches of paper does she need?

8. A farmer paid $40,000 for 25 acres of farmland. The farmer wants to buy 15 acres more at the same price per acre. How much will 15 acres cost?

9. Kyle has an average of 88 on four tests. What must he get on his fifth test to have an average of 90?

The histogram shows how many salespeople sold different numbers of cars in the first month of work. Use the histogram for 10–15.

10. How many salespeople sold more than 40 cars in the first month?

11. How many salespeople are represented by the histogram?

12. Did more salespeople sell 0–30 cars or 31–60 cars? How many more?

13. What is the greatest possible number of cars any one salesperson sold?

14. What if no salespeople had sold 31–40 cars? How would the histogram change?

15. **? What's the Error?** Al said that four people sold exactly 21 cars each. What was his mistake?

SALES IN FIRST MONTH

LESSON 20.4

ALGEBRA
Ratios and Similar Figures

Learn how to use ratios to identify similar figures.

QUICK REVIEW

Simplify.

1. $\frac{6}{8}$ 2. $\frac{6}{12}$ 3. $\frac{8}{12}$ 4. $\frac{15}{18}$ 5. $\frac{9}{15}$

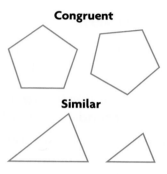

Congruent

Similar

Vocabulary

similar figures

corresponding sides

corresponding angles

Congruent figures have the same shape and the same size. These pentagons are congruent.

Similar figures have the same shape. These triangles are similar. The pentagons are similar, too, because they have the same shape. Similar figures may be the same size, like the pentagons, or different sizes, like the triangles.

EXAMPLE 1

Decide whether the figures in each pair appear to be similar, congruent, both, or neither.

A
same shape, not same size
The figures are similar but not congruent.

B
same shape, same size
The figures are both similar and congruent.

C
not same shape, not same size
The figures are neither similar nor congruent.

Activity

You need: metric ruler

- Do these triangles appear to be congruent? Do they appear to be similar? Explain.

- Measure and record the lengths of the sides of both triangles in cm.

- Write the ratios *AB* to *DE*, *BC* to *EF*, and *AC* to *DF*.

- Write each of the ratios in simplest form. What do you notice about the ratios?

380 Chapter 20

CALIFORNIA STANDARDS O━ⁿNS 1.0 Students compare and order positive and negative fractions, decimals, and mixed numbers. Students solve problems involving fractions, ratios, proportions, and percentages. *also* **AF 2.0, MR 3.2**

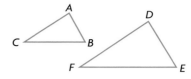
Sometimes it is easier to find the **corresponding sides** and **corresponding angles** of similar figures if they are oriented the same way on the page.

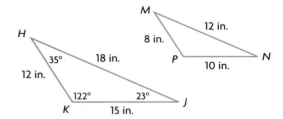

$\angle A$ corresponds to $\angle D$. $\overline{AB}$ corresponds to $\overline{DE}$.

$\angle B$ corresponds to $\angle E$. $\overline{BC}$ corresponds to $\overline{EF}$.

$\angle C$ corresponds to $\angle F$. $\overline{CA}$ corresponds to $\overline{FD}$.

Math Idea ▶ If figures are similar, their corresponding angles are congruent and the lengths of their corresponding sides have the same ratio.

EXAMPLE 2

$\triangle MNP$ and $\triangle HJK$ are similar. Find the measures of $\angle M$, $\angle N$, and $\angle P$ and the ratio of the lengths of the corresponding sides.

$\angle M$ corresponds to $\angle H$, so $\angle M$ measures 35°.

$\angle N$ corresponds to $\angle J$, so $\angle N$ measures 23°.

$\angle P$ corresponds to $\angle K$, so $\angle P$ measures 122°.

$\overline{MN}$ corresponds to $\overline{HJ}$, $\overline{NP}$ corresponds to $\overline{JK}$, and $\overline{PM}$ corresponds to $\overline{KH}$.

$\dfrac{MN}{HJ} \rightarrow \dfrac{12}{18} = \dfrac{2}{3}$ $\dfrac{NP}{JK} \rightarrow \dfrac{10}{15} = \dfrac{2}{3}$ $\dfrac{PM}{KH} \rightarrow \dfrac{8}{12} = \dfrac{2}{3}$

So, the ratio of the lengths of corresponding sides is $\frac{2}{3}$.

For two polygons to be similar, the ratios of the corresponding sides must be equal, and the corresponding angles must be congruent.

EXAMPLE 3

Sean used a copy machine to make an enlargement of $\triangle ABC$. He labeled his new triangle $\triangle DEF$. Are the triangles similar?

$\dfrac{DE}{AB} \rightarrow \dfrac{30}{20} = \dfrac{3}{2}$ $\dfrac{EF}{BC} \rightarrow \dfrac{33}{22} = \dfrac{3}{2}$ *Write the ratios of the corresponding sides in simplest form.*

$\dfrac{FD}{CA} \rightarrow \dfrac{18}{12} = \dfrac{3}{2}$

$\dfrac{3}{2} = \dfrac{3}{2} = \dfrac{3}{2}$ *Compare the ratios.*

$\angle D$ and $\angle A$ are congruent.   *Check whether corresponding*
$\angle E$ and $\angle B$ are congruent.   *angles are congruent.*
$\angle F$ and $\angle C$ are congruent.

Since the ratios of the corresponding sides are equal and the corresponding angles are congruent, $\triangle ABC$ is similar to $\triangle DEF$.

Think and ▶ Discuss

Look back at the lesson to answer each question.

1. **Tell** whether the ratio for corresponding sides in Example 3 could be $\frac{2}{3}$. Explain your decision.

2. **Choose** the one correct statement:
   A All similar figures are also congruent.
   B All congruent figures are also similar.

Guided ▶ Practice

Tell whether the figures appear to be *similar, congruent, both,* or *neither.*

3.   4.   5.

6. Name the corresponding sides and angles. Write the ratio of the corresponding sides in simplest form.

Independent ▶ Practice

Tell whether the figures appear to be *similar, congruent, both,* or *neither.*

7.   8.   9.

Name the corresponding sides and angles. Write the ratio of the corresponding sides in simplest form.

10.

11.

Tell whether the figures in each pair are similar. Write *yes* or *no*. If you write *no*, explain.

12.

13.

14.

15.

The figures in each pair are similar. Find the missing measures.

16.
B
6 ft ? 6 ft
48° 48°
A   8 ft   C

E
? ? 12 ft
48°   ?
D   16 ft   F

17.
R 6 m S
3 m 110° ?
Q   9 m   T

Y   12 m   X
7 m ?   6 m
70°
Z   18 m   W

Write *yes* or *no*. Explain your answers.

18. Are all squares similar?

19. Are all rectangles similar?

20. Are all rhombi similar?

21. Are all right triangles similar?

Problem Solving ▶ Applications

Use the following information for 22–23: The rectangular front of the North End Mall is 60 ft tall and 80 ft wide. The front of the Eastside Mall has sides that are $\frac{3}{4}$ the lengths of the sides of the North End Mall.

22. Are the fronts similar?

23. Find the dimensions of the front of the Eastside Mall.

24. Most basketball courts are 94 ft long and 50 ft wide. A half court is 47 ft by 50 ft. Is a half court similar to a full court? Explain.

25. ✎ **Write About It** Explain how you know that *ABCD* is similar to *AEFG*.

G
80° 24 m
36 m   125° F
B   20 m   A   119° 18 m
12 m 119°   30 m   E
C 125°
16 m 80° 24 m
D

MIXED REVIEW AND TEST PREP

26. What is the unit rate? 5 cans for $3 (p. 374)

27. If $5y = 15$, what does *y* equal? (p. 301)

28. How many faces does a rectangular pyramid have? (p. 354)

29. TEST PREP Evaluate $4x + x + 9$ for $x = 3$. (p. 276)

   A 19   **B** 21   **C** 24   **D** 25

30. TEST PREP Evaluate $7^2 - \sqrt{49}$. (p. 282)

   F ⁻42   **G** 0   **H** 42   **J** 56

LiNKUP to Careers

Careers Artists, photographers, designers, and architects are among the many professionals who use similar figures in their occupations.

1. ❓ **What's the Error?** A photographer took the photo at the right. The photo was then reduced in size. What error did the photographer make in reducing the photo?

2. Suppose that an artist wants to enlarge a drawing that is octagonal in shape. How will the angles of the enlargement be related to those of the original drawing? How will the sides be related?

ALGEBRA
Proportions and Similar Figures

Learn how to use proportions and similar figures to find unknown measures.

Arizona's Grand Canyon is more than 1 mile deep and 217 miles long. It has been called one of the Seven Natural Wonders of the World. Each year, tourists snap millions of photos of the canyon. A photo is similar to the scene it depicts.

If two figures are similar and you know the length of one side, you can use a proportion to find the length of another side.

EXAMPLE 1

Remember that a proportion is an equation which states that two ratios are equivalent.

Armando made an enlargement that is similar to a 4-in. × 6-in. photo he took of the Grand Canyon. The enlargement is 12 in. wide. How long is it?

Write the ratios of the corresponding sides. Let x = the length of $\overline{EF}$.

$$\frac{AB}{EF} \rightarrow \frac{6}{x} \qquad \frac{AD}{EH} \rightarrow \frac{4}{12}$$

$$\frac{6}{x} = \frac{4}{12} \qquad \text{\textit{Use the ratios to write a proportion.}}$$

$$4 \cdot x = 6 \cdot 12 \qquad \text{\textit{Find the cross products.}}$$

$$4x = 72 \qquad \text{\textit{Solve the equation.}}$$

$$\frac{4x}{4} = \frac{72}{4}$$

$$x = 18$$

So, the enlargement is 18 in. long.

Diagram:
Rectangle ABCD: A — 6 in. — B, 4 in. (side AD), D, C
Rectangle EFGH: E — x — F, 12 in. (side EH), H, G

Math Idea ▶ In Example 1, similar figures and a proportion were used to find an unknown length. This method of finding a measurement is called **indirect measurement**. It can be used to find lengths and distances that are too great to be measured with a ruler.

CALIFORNIA STANDARDS ⊶ **NS 1.3** Use proportions to solve problems. Use cross-multiplication as a method for solving such problems, understanding it as the multiplication of both sides of an equation by a multiplicative inverse. ⊶**NS 1.0** Students compare and order positive and negative fractions, decimals, and mixed numbers. Students solve problems involving fractions, ratios, proportions, and percentages. *also* ⊶**AF 2.2, MR 3.2**

384 Chapter 20

EXAMPLE 2

On a sunny day, one saguaro cactus casts a shadow that is 56 ft long. At the same time, a yardstick casts a shadow that is 4 ft long. Use the similar right triangles shown below to find the height of the cactus.

$$\frac{3}{h} = \frac{4}{56} \quad \begin{matrix} \leftarrow \text{ small triangle} \\ \leftarrow \text{ large triangle} \end{matrix}$$ *Write a proportion.*

$$4 \times h = 3 \times 56$$ *Find the cross products.*
Solve the equation.

$$4h = 168$$

$$\frac{4h}{4} = \frac{168}{4}$$

$$h = 42$$

So, the cactus is 42 ft tall.

The saguaro cactus, the world's tallest cactus, is native to Arizona.

CHECK FOR UNDERSTANDING

Think and Discuss ▶ **Look back at the lesson to answer each question.**

1. *REASONING* **Explain** why you can use similar figures to measure objects indirectly.

2. **Name** three objects that would be difficult to measure directly.

3. **Determine** how long the picture in Example 1 would be if it were 8 in. wide.

Guided Practice ▶ **The figures are similar. Write a proportion. Then find the unknown length.**

4.

5.

Independent ▶ Practice

The figures are similar. Write a proportion. Then find the unknown length.

6.
n
4 m
10 m
8 m

7.
8 in.
8 in.
n
3 in.

8.
40 ft
60 ft
100 ft
n

9.
30 cm
n
18 cm
24 cm

10.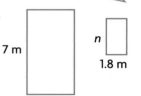
7 m
n
1.8 m
4.5 m

11.
6.2 cm
5 cm
n
2.5 cm

Problem Solving ▶ Applications

12. *REASONING* Two common envelope sizes are $3\frac{1}{2}$ in. $\times$ $6\frac{1}{2}$ in. and 4 in. $\times$ $9\frac{1}{2}$ in. Are these envelopes similar? Explain.

13. Art The famous painting known as "Red Interior, Still-life on a Blue Table" was painted by Henri Matisse in 1947. The copy of the painting at the right measures 1 in. tall by $\frac{3}{4}$ in. wide. The height of the real painting is 32 in. Find the width of the real painting.

14. ? What's the Error? The shadow of a 6-ft man measures 5 ft at the same time his son's shadow is 2 ft. Taylor used a proportion to find the son's height as $1\frac{2}{3}$ ft. Explain Taylor's mistake.

MIXED REVIEW AND TEST PREP

15. Triangle *A*'s sides measure 2 cm, 3 cm, and 4 cm. Triangle *B*'s sides measure 6 cm, 7 cm, and 8 cm. Are the triangles similar? Explain. (p. 380)

16. The base of a prism has 20 sides. How many vertices does the prism have? (p. 354)

For 17–18, use the box-and-whisker graph.

17. What is the median? (p. 130)

18. What is the range? (p. 130)

10 12 14 16 18 20 22

19. TEST PREP Which value for *p* solves the equation $0.4p = 20$? (p. 301)

A 5 **B** 19.6 **C** 20.4 **D** 50

Extra Practice page H51, Set C

ALGEBRA
Scale Drawings

Learn how to use scale to find the dimensions for a drawing or the actual dimensions of an object.

QUICK REVIEW

1. 490 ÷ 7   **2.** 800 ÷ 10   **3.** 540 ÷ 6   **4.** 3 × 120   **5.** 70 × 11

Vocabulary

scale drawing

scale

Floor plans, road maps, and diagrams are examples of scale drawings. A **scale drawing** is a drawing of a real object that is smaller than (a reduction of) or larger than (an enlargement of) the real object. Measurements on a scale drawing are proportional to measurements of the real object.

The above floor plan and other types of architectural drawings are called blueprints. A blueprint can be a reduction or an enlargement of an object.

More Practice: Use **Mighty Math Cosmic Geometry**, *Robot Studio*, Level F.

Activity

You need: centimeter graph paper, metric ruler

- Measure the length and width of the scale drawing above in centimeters.

- On graph paper, make a drawing with measurements that are twice those in the scale drawing.

- Write the ratio of length to width for the original drawing.

- Write the ratio of length to width in your drawing. Are the two ratios equal? Are the drawings similar? Explain.

- Make a drawing with measurements that are half those in the original scale drawing.

- Write the ratio of length to width for your new drawing. Is the ratio equal to the ratio for the original scale drawing? Are the drawings similar? Explain.

- Suppose you made a drawing of the floor plan that was 21 cm long. If your drawing was similar to the floor plan, how wide would it be? Explain.

CALIFORNIA STANDARDS ⊙┓NS **1.2** Interpret and use ratios in different contexts to show the relative sizes of two quantities, using appropriate notations. ⊙┓NS **1.3** Use proportions to solve problems. Use cross-multiplication as a method for solving such problems, understanding it as the multiplication of both sides of an equation by a multiplicative inverse. *also* ⊙┓NS **1.0,** AF **1.0,** ⊙┓AF **2.2,** MR **2.2**

A **scale** is a ratio between two sets of measurements. The scale 1 in. = 3 ft on this drawing of a bicycle means that 1 in. on the drawing represents 3 ft on the actual bike.

scale: 1 in. = 3 ft

EXAMPLE 1

TECHNOLOGY LINK

More Practice: Use E-Lab, *Stretching and Shrinking*. **www.harcourtschool.com/ elab2002**

Measure the length of the bike in the scale drawing. Then use the scale to find the actual length of the bike.

scale drawing length: 2 in.

$$\underset{\text{actual (ft)}}{\overset{\text{drawing (in.)}}{}} \rightarrow \frac{1}{3} = \frac{2}{n}$$

Write a proportion. Let n represent the actual length of the bike.

$$1 \times n = 3 \times 2$$

Find the cross products and solve.

$$n = 6$$

So, the actual length of the bike is 6 ft.

If an object is very small, a scale drawing might be larger than the object.

scale: 8 cm = 1 mm

EXAMPLE 2

A paramecium is a tiny organism whose entire body consists of a single cell. Use the scale and the scale drawing to find the length of an average-size paramecium.

scale drawing length: 2.4 cm

$$\underset{\text{actual (mm)}}{\overset{\text{drawing (cm)}}{}} \rightarrow \frac{8}{1} = \frac{2.4}{n}$$

Write a proportion. Let n represent the actual length of the paramecium.

$$8 \times n = 1 \times 2.4$$

Find the cross products and solve.

$$n = \frac{2.4}{8}$$

$$n = 0.3$$

So, an average paramecium is about 0.3 mm long.

CHECK FOR UNDERSTANDING

Think and ▶
Discuss

Look back at the lesson to answer each question.

1. **Compare** the actual length of the bike in Example 1 with the length of the bike in the scale drawing.

2. **Make** a scale drawing of a square microchip that has a side of 2 mm. Use a scale of 4 cm:1 mm. How long will the side on your drawing be?

3. **Explain** how the scale drawing of a microchip is different from the scale drawing of the bicycle.

Find the unknown dimension.

4. scale: 1 in.:8 ft
drawing length: 3 in.
actual length: ■ ft

5. scale: 2 in.:7 yd
drawing length: 8 in.
actual length: ■ yd

PRACTICE AND PROBLEM SOLVING

**Independent ►
Practice**

Find the unknown dimension.

6. scale: 1 in.:4 ft
drawing length: ■ in.
actual length: 28 ft

7. scale: 1 in.:6 ft
drawing length: 5 in.
actual length: ■ ft

8. scale: 3 in.:4 ft
drawing length: ■ in.
actual length: 14 ft

9. scale: 4 cm:3 mm
drawing length: ■ cm
actual length: 15 mm

10. scale: 2 in.:5 ft
drawing length: $\frac{1}{2}$ in.
actual length: ■ ft

11. scale: 6 cm:3 mm
drawing length: ■ cm
actual length: 2 mm

**Problem Solving ►
Applications**

12. A hallway is 15 ft long. How long is it on a floor plan drawn to the scale 2 in. = 3 yd?

13. A model of a storage shed is shown at the right. The model measures 8 in. × 3 in. × 4 in. How many cubical boxes measuring 2 ft on a side can be stacked in the actual shed?

scale: 1 in. = 2 ft

14. Astronomy Gail wanted to make a scale drawing of the solar system. She drew the sun in the center with a diameter of 1 in. The actual diameter of the sun is about 900,000 miles. What problem did Gail run into in drawing the planet Pluto at its average distance of 3.6 billion miles from the sun?

15. ✎ **Write About It** Explain what the scale 1 cm = 2 m on a scale drawing means.

MIXED REVIEW AND TEST PREP

16. Two rectangles measure 9 cm by 12 cm and 12 cm by 16 cm. Are the rectangles similar? (p. 380)

17. True or false? A cone is a polyhedron. (p. 354)

For 18–19, use the stem-and-leaf plot at the right.

18. What is the greatest data value shown? (p. 126)

19. What is the mode of the data? (p. 126)

Stem	Leaves
3	0 0 3 6 9
4	2 3 8
5	3 6 7

20. TEST PREP Ted recorded these scores on 10-point history quizzes: 8, 7, 10, 9, 7, 7. What was his median score? (p. 106)

A 3　　　**B** 7　　　**C** 7.5　　　**D** 8

ALGEBRA
Maps

Learn how to read and use map scales.

Mapmakers are called cartographers. Cartographers use ratios, proportions, and scales to convert actual distances on Earth to map distances. You can use the same method to convert map distances into real distances.

EXAMPLE

Use the map of California to determine the actual straight-line distance from Los Angeles to Malibu. The map scale is 1 in. = 20 mi.

Use a ruler to measure the distance from Los Angeles to Malibu.

map distance → $1\frac{1}{4}$, or 1.25 in.

Let n represent the actual distance in miles. Write a proportion.

$$\frac{\text{in.}}{\text{mi}} \rightarrow \frac{1}{20} = \frac{1.25}{n}$$

Find the cross products.
$1 \times n = 1.25 \times 20$

Multiply.
$n = 25$

So, it is 25 mi from Los Angeles to Malibu.

• Riverside, California, is 50 mi from Los Angeles. If you were a cartographer, how far from Los Angeles would you place Riverside?

CHECK FOR UNDERSTANDING

Think and ▶
Discuss

Look back at the lesson to answer each question.

1. **Find** the actual straight-line distance from Los Angeles to Long Beach.

Guided ▶
Practice

2. **Tell** the map distance that represents 90 mi on the California map.

The map distance is given. Find the actual distance. Use a map scale of 1 in. = 100 mi.

3. $\frac{1}{2}$ in. 4. 5 in. 5. $4\frac{1}{4}$ in.

CALIFORNIA STANDARDS O—ᴨ NS 1.2 Interpret and use ratios in different contexts to show the relative sizes of two quantities, using appropriate notations. O—ᴨ NS 1.3 Use proportions to solve problems. Use cross-multiplication as a method for solving such problems, understanding it as the multiplication of both sides of an equation by a multiplicative inverse. *also* AF 1.0, O—ᴨ AF 2.2, MR 2.2

Independent ▶
Practice

The map distance is given. Find the actual distance. The scale is 1 in. = 80 mi.

6. $1\frac{1}{2}$ in.   **7.** 3 in.   **8.** $2\frac{1}{2}$ in.   **9.** $5\frac{1}{4}$ in.

10. $\frac{1}{2}$ in.   **11.** 18 in.   **12.** $2\frac{3}{4}$ in.   **13.** $4\frac{5}{8}$ in.

The actual distance is given. Find the map distance. The scale is 1 in. = 25 mi.

14. 100 mi   **15.** 150 mi   **16.** $12\frac{1}{2}$ mi   **17.** $62\frac{1}{2}$ mi

Problem Solving ▶
Applications

For 18–20, use the map of Colorado. The scale is $\frac{1}{2}$ in. = 46 mi.

18. Find the straight-line distance in miles from Pueblo to Denver and then from Denver to Steamboat Springs.

19. Stefano drove from Denver to Vail. If he drove 55 mi per hour, about how long did the trip take him?

20. Fran drove for 2.3 hours at 50 mi per hour. If you drew her route as a line on the Colorado map, how long would it be?

21. *REASONING* The auto club map of California has a scale of 1 in. = 20 mi. The tourist map has a scale of 1 in. = 40 mi. On which map does the state of California appear larger?

22. ✎ **Write About It** Explain how you can find an actual distance between two cities if you know the map distance and the scale.

MIXED REVIEW AND TEST PREP

23. Scale: 1 in.:5 ft, actual length: 20 ft, drawing length: ___?___. (p. 387)

24. How many bases does a cone have? (p. 354)

For 25–26, use the line plot at the right. (p. 102)

25. What is the mode of the data?

26. What is the median of the data?

27. TEST PREP Which fraction is greatest: $\frac{3}{8}, \frac{7}{24}, \frac{5}{12}, \frac{1}{3}$? (p. 166)

A $\frac{3}{8}$ **B** $\frac{7}{24}$ **C** $\frac{5}{12}$ **D** $\frac{1}{3}$

1. **VOCABULARY** A comparison of two numbers is a(n) __?__. (p. 374)

2. **VOCABULARY** An equation that shows two equivalent ratios is a(n) __?__. (p. 377)

3. **VOCABULARY** __?__ figures have the same shape but not necessarily the same size. (p. 380)

Write each ratio in fraction form. (pp. 374–376)

4. white sections to all sections

5. yellow sections to white sections

6. all sections to yellow sections

Write each ratio in fraction form. Then find the unit rate. (pp. 374–376)

7. $0.75 for 3

8. 210 km in 3 hr

9. $4.00 for 10

10. 20 mi in 5 hr

Solve. (pp. 374–376)

11. Beach towels are on sale at 2 for $18. Rich buys 5 of them. How much does he pay?

12. Oranges are selling for $3.00 a dozen. Find the cost of 2 oranges.

Tell whether the figures in each pair are similar. Write *yes* or *no*. If you write *no*, explain. (pp. 380–383)

13.

15 cm 6 cm   5 cm 3 cm

14.

10 ft 99° 8 ft 34° 47° 14 ft   25 ft 99° 20 ft 34° 47° 35 ft

The figures are similar. Write a proportion. Then find the unknown length. (pp. 384–386)

15.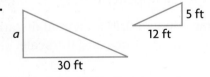

a 5 ft   12 ft   30 ft

16.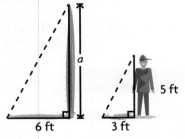

a 5 ft   6 ft   3 ft

Find the unknown dimension. (pp. 387–389)

17. scale: 2 in. = 5 ft
drawing length: 8 in.
actual length: ▇ ft

18. scale: 2 in. = 5 ft
drawing length: ▇ in.
actual length: 40 ft

19. scale: 2 in. = 5 ft
drawing length: 13 in.
actual length: ▇ ft

Use a map scale of 1 in. = 150 mi. Find the actual distance. (pp. 390–391)

20. 5 in.

21. 12 in.

22. $3\frac{1}{2}$ in.

23. $6\frac{3}{4}$ in.

24. $\frac{1}{2}$ in.

25. 0.4 in.

Decide on a plan.

See item **6**.

Use the ratios of width to height of the model and width to the unknown height to write a proportion. Find the cross products and solve the equation.

Also see problem **4**, p. H63.

For 1–10, choose the best answer.

1. Tracy drove 400 kilometers in 8 hours. How many kilometers per hour is that?

   A 5 km per hr   **C** 392 km per hr

   B 50 km per hr   **D** 1,200 km per hr

2. Diane can type 248 words in 4 minutes. At this rate, how long would it take her to type 558 words?

   F 9 min   **H** 14 min

   G 12 min   **J** Not here

3. Sonny uses 68 strawberry plants to fill 4 rows in his garden. Which proportion could he use to find the number of plants he needs for 3 more of these rows?

   A $\frac{4}{68} = \frac{1}{x}$ **C** $\frac{4}{68} = \frac{3}{x}$

   B $\frac{4}{3} = \frac{x}{68}$ **D** $\frac{4}{68} = \frac{x}{3}$

4. If 6 pairs of socks cost $4.50, how much will 9 pairs cost?

   F $6.75   **H** $9.00

   G $7.25   **J** Not here

5. Rectangles *ABCD* and *GHJK* are similar. Which of the following ratios is equivalent to $\frac{AB}{GH}$?

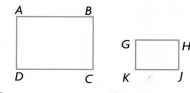

   A $\frac{CD}{JK}$ **C** $\frac{AD}{KJ}$

   B $\frac{AB}{BC}$ **D** $\frac{JK}{CB}$

6. A model of a building is 5 inches wide and 12 inches high. The actual building is 120 feet wide. How high is the actual building?

   F 24 ft   **H** 192 ft

   G 50 ft   **J** 288 ft

7. The diagram below shows similar figures. Which proportion could be used to find *x*?

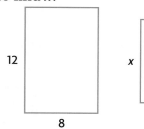

   A $\frac{8}{x} = \frac{6}{12}$ **C** $\frac{x}{12} = \frac{8}{6}$

   B $\frac{6}{8} = \frac{x}{12}$ **D** $\frac{x}{6} = \frac{8}{12}$

8. At 6:00 A.M., the temperature was ⁻4°F. By 6:00 P.M., the temperature had risen 17°. What was the temperature at 6:00 P.M.?

   F ⁻17°F   **H** 3°F

   G ⁻13°F   **J** 13°F

9. The field goal averages of four players are 0.309, 0.234, 0.312, and 0.289. Which shows these averages in order from least to greatest?

   A 0.234, 0.289, 0.312, 0.309

   B 0.312, 0.309, 0.289, 0.234

   C 0.234, 0.289, 0.309, 0.312

   D 0.289, 0.312, 0.234, 0.309

10. What is the range of the data shown below?

PRICES OF DAYPACKS ($)							
21	27	31	26	22	27	20	19
28	31	25	32	20	28	25	20

   F 8   **H** 15

   G 13   **J** 17

CHAPTER 21

Percent and Change

Most states, counties, and cities use sales tax and income tax to raise money. Sales tax is given as a percent. Five states—Alaska, Delaware, Montana, New Hampshire, and Oregon—have no sales tax. All other states have sales taxes varying from 3% to 7.25%. Suppose you have $65 and you want to buy a dictionary for $18.75, a map for $6, a magazine for $4.50, and a book on tape for $34. The sales tax rate is 7.25%. Do you have enough money to buy all of the items? Why or why not?

SALES TAX FOR SELECTED STATES

Sales Tax Rate (%)

State: California, Idaho, Louisiana, Oklahoma, Rhode Island, Virginia

Check What You Know

Use this page to help you review and remember important skills needed for Chapter 21.

 Vocabulary

Choose the best term from the box.

> equivalent
> proportion
> ratio

1. A comparison of two numbers is a(n) __?__ .

2. An equation that shows two equivalent ratios is a(n) __?__ .

 Write Fractions as Decimals (See p.H22.)

Write each fraction as a decimal.

3. $\frac{1}{4}$
4. $\frac{3}{5}$
5. $\frac{1}{10}$
6. $\frac{7}{20}$
7. $\frac{4}{5}$

8. $\frac{9}{10}$
9. $\frac{1}{5}$
10. $\frac{13}{25}$
11. $\frac{81}{100}$
12. $\frac{31}{50}$

 Multiply with Decimals and Fractions (See p.H23.)

Find the product.

13. $\begin{array}{r} 3.6 \\ \times 4.8 \end{array}$
14. $\begin{array}{r} 9.2 \\ \times 4.9 \end{array}$
15. $\begin{array}{r} 6.5 \\ \times 4.1 \end{array}$
16. $\begin{array}{r} 2.7 \\ \times 4.3 \end{array}$

17. $\begin{array}{r} 72 \\ \times 0.4 \end{array}$
18. $\begin{array}{r} 22.4 \\ \times 0.16 \end{array}$
19. $\begin{array}{r} 19 \\ \times 1.8 \end{array}$
20. $\begin{array}{r} 0.043 \\ \times \ 240 \end{array}$

Find the product. Write the answer in simplest form.

21. $\frac{1}{2} \times \frac{1}{3}$
22. $\frac{3}{8} \times \frac{3}{5}$
23. $\frac{2}{3} \times \frac{1}{7}$
24. $\frac{4}{5} \times \frac{1}{6}$

25. $\frac{3}{4} \times \frac{4}{5}$
26. $\frac{4}{9} \times \frac{3}{8}$
27. $\frac{6}{7} \times \frac{8}{9}$
28. $\frac{5}{12} \times \frac{3}{10}$

 Percents and Decimals (See p.H23.)

Write the percent as a decimal or the decimal as a percent.

29. 45%
30. 0.8

31. 0.14
32. 73%

33. 40%
34. 0.39

35. 53%
36. 125%

37. 100%
38. 1.4

39. 3%
40. 0.007

> **LOOK AHEAD**
>
> **In Chapter 21 you will**
> - write ratios as percents
> - convert among percents, decimals, and fractions
> - find percents
> - construct circle graphs
> - find discounts, sales tax, and interest

Percent

Learn how to write ratios as percents and express parts of a whole area as percents.

Complete to find an equivalent fraction.

1. $\frac{1}{2} = \frac{\blacksquare}{100}$ 2. $\frac{1}{10} = \frac{\blacksquare}{100}$ 3. $\frac{7}{10} = \frac{\blacksquare}{100}$

4. $\frac{2}{25} = \frac{\blacksquare}{100}$ 5. $\frac{9}{50} = \frac{\blacksquare}{100}$

Twenty out of 100 teens have a checking account, according to a survey. You can express the ratio 20 out of 100 as a percent.

Math Idea ▶ A percent is the ratio of a number to 100. Percent, %, means "per hundred." So, 20% of teens have a checking account.

EXAMPLE 1

What percent of the squares are shaded?

$\frac{\text{shaded} \rightarrow}{\text{total} \rightarrow} \frac{46}{100}$ *Write the ratio of shaded squares to total squares.*

$\frac{46}{100} = 46\%$ *Write the ratio as a percent.*

So, 46% of the squares are shaded.

You can write a fraction as a percent.

EXAMPLE 2

Sheri has $25 in her checking account. She writes a check for $9. What percent of the total is the check amount?

$\frac{\text{check amount} \rightarrow}{\text{total amount} \rightarrow} \frac{9}{25}$ *Write a ratio in fraction form.*

$\frac{9}{25} \times \frac{4}{4} = \frac{36}{100} = 36\%$ *Write an equivalent fraction with a denominator of 100. Then write the percent.*

So, the check amount is 36% of the total in the checking account.

You can compare percents just as you compare other numbers.

EXAMPLE 3

The average weekly allowance for 9–11-year-olds in a recent survey was $5. Gina gets 85% of the average. Ricky gets 60% of the average. Compare the percents.

85 > 60, so 85% > 60%.

Allowance Amounts, 9–11-year-olds

CALIFORNIA STANDARDS ○━┓NS 1.0 Students compare and order positive and negative fractions, decimals, and mixed numbers. Students solve problems involving fractions, ratios, proportions, and percentages. *also* MR 2.5, MR 3.2

Think and ▶
Discuss

Look back at the lesson to answer the question.

1. **Draw** a model to show 63 out of 100. Then write the ratio as a percent.

Guided ▶
Practice

Write the percent that is shaded.

2.

3.

4.

Write as a percent.

5. $\frac{31}{100}$ 6. $\frac{125}{100}$ 7. $\frac{6}{10}$ 8. $\frac{43}{50}$

Compare. Write <, >, or = for each ●.

9. 30% ● 25%   10. 99% ● 100%   11. 54% ● 54.0%

PRACTICE AND PROBLEM SOLVING

Independent ▶
Practice

Write the percent that is shaded.

12.

13.

14.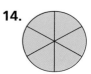

Write as a percent.

15. $\frac{18}{100}$ 16. $\frac{150}{100}$ 17. $\frac{21}{25}$ 18. $\frac{13}{20}$

Compare. Write <, >, or = for each ●.

19. 6% ● 12%   20. 50.0% ● 50%   21. 0.4% ● 0.04%

Problem Solving ▶
Applications

22. Out of 100 students in the cafeteria, 38 wanted chicken and 29 wanted roast beef. What percent did not want either one?

23. The Geckos won 4 out of their 8 games in April, 7 out of 9 in May, and 5 out of 8 in June. What percent of all their games did they win?

24. ✎ **Write a problem** that involves money and can be solved using percents.

MIXED REVIEW AND TEST PREP

25. If a map scale is 1 in. = 60 mi, what map distance represents 150 mi? (p. 390)

26. Write an expression for the phrase "21 decreased by *y*." (p. 274)

27. 1.5 + 0.51 + 11 + 0.005 (p. 66)

28. $\frac{9}{16} - \frac{3}{8}$ (p. 182)

29. **TEST PREP** What is the solution of the equation 2 = 8 + *n*? (p. 291)

A $n = \frac{1}{4}$ B $n = 4$ C $n = 6$ D $n = {}^-6$

Percents, Decimals, and Fractions

Learn how to convert among percents, decimals, and fractions.

Every day, data appear in newspapers, on television, and on the Internet. Some data are given in decimal form, some in fraction form, and some in percent form. To fully understand the data you see in your everyday life, you should be able to change from one number form to the others.

Below are two ways to write a decimal as a percent.

One Way   Use place value.

EXAMPLE 1

Only about 0.2 of the people who use the Internet are younger than 18. Write 0.2 as a percent.

$0.2 = \frac{2}{10}$ — *Use place value to express the decimal as a ratio in fraction form.*

$= \frac{2 \times 10}{10 \times 10} = \frac{20}{100}$ — *Write an equivalent fraction with a denominator of 100.*

$= 20\%.$ — *Since percent is the ratio of a number to 100, write the ratio as a percent.*

• Write 0.6 as a percent.

Another Way   Multiply by 100.

EXAMPLE 2

Write the decimal as a percent.

A. 0.2
   $0.2 = 0.20 = 20\%$ — *Multiply by 100 by moving the decimal point two places to the right.*

B. 0.87
   $0.87 = 0.87 = 87\%$ — *Multiply by 100 by moving the decimal point two places to the right.*

• Write 0.3 and 0.94 as percents.

Remember that when multiplying decimal numbers by powers of 10, you move the decimal point one place to the right for each factor of 10.

CALIFORNIA STANDARDS O—NS 1.0 Students compare and order positive and negative fractions, decimals, and mixed numbers. Students solve problems involving fractions, ratios, proportions, and percentages. *also* **MR 1.1, MR 2.2, MR 2.5, MR 3.2**

Write Fractions as Percents

Below are two ways to write a fraction as a percent.

One Way Write an equivalent fraction with a denominator of 100.

In 1999, about $\frac{2}{5}$ of small businesses in the United States had websites on the Internet. What percent of small businesses had websites?

$$\frac{2}{5} = \frac{2 \times 20}{5 \times 20} = \frac{40}{100}$$ *Write an equivalent fraction with a denominator of 100.*

$$= 40\%$$ *Since percent is the ratio of a number to 100, write the ratio as a percent.*

So, about 40% of small businesses had websites in 1999.

Another Way Use division to write the fraction as a decimal.

$$\frac{2}{5} \rightarrow 5\overline{)2.00}^{\,0.40}$$ *Divide the numerator by the denominator.*

$$0.40 = 40\%$$ *Multiply by 100 by moving the decimal point two places to the right.*

You can also write a percent as a fraction.

Remember that *percent* means "per hundred."

In 1999, about 52% of the world's Internet users lived in the United States. Write 52% as a fraction.

$$52\% = \frac{52}{100}$$ *Write the percent as a fraction with a denominator of 100.*

$$\frac{52}{100} = \frac{52 \div 4}{100 \div 4} = \frac{13}{25}$$ *Write the fraction in simplest form.*

So, 52% written as a fraction is $\frac{13}{25}$.

You can write a percent as an equivalent decimal.

The number of websites on the Internet in May 1999 was 218% of the websites two years earlier. Write 218% as a decimal.

$$218\% = \frac{218}{100}$$ *Write the percent as a fraction with a denominator of 100.*

$$= 2.18$$ *Write the fraction as a decimal.*

So, 218% written as a decimal is 2.18.

The key sequence below shows how to change a percent to a decimal or a fraction by using the *TI-Explorer Plus*.

70 ⬭ 2nd ⬭ % □ 0.7 F◐D 7/10

399

Sometimes it takes several steps to rewrite in a different form a percent that is less than 1.

EXAMPLE 6

In 2000, about 30% of people in the United States used the Internet. In China, about 0.8% of the people were Internet users. Write 0.8% as a fraction and as a decimal.

To write 0.8% as a fraction, recall that *percent* means "per hundred."

$$0.8\% = \frac{0.8}{100}$$ *Write the percent as a fraction with a denominator of 100.*

$$\frac{0.8}{100} \times \frac{10}{10} = \frac{8}{1,000}$$ *Multiply numerator and denominator by 10 to remove decimals from the fraction.*

$$= \frac{8 \div 8}{1,000 \div 8}$$ *Simplify.*

$$= \frac{1}{125}$$

So, about $\frac{1}{125}$ of the Chinese people used the Internet.

To write 0.8% as a decimal, write it as a fraction and then as a division problem.

$$0.8\% = \frac{0.8}{100} = 0.8 \div 100$$ *Write the percent as a fraction and then as a division problem.*

$$0.8 \div 100 = 0.008$$ *Divide by 100. This moves the decimal point two places to the left.*

So, about 0.008 of the Chinese people use the Internet.

CHECK FOR UNDERSTANDING

**Think and ▶
Discuss**

Look back at the lesson to answer the question.

1. **Explain** how to move the decimal point to write a percent as a decimal and a decimal as a percent.

**Guided ▶
Practice**

Write as a percent.

2. 0.8   3. 0.25   4. 1.2   5. $\frac{3}{5}$ 6. $1\frac{1}{2}$

Write each percent as a fraction or mixed number in simplest form.

7. 50%   8. 40%   9. 75%   10. 15%   11. 180%

Write each percent as a decimal.

12. 33%   13. 8%   14. 19%   15. 0.2%   16. 240%

PRACTICE AND PROBLEM SOLVING

**Independent ▶
Practice**

Write as a percent.

17. 0.15   18. 0.9   19. 0.09   20. 0.483   21. 0.007

22. 1   23. 2.5   24. $\frac{1}{5}$ 25. $\frac{1}{8}$ 26. $\frac{1}{400}$

Write each percent as a fraction or mixed number in simplest form.

27. 88%   **28.** 73%   **29.** 4%   **30.** 700%   **31.** $33\frac{1}{3}$%

Write each percent as a decimal.

32. 58%   **33.** 7%   **34.** 2.5%   **35.** 220%   **36.** 0.03%

Compare. Write <, >, or = for each ⬤.

37. 3 ⬤ 3%   **38.** $\frac{1}{2}$ ⬤ 50%   **39.** 7% ⬤ 0.7   **40.** 12% ⬤ 1.2

Problem Solving ▶
Applications

41. Out of 30 students, 15 prefer dogs as pets and 9 prefer cats. The rest prefer other animals. What percent prefer other animals as pets?

42. Carlo has $10 more than Erin, and Erin has $5 more than Minh. Altogether they have $50. What percent of the money does Erin have?

43. ❓ **What's the Question?** If the answer is "32% of the instruments," what is the question? Use the table.

Central Avenue School Band	
Instruments	Number
Brass	28
Woodwinds	16
Percussion	6

MIXED REVIEW AND TEST PREP

44. If 4 of 25 students are absent, what percent are absent? (p. 396)

45. Joe read 78 pages in 3 hours. What was his unit rate? (p. 374)

46. Estimate the quotient: 47.96 ÷ 6.03. (p. 76)

47. TEST PREP Which is the prime factorization of 45? (p. 148)

   A 3 × 3 × 5   **B** 9 × 5   **C** 45 × 1   **D** 3 × 5

48. TEST PREP Kendra counted cars passing her school each hour. She made this stem-and-leaf plot. What is the median number? (p. 106)

   F 55   **G** 45   **H** 44   **J** 18

Number of Cars

Stem	Leaves
3	3 8
4	0 1 7
5	1 5 5

Thinker's CORNER

Math Match Play this game with a partner. Write each percent shown at the right on an index card, and put the cards down as shown. Then write each fraction and decimal shown below on an index card.

| 50% |
| 60% |
| 5% |
| 4% |
| 25% |
| 16% |

$\frac{1}{20}$ $\frac{1}{2}$ $\frac{3}{5}$ $\frac{4}{100}$ $\frac{25}{100}$ $\frac{8}{50}$ $\frac{10}{200}$ $\frac{5}{10}$ $\frac{6}{10}$ 0.040

0.05   0.50   0.60   0.04   0.25   0.16   0.050   0.5   0.6   0.250

$\frac{1}{25}$ $\frac{1}{4}$ $\frac{4}{25}$ $\frac{5}{100}$ $\frac{5}{10}$ $\frac{60}{100}$ $\frac{2}{50}$ $\frac{4}{16}$ $\frac{12}{75}$ 0.160

Shuffle the fraction and decimal cards. Place them in two equal stacks. Each player draws a card. The first player to place his or her card on the correct equivalent percent card earns 1 point. Continue until all cards have been used. The player with more points is the winner.

Extra Practice page H52, Set B

Estimate and Find Percent of a Number

Learn how to estimate and find a percent of a number.

Remember that to change a percent to a ratio in fraction form, write the percent over 100. Then simplify.

$$36\% = \frac{36}{100}$$
$$= \frac{9}{25}$$

In 1999, basketball star Grant Hill made about 75% of his free throws. At that rate, how many out of 20 free throws would he make?

To find the answer, you need to find 75% of 20.

Activity

You need: two-color counters or colored paper squares

- To find 75% of 20, use 20 yellow counters.

- 75% equals $\frac{3}{4}$, so separate the counters into four groups. Change the color of three groups to red.

- Count the number of red counters. How many free throws would Grant Hill make?

- Use two-color counters to find 50% of 16.

One way to find a percent of a number is to change the percent to a fraction and multiply.

EXAMPLE 1

Due to an injury, Josh played in only 60% of his team's 25 hockey games. In how many games did he play?

Find 60% of 25.

$60\% = \frac{60}{100}$ *Write the percent as a ratio in fraction form.*

$\quad\quad = \frac{3}{5}$ *Write the ratio in simplest form.*

$\frac{3}{5} \times \frac{25}{1} = \frac{75}{5} = 15$ *Multiply the ratio by the number.*

So, Josh played in 15 games.

CALIFORNIA STANDARDS O—n NS 1.0 Students compare and order positive and negative fractions, decimals, and mixed numbers. Students solve problems involving fractions, ratios, proportions, and percentages. O—n NS 1.4 Calculate given percentages of quantities and solve problems involving discounts at sales, interest earned, and tips. *also* MR 2.5, MR 3.1, MR 3.2

Sometimes it is easier to change a percent to a decimal.

EXAMPLE 2

In 1998, Randall Cunningham threw 425 passes, and 8% of them resulted in touchdowns. How many touchdown passes did he throw?

Find 8% of 425.

8% = 0.08   *Change the percent to a decimal.*

0.08 × 425 = 34   *Multiply the number by the decimal.*

So, Randall Cunningham threw 34 touchdown passes.

It is customary to leave a tip when you eat in a restaurant.

EXAMPLE 3

Katie's lunch bill came to $13.79. She wants to leave a 15% tip. How much should she leave?

You can estimate the amount of the tip.

Estimate. $13.79 is about $14.
THINK: 15% = 10% + 5%

10% of $14 is $1.40.
5% of $14 is half of $1.40, or $0.70.

So, 15% of $13.79 is about $1.40 + $0.70, or $2.10.

To find the amount of the tip to the nearest cent, use a proportion.

Let t = amount of tip.   **THINK:** $15\% = \frac{15}{100}$

So, the tip (t) is to the total bill ($13.79) as 15 is to 100.

$$\frac{t}{13.79} = \frac{15}{100}$$
$$100 \times t = 15 \times 13.79$$
$$100t = 206.85$$
$$t = 206.85 \div 100 = 2.0685$$

To the nearest cent, a 15% tip should be $2.07.

You can use a calculator to find a percent of a number.

EXAMPLE 4

Use a calculator to find 15% of 13.79.

Use the key sequence below.

13.79 15   2.0685

So, 2.0685 is 15% of 13.79.

403

You can use mental math to estimate the percent of a number.

Estimate 15% of 81.

THINK: 81 is about 80.
15% = 10% + 5%

10% of 80 = 8.
5% of 80 is half of 10% of 80, or half of 8, or 4.

So, 15% of 81 is about 8 + 4, or 12.

- In Example 2, 8% of 425 is given as 34. Use estimation to determine if 34 is reasonable.

CHECK FOR UNDERSTANDING

Think and ▶ Discuss

Look back at the lesson to answer each question.

1. **Tell** how you can easily find 50% of a number if you know 25% of the number.

2. **Explain** how you can tell whether a percent of a number should be less than or greater than the number.

Guided ▶ Practice

Use a fraction in simplest form to find the percent of the number.

3. 20% of 30   **4.** 50% of 80   **5.** 75% of 48   **6.** 15% of 20

Use a decimal to find the percent of the number.

7. 52% of 15   **8.** 40% of 90   **9.** 28% of 75   **10.** 76% of 25

PRACTICE AND PROBLEM SOLVING

Independent ▶ Practice

Use a fraction in simplest form to find the percent of the number.

11. 25% of 80   **12.** 45% of 10   **13.** 80% of 55   **14.** 90% of 120

Use a decimal to find the percent of the number.

15. 12% of 9   **16.** 28% of 20   **17.** 71% of 84   **18.** 95% of 18

Use the method of your choice to find the percent of the number.

19. 21% of 88   **20.** 35% of 92   **21.** 40% of 106   **22.** 2% of 12

23. 51% of 30   **24.** 99% of 99   **25.** 82% of 150   **26.** 63% of 85

27. 5.5% of 70   **28.** 200% of 100   **29.** 150% of 38   **30.** 0.5% of 400

Estimate a 15% tip for each amount.

31. $8.00   **32.** $12.25   **33.** $15.75   **34.** $18.09   **35.** $32.18

Estimate a 20% tip for each amount.

36. $12.00   **37.** $25.00   **38.** $18.14   **39.** $3.95   **40.** $46.12

Each proportion shows *n* as a percent of a number. What is the percent? What is *n*?

41. $\frac{25}{100} = \frac{n}{60}$ **42.** $\frac{40}{100} = \frac{n}{280}$ **43.** $\frac{n}{6} = \frac{10}{100}$ **44.** $\frac{n}{240} = \frac{75}{100}$

Problem Solving ▶
Applications

Yosemite Falls

45. Geography   The surface area of Earth is about 200,000,000 mi². If 70% of the surface is water, about how many square miles of Earth's surface are water?

46. Geography   California's Yosemite Falls, the tallest waterfall in North America, actually consists of three falls. The Upper Fall is about 210% of the height of the Middle Cascades. The Middle Cascades are about 210% of the height of the Lower Fall. If the Lower Fall is 320 ft tall, about how tall is Yosemite Falls?

47. Arnold wants to leave a 15% tip on a bill of $32. He has $35. Does he have enough money for the tip? Explain.

48. *REASONING*   A diner left a $7.20 tip on a $40 meal. What percent of the meal price was the tip?

MIXED REVIEW AND TEST PREP

49. Write 0.001 as a percent. (p. 398)

50. $3\frac{2}{5} - 1\frac{7}{10}$ (p. 186)

51. Do the figures appear to be *similar, congruent, both,* or *neither*? (p. 380)

52. TEST PREP   An obtuse triangle contains an angle ___?___ 90°. (p. 336)

A greater than   **B** less than   **C** equal to   **D** less than or equal to

53. TEST PREP   Two perpendicular lines form an angle which is ___?___. (p. 330)

F straight   **G** right   **H** obtuse   **J** acute

LiNKUP to Reading

Strategy • Sequence   Sometimes the events in a problem are given in an order which is different from the order in which they occurred. To solve the problem, you must first *sequence* the events in the correct order.

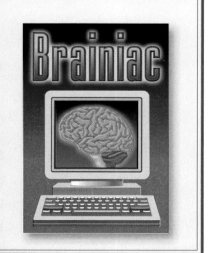

The Brainiac computer was introduced in 1981. Today, a Brainiac sells for 40% of its original price. The price increased an average of $200 per year during the first 6 years. The price topped out at $3,600 in 1987 and has dropped steadily ever since. How much does a Brainiac sell for today?

• Use the words *increased* and *decreased* to sequence the changes in the price of a Brainiac from 1981 till today.

Extra Practice   page H52, Set C

Construct Circle Graphs

Explore how to construct circle graphs.

You need compass, protractor, straightedge.

A circle graph shows parts of a whole. If you think of a complete circle as 100%, you can express portions of a circle graph as percents.

- Ms. Shipley's class earned $400 at the school fair. What fraction of the $400 did the class earn at the bake sale?

- What percent of the $400 did the class earn at the bake sale?

MONEY RAISED AT SCHOOL FAIR

Activity

ANNUAL PASTA SALES	
Shape	**Pounds Sold (millions)**
Spaghetti	310
Elbows	115
Noodles	75

People in the United States spend over $1.1 billion per year on pasta. The table gives the approximate total weights of the three most popular shapes sold.

- Use a compass to draw a circle on a piece of paper. Mark the center.

- Find the combined weight of the three shapes of pasta

- Find the percent of the total weight represented by each shape.

    spaghetti: $\frac{310}{\text{total weight}}$ = ■%

    elbows: $\frac{115}{\text{total weight}}$ = ■%

    noodles: $\frac{75}{\text{total weight}}$ = ■%

- Since there are 360° in a circle, multiply each of the percents by 360° to find the degrees in each sector of the circle graph. Round your answers to the nearest degree.

- Use a protractor to draw each angle of 180° or less. Angles greater than 180° remain.

- Label the sectors and write a title.

TOP-SELLING PASTA (in millions of lb)

CALIFORNIA STANDARDS ○━▪NS 1.0 Students compare and order positive and negative fractions, decimals, and mixed numbers. Students solve problems involving fractions, ratios, proportions, and percentages. ○━▪NS 1.4 Calculate given percentages of quantities and solve problems involving discounts at sales, interest earned, and tips. *also* MR 2.4, MR 2.5, MR 3.2

Think and Discuss

- In the circle graph you drew, what does the whole circle represent?

- Suppose the combined weight of the three shapes of pasta was 930 million pounds but the weight of the spaghetti remained the same. Find the number of degrees in the sector representing spaghetti.

Practice

Use Data Make a circle graph of the data.

1. The table shows how Shari budgets her earnings from her job.

MONTHLY BUDGET	
Item	**Percent**
Rent	30%
Food	25%
Clothes	12.5%
Other	32.5%

2. The table shows the number of videos rented in one week.

VIDEOS RENTED			
Drama	**Comedy**	**Kids**	**Classics**
250	250	300	200

3. The table records the results of a survey.

FAVORITE SPORTS SURVEY				
Tennis	**Soccer**	**Baseball**	**Golf**	**Other**
50	20	30	80	20

4. Collect data from your classmates about their favorite color. Use the data to make a circle graph with no more than five categories. Compare your graph with those of your classmates.

MIXED REVIEW AND TEST PREP

5. Find 20% of 30. (p. 402)

6. Evaluate $(m - 2m)^2$ for $m = {}^-2$. (p. 276)

7. A rectangle is 16 in. long and 12 in. wide. A similar rectangle is 4 in. long. How wide is it? (p. 384)

8. Is this survey question biased? "Do you agree that our excellent mayor should be re-elected?" Explain. (p. 98)

9. **TEST PREP** Which shows the value of y for the equation ${}^-3y = 15$? (p. 301)

   A $y = {}^-6$ **B** $y = {}^-5$ **C** $y = 5$ **D** $y = 6$

Discount and Sales Tax

Learn how to solve problems involving discounts and sales tax.

Vocabulary

discount
sales tax

Anthony wants to buy a backpack for summer camp. He saw the newspaper ad shown below. How much will the pack cost with a 30% discount?

Backpack Sale!
Regular Price $129
NOW 30% OFF!!

Math Idea ▶ To find the sale price, you can first find the discount. A **discount** is an amount that is subtracted from the regular price of an item.

discount = regular price × discount rate
= $129 × 30%
= $129 × 0.30   *Write the percent as a decimal.*
= $38.70

So, the amount of the discount is $38.70.

To find the sale price of the backpack, subtract the discount from the regular price.

Regular Price

Discount

Sale Price

sale price = regular price − discount
= $129.00 − $38.70
= $90.30

So, the sale price of the backpack is $90.30.

CALIFORNIA STANDARDS O─ɴNS 1.0 Students compare and order positive and negative fractions, decimals, and mixed numbers. Students solve problems involving fractions, ratios, proportions, and percentages. O─ɴNS 1.4 Calculate given percentages of quantities and solve problems involving discounts at sales, interest earned, and tips. *also* AF 1.0, MR 2.2, MR 2.4, MR 2.5, MR 3.2

EXAMPLE 1

A volleyball set that regularly sells for $32 is on sale at a 25% discount. Find the amount of the discount and the sale price of the set.

You can use mental math to find the discount if you change 25% to the fraction $\frac{1}{4}$. **THINK:** $\frac{1}{4}$ of $32 is $8.

You can also find the discount by writing the percent as a decimal.

discount = $32 × 25%
= $32 × 0.25
= $8

sale price = regular price − discount
= $32 − $8
= $24

So, the discount is $8 and the sale price is $24.

In Example 1, you found the sale price when you knew the regular price and the discount rate. Sometimes you may want to find the regular price when you know the sale price and the discount rate.

EXAMPLE 2

Find the regular price of the in-line skates.

The regular price has been discounted 40%. That means the sale price must be 60% of the regular price, since 100% − 40% = 60%.

Regular Price
|←——— 100% ———→|
| 60% | 40% |
Sale Price   Discount

THINK: sale price = 60% × regular price
Let n = the regular price.

$72 = 60% × n   *Write an equation relating the sale price and the regular price.*

$72 = 0.6 × n   *Write the percent as a decimal.*

$\frac{\$72}{0.6} = \frac{0.6 × n}{0.6}$ *Divide both sides of the equation by 0.6.*

$120 = n

So, the skates regularly sell for $120.

409

Most states charge a **sales tax** on purchases. A sales tax is calculated as a percent of the cost of an item. To find the amount of the sales tax, multiply the amount of the purchase by the sales tax rate.

EXAMPLE 3

Martina bought a table-tennis table for $119. The sales tax rate was 5%. How much did she pay in sales tax?

Estimate. **THINK:** 10% of $119 is about $12.00, so 5% of $119 is about $6.

$119 × 5% = $119 × 0.05   *Multiply the price by the percent.*
          = $5.95

So, she paid $5.95 in sales tax. This is close to the estimate of $6, so the answer is reasonable.

You can find the total cost of a purchase that includes sales tax.

EXAMPLE 4

The Carduso family purchased a pool table for $825. The sales tax rate was 7%. What was the total cost of the purchase?

One Way Add the sales tax to the price.

$825 × 7% = $825 × 0.07
          = $57.75

$825 + $57.75 = $882.75

Another Way Multiply the price by 107% since 7% is added to the cost.

$825 × 107% = $825 × 1.07
            = $882.75

So, the total cost of the purchase was $882.75.

CHECK FOR UNDERSTANDING

Think and Discuss Look back at the lesson to answer each question.

1. **Explain** how you can estimate the 25% discount on a $96.50 coat.

2. **Describe** an easy way to find the regular price of an item that has been discounted 50%, if you know the discount price.

Guided Practice Find the sale price.

3. regular price: $80

   SALE 25% OFF

4. regular price: $55

   10% OFF

5. regular price: $96

   20% Discount

6. regular price: $110

   SAVE 75%

Find the regular price.

7. sale price: $80
   discount rate: 20%

8. sale price: $150
   discount rate: 40%

PRACTICE AND PROBLEM SOLVING

Independent ▶
Practice

Find the sale price.

9. regular
price:
$47.90

DISCOUNT
30%

10. regular
price:
$120.00

SAVE
40%

11. regular
price:
$51.80

SALE
10% OFF

12. regular
price:
$88.50

20% OFF

Find the regular price.

13. sale price: $28.56
discount rate: 25%

14. sale price: $595.63
discount rate: 30%

Find the sales tax for the given price. Round to the nearest cent.

15. $512.00
tax: 6%

16. $24.95
tax: 7.5%

17. $84.50
tax: 5.5%

18. $260.00
tax: 6.75%

Find the total cost of the purchase. Round to the nearest cent.

19. price:
$44.00
tax: 5%

20. price:
$125.00
tax: 7%

21. price:
$56.95
tax: 8%

22. price:
$899.99
tax: 8.75%

Problem Solving ▶
Applications

Use Data **For 23–24, use the graph.**

23. In February, apples cost 12% less per
pound than they cost in January.
Estimate the cost per pound in
February.

24. Maureen bought 15 lb of apples in
January and paid 4% sales tax.
What was the cost of her purchase?

25. *REASONING* A refrigerator that
regularly sold for $800 was discounted
25%. After the sale, the sale price was raised 25%. Was the new
price less than, greater than, or equal to $800? Explain.

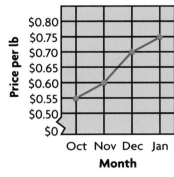

PRICE OF APPLES

MIXED REVIEW AND TEST PREP

26. Find 80% of 12.5. (p. 402)

27. Solve for b: $2 = b - 15$. (p. 294)

28. On a scale of 2 in.:3 ft, what is the actual length of a room that is 12
in. long in a scale drawing? (p. 387)

29. **TEST PREP** Which is $\frac{37}{8}$ written as a mixed number? (p. 164)

A $3\frac{5}{8}$ **B** $4\frac{3}{8}$ **C** $4\frac{5}{8}$ **D** $8\frac{29}{37}$

30. **TEST PREP** The part of a circle shown in red is a(n) __?__ . (p. 348)

F chord   **G** arc   **H** sector   **J** radius

Simple Interest

Learn how to find simple interest.

Vocabulary

principal

simple interest

QUICK REVIEW

Write each percent as a decimal.

1. 32%   2. 8%   3. 10%
4. 6.9%   5. 0.5%

When money is in a savings account, the bank regularly adds money to the account. The original amount you put in the account is called the **principal**. The amount the bank adds is called interest.

Simple interest is a fixed percent of the principal and is paid yearly. Use the formula $I = prt$ to calculate simple interest, where I = interest earned, p = principal, r = interest rate per year, and t = time in years.

EXAMPLE 1

Katrina put $500 in a savings account at a simple interest rate of 6% per year for 2 years. How much interest will she earn?

$I = prt$	$p = \$500, r = 6\%, t = 2$ years
$I = \$500 \times 0.06 \times 2$	Multiply.
$I = \$60$	

So, she will earn $60 in interest.

Activity

You need: 12 index cards, as shown

Shuffle and stack the principal cards and turn them face down. Do the same with the interest-rate and time cards.

p	r	t
$50	4%	1 yr
$125	5%	2 yr
$200	8%	3 yr
$300	$6\frac{1}{2}\%$	4 yr

- Each player draws one card from each pile and computes the interest earned. The player with the most interest earns points equal to the total interest of all the players.

- Play for five rounds. What is the most interest a player could earn in one round? What is the least?

EXAMPLE 2

Bill borrows $500. He will repay it in 2 years at a simple interest rate of 8%. How much will he have to repay at the end of 2 years?

$I = prt$	Find the amount of interest.
$I = \$500 \times 0.08 \times 2 = \$80$	
$\$80 + \$500 = \$580$	Add the interest to the loan.

So, Bill will have to repay $580.

CALIFORNIA STANDARDS O—nNS 1.0 Students compare and order positive and negative fractions, decimals, and mixed numbers. Students solve problems involving fractions, ratios, proportions, and percentages. O—nNS 1.4 Calculate given percentages of quantities and solve problems involving discounts at sales, interest earned, and tips. *also* **MR 2.5, MR 3.2**

Think and ▶
Discuss

Look back at the lesson to answer each question.

1. **Explain** what $I = prt$ means.

2. **Tell** how to find the simple interest earned on $500 for 6 months at an annual interest rate of 5%. Then find the simple interest.

Guided ▶
Practice

Find the simple interest.

3. principal: $100
   rate: 4%
   time: 1 year

4. principal: $750
   rate: 8%
   time: 4 years

5. principal: $2,400
   rate: 12%
   time: 6 years

PRACTICE AND PROBLEM SOLVING

Independent ▶
Practice

Find the simple interest.

6. principal: $12,000
   rate: 8%
   time: 2 years

7. principal: $640
   rate: 6.5%
   time: 5 years

8. principal: $21,000
   rate: 5.25%
   time: 18 years

Find the simple interest.

	Principal	Rate	1 Year	5 Years
9.	$70	2%	■	■
10.	$250	3.5%	■	■
11.	$1,200	6.9%	■	■

Problem Solving ▶
Applications

12. Amber put $2,000 in a savings account for 6 years at a simple interest rate of 6%. Haines put $2,400 in an account for 5 years at a rate of 5.5%. Who earned more interest? How much more?

13. Ray needs to borrow $1,000. Bank A offers a 3-year loan with a simple interest rate of 8.5%. Bank B offers a 5-year loan at a simple interest rate of 4%. Which bank should Ray borrow from? Explain.

14. **REASONING** Jake earned $54 in simple interest on a principal of $900 invested for 1 year. What was the interest rate?

15. **?** **What's the Error?** Jeff found simple interest of $360 for a principal of $200, interest rate of 6%, and a time of 3 years. Find his error and the correct amount of simple interest.

MIXED REVIEW AND TEST PREP

16. What is the sale price on a $150 item discounted 20%? (p. 408)

17. Two towns are 125 mi apart. How far apart do they appear on a map with a scale of 2 in. = 50 mi? (p. 390)

18. Solve $4x = 96$. (p. 301)

19. Evaluate $12 + 24 \div 12 - 4$. (p. 44)

20. **TEST PREP** Which is the best estimate of the quotient $618.6 \div 5.914$? (p. 76)

   A 1   **B** 10   **C** 100   **D** 1,000

1. **VOCABULARY**   An amount subtracted from the regular price of an item to give the sale price is called the ___?___. (pp. 408–411)

2. **VOCABULARY**   A fixed percent of the principal, paid yearly, is the ___?___. (p. 412)

Write the percent that is shaded. (p. 396)

3.

4.

5.

Write as a percent. (pp. 398–401)

6. 0.38   7. $\frac{1}{4}$ 8. 0.09   9. $\frac{2}{5}$

Write each percent as a decimal and as a fraction in simplest form. (pp. 398–401)

10. 20%   11. 18%   12. 75%   13. 10%

Find the percent. (pp. 402–405)

14. 40% of 300   15. 25% of 24   16. 1% of 350   17. 100% of 80

18. 200% of 7   19. 15% of 24   20. 11% of 80   21. 90% of 300

22. Use these data to make a circle graph of the Vernon family's monthly budget of $2,000: clothing and entertainment, 25%; food, 10%; rent, 35%; other, 30%. (pp. 406–407)

23. Find the amount spent on food in the Vernon family's budget.

24. Find the amount spent on rent in the Vernon family's budget.

Find the sale price. (pp. 408–411)

25. regular price: $40

26. regular price: $79

Find the regular price. (pp. 408–411)

27. sale price: $36
    20% off

28. sale price: $95
    50% off

29. What is the sales tax on an item that costs $33.50 if the sales tax rate is 6%?

30. What is the total cost of an item that costs $98 if the sales tax rate is 8%?

Find the simple interest. (p. 412)

31. principal: $200
    rate: 6%
    time: 1 year

32. principal: $8,000
    rate: 7.5%
    time: 4 years

33. Julia put $1,500 in a savings account. The yearly simple interest rate was 5%. How much did she have in her account after 3 years?

Eliminate choices.

See item **3**.

Estimate the simple interest at 10% per year for 2 years. Interest at 5% would be half of that amount, so interest at 7% must be more than at 5% and less than at 10%.

Also see problem **5**, p. H64.

Choose the best answer.

1. Shannon correctly answered 17 out of 20 questions on a math test. What is her score written as a percent?

   A 70%   **C** 85%

   B 80%   **D** 92%

2. The Fremont Tennis Club has 120 members. Teenagers make up 85% of the club. How many teens are in the club?

   F 18   **H** 96

   G 24   **J** 102

3. Tonya put $1,800 in the bank and left it for 2 years. The account earned simple interest at the rate of 7% per year. How much interest did Tonya get?

   A $70   **C** $252

   B $140   **D** $360

4. How is 46% written as a decimal?

   F 0.46%   **H** 4.6

   G 0.46   **J** 46

5. The regular price of a coat is $118.00. It is discounted 25%. What is the discounted price?

   A $2.95   **C** $88.50

   B $29.50   **D** $93.00

6. Which pair shows two numbers equivalent to $\frac{3}{5}$?

   F 66% and 6.6   **H** 40% and 0.4

   G 0.6 and 60%   **J** 3.5 and 35%

7. What is 18% of 250?

   A 45   **C** 18

   B 36   **D** Not here

8. If 6 pens cost $2.50, how much will 9 pens cost at the same rate?

   F $2.75   **H** $3.25

   G $2.95   **J** $3.75

9. A triangle has angles that measure 20° and 75°. What is the measure of the third angle?

   A 55°   **C** 95°

   B 85°   **D** 180°

10. What is the value of $s - t + v$ for $s = 25$, $t = 12$, and $v = 3$?

    F 16   **H** 10

    G 13   **J** Not here

11. Which group shows an equivalent fraction, decimal, and percent?

    A $\frac{11}{25}$, 0.44, 44%   **C** $\frac{11}{25}$, 4.4, 44%

    B $\frac{11}{25}$, 0.44, 4.4%   **D** $\frac{11}{25}$, 44, 44%

12. $1\frac{3}{4} \times 9\frac{3}{5}$

    F $17\frac{3}{5}$ **H** $18\frac{4}{5}$

    G $18\frac{2}{5}$ **J** Not here

CHAPTER 22 Probability of Simple Events

Many states require bicyclists to wear helmets and have lights and reflectors on their bikes for safety. Suppose your school does random safety checks on the bikes in the bike rack. The graph shows data for the bike rack on a day when you ride your bike to school. What is the probability that your bike will be the one chosen for inspection? Which color has the greatest chance of being chosen?

BICYCLES IN RACK

Bicycle Colors	
Green	
Pink	
Blue	
Red	
Black	

key: each ◯ = 4 bikes

Check What You Know

Use this page to help you review and remember important skills needed for Chapter 22.

✓ Vocabulary

Choose the best term from the box.

1. A(n) __?__ event will never happen.

2. A(n) __?__ is a specific set of outcomes in a probability experiment.

3. Events that have the same probability of occurring are __?__ .

> equally likely
> event
> impossible
> outcome

✓ Fractions, Decimals, and Percents (See p. H23.)

Write each fraction as a decimal and a percent.

4. $\frac{2}{5}$
5. $\frac{1}{2}$
6. $\frac{7}{10}$
7. $\frac{3}{4}$
8. $\frac{9}{20}$

9. $\frac{1}{10}$
10. $\frac{1}{4}$
11. $\frac{3}{5}$
12. $\frac{1}{5}$
13. $\frac{18}{25}$

14. $\frac{3}{8}$
15. $\frac{3}{25}$
16. $\frac{5}{8}$
17. $\frac{19}{20}$
18. $\frac{3}{50}$

✓ Simplest Form of Fractions (See p. H9.)

Write each fraction in simplest form.

19. $\frac{4}{12}$
20. $\frac{9}{6}$
21. $\frac{32}{64}$
22. $\frac{40}{15}$
23. $\frac{36}{54}$

24. $\frac{9}{75}$
25. $\frac{22}{55}$
26. $\frac{96}{60}$
27. $\frac{49}{56}$
28. $\frac{81}{120}$

29. $\frac{25}{75}$
30. $\frac{72}{60}$
31. $\frac{70}{120}$
32. $\frac{155}{200}$
33. $\frac{200}{800}$

✓ Certain, Impossible, Likely, Unlikely (See p. H24.)

Tell if the event is *certain*, *impossible*, *likely*, or *unlikely*.

34. Having homework this week

35. The sun rising today

36. Spinning an odd number on a spinner that is numbered 1, 3, 5, 6, 7, and 9

37. The month of January having 34 days next year

38. Reading 3 novels in one day

> **LOOK AHEAD**
>
> **In Chapter 22 you will**
> - find the theoretical probability of an event
> - use a simulation to model an experiment
> - find the experimental probability of an event

Theoretical Probability

Learn how to find the theoretical probability of an event.

Vocabulary

outcome

sample space

theoretical probability

Each color on the spinner shown represents one equally likely **outcome**, or possible result, of spinning. The **sample space**, or set of all possible outcomes, is white, purple, red, yellow, and green.

Math Idea ▶ The **theoretical probability**, P, of an event is a comparison of the number of favorable outcomes to the number of possible, equally likely outcomes. It can be written as a ratio.

$$P(\text{event}) = \frac{\text{number of favorable outcomes}}{\text{number of possible equally likely outcomes}}$$

EXAMPLE 1

In a game students created, if the pointer on the spinner above lands on red, you get an extra turn. What is the probability that you get an extra turn?

1 favorable outcome: red   *List the favorable outcomes.*

5 possible outcomes.   *Count the possible outcomes.*

$P(\text{red}) = \frac{1 \text{ favorable outcome}}{5 \text{ possible outcomes}} = \frac{1}{5}$ *Write the probability as a ratio.*

So, the probability that you get an extra turn in the game is $\frac{1}{5}$.

Probabilities can be expressed as fractions, decimals, and percents.

EXAMPLE 2

Miyoko rolled a number cube labeled 1 to 6. Find each probability. Express your answers as fractions, decimals, and percents.

A. $P(2) = \frac{1}{6}$, $0.1\overline{6}$, or $16\frac{2}{3}\%$ ← 1 choice out of 6

B. $P(2 \text{ or } 3) = \frac{1+1}{6}$ ← 2 choices out of 6

$= \frac{2}{6} = \frac{1}{3}$, $0.\overline{3}$, or $33\frac{1}{3}\%$

C. $P(\text{greater than } 3) = \frac{1+1+1}{6}$ ← 3 choices out of 6

$= \frac{3}{6} = \frac{1}{2}$, 0.50, or 50%

CALIFORNIA STANDARDS O—n **SDAP 3.3** Represent probabilities as ratios, proportions, decimals between 0 and 1, and percentages between 0 and 100 and verify that the probabilities computed are reasonable; know that if *P* is the probability of an event, 1–*P* is the probability of an event not occurring. *also* O—n NS 1.0, O—n NS 2.0, SDAP 3.0, SDAP 3.4, MR 2.5, MR 2.7

The number line below shows that the probability of an event ranges from 0, or impossible, to 1, or certain. The closer a probability is to 1, the more likely the event is to occur.

EXAMPLE 3

Each letter of the word *OUTCOMES* is written on a card and placed in a bag. One card is chosen at random. Which outcome is more likely to occur, an *M* or an *O*?

$P(M) = \frac{1}{8}$ *Find each probability.*

$P(O) = \frac{2}{8} = \frac{1}{4}$

$\frac{1}{4} > \frac{1}{8}$ *Compare the probabilities.*

So, the outcome *O* is more likely to occur than the outcome *M*.

EXAMPLE 4

Use the spinner at the right to find each probability.

A. $P(blue) = \frac{0}{4}$. None of the sections are blue.

B. $P(green) = \frac{1}{4}$

C. $P(not\ green) = \frac{1 + 1 + 1}{4} = \frac{3}{4}$

Look at P(green) and P(not green) in Example 4.
The sum of those probabilities is $\frac{1}{4} + \frac{3}{4}$, or 1. If P is the probability of an event occurring, $1 - P$ is the probability of an event not occurring.

EXAMPLE 5

Use the spinner above to find each probability. Write each answer as a fraction, a decimal, and a percent.

A. $P(not\ yellow) = 1 - P(yellow)$

$= 1 - \frac{1}{4}$

$= \frac{3}{4}$

So, P(not yellow) is $\frac{3}{4}$, 0.75, or 75%.

B. $P(not\ orange) = 1 - P(orange)$

$= 1 - \frac{1}{2}$

$= \frac{1}{2}$

So, P(not orange) is $\frac{1}{2}$, 0.50, or 50%.

Think and ▶ Discuss

Look back at the lesson to answer each question.

1. **Write** the sample space for the number cube in Example 2.

2. **Explain** why the probabilities computed in Example 4 are reasonable.

Guided ▶ Practice

Use the spinner at the right to find each probability. Write each answer as a fraction, a decimal, and a percent.

3. P(yellow)

4. P(purple)

5. P(red)

6. P(red or purple)

7. P(not yellow)

8. P(green)

Independent ▶ Practice

Use the spinner at the right to find each probability. Write each answer as a fraction, a decimal, and a percent.

9. P(*L*)

10. P(*T*)

11. P(*B*)

12. P(*E* or *R*)

13. P(*L*, *F*, or *T*)

14. P(vowel)

A number cube is labeled 1, 2, 3, 4, 5, and 6. Find each probability. Write each answer as a fraction.

15. P(2)

16. P(not 5)

17. P(4 or 6)

18. P(less than 4)

19. P(1, 2, or 3)

20. P(even)

21. P(multiple of 3)

22. P(divisible by 6)

23. P(integer)

Cards numbered 1, 1, 2, 2, 3, 4, 5, and 6 are placed in a box. You choose one card without looking. Compare the probabilities. Write <, >, or = for each ●.

24. P(1) ● P(2)

25. P(1) ● P(5 or 6)

26. P(6) ● P(2)

27. P(1 or 2) ● P(3 or 4)

28. P(even) ● P(odd)

29. P(multiple of 3) ● P(greater than 2)

Use P(A), the probability of event A, to find P(not A).

30. P(A) = $\frac{1}{2}$

31. P(A) = 0.4

32. P(A) = 80%

33. P(A) = $\frac{7}{12}$

34. P(A) = 0.61

35. P(A) = 23%

Problem Solving ▶
Applications

36. A rental car agency has 55 blue cars, 32 red cars, and 70 white cars. If a customer selects at random, what is the probability that she will select a green car? Write your answer as a percent.

37. Suppose the probability of an event is $\frac{5}{12}$. Which is greater, the probability that the event will occur or the probability that it will not occur?

38. You are given a choice of three answers to a trivia question. If you don't know the answer, what is the probability of guessing the correct answer?

39. **(?)** **What's the Question?** Shalonda has a number cube labeled 1 to 6. The answer is 1.

40. ***REASONING*** Brenda has a bag with 8 cubes all the same size: 1 red, 4 yellow, and 3 blue. She chooses a blue cube without looking and does not put it back in the bag. What is the probability of choosing a yellow cube after the blue cube has been removed?

41. A box has 24 marbles. If $\frac{3}{8}$ of the marbles are blue and $\frac{5}{8}$ of the marbles are red, how many of each color are in the box?

MIXED REVIEW AND TEST PREP

42. Tong invested $420 at a simple interest rate of $3\frac{1}{2}\%$. How much interest will he earn in 5 years? (p. 412)

43. Order $7\frac{3}{8}$, $^-7.75$, $7\frac{1}{4}$, and 7.5 from least to greatest. (p. 234)

44. Divide. $135.892 \div 6.41$ (p. 76)

45. **TEST PREP** Evaluate $3p$ for $p = 7.8$.
(p. 82)

  A 2.6   **B** 4.8   **C** 10.8   **D** 23.4

46. **TEST PREP** Find the product.
$^-10 \times {}^-12$ (p. 258)

  F $^-120$ **G** $^-22$ **H** 2   **J** 120

Thinker's CORNER **Geometric Probability**

You can use the areas of geometric figures to find probabilities.

In a certain game, students throw a quarter onto a target that looks like the square at the right. To find the probability that the quarter lands on the green section, compare the areas. Assume that the center point of the quarter lands on the target at some random point and that all points are equally likely.

area of green section

5 in. × 5 in. = 25 in.^{2}

area of target

10 in. × 10 in. = 100 in.^{2}

$P(green) = \dfrac{\text{area of green section}}{\text{area of target}} = \dfrac{25}{100} = \dfrac{1}{4}$

• If the center point of a randomly thrown quarter lands on the target at the right, what is the probability of its being in the blue section?

PROBLEM SOLVING SKILL
Too Much or
Too Little Information

Analyze
Choose
Solve
Check

Learn how to use the skill of determining whether there is *too much or too little information* to solve problems.

Karyn is waiting in line with 14 other people to ride the bumper cars at a carnival. Each person in line is randomly assigned to one of 15 bumper cars. There are 3 blue, 5 red, 4 yellow, and 3 green bumper cars. What is the probability that Karyn is assigned to a red bumper car?

If a problem has too much information, you must decide what to use to solve the problem.

1. Want to know: P(red bumper car)

2. Know: number of red bumper cars, total number of bumper cars

3. Don't need to know: number of people waiting in line; number of blue, yellow, and green bumper cars

4. Missing and need to know: none

Decision: Too much information

Solve: 5 red bumper cars, 15 bumper cars total; P(red) $= \frac{5}{15} = \frac{1}{3}$

So, the probability that Karyn is assigned to a red bumper car is $\frac{1}{3}$.

Sometimes there is too little information to solve a problem.

What fraction of the people waiting in line to ride the bumper cars are females?

1. Want to know: fraction of people in line who are females

2. Know: 15 people waiting in line

3. Don't need to know: number of blue, yellow, green, and red bumper cars; number of bumper cars

4. Missing and need to know: number of females in line

Decision: Too little information

Solve: can't; need more information

CALIFORNIA STANDARDS SDAP 3.0 Students determine theoretical and experimental probabilities and use these to make predictions about events. **MR 1.1** Analyze problems by identifying relationships, distinguishing relevant from irrelevant information, identifying missing information, sequencing and prioritizing information, and observing patterns. *also* o—¬ NS 2.0, MR 1.3, MR 2.0, AF 2.1, o—¬ SDAP 3.3

Tell if each problem has *too much, too little,* or *the right amount* of information. Then solve the problem if possible, or describe the information needed to solve it.

1. A club has 32 female members and 24 male members. The members range in age from 22 years to 51 years. If a club member is randomly selected to be president, what is the probability that a female is chosen?

2. The Jenkins family is trying to decide on a name for their dog. They place the names Rusty, Sporto, and Buddy in a box and randomly select one name. What is the probability that they select the name Rusty or Buddy?

3. G. de Chasseloup-Laubat held the official land speed record in 1898. He drove a vehicle, the Jeantaud, at a speed of 39.24 mi/hr. How many times as fast was the land speed record in 1970?

4. Lina is saving to buy a bicycle and a helmet. The price of the bike is $199. So far, she has $155 saved. How much more does she need to save?

Use Data   For 5–6, use this table.

5. What is the difference between the average high temperature and the average low temperature for the 4 cities?

   A ⁻22°F   **B** ⁻16°F   **C** 16°F   **D** 22°F

6. Use the formula $C = \frac{5}{9} \times (F - 32)$ to choose the high temperature in San Diego in °C.

   F ⁻18°C   **G** ⁻10°C   **H** 10°C   **J** 18°C

TEMPERATURES (°F)		
City	Low	High
Los Angeles, CA	45	67
Sacramento, CA	42	58
San Diego, CA	50	64
San Francisco, CA	47	59

7. Lydia studies twice as long as Jorge. Jorge and Olga study a total of 7 hr, but Olga studies 3 hr longer than Jorge. How long does Lydia study?

8. A bus travels 55 mi per hour. Misty starts out by bus at 11:45 A.M. to go to the carnival. How many miles does she travel if she arrives at the carnival at 1:15 P.M.?

9. There are 164 steps in the Astoria Column on Coxcomb Hill in Astoria, Oregon. Rick decided to climb to the top once a week for a year. How many steps must he climb up to do that?

10. The Stefan family spent a total of $42.75 at the carnival. They bought food for $12.50, souvenirs for $5.25, and show tickets for $5.00 each. How many tickets did they buy?

11. ✏️ **Write a problem** about probability and include too much information. Then write a related problem, giving too little information.

Astoria Column

Simulations

Explore how to use a simulation to model an experiment.

Find the mean.

1. 4, 5, 4, 6, 6   **2.** 2, 3, 1, 2, 3, 1

3. 10, 9, 17, 10   **4.** 15, 10, 14

5. 250, 200, 150

You need a 5-section spinner and a calculator.

A cereal company is having a contest. To win a prize, you have to collect five cards that spell *PRIZE*. One of the five letters is put into each cereal box when the cereal is produced. The letters are divided equally among the cereal boxes.

You can conduct an experiment to simulate how many boxes of cereal you have to buy to get all five letters.

Activity 1

• Use a spinner to generate random numbers. Each of the numbers 1 to 5 will represent one of the letters.

• Spin the pointer on the spinner, and tally the numbers you get.

• Continue to spin the pointer until you get every number at least once.

• Repeat the experiment.

P	R	I	Z	E
1	2	3	4	5

NUMBER	TALLY
1	III
2	III
3	I
4	I
5	III

Think and Discuss

• How many spins did it take in the first experiment to get all five numbers? in the second experiment?

• What is the mean of the spins in your two experiments?

• How many boxes of cereal do you expect you will have to buy to get all five letters? If you bought this many boxes, would you be sure to win? Explain.

Practice

• Repeat the experiment three more times.

• Combine your data with data from 5 classmates. Find the mean of the spins from all five sets of data.

• How many boxes of cereal do you expect you will have to buy to get all five letters? How does this differ from what you expected after the first two experiments?

CALIFORNIA STANDARDS SDAP 3.0 Students determine theoretical and experimental probabilities and use these to make predictions about events. **SDAP 1.1** Compute the range, mean, median, and mode of data sets. *also* MR 2.0, MR 2.4

You can produce random numbers by using a calculator.

Activity 2

- Use this key sequence on some calculators to produce random numbers from 1 to 15.

    1   2nd   RAND   15   =

- Record the number you get each time.

- Repeat the key sequence until you have gotten each of the numbers 1 to 15.

- Make a bar graph to show the results.

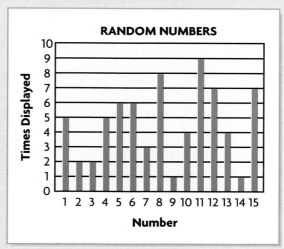

Think and Discuss

- Compare the numbers of times you and your classmates used the key sequence to get all 15 numbers.

- Look at the results in the graph above. What was the last random number produced?

Practice

- Use a calculator to produce random numbers from 1 to 25. Record the number of times you get each number. Continue until you get each number.

MIXED REVIEW AND TEST PREP

1. A spinner has 8 equal sections numbered 1, 2, 1, 3, 2, 4, 1, 5. Find P(1 or 2). (p. 418)

2. The regular price of an item is $140. The discount rate is 25%. Find the amount of discount and the sale price. (p. 408)

3. Evaluate $y^2 + 8$, for $y = 9$. (p. 276)

4. Evaluate $b^2 \div (6 + 4)$, for $b = 8$. (p. 276)

5. **TEST PREP** Find the product. 6.507×14 (p. 70)

    A 91.098   **B** 71.098   **C** 26.028   **D** 20.507

Experimental Probability

Learn how to find the experimental probability of an event.

Vocabulary

experimental probability

Mateo put some red, yellow, and green marbles in a box. He randomly selected a marble and recorded the color. He returned the marble to the box each time. Mateo did this a total of 25 times.

By performing an experiment, Mateo can find the experimental probability of selecting each marble color.

Math Idea ▶ The **experimental probability** of an event is the number of times a certain outcome actually occurs compared with the total number of trials, or the total number of times you do the activity.

$$\textbf{experimental probability} = \frac{\text{number of times outcome occurs}}{\text{total number of trials}}$$

EXAMPLE 1

Mateo recorded his results in a table. Use these results to find

MARBLE COLOR	Green	Red	Yellow
	2	10	13

the experimental probability of selecting each color. Express your answers as fractions, decimals, and percents.

P(green) $= \frac{2}{25}$, 0.08, 8%   P(red) $= \frac{10}{25} = \frac{2}{5}$, 0.40, 40%

P(yellow) $= \frac{13}{25}$, 0.52, 52%

You can use the experimental probability to predict future events.

Based upon his experimental results, how many times can Mateo expect to select a red marble in his next 10 selections?

EXAMPLE 2

P(red) $= \frac{2}{5}$ *Use the experimental probability.*

$\frac{2}{5} \times 10 = 4$ *Multiply 10 selections by $\frac{2}{5}$.*

So, he can expect to select a red marble 4 times.

CHECK FOR UNDERSTANDING

Think and ▶ **Discuss**

Look back at the lesson to answer the question.

1. **What if** Mateo has 10 green, 50 red, and 40 yellow marbles? Tell how the experimental probability of selecting each color compares with the theoretical probability.

CALIFORNIA STANDARDS SDAP 3.0 Students determine theoretical and experimental probabilities and use these to make predictions about events. **SDAP 3.2** Use data to estimate the probability of future events (e.g., batting averages or number of accidents per mile driven). *also* O—n NS 1.2, O—n NS 2.0, O—n SDAP 3.3

Guided Practice

Celia rolled a number cube 30 times. For 2–7, use her results shown in the table below to find the experimental probability of each event. Write the answer in simplest form.

NUMBER	1	2	3	4	5	6
TIMES ROLLED	2	6	7	3	8	4

2. P(1)　　　　**3.** P(2)　　　　**4.** P(3)

5. P(4)　　　　**6.** P(5)　　　　**7.** P(6)

8. What is the theoretical probability of rolling each number?

PRACTICE AND PROBLEM SOLVING

Independent Practice

For 9–15, use the table.
Vincent spun the pointer of this spinner 40 times.

COLOR	Red	Yellow	Blue	Green
SPINS	10	6	16	8

Find the experimental probability.

9. P(red)　　　　**10.** P(yellow)

11. P(blue)　　　　**12.** P(green)

13. P(yellow or red)　　　　**14.** P(not red)

15. Based on his experimental results, how many times can Vincent expect a result of yellow in the next 100 spins?

Problem Solving Applications

16. Conduct an experiment in which you toss a coin 50 times. Keep a tally of the number of times you toss heads. Combine your results with those of five other students to find the experimental probability of tossing heads. How does it compare to the theoretical probability?

17. ✎ **Write About It** Explain how to find the experimental probability of rolling a 4 using a number cube labeled 1 to 6.

18. **Science** On a plant called the four o'clock, the flowers may be red, white, or pink. Pink flowers are twice as common as either red or white flowers. If a four o'clock has 140 flowers, how many should be red? white? pink?

MIXED REVIEW AND TEST PREP

19. For a certain experiment, P(vowel) = $\frac{3}{7}$. Find P(consonant). (p. 418)

Find the percent of the number. (p. 402)

20. 82% of 150　　　　**21.** 7.5% of 130　　　　**22.** 200% of 95

23. **TEST PREP** Which type of graph would you use to show the lengths of the five longest rivers? (p. 120)

A circle　　　**B** bar　　　**C** histogram　　　**D** line

1. **VOCABULARY** The number of times a certain outcome occurs compared with the total number of trials is called the ___?___ . (p. 426)

2. **VOCABULARY** A comparison of the number of favorable outcomes to the number of possible equally likely outcomes is called the ___?___ . (p. 418)

A bag has 10 new pencils: 3 red, 4 yellow, 1 blue, and 2 green. Find the probability of randomly choosing the given color from the bag. Write each answer as a fraction, a decimal, and a percent. (pp. 418–421)

3. P(green)

4. P(yellow)

5. P(orange)

6. P(not yellow)

7. P(red or green)

8. P(red, yellow, or blue)

A bag has slips of paper numbered 2–10. Find the probability of randomly drawing the number from the bag. Write each answer as a fraction. (pp. 418–421)

9. P(5)

10. P(7 or 9)

11. P(1)

12. P(2, 5, or 9)

13. P(even number)

14. P(not 8)

15. P(5 or 6)

16. P(odd number)

17. P(a number less than 10)

18. P(multiple of 5)

19. P(integer)

For 20–23, use the table. The table shows the results of rolling a number cube 100 times. Find the experimental probability. (426–427)

NUMBER	1	2	3	4	5	6
TIMES ROLLED	8	12	24	13	17	26

20. P(3)

21. P(6)

22. P(2 or 5)

23. Suppose the number cube is rolled 500 times. Based on the experimental results, how many times can you expect to roll a 1 or a 2?

Tell if each problem has *too much*, *too little*, or *the right amount* of information. Then solve the problem if possible, or describe the information needed to solve it. (pp. 422–423)

Forty-two people have signed up for a race. Sixteen of them are male. Five of the people are wearing red running shoes.

24. What is the probability that the winner of the race will be a female?

25. What is the probability that the winner of the race will be a male wearing red running shoes?

Look for important words.
See item **7.**

Important words are *that is not green*. Use the spinner to determine the probability of spinning any color that is not green before you choose the expression that shows it.

Also see problem **2**, p. H62.

Choose the best answer.

1. Tong has a spinner with 5 equal sections labeled 1 to 5. What is the probability of spinning a number that is not 2?

   A $\frac{1}{5}$ **C** $\frac{4}{5}$

   B $\frac{2}{5}$ **D** 1

2. A spinner is divided into 8 equal sections. Of the sections, 4 are colored white, 3 are colored blue, and 1 is colored yellow. What is the probability that the pointer will stop on a white section?

   F $\frac{1}{8}$ **H** $\frac{3}{5}$

   G $\frac{1}{2}$ **J** $\frac{7}{8}$

3. Stan has a number cube labeled 1 to 6. What is the probability that he will roll a number less than 3?

   A $\frac{1}{6}$ **C** $\frac{1}{2}$

   B $\frac{1}{3}$ **D** $\frac{2}{3}$

4. A bag contains 4 blue marbles, 2 green marbles, 3 yellow marbles, and 6 red marbles. What is the probability of randomly pulling either a blue or a red marble from the bag?

   F $\frac{1}{10}$ **H** $\frac{1}{2}$

   G $\frac{2}{5}$ **J** $\frac{2}{3}$

5. Alice's class toured a toy factory. At the end of the tour, one person's name was drawn to win a toy. What information do you need to know to determine the probability that Alice won?

   A The kind of toy to be given away

   B The number of toys made at the factory

   C The name of the person drawing

   D The number of names in the drawing

Use the spinner for 6–7.

6. Which is the probability of spinning yellow?

   F $\frac{1}{2}$ **H** $\frac{1}{4}$

   G $\frac{1}{3}$ **J** $\frac{1}{6}$

7. Which expression shows the probability of spinning a color that is not green?

   A $1 + \frac{1}{6}$ **C** $1 - \frac{1}{3}$

   B $1 - \frac{1}{6}$ **D** $1 - \frac{1}{2}$

8. If two angles are supplementary and the measure of one of the angles is 63°, what is the measure of the second angle?

   F 27°   **H** 137°

   G 117°   **J** Not here

9. A store manager decided to survey 1 out of every 10 shoppers. How many would be surveyed out of 350 shoppers?

   A 35   **C** 70

   B 50   **D** Not here

10. Estimate. 629 × 37

   F 18,000   **H** 24,000

   G 21,000   **J** 28,000

CHAPTER 23

Probability of Compound Events

The yo-yo is the second oldest toy in the world, dating from the ancient Greeks 3,000 years ago. Today, yo-yos are popular toys among collectors. Suppose you are a yo-yo collector. At a flea market, one vendor has grab bags. Each bag contains a yo-yo. You want to get a yo-yo with plastic sides and a metal axle. How can a chart like the one below help you figure out the probability of getting the yo-yo you want?

GRAB BAG YO-YOS				
Yo-Yo	Sides		Axle	
	Wooden	Plastic	Wooden	Metal
1	X		X	
2		X	X	
3	X		X	
4		X		X
5		X		X
6	X		X	

Check What You Know

Use this page to help you review and remember important skills needed for Chapter 23.

 Vocabulary

Choose the best term from the box.

1. The set of all possible outcomes in a probability experiment is called the __?__ .

2. The ratio of the number of favorable outcomes to the number of possible outcomes is the __?__ .

3. A possible result in a probability experiment is a(n) __?__ .

> event
> outcome
> theoretical
> probability
> sample space

 Multiply Fractions (See p. H23.)

Multiply. Write your answer in simplest form.

4. $\frac{1}{3} \times \frac{1}{2}$ 5. $\frac{1}{2} \times \frac{2}{3}$ 6. $\frac{1}{4} \times \frac{1}{8}$ 7. $\frac{3}{8} \times \frac{1}{2}$ 8. $\frac{1}{3} \times \frac{3}{7}$

9. $\frac{1}{2} \times \frac{4}{5}$ 10. $\frac{1}{9} \times \frac{2}{5}$ 11. $\frac{3}{4} \times \frac{1}{6}$ 12. $\frac{3}{5} \times \frac{1}{5}$ 13. $\frac{1}{10} \times \frac{2}{5}$

 Analyze Data (See p. H24.)

Use the line graph for 14–16.

14. During which month did the greatest amount of snow fall?

15. About how many inches of snow fell in all five months?

16. In which two months did 6 inches of snow fall?

MEASURED SNOWFALL

Inches of Snow

Nov Dec Jan Feb Mar

Month

Use the circle graph for 17–20.

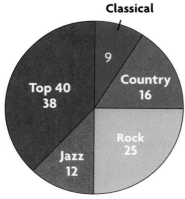

MUSIC PREFERRED by Students

Classical

9

Country 16

Top 40 38

Rock 25

Jazz 12

17. How many students were surveyed?

18. What was the most popular music?

19. How many more students preferred rock than preferred classical?

20. How many students preferred rock or jazz?

> **LOOK AHEAD**
>
> **In Chapter 23 you will**
> - find outcomes of compound events
> - identify and find probabilities of independent and dependent events
> - find probabilities and make predictions from sample data

PROBLEM SOLVING STRATEGY
Make an Organized List

Analyze
Choose
Solve
Check

Learn how to use the strategy *make an organized list* to solve problems.

QUICK REVIEW

A number cube labeled 1–6 is rolled. Find each probability.

1. P(2) **2.** P(4)

3. P(1 or 3) **4.** P(not 5)

5. P(even)

The 5 members of the Montoya family are planning their next vacation. They will travel in June or July to San Diego, Lake Tahoe, Los Angeles, or San Francisco. How many different vacations involving one place and one month are possible?

Analyze

What are you asked to find?

What facts are given?

Is there any numerical information you will not use? If so, what?

Choose

What strategy will you use?

You can use the strategy *make an organized list* to show the sample space.

Solve

How will you solve the problem?

List the two months and pair each with the four places the Montoyas might visit.

June, San Diego	**July, San Diego**
June, Lake Tahoe	**July, Lake Tahoe**
June, Los Angeles	**July, Los Angeles**
June, San Francisco	**July, San Francisco**

So, there are 8 possible choices for the Montoyas' next vacation.

Check

How can you check your answer?

What if the Montoya family adds the city of Sacramento as a choice? How many different vacations are possible?

CALIFORNIA STANDARDS O━┓SDAP 3.1 Represent all possible outcomes for compound events in an organized way (e.g., tables, grids, tree diagrams) and express the theoretical probability of each outcome. **MR 2.2** Apply strategies and results from simpler problems to more complex problems. *also* **MR 1.0, MR 1.3, MR 2.1, MR 3.2**

PROBLEM SOLVING PRACTICE

PROBLEM SOLVING STRATEGIES

Draw a Diagram or Picture

Make a Model

Predict and Test

Work Backward

▶ **Make an Organized List**

Find a Pattern

Make a Table or Graph

Solve a Simpler Problem

Write an Equation

Use Logical Reasoning

Solve by making an organized list.

1. Mrs. Chen is making a dental appointment. It can be on Monday, Tuesday, Wednesday, or Thursday, at 9:30 A.M., 11:00 A.M., or 2:00 P.M. How many choices does Mrs. Chen have?

2. Gilbert is packing for a ski trip. He has 2 jackets, 2 pairs of ski pants, and 2 hats. How many different outfits can Gilbert make if each outfit consists of a jacket, pants, and a hat?

MIXED STRATEGY PRACTICE

For 3–4, use this information. Chad spends $44.39 on two packages of computer disks and a cartridge for his printer.

3. If each package of disks costs $9.95, what equation would you use to find the cost of the cartridge for his printer?
   A $9.95 + c = 44.39$
   B $2(9.95) + c = 44.39$
   C $44.39 - 9.95 = c$
   D $9.95 + 2c = 44.39$

4. Chad's total includes $2.88 in sales tax. What would you do first to find the cost of the items he buys?
   F Add $2.88 to $44.39.
   G Subtract $2.88 from $44.39.
   H Multiply $44.39 by $2.88.
   J Divide $44.39 by $2.88.

5. **Use Data** Use the graph below. How many more visits did Great Smoky Mountains National Park have than Olympic National Park and Yosemite National Park combined? Estimate to determine if your answer is reasonable.

6. Yumi arranged five of her friends in line for a photograph. Alberto stood between Harley and Joshua. Eleni stood between Joshua and Merika. Merika was on the left. Starting from the left, write the order in which Yumi's friends were arranged for the photo.

7. A dance company performs in nonoverlapping groups of 8 and 10. There are 64 dancers in all. There are more groups of 10 than groups of 8. How many groups of each size are there?

8. ❓ **What's the Error?** Jay's lunch cost $7.55. He left $0.15 for a 20% tip. Describe his error. How much should a 20% tip be?

**MOST-VISITED
NATIONAL PARKS IN THE U.S.**
(average yearly visits)

Great Smoky Mountains (NC, TN)	9,265,667
Grand Canyon (AZ)	4,537,703
Yosemite (CA)	4,046,207
Olympic (WA)	3,348,723

Visits (in millions)

Compound Events

Learn how to find all possible outcomes for compound events.

Vocabulary

compound event

tree diagram

Fundamental Counting Principle

How many choices of a beverage, a sandwich, and a dessert does Carmella have for lunch?

Choosing three different items for lunch is an example of a **compound event**. A compound event includes two or more simple events.

One Way   Find the number of possible outcomes for a compound event by drawing a **tree diagram**.

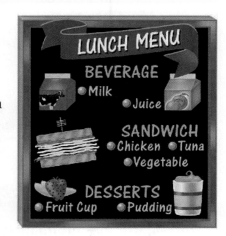

Beverage	Sandwich	Dessert		Outcome
milk (M)	chicken (C)	fruit cup (F)	→	M, C, F
		pudding (P)	→	M, C, P
	tuna (T)	fruit cup (F)	→	M, T, F
		pudding (P)	→	M, T, P
	vegetable (V)	fruit cup (F)	→	M, V, F
		pudding (P)	→	M, V, P
juice (J)	chicken (C)	fruit cup (F)	→	J, C, F
		pudding (P)	→	J, C, P
	tuna (T)	fruit cup (F)	→	J, T, F
		pudding (P)	→	J, T, P
	vegetable (V)	fruit cup (F)	→	J, V, F
		pudding (P)	→	J, V, P

So, Carmella has 12 choices of a beverage, a sandwich, and a dessert.

Another Way   Find the number of possible outcomes for 2 or more events by using the **Fundamental Counting Principle**.

Fundamental Counting Principle	If one event has m possible outcomes and a second independent event has n possible outcomes, then there are $m \times n$ total possible outcomes for the two events together.

EXAMPLE 1

What if the school cafeteria adds 2 more desserts to the menu? Find the total number of beverage-sandwich-dessert selections.

Beverages		Sandwiches		Desserts		
2	×	3	×	4	=	24

So, there are 24 beverage-sandwich-dessert selections.

CALIFORNIA STANDARDS O—nSDAP 3.1 Represent all possible outcomes for compound events in an organized way (e.g., tables, grids, tree diagrams) and express the theoretical probability of each outcome. *also* O—nNS 2.0, MR 2.2, MR 2.5

You can also show the sample space by using a grid to make a table.

EXAMPLE 2

Find the number of possible outcomes if Brianna selects one card at random and spins the pointer on the spinner.

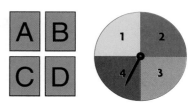

Spinner

Card	1	2	3	4
A	(A,1)	(A,2)	(A,3)	(A,4)
B	(B,1)	(B,2)	(B,3)	(B,4)
C	(C,1)	(C,2)	(C,3)	(C,4)
D	(D,1)	(D,2)	(D,3)	(D,4)

List the card choices in rows and the spinner choices in columns.

So, there are 16 possible outcomes if Brianna selects one card at random and spins the pointer on the spinner.

CHECK FOR UNDERSTANDING

Think and ▶ Discuss

Look back at the lesson to answer each question.

1. **What if** Brianna spins the pointer on the spinner in Example 2 and rolls a number cube labeled 1 to 6? Find the number of possible outcomes.

2. **Draw** a tree diagram to show all possible outcomes for tossing three coins.

Guided ▶ Practice

Draw a tree diagram or make a table to find the number of possible outcomes for each situation.

3. a choice of vanilla, chocolate, or strawberry yogurt in a small, medium, or large cup

4. spinning the pointers on these two spinners

Use the Fundamental Counting Principle to find the number of possible outcomes for each situation.

5. rolling two number cubes labeled 1 to 6

6. a choice of 5 cards, 3 envelopes, and 2 stickers

PRACTICE AND PROBLEM SOLVING

Independent ▶ Practice

Draw a tree diagram or make a table to find the number of possible outcomes for each situation.

7. a choice of a car or a van with a blue, black, white, or silver exterior, and a black or gray interior

8. tossing a coin and rolling a number cube labeled 1 to 6

Use the Fundamental Counting Principle to find the number of possible outcomes for each situation.

9. a choice of 3 pizza crusts and 6 toppings

10. a choice of 2 ties, 5 shirts, 3 trousers, 2 belts

11. rolling 3 number cubes

12. tossing a coin 4 times

Problem Solving ▶ Applications

13. A cafe offers 12 different sandwiches, 4 different salads, and 8 different soups. If the cafe is open every day of the year, is it possible to have a different combination of sandwich, salad, and soup for lunch every day of the year? Explain.

14. ✎ **Write About It**   Is it easier to draw a tree diagram, make a table, or use the Fundamental Counting Principle to find the number of possible outcomes for a compound event? Explain.

MIXED REVIEW AND TEST PREP

15. Out of 60 rolls of a number cube, Alicia rolls a six 12 times. What is the experimental probability of rolling a six? (p. 426)

Find the measure of the supplement of each angle. (p. 326)

16. 86°   **17.** 17°   **18.** 39°

19. TEST PREP   Which fraction is greater than $\frac{7}{16}$? (p. 166)

   A $\frac{1}{4}$ **B** $\frac{1}{3}$ **C** $\frac{3}{8}$ **D** $\frac{5}{8}$

Explore Combinations Suppose each student is to write a report on 2 of the following careers in the television industry: director, actor, producer, and sound mixer. How many combinations of 2 careers are possible?

A **combination** is a selection of objects or individuals in which the order is not important. You can make an organized list to find the number of combinations.

Choose a career. Pair it with each of the other careers. Do not use any combinations that reverse the pairs already listed. The combination "director, actor" is the same as "actor, director."

director, actor   actor, producer   producer, sound mixer
director, producer   actor, sound mixer
director, sound mixer

So, there are 6 possible combinations of 2 careers.

• How many combinations of 2 careers are possible if another career, makeup artist, is added to the list of report topics?

Independent and Dependent Events

Learn how to identify and find probabilities of dependent and independent events.

QUICK REVIEW

1. $\frac{1}{4} \times \frac{1}{2}$ 2. $\frac{2}{3} \times \frac{1}{6}$ 3. $\frac{1}{5} \times \frac{1}{4}$ 4. $\frac{1}{2} \times \frac{1}{2}$ 5. $\frac{3}{5} \times \frac{1}{6}$

The table below shows the contents of a bag of marbles. Juan randomly selects a marble from the bag, replaces it, and selects again.

Marbles have been around for at least 4,000 years.

RED	BLUE	GREEN	YELLOW
3	8	5	4

Because Juan replaces the marble after the first selection, the outcome of the second event does not depend on the outcome of the first event. These are **independent events**.

To find the probability of two independent events, multiply their probabilities.

> If A and B are independent events, then
> $P(A, B) = P(A) \times P(B)$.

EXAMPLE 1

Remember that to find the theoretical probability of an event with equally likely outcomes, you write the ratio of the number of favorable outcomes to the number of possible outcomes.

What is the probability that Juan first selects a blue marble, replaces it, and then selects a yellow marble?

Find P(blue, yellow).   *First blue, then yellow.*

first selection: $P(\text{blue}) = \frac{8}{20}$, or $\frac{2}{5}$ *Find the probability of blue.*

second selection: $P(\text{yellow}) = \frac{4}{20}$, or $\frac{1}{5}$ *Find the probability of yellow.*

$P(\text{blue, yellow}) = \frac{2}{5} \times \frac{1}{5} = \frac{2}{25}$ *Multiply the probabilities.*

So, the probability is $\frac{2}{25}$, 0.08, or 8%.

EXAMPLE 2

Lisa spins the pointer on the spinner two times.

Find the probability of each event.

A. $P(1, 2) = P(1) \times P(2) = \frac{1}{5} \times \frac{1}{5} = \frac{1}{25}$

B. $P(\text{not } 3, \text{not } 4) = P(\text{not } 3) \times P(\text{not } 4) = \frac{4}{5} \times \frac{4}{5} = \frac{16}{25}$

C. $P(1 \text{ or } 2, 2) = P(1 \text{ or } 2) \times P(2) = \frac{1 + 1}{5} \times \frac{1}{5} = \frac{2}{25}$

CALIFORNIA STANDARDS **SDAP 3.4** Understand that the probability of either of two disjoint events occurring is the sum of the two individual probabilities and that the probability of one event following another, in independent trials, is the product of the two probabilities. **O—n SDAP 3.5** Understand the difference between independent and dependent events. *also* **NS 2.1, O—n SDAP 3.1 O—n SDAP 3.3, MR 2.5, MR 2.7**

TECHNOLOGY LINK

More Practice: Use **Mighty Math Number Heroes**, *Probability*, Level N.

Dependent Events

Without looking, Isabella selects a magnet from a box, does not replace it, and then passes the box to Jason, who also selects a magnet. Since Isabella does not replace the magnet, Jason's magnet selection depends on the result of Isabella's selection.

If the outcome of the second event depends on the outcome of the first event, the events are called **dependent events**. To find the probability of two dependent events, use the formula below.

> If A and B are dependent events, then
> $P(A, B) = P(A) \times P(B \text{ after } A)$.

EXAMPLE 3

The contents of Isabella's magnet collection are shown in the table. Assuming that Jason and Isabella select without replacement, find the probability that they both select animal magnets.

FLOWERS	ANIMALS	CITIES
4	5	16

Find P(animal, animal).

Isabella's selection:
$P(\text{animal}) = \frac{5}{25} = \frac{1}{5}$

There are 5 animal magnets and 25 equally likely outcomes.

Jason's selection:
$P(\text{animal after animal}) = \frac{4}{24} = \frac{1}{6}$

There are now 4 animal magnets and 24 equally likely outcomes.

$P(\text{animal, animal}) = \frac{1}{5} \times \frac{1}{6} = \frac{1}{30}$

Multiply the probabilities.

So, the probability of Isabella and Jason both selecting animal magnets is $\frac{1}{30}$, $0.0\overline{3}$, or $3\frac{1}{3}\%$.

Math Idea ▶ Compound events consist of either independent events or dependent events. To find the probability of a compound event, multiply the probabilities of the simple events.

CHECK FOR UNDERSTANDING

Think and Discuss ▶ Look back at the lesson to answer each question.

1. **Find** the probability of spinning two even numbers in Example 2. Find the probability of spinning two odd numbers. Express your answers as percents. Are your computed probabilities reasonable? Explain.

2. **Give an example** of two events that are dependent.

Guided Practice ▶ Write *independent* or *dependent* to describe the events.

3. Draw one card from a box, do not replace it, and draw another card.

4. Choose a sandwich and a soup from a menu.

John rolls two number cubes labeled 1 to 6. Find the probability of each event.

5. P(1, 3)   **6.** P(not 2, 4)   **7.** P(7, 3)   **8.** P(1 or 2, 4)

A bag contains five lettered tiles labeled *P, I, Z, Z, A*. Without looking, Kazuhiro selects a tile, does not replace it, and selects another tile. Find the probability of each event.

9. P(*P, A*)   **10.** P(*Z,* not *I*)   **11.** P(*A* or *P, Z*)   **12.** P(vowel, *U*)

PRACTICE AND PROBLEM SOLVING

Independent ▶ Practice

Write *independent* or *dependent* to describe the events.

13. Draw a card from a deck of cards, do not replace it, and draw a second card.

14. Toss a coin three times.

Without looking, you take a card out of the jar and replace it before selecting again. Find the probability of each event. Then find the probability, assuming the card is not replaced after each selection.

15. P(1, 3)   **16.** P(2, 4)   **17.** P(4, 1 or 2)

18. P(4, not 2)   **19.** P(1, 2 or 4)   **20.** P(3, odd)

21. P(2, 2, 2)   **22.** P(2, 1, 2)   **23.** P(4, 1, 1)

Problem Solving ▶ Applications

24. *REASONING*   Mong has two numbered spinners. The probability of spinning a 5 on both spinners is 0.075. If the probability of spinning a 5 on the first spinner alone is 0.375, what is the probability of spinning a 5 on the second spinner alone?

25. ❓ **What's the Question?**   Owen rolls two number cubes labeled 1 to 6. The answer is $\frac{1}{9}$.

26. Donte and his two sisters collect stamps. Donte has twice as many as his older sister, who has 21 stamps, and three times as many as his younger sister. How many stamps do they have in all?

MIXED REVIEW AND TEST PREP

Find the number of possible outcomes of making choices for each situation. (p. 434)

27. 5 shirts, 3 pairs of slacks   **28.** 4 cards, 6 envelopes   **29.** 3 beverages, 6 entrees

30. A bag contains 3 blue, 4 yellow, 5 purple, and 3 red marbles. One marble is chosen at random. Find P(not yellow). (p. 418)

31. TEST PREP   Which figure can always be described as a rectangle? (p. 342)

A rhombus   **B** square   **C** parallelogram   **D** trapezoid

Extra Practice page H54, Set B

Make Predictions

Learn how to use sample data to make predictions about a population.

Kevin and Shaudra conducted a survey of a random sample of sixth graders at Washington Middle School. The table below shows the results of the survey.

Remember that a population is a group being studied. A sample is a representative part of that group.

FAVORITE PETS	
Pet	**Number of Students**
Cat	30
Dog	25
Hamster	9
Rabbit	5
Bird	3
Other	3

Math Idea ▶ You can use the results from a sample to make predictions about preferences or actions of the population.

EXAMPLE 1

What is the probability that a randomly selected sixth grader at Washington Middle School chooses a cat as his or her favorite pet?

$P(\text{cat}) = \frac{\text{number of students who prefer cats}}{\text{number of students surveyed}} = \frac{30}{75}$, or $\frac{2}{5}$

So, the probability that a sixth grader chooses a cat as his or her favorite pet is about $\frac{2}{5}$, 0.40, or 40%.

EXAMPLE 2

There are 210 sixth graders at Washington Middle School. Predict about how many of them would choose cats as their favorite pets.

$\frac{2}{5} = \frac{n}{210}$ *Write and solve a proportion.*

$2 \times 210 = 5 \times n$ *Find the cross products.*

$420 = 5n$ *Solve the equation.*

$\frac{420}{5} = \frac{5n}{5}$

$84 = n$

So, about 84 sixth graders would choose cats as their favorite pets.

CALIFORNIA STANDARDS SDAP 3.0 Students determine theoretical and experimental probabilities and use these to make predictions about events. **SDAP 3.2** Use data to estimate the probability of future events. *also* **NS 1.3, AF 1.1, SDAP 3.3, MR 2.4, MR 3.2**

Think and ▶
Discuss

Look back at the lesson to answer each question.

1. Explain how to predict how many sixth graders from Washington Middle School would choose rabbits as their favorite pets.

2. What if there were 300 sixth graders at Washington Middle School? About how many of them would you predict would choose cats as their favorite pets?

Guided ▶
Practice

The results of a survey of 500 randomly selected teenagers in California indicate that 175 of them use the Internet regularly.

3. What is the probability that a randomly selected teenager in California uses the Internet on a regular basis?

4. Predict about how many California teenagers out of 4,500 use the Internet on a regular basis.

PRACTICE AND PROBLEM SOLVING

Independent ▶
Practice

Use Data For 5–6, use the graph.
The graph shows the favorite colors of 100 randomly selected sixth graders from Glenville Middle School.

FAVORITE COLORS

Other 12 · Purple 15 · Yellow 8 · Green 13 · Red 7 · Blue 45

5. If there are 340 sixth graders at Glenville Middle School, about how many would prefer the color blue?

6. If there are 400 sixth graders at Glenville Middle School, about how many will prefer a color that is not green?

Problem Solving ▶
Applications

7. In a random sample of 400 remote control cars, the quality control department found that 24 were defective. If the company manufactures 10,000 remote control cars, about how many of them would you predict would be defective?

8. **Write a problem** about making a prediction about a population of 150 sixth graders based on a sample.

MIXED REVIEW AND TEST PREP

There are 6 cards that spell out *DIVIDE*. Suppose you choose one card at random. Find the probability of each event. (p. 418)

9. P(*D*)

10. P(*I* or *D*)

11. P(vowel)

12. Two number cubes labeled 1 to 6 are rolled. Find P(3, not 4). (p. 437)

13. TEST PREP Find $4^3 \div 2^2 + 4$. (p. 282)

A 6   **B** 8   **C** 20   **D** 64

1. **VOCABULARY** If the outcome of the second event depends on the outcome of the first event, the events are called __?__ events. (p. 438)

2. **VOCABULARY** If the outcome of the second event does not depend on the outcome of the first event, the events are called __?__ events. (p. 437)

Draw a tree diagram or make a table to find the number of possible outcomes for each situation. (pp. 434–436)

3. Miguel tosses a coin and selects 1 marble at random from a jar containing 1 red, 1 blue, 1 yellow, and 1 green marble.

4. To get to work, Lea can walk or ride the subway to Central Station. From there, she takes either a bus, a cab, or a train.

Use the Fundamental Counting Principle to find the number of possible outcomes for each situation. (pp. 434–436)

5. A choice of 4 entrees, 3 drinks, and 2 desserts

6. A choice of 5 ties, 6 shirts, and 2 jackets

For 7–10, suppose you spin the pointer and select a card at random. (pp. 437–439)

7. Find P(green, 1).

8. Find P(red, 2).

9. Find P(red, odd).

10. Find P(blue, 4).

A bag contains 1 green, 2 purple, and 2 white blocks. Without looking in the bag, you choose a block, do not replace it, and then choose another block. **For 11–14, find the probability. Write your answer as a fraction, a decimal, and a percent.** (pp. 437–439)

11. P(green, white)

12. P(purple, black)

13. P(white or green, purple)

14. P(purple, not white)

Sixty students from Center Middle School were randomly surveyed about their favorite types of movies. The results are shown in the table. **For 15–18, find the probability that a student picked at random from Center Middle School prefers each type of movie.** (pp. 440–441)

15. drama

16. comedy

17. action

18. horror

19. Suppose there are 300 students at Center Middle School. Predict the number of students who prefer horror movies.

Favorite Types of Movies

Movies	Number of Students
Action	18
Comedy	12
Drama	15
Horror	7
Other	8

Make an organized list to solve. (pp. 432–433)

20. Armand is making a haircut appointment. It can be on Tuesday, Wednesday, or Friday at 11:00 A.M., 1:00 A.M., 2:30 P.M., or 4:00 P.M. How many possible appointments are there?

Understand the problem.

See item **3.**

A through F describes the letters A, B, C, D, E, and F. If a card is not replaced, there is one fewer letter. You also need to remember which letters are vowels.

Also see problem **1**, p. H62.

Choose the best answer.

1. Shaun wants to buy an ice-cream cone. He can have vanilla, chocolate, blueberry, peach, or strawberry ice cream. He can get sprinkles, candy pieces, or peanuts for the topping. How many different kinds of cones made with one flavor and one topping are possible?

   A 3   **B** 4   **C** 12   **D** 15

2. Stan has a spinner with 5 equal sections labeled 1 through 5. The results of spinning the pointer 100 times are shown.

NUMBER	1	2	3	4	5
TIMES SPUN	20	18	23	19	20

   What was the experimental probability of spinning an odd number?

   F $\frac{9}{100}$ **H** $\frac{50}{100}$

   G $\frac{20}{100}$ **J** $\frac{63}{100}$

3. Cards labeled *A* through *F* are placed in a bag. If the card is not replaced after each selection, what is the probability that the first card drawn is C and the second card drawn is a vowel?

   A $\frac{1}{18}$ **C** $\frac{1}{6}$

   B $\frac{1}{15}$ **D** $\frac{17}{30}$

4. For dessert, there are choices of four kinds of pies and three flavors of ice cream. How many choices of one kind of pie and one flavor of ice cream are there?

   F 24   **H** 7

   G 12   **J** Not here

5. There are 9 boxes of cereal. Three boxes are shredded wheat, 4 are bran flakes, and 2 are oatmeal. What is the probability that you will choose a box of oatmeal at random?

   A $\frac{2}{9}$ **C** $\frac{4}{9}$

   B $\frac{3}{9}$ **D** $\frac{7}{9}$

6. Cody can choose either pancakes, cereal, or eggs and either milk or orange juice for breakfast. How many different breakfasts consisting of 1 entree and 1 drink are possible?

   F 4   **G** 5   **H** 6   **J** 7

7. A spinner is divided into 10 equal sections; 3 are black, 2 are blue, 4 are red, and 1 is white. What is the probability of the pointer stopping on either a blue or a red section?

   A $\frac{1}{10}$ **C** $\frac{3}{10}$

   B $\frac{1}{8}$ **D** $\frac{3}{5}$

8. Jake measured $2\frac{1}{2}$ inches between two cities on a map. On the map scale, 1 inch represented 120 miles. What was the actual distance between the two cities?

   F 48 mi   **H** 300 mi

   G 180 mi   **J** 360 mi

9. What is the mode for this data set?
   30, 25, 40, 35, 40

   A 15   **C** 35

   B 34   **D** 40

10. 48.96 ÷ 16

   F 3.06

   G 3.6

   H 30.6

   J Not here

MATH DETECTIVE

You Be the Judge

REASONING   Use what you know about percent to decide who is correct. Be prepared to explain your decision.

Disagreement 1

Keisha and Rashad are shopping for a CD player. At the local electronics store, CD players that are marked down 20% have an extra 10% off the sale price for today only. Keisha says, "Great, that means I'll save 30% off the original price." Rashad disagrees. He says she'll save less than 30% off the original price.

You be the judge.

Who is correct? Explain your decision.

Disagreement 2

José and Mohammed are shopping at the bookstore. José is interested in buying a novel in hardcover as a gift for his father. The bookstore discounts hardcover books 30% from the full price. José also has a coupon good for 10% off the full price of any hardcover book. José wants the clerk to take the 30% off first and then the 10% rather than taking 40% off. Mohammed says it won't matter.

You be the judge.

Who is correct? Explain your decision.

Think It Over!

* **Write About It**   Explain how you made your decision for each disagreement.

* **Stretch Your Thinking**   If an item already marked down 25% is then marked down an additional 20%, what percent of the original price will you pay? Explain.

Percent of Increase and Decrease

Learn how to find the percent of increase or decrease when given the amount of increase or decrease and the original amount.

The city of Sandville is increasing bus fares. The cost of a monthly bus pass for students will increase from $25 to $26.

Use this formula to find the percent of increase.

$$\% \text{ increase} = \frac{\text{amount of increase}}{\text{original amount}}$$

You divide by the original amount since you are comparing the increase to the original amount.

EXAMPLE 1

Find the percent of increase in a student's bus pass.

$\% \text{ increase} = \dfrac{\text{amount of increase}}{\text{original amount}}$ *Write the formula.*

$\% \text{ increase} = \dfrac{26 - 25}{25} = \dfrac{1}{25}$ *Amount of increase: 26 − 25 = 1 original amount: 25*

$\% \text{ increase} = 0.04, \text{ or } 4\%$ *Write the fraction as a percent.*

So, the percent of increase is 4%.

Use this formula to find the percent of decrease.

$$\% \text{ decrease} = \frac{\text{amount of decrease}}{\text{original amount}}$$

You divide by the original amount since you are comparing the decrease to the original amount.

EXAMPLE 2

Beth observed that 64 elephants drank from a water hole this week. Last week, 80 elephants drank from the water hole. What is the percent of decrease?

$\% \text{ decrease} = \dfrac{\text{amount of decrease}}{\text{original amount}}$ *Write the formula.*

$\% \text{ decrease} = \dfrac{80 - 64}{80} = \dfrac{16}{80}$ *Amount of decrease: 80 − 64 = 16 Original amount: 80*

$\% \text{ decrease} = 0.20, \text{ or } 20\%$ *Write the fraction as a percent.*

So, the percent of decrease is 20%.

TRY IT

Find the percent of increase or decrease.

1. 30 is increased to 39.

2. 88 is decreased to 66.

3. $72 is increased to $99.

4. 50 is increased to 169.

VOCABULARY

1. A ratio that compares two quantities having different units of measure is a ___?___ . (p. 375)

2. Figures that have the same shape are ___?___ figures. (p. 380)

3. The set of all possible outcomes is called the ___?___ . (p. 418)

EXAMPLES

EXERCISES

Chapter 20

• **Find the unit rate.** (pp. 374–376)
  Apples cost $2.97 for 3 lb. What is the cost per pound?

  $\frac{\text{cost}}{\text{pounds}} = \frac{\$2.97}{3}$ *Write a ratio to compare cost to pounds.*

  $\frac{2.97 \div 3}{3 \div 3} = \frac{0.99}{1}$ *Find the cost per pound.*

  So, the cost is $0.99 per pound.

Find the unit rate.

4. 480 mi in 6 hr

5. $51.96 for 4 CDs

6. 15 books in 5 months

7. 540 mi on 30 gal

8. $1.80 for a dozen rolls

9. 468 tickets for 3 games

• **Use proportions to solve problems using scale drawings.** (pp. 387–389)

  Find the actual length if the drawing length is 6 cm and the scale is 1 cm:5 m.

  $\frac{\text{drawing (cm)}}{\text{actual (m)}} \rightarrow \frac{6}{n} = \frac{1}{5}$ *Write a proportion. Let n represent the actual length.*

  $6 \times 5 = n \times 1$ *Find the cross products.*
  $n = 30$

  The actual length is 30 m.

Find the unknown dimension.

10. scale: 3 cm:5 mm
    drawing length: 12 cm
    actual length: ■ mm

11. scale: 1 in.:4 ft
    drawing length: 6 in.
    actual length: ■ ft

12. scale: 2 in.:10 ft
    drawing length: ■ in.
    actual length: 5 ft

13. scale: 4 cm: 1 mm
    drawing length: ■ cm
    actual length: 12 mm

Chapter 21

• **Convert among percents, decimals, and fractions.** (pp. 398–401)

  Write 0.4 as a percent.

  0.4 = 40%

  Write $\frac{3}{4}$ as a percent.

  $\frac{3}{4} = \frac{75}{100} = 75\%$

  Move the decimal point two places to the right. Write an equivalent fraction with a denominator of 100.

Write as a percent.

14. 0.9   15. 0.34   16. 2.5

17. $\frac{4}{5}$ 18. $\frac{1}{200}$ 19. $\frac{5}{8}$

Write as a decimal.

20. 45%   21. $\frac{1}{4}$ 22. 135%

Chapter 22

- **Find the theoretical probability of a simple event.** (pp. 418–421)

  A cube is numbered 1 to 6. What is the probability you will roll a number greater than 4?

  $$P(event) = \frac{number\ of\ favorable\ outcomes}{number\ of\ possible\ outcomes}$$

  $$P(number\ greater\ than\ 4) = \frac{2}{6} = \frac{1}{3}$$

A bag has 10 red marbles, 5 blue marbles, 6 yellow marbles, and 4 green marbles. Find each probability. Express as a fraction, a decimal, and a percent.

23. P(green)

24. P(blue, yellow, or red)

25. P(white)

- **Make predictions based on experimental probabilities.** (pp. 426–427)

COIN	Heads	Tails
TOSS	15	35

  Based upon the experimental results, how many times would you expect to get heads if you tossed the coin 80 times?

  $$P(heads) = \frac{15}{50} = \frac{3}{10} \qquad \frac{3}{10} \times 80 = 24$$

  You would expect to get heads 24 times.

For 26–27, use the table to find the experimental probability.

26. If you tossed the coin 120 times, how many times would you expect to get tails?

27. If you tossed the coin 30 times, how many heads would you expect to get?

28. A baseball player hit the ball 42 times out of the last 150 times at bat. How many hits can he expect to get during his next 100 times at bat?

Chapter 23

- **Find the number of possible outcomes of a compound event.** (pp. 434–436)

  Find the number of possible outcomes for tossing a coin and selecting a card from A, B, C, or D.

  Coin: heads or tails—2 choices
  Card: A, B, C, D—4 choices

  $2 \times 4 = 8$; 8 possible outcomes

Find the number of possible outcomes for each situation.

29. picking chocolate or vanilla ice cream and sprinkles, hot fudge, nuts, or whipped cream toppings.

30. picking a white, blue, striped, or plaid shirt and black, khaki, or navy pants.

- **Find the probability of a dependent event.** (pp. 437–439)

  Find the probability of picking an X and then a Y from a bag of cards labeled X, Y, Z, W, Q if you do not replace the cards drawn.

  $$P(X) = \frac{1}{5} \qquad P(Y\ after\ X) = \frac{1}{4}$$

  $$P(X,Y) = \frac{1}{5} \times \frac{1}{4} = \frac{1}{20}$$

Six cards labeled *R*, *E*, *V*, *I*, *E*, and *W* are in a jar. Suppose you pick a card, do not replace it, and then pick a second card. Find the probability of each event.

31. P(R,V)   32. P(W,E)

33. P(V,E)   34. P(E,E)

PROBLEM SOLVING APPLICATION

35. A drawing of a birdhouse has a scale of 1 in. = 4 in. If the height of the birdhouse on the drawing is 8 in., how high is the actual birdhouse? (pp. 387–389)

San Diego

San Diego, California's second-largest city, is both an important business center and a popular year-round resort. Its location on the southern coast of California is one reason for its very mild climate.

There are many things to do in San Diego's Balboa Park. The park is home to the San Diego Zoo, sports facilities, and many museums.

AVERAGE HIGH/LOW TEMPERATURE (IN °F) FOR SAN DIEGO											
	Jan	Feb	Mar	Apr	May	Jun	Jul	Aug	Sep	Oct	Nov	Dec
High	65	65	67	68	70	72	76	77	77	73	71	67
Low	46	48	50	54	57	60	63	65	62	58	51	48

Use Data **For 1–4, use the table.**

1. Write the ratio of the number of months with an average high temperature of 70°F or greater to the number of months with an average high temperature less than 70°.

2. Write the ratio of the number of months with an average low temperature of 50°F or less to the total number of months.

3. What percent of the months have an average high temperature greater than 75°F?

4. What percent of the months have an average low temperature less than 66°F?

Suppose you randomly choose one of these Balboa Park museums to visit.

MUSEUMS IN BALBOA PARK
Natural History Museum
Aerospace Museum
Museum of Man
Automotive Museum
Historical Society Museum

5. Find the probability of choosing a museum whose name begins with the letter A. Write the answer as a fraction, a decimal, and a percent.

6. Find the probability that you will *not* choose the Museum of Man.

7. Is it certain, impossible, likely, or unlikely that the museum you choose will be in San Diego? Explain.

Palomar Observatory

The Palomar Observatory is about 2 hours northeast of San Diego. The observatory consists of four domes that house telescopes used to study the planets, stars, comets, and other celestial bodies.

The largest dome at the observatory houses the 200-inch Hale Telescope.

1. Find the actual straight-line distance from San Diego to the Palomar Observatory. Explain how you found the distance.

2. Determine the map distance that represents 75 mi on the California map.

3. The actual driving distance from San Diego to the Palomar Observatory is 74 miles. If you take 2 hours to drive this distance, what is your average rate of speed?

4. Suppose you plan a day trip from San Diego to the Palomar Observatory and decide you will stop for a picnic lunch in Escondido, Pala, or Pauma Valley. You can stop on your way to the observatory or on your way home. How many choices do you have? Draw a tree diagram or make a table to find the possible outcomes.

5. The California Institute of Technology (Caltech) shares the use of the Hale Telescope with Cornell University and the Jet Propulsion Laboratory. Cornell University and the Jet Propulsion Laboratory each have 25% of the time. Caltech has the remaining time. Make a circle graph to show how the three institutions share the time.

Football, soccer, lacrosse, tennis, hockey, and curling are some of the sports that have carefully defined rules about the sizes of the courts or fields on which they are played. Lacrosse is played on a field that is 300 ft long. It may be anywhere from 160 to 180 ft wide. Approximately how many times as long is the field as it is wide?

DIMENSIONS OF PLAYING AREAS		
Sport	**Length (in feet)**	**Width (in feet)**
Tennis		
Singles	78	27
Doubles	78	36
Football	360	160
Soccer	300 to 390	150 to 300
Curling	146	14
Ice Hockey	200	85

Check What You Know

Use this page to help you review and remember important skills needed for Chapter 24.

✓ Vocabulary

Choose the best term from the box.

multiply
divide
feet
yards

1. You can change __?__ to inches by multiplying by 12.

2. To change meters to kilometers, you __?__ by 1,000.

3. To change pints to ounces, you __?__ by 16.

✓ Customary Units (See p. H25.)

Change to the given unit.

4. 1 yd = ■ ft

5. 9 ft = ■ yd

6. 24 in. = ■ ft

7. $4\frac{2}{3}$ ft = ■ in.

8. 6 yd = ■ ft

9. 1 gal = ■ qt

10. 3 gal = ■ qt

11. $1\frac{1}{2}$ lb = ■ oz

12. 5 lb = ■ oz

13. 36 ft = ■ yd

14. 8 qt = ■ gal

15. 60 in. = ■ ft

✓ Metric Units (See p. H25.)

Change to the given unit.

16. 1 m = ■ cm

17. 1 m = ■ mm

18. 2,000 g = ■ kg

19. 1 cm = ■ mm

20. 30 mm = ■ cm

21. 1 km = ■ m

22. 3,000 mL = ■ L

23. 2,000 mm = ■ m

24. 2 cm = ■ mm

25. 4 L = ■ mL

26. 9 kg = ■ g

27. 900 cm = ■ m

✓ Solve Proportions (See p. H26.)

Solve for n.

28. $\frac{2}{n} = \frac{6}{84}$

29. $\frac{8}{15} = \frac{24}{n}$

30. $\frac{3}{5} = \frac{30}{n}$

31. $\frac{3}{8} = \frac{n}{24}$

32. $\frac{n}{6} = \frac{8}{12}$

33. $\frac{4}{10} = \frac{n}{25}$

LOOK AHEAD

In Chapter 24 you will

• use and relate customary and metric measurements

• use appropriate tools and units to solve problems

ALGEBRA
Customary Measurements

Learn how to use proportions to convert one customary unit of measurement to another.

Solve for *n*.

1. $\frac{3}{4} = \frac{n}{8}$ **2.** $\frac{1}{3} = \frac{3}{n}$ **3.** $\frac{4}{n} = \frac{2}{5}$

4. $\frac{n}{18} = \frac{1}{6}$ **5.** $\frac{10}{24} = \frac{n}{36}$

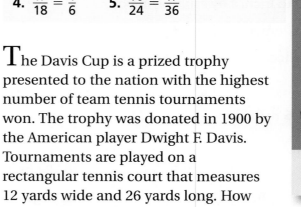

The Davis Cup is a prized trophy presented to the nation with the highest number of team tennis tournaments won. The trophy was donated in 1900 by the American player Dwight F. Davis. Tournaments are played on a rectangular tennis court that measures 12 yards wide and 26 yards long. How many feet long is the tennis court?

The Davis Cup is 13 in. high and 18 in. across the top.

You can write a proportion to convert units of length.

Math Idea ▶ In a proportion, both numerators must have the same units and both denominators must have the same units.

$$\frac{1 \text{ yd}}{3 \text{ ft}} = \frac{26 \text{ yd}}{x \text{ ft}} \rightarrow \frac{1}{3} = \frac{26}{x}$$ *Use the relationship of 1 yard to 3 feet to write a proportion.*

$$1 \cdot x = 3 \cdot 26$$ *Find the cross products.*

$$x = 78$$ *Multiply.*

Remember that multiplication problems with variables can be written in different ways.

$7 \times a$ $7(a)$
$7a$ $7 \cdot a$

So, the tennis court is 78 ft long.

• How many feet wide is the tennis court?

You can also use proportions to convert units of weight.

EXAMPLE

In the powerlifting competition held April 5–6, 1997, Chris Lawton lifted a total of 992,288 lb. How many tons is 992,288 lb?

$$\frac{1 \text{ T}}{2{,}000 \text{ lb}} = \frac{x \text{ T}}{992{,}288 \text{ lb}}$$ *Use the relationship of 1 ton to 2,000 pounds to write a proportion.*

$$\frac{1}{2{,}000} = \frac{x}{992{,}288}$$ *Find the cross products.*

$$2{,}000x = 992{,}288$$ *Solve for x.*

$$\frac{2{,}000x}{2{,}000} = \frac{992{,}288}{2{,}000}$$

$$x = 496.144$$

So, Chris Lawton lifted 496.144 T or about 496T.

• How many tons is 630,932 lb?

CALIFORNIA STANDARDS AF 2.0 Students analyze and use tables, graphs, and rules to solve problems involving rates and proportions. **AF 2.1** Convert one unit of measurement to another (e.g., from feet to miles, from centimeters to inches). *also* O━┓NS 1.3, MR 1.1, MR 2.5, MR 2.7

Think and ▶
Discuss

Look back at the lesson to answer each question.

1. **Describe** how you could use a proportion to convert quarts into gallons.

2. **Explain** why sometimes it is important for a proportion to have both numerators with the same units and both denominators with the same units.

Guided ▶
Practice

Use a proportion to convert to the given unit.

3. 2 qt = ■ pt   **4.** 84 in. = ■ ft   **5.** 8 lb = ■ oz

PRACTICE AND PROBLEM SOLVING

Independent ▶
Practice

Use a proportion to convert to the given unit.

6. 10 c = ■ fl oz   **7.** 36 ft = ■ yd   **8.** 8 days = ■ hr

9. 8 oz = ■ lb   **10.** 3.5 T = ■ lb   **11.** 54 in. = ■ yd

12. 20 pt = ■ fl oz   **13.** 15 gal = ■ qt   **14.** 34 pt = ■ qt

15. 36 oz = ■ lb   **16.** 27 c = ■ qt   **17.** 39 in. = ■ ft

Compare. Write <, >, or = for each ●.

18. 31 ft ● 16 yd   **19.** 144 hr ● 6 days   **20.** 77 fl oz ● 9 c

Problem Solving ▶
Applications

78 ft

27 ft

21. The width of the singles half-court section on a tennis court is $13\frac{1}{2}$ ft. How many yards wide is the singles half-court section?

22. Juan says that each guest will drink $2\frac{1}{2}$ cups of punch. How many gallons of punch will Juan need for 80 guests to each have $2\frac{1}{2}$ cups?

23. **(?)** **What's the Error?** A sign on a bridge says, "Weight limit 12,000 lb." Eric's truck weighs 12 tons. He says he will be able to drive across the bridge. What's his error? How many pounds does his truck weigh?

MIXED REVIEW AND TEST PREP

24. Write $\frac{4}{5}$ as a percent. (p. 398)

25. $3\frac{2}{5} \div 2\frac{3}{4}$ (p. 210)

26. $\frac{5}{9} - \frac{1}{2}$ (p. 182)

27. 0.218 + 2.143. (p. 66)

28. **TEST PREP** Martha had 132 trading cards. She gave $\frac{1}{4}$ to her sister and $\frac{1}{3}$ to her brother. How many trading cards does she have left? (p. 202)

A 33   **B** 44   **C** 55   **D** 77

ALGEBRA
Metric Measurements

Learn how to use proportions to convert one metric unit of measurement to another.

QUICK REVIEW

Solve for *n*.

1. $\frac{1}{4} = \frac{n}{8}$ **2.** $\frac{3}{8} = \frac{n}{16}$ **3.** $\frac{n}{9} = \frac{2}{3}$

4. $\frac{n}{12} = \frac{1}{6}$ **5.** $\frac{10}{n} = \frac{15}{21}$

Remember

kilo- = 1,000
hecto- = 100
deka- = 10
deci- = 0.1
centi- = 0.01
milli- = 0.001

The Stanley Cup is the oldest trophy competed for by professional ice hockey teams in North America. The trophy was donated in 1892 by Sir Frederick Arthur, Lord Stanley of Preston. Ice hockey games are played on an ice rink 61 meters long and 26 meters wide. How many kilometers long is the ice rink?

The original Stanley Cup is displayed in the Hockey Hall of Fame in Toronto, Canada.

You can write a proportion to convert metric units of length.

$\frac{1 \text{ km}}{1,000 \text{ m}} = \frac{x \text{ km}}{61 \text{ m}}$ *Use the relationship of kilometers to meters to write a proportion.*

$\frac{1}{1,000} = \frac{x}{61}$ *Find the cross products.*

$1 \cdot 61 = 1,000x$ *Solve for x.*

$x = 0.061$

So, the ice rink is 0.061 km long.

• How many kilometers wide is the ice rink?

You can also use proportions to convert metric units of liquid.

EXAMPLE

It takes about 30,200 liters of water to make the ice for the ice rink. How many kiloliters of water does it take to make the ice?

$\frac{1 \text{ kL}}{1,000 \text{ L}} = \frac{x \text{ kL}}{30,200 \text{ L}}$ *Use the relationship of kiloliters to liters to write a proportion.*

$\frac{1}{1,000} = \frac{x}{30,200}$ *Find the cross products.*

$1,000x = 30,200$ *Solve for x.*

$\frac{1,000x}{1,000} = \frac{30,200}{1,000}$

$x = 30.2$

So, it takes 30.2 kiloliters of water.

• How many kiloliters are in 1,800 L?

CALIFORNIA STANDARDS AF 2.0 Students analyze and use tables, graphs, and rules to solve problems involving rates and proportions. **AF 2.1** Convert one unit of measurement to another (e.g., from feet to miles, from centimeters to inches). *also* **NS 1.3, MR 1.1, MR 2.5, MR 2.7**

**Think and ▶
Discuss**

Look back at the lesson to answer each question.

1. **Describe** how you could use a proportion to convert decimeters to meters.

2. **Explain** what you could multiply or divide by to convert one metric measurement to another.

**Guided ▶
Practice**

Use a proportion to convert to the given unit.

3. 5 m = ■ km **4.** 35 cm = ■ mm **5.** 9 L = ■ mL **6.** 2 g = ■ kg

PRACTICE AND PROBLEM SOLVING

**Independent ▶
Practice**

Use a proportion to convert to the given unit.

7. 200 kL = ■ L **8.** 40 dm = ■ cm **9.** 500 m = ■ km

10. 10 g = ■ kg **11.** 1,200 m = ■ km **12.** 12 mm = ■ m

13. 440 g = ■ mg **14.** 9 L = ■ mL **15.** 18 kL = ■ L

16. 0.34 m = ■ km **17.** 425,000 mg = ■ g **18.** 320 mm = ■ cm

Compare. Write <, >, or = for each ●.

19. 40 m ● 400 cm **20.** 200 kL ● 2,000,000 L **21.** 10 g ● 0.01 kg

**Problem Solving ▶
Applications**

22. Karen carries her hockey equipment in her gym bag when she goes to practice. When the bag is empty its mass is 2,000 g. What is the mass of the empty bag in kilograms?

23. ❓ **What's the Question?** Marie needs 2,500 mL of soup for a dinner party. The answer is 2.5 L.

24. Jason made a scale drawing of his room. The drawing is 5.5 in. by 6 in. If the scale is 1 in.:2 ft, what are the dimensions of the room?

25. *REASONING* Shawna bought a 3-m length of tissue paper to use for an art project. She used a piece 200 cm long. What fraction of the paper is left over?

MIXED REVIEW AND TEST PREP

26. Solve for *n*. 8 days = *n* hours (p. 452)

27. Write $\frac{2}{5}$ as a percent. (p. 398)

28. Solve for *x*. 8.2*x* = 131.2 (p. 82)

29. What is the opposite of 9? (p. 228)

30. **TEST PREP** Mina spends $7.60 for $9\frac{1}{2}$ lb of apples. What is the cost per pound? (p. 210)

   A $0.80   **B** $0.84   **C** $0.95   **D** $1.29

Relate Customary and Metric

Learn how to convert between customary and metric measurements.

At least 100,000 cubic feet of water passes over the falls every second during daylight hours between April 1 and October 31.

Niagara Falls consists of two waterfalls, the Horseshoe Falls and the American Falls. While Catherine is visiting the Horseshoe Falls, she does some sightseeing in Ontario, Canada. Catherine notices that Canada uses the metric system. She reads that Horseshoe Falls is 792 m wide at its widest point. Catherine wants to find how many feet 792 m is.

Use the table of measures below.

1 in. ≈ 2.54 cm	1 m ≈ 3.28 ft	1 lb ≈ 0.45 kg
1 ft ≈ 30.48 cm	1 mi ≈ 1.61 km	1 qt ≈ 0.95 L
1 yd ≈ 0.91 m	1 oz ≈ 28.35 g	1 gal ≈ 3.79 L

$$\frac{1 \text{ m}}{3.28 \text{ ft}} = \frac{792 \text{ m}}{x \text{ ft}}$$ *Use the relationship of meters to feet to write a proportion.*

$$\frac{1}{3.28} = \frac{792}{x}$$ *Find the cross products.*

$$1 \cdot x = 3.28 \cdot 792$$ *Solve for x.*

$$x = 2{,}597.76$$

So, 792 m ≈ 2,597.76 ft.

EXAMPLES

A. How many kilograms is 365 lb?

THINK: A kilogram is heavier than a pound. So the number of kilograms in the answer should be *less than* 365.

$$\frac{1 \text{ lb}}{0.45 \text{ kg}} = \frac{365 \text{ lb}}{x \text{ kg}}$$

$$\frac{1}{0.45} = \frac{365}{x}$$

$$1 \cdot x = 0.45 \cdot 365$$

$$x = 164.25$$

So, 365 lb ≈ 164.25 kg.

B. How many quarts is 24 L?

THINK: A quart is smaller than a liter. So the number of quarts in the answer should be *greater than* 24.

$$\frac{1 \text{ qt}}{0.95 \text{ L}} = \frac{x \text{ qt}}{24 \text{ L}}$$

$$\frac{1}{0.95} = \frac{x}{24}$$

$$24 = 0.95x$$

$$\frac{24}{0.95} = \frac{0.95x}{0.95}$$

$$25.26 \approx x$$

So, 24 L ≈ 25.26 qt.

CALIFORNIA STANDARDS **AF 2.0** Students analyze and use tables, graphs, and rules to solve problems involving rates and proportions. **AF 2.1** Convert one unit of measurement to another (e.g., from feet to miles, from centimeters to inches). *also* O⁻¬NS 1.2, NS 1.3, MR 1.1, MR 2.5

Think and ▶ Discuss

Look back at the lesson to answer each question.

1. **Tell** if you get a greater or smaller number when you convert miles to kilometers. Explain your reasoning.

2. **Write** the proportion you would use to convert 6 lb to kilograms.

Guided ▶ Practice

Use a proportion to convert to the given unit.

3. 1 in. ≈ ▦ mm   **4.** 4 in. ≈ ▦ cm   **5.** 470 yd ≈ ▦ m

6. 70.5 L ≈ ▦ gal   **7.** 25.4 cm ≈ ▦ in.   **8.** 1,000 g ≈ ▦ oz

PRACTICE AND PROBLEM SOLVING

Independent ▶ Practice

Use a proportion to convert to the given unit.

9. 2.5 ft ≈ ▦ cm   **10.** 1 lb ≈ ▦ g   **11.** 5 qt ≈ ▦ L

12. 10 kg ≈ ▦ lb   **13.** 24 m ≈ ▦ yd   **14.** $\frac{1}{2}$ yd ≈ ▦ cm

Compare. Write < , >, or ≈ for each ●.

15. 12 oz ● 737.10 g   **16.** 33 km ● 20.50 mi   **17.** 89.85 L ● 15 gal

Problem Solving ▶ Applications

18. Ramona saw a road sign near the Canadian border stating the speed limit was 80 km per hour. To the nearest mile per hour, how many miles per hour is 80 km per hour?

19. Some cameras use film that is 35 mm wide. How many inches wide is 35 mm?

20. 📓 **Write a problem** in which you have to change 137 grams to pounds.

21. More than $\frac{2}{3}$ of Canada has an average January temperature of ⁻18°C. The coastal areas of British Columbia have average January temperatures of 0°C. What is the difference in the average January temperatures?

MIXED REVIEW AND TEST PREP

22. Change 9 L to kL. (p. 454)

23. Write an integer to show a 10 ft decline. (p. 228)

24. Write the ratio one to eight as a fraction. (p. 374)

25. Simplify $\frac{21}{18}$. (p. 160)

26. TEST PREP A shirt originally cost $13.65. It was discounted 40% during a sale. What was the sale price of the shirt? (p. 408)

A $5.46   **B** $8.19   **C** $9.10   **D** $10.19

Appropriate Tools and Units

Learn how to measure to a given degree of precision by using the appropriate mathematical tools and units of measure.

QUICK REVIEW

Which number is greater?

1. $\frac{1}{8}$, $\frac{1}{4}$ 2. $\frac{5}{16}$, $\frac{3}{4}$ 3. $\frac{7}{8}$, $\frac{6}{40}$ 4. 0.2, 0.02   5. 1.4, 1.41

When Doris runs to train for the track and field competition, she measures her time to the nearest minute. When she runs in the competition, the time is measured to the nearest hundredth of a second.

Doris's time is measured more precisely when seconds are used instead of minutes.

The smaller the unit of measure used, the more precise the measurement is.

Math Idea ▶ Metric and customary units are used to measure length, capacity, temperature, and weight (or mass). When you measure quantities, the measurement you get is an approximation.

Activity

You need: 2 different-size bags of rice; quart, pint, and cup measuring tools

Copy the table and record your findings.

	CUP	PINT	QUART
LARGER BAG			
SMALLER BAG			

Work with a partner, and begin with the larger bag of rice.

• Use the quart measure. Record the amount of rice you have to the nearest quart.

• Use the pint measure. Record the amount of rice you have to the nearest pint.

• Use the one-cup measure. Record the amount of rice you have to the nearest cup.

• Repeat the activity with the smaller bag of rice.

Think and Discuss

• Which measurement for each bag of rice is most precise? Why?

• How could you get a more precise measurement?

CALIFORNIA STANDARDS MG 1.0 Students deepen their understanding of the measurement of plane and solid shapes and use this understanding to solve problems. MR 1.0 Students make decisions about how to approach problems. *also* MR 2.6

Measured to the nearest centimeter, this segment is about 4 cm.

You can measure the length of the same line segment more precisely with a ruler that is divided into millimeters.

A more precise measurement of the length is 38 mm.

EXAMPLE 1

Measure the segment to the nearest $\frac{1}{4}$ in., $\frac{1}{8}$ in., and $\frac{1}{16}$ in. Which measurement is most precise?

To the nearest $\frac{1}{4}$ in., the line segment is $2\frac{3}{4}$ in. long.

To the nearest $\frac{1}{8}$ in., the line segment is $2\frac{5}{8}$ in. long.

To the nearest $\frac{1}{16}$ in., the line segment is $2\frac{11}{16}$ in. long.

So, $2\frac{11}{16}$ in. is most precise.

When you measure an object, you must decide which unit of measurement is most appropriate. If you wanted to measure the amount of water in a bathtub, for example, you would use liters or gallons. It would not be as appropriate to use milliliters or cups.

EXAMPLE 2

What is an appropriate unit of measure for each item?

A. The mass of two small apples *gram* or *kilogram* Gram is more appropriate.	**B.** The amount of punch per serving *quart* or *cup* Cup is more appropriate.
C. The length of your fingernail *millimeter* or *centimeter* Millimeter is more appropriate.	**D.** The weight of a math book *ounce* or *pound* Pound is more appropriate.

Think and ▶
Discuss

Look back at the lesson to answer each question.

1. **Tell** whether the mass of a textbook is more precise measured to the nearest gram or kilogram. Explain your reasoning.

2. **Give** a more precise measurement for a room that is about 12 ft long.

Guided ▶
Practice

Measure the line segment to the given length.

3. nearest centimeter; nearest millimeter

4. nearest inch; nearest half inch

Tell which measurement is more precise.

5. 7 ft or 85 in.   6. $\frac{1}{2}$ c or $\frac{1}{4}$ c   7. 4 lb or 65 oz   8. 6 cm or 61 mm

Name an appropriate customary or metric unit of measure for each item.

9. length of a classroom

10. liquid in a soft-drink can

11. mass or weight of a new pencil

PRACTICE AND PROBLEM SOLVING

Independent ▶
Practice

Measure the line segment to the given length.

12. nearest centimeter; nearest millimeter

13. nearest half inch, nearest quarter inch

14. nearest centimeter; nearest millimeter

15. nearest quarter inch; nearest eighth inch

Tell which measurement is more precise.

16. 71 mm or 7 cm   17. 3 yd or 8 ft   18. 3 pt or 5 c   19. $\frac{1}{2}$ lb or 9 oz

Name an appropriate customary or metric unit of measure for each item.

20. weight or mass of a chair

21. thickness of a magazine

22. distance from one city to another

Compare. Write <, >, or = for each ●.

23. 8 qt ● 16 pt   24. 9,000 lb ● 6 T   25. 739 mL ● 0.739 L

Problem Solving ▶ Applications

26. Chris said the county repaired 2,650 yd of highway in her neighborhood. Her brother said the county repaired 1.5 mi of highway. Which measurement is more precise?

27. Which is more precise, measuring to the nearest half centimeter or to the nearest millimeter? Explain.

28. **Write About It** Explain the most precise unit of measurement you might use to measure the length and width of a photograph.

Use Data **For 29–30, use the table below.**

29. Write an expression to find how much shorter in length the Peace Bridge is than the Brooklyn Bridge and the Golden Gate Bridge combined.

BRIDGES	
Bridge	**Length in feet**
Golden Gate	8,981
Brooklyn Bridge	$1,595\frac{1}{2}$
The Peace Bridge	5,800

30. What is the length of the Golden Gate Bridge in miles?

MIXED REVIEW AND TEST PREP

31. Change 4.5 gal to L. (p. 456) **32.** $^-7 - {}^-12$ (p. 250) **33.** $^-12 \times 8$ (p. 258)

34. **TEST PREP** Kim has to allow 45 min to practice her cello before she leaves for soccer practice. She also wants to study for 1 hr 10 min and spend 25 min eating breakfast and getting dressed. If she has to be at soccer practice at 1 P.M., what time should she get up? (p. 452)

    A 9:40 A.M.   **B** 10:40 A.M.   **C** 11:00 A.M.   **D** 10:40 P.M.

35. **TEST PREP** Which shows 34% of 124? (p. 402)

    F 4.216   **G** 42.16   **H** 421.6   **J** 4,216

Thinker's CORNER

WHAT'S THE UNIT? Choose the appropriate unit of measure from the column on the right. Write the letters in the order shown below the blanks to learn the name of a customary measure that is equal to 4 pecks.

1. Volume of gasoline   **B** mile

2. Length of a football field   **E** yard

3. Distance from Chicago to New York **H** gram

4. Weight of a hummingbird   **L** gallon

5. Weight of a herd of elephants   **S** pound

6. Weight of a pork roast   **U** ton

   ___   ___   ___   ___   ___   ___
    3   5   6   4   2   1

Extra Practice page H55, Set D

PROBLEM SOLVING SKILL
Estimate or Find Exact Answer

Analyze
Choose
Solve
Check

Learn how to decide when to estimate and when to find an exact answer.

Alex is a radio-station disk jockey. He is making a list of songs that should last about a half hour, but no longer. Below is his first list of songs and their playing times. Does he have enough music?

SONG	LENGTH	SONG	LENGTH
Color Me Blue	4.5 min	Smile on Me	5.7 min
Top Dog	3.6 min	Kelso Blues	4.3 min
Hittin' the Road	7.2 min	A Long Time Ago	6.4 min
Stand Up and Shout	2.6 min		

To find out if he has enough music, Alex can estimate the total amount of time for the songs in his list. To estimate, he rounds each amount of time to the nearest minute.

$$5 + 4 + 7 + 3 + 6 + 4 + 6 = 35$$

Since his estimate is 35 min, he has more music than he needs.

Alex also has to play 5.7 min of commercials. How much music time does he need to cut?

Now Alex needs to know the exact amount of time. He adds the lengths of the songs and the lengths of the commercials.

$$4.5 + 3.6 + 7.2 + 2.6 + 5.7 + 4.3 + 6.4 + 5.7 = 40$$

He has 40 min of music and commercials, so he needs to cut the music by 10 min.

Math Idea ▶ Sometimes an estimate is all you need to answer a problem and sometimes you need to find an exact answer.

Talk About It ▶ • Which songs could Alex remove from his list so he has 30 min of music and commercials? Is an estimate or exact answer needed?

CALIFORNIA STANDARDS O─┐ **NS 2.0** Students calculate and solve problems involving addition, subtraction, multiplication, and division. **MR 2.6** Indicate the relative advantages of exact and approximate solutions to problems and give answers to a specified degree of accuracy. *also* **NS 2.1,** O─┐ **NS 2.4, MR 1.3**

Decide whether you need an estimate or an exact answer. Solve.

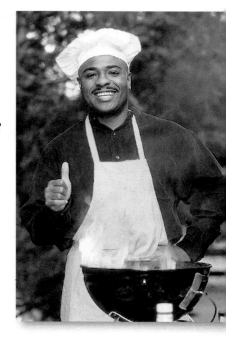

1. Eric is having a party. He is planning for each of his six guests to have two 0.25-lb meat patties. He buys four packages of meat weighing 6 oz, 9 oz, 4.5 oz, and 11.5 oz. Does he have enough meat?

2. Sally uses cereal, $\frac{1}{2}$ c of nuts, $\frac{1}{4}$ c of coconut, and 1 c of raisins to make trail mix. She wants to make a total of 1 qt of mix. How many cups of cereal does she need to add?

3. Sherry took $20.00 to spend at the mall. She paid $8.95 for a blouse and $5.25 for lunch at the food court. Will she be able to spend $10.00 at another store?

4. Ronny takes $50.00 to the music store to buy three books of music. He spends $13.95, $18.99, and $15.75 on the music. How much money does Ronny have left?

MIXED APPLICATIONS

Karen has a checking account at her bank. The bank charges $4 per month plus $0.25 for each check she writes.

5. Which of the following equations can be used to determine the amount Karen pays each month? Let c = the number of checks.

    A $4 \times 0.25c$ **C** $4 + 0.25c$

    B $4c + 0.25$ **D** $4 - 0.25c$

6. Karen is charged $6.00 by the bank during the month of September. How many checks did she write in September?

    F 2   **H** 8

    G 6   **J** 24

Use Data For 7–9, use the table at the right.

Precipitation In Inches

Nov.	Dec.	Jan.	Feb.	Mar.	Apr.
$5\frac{2}{5}$	$6\frac{3}{4}$	$8\frac{1}{2}$	$10\frac{1}{4}$	$5\frac{5}{8}$	$1\frac{1}{10}$

7. How much more rain fell during the month with the most rainfall than the month with the least rainfall?

8. Jon kept records of the rainfall at his home for six months. Estimate the total amount of precipitation from November through April.

9. The record rainfall for the months November through April is 40 inches. How much above or below the record is this year's rainfall?

10. The month of April has 30 days. What percent of the dates in April are prime numbers?

11. The sum of Maris's and Ramona's ages is 40 yr. Maris is 6 yr older than Ramona. How old is Maris?

12. Thirty-six students try out for baseball and basketball. If 20 students try out only for baseball and 8 try out for both sports, how many try out only for basketball?

13. **Write a problem** about measurement for which an estimate would be appropriate as an answer. Then write a related problem for which an exact answer is needed.

Use a proportion to convert to the given unit. (pp. 452–453)

1. 5 ft = ■ in.

2. 4 yd = ■ in.

3. 8 qt = ■ gal

4. 40 oz = ■ lb

5. 6 c = ■ pt

6. 2.3 T = ■ lb

Use a proportion to convert to the given unit. (pp. 454–455)

7. 50 mm = ■ cm

8. 3,000 mg = ■ g

9. 100 g = ■ kg

10. 3.7 L = ■ mL

11. 1.25 km = ■ m

12. 4,700 mm = ■ m

Use a proportion to convert to the given unit. Use the table on page 456. Round to the nearest hundredth if necessary. (pp. 456–457)

13. 2 ft ≈ ■ cm

14. 4 oz ≈ ■ g

15. 3.2 gal ≈ ■ L

16. 70 cm ≈ ■ in.

17. 200 g ≈ ■ oz

18. 6 L ≈ ■ qt

19. 7 in. ≈ ■ mm

20. 190 m ≈ ■ yd

21. 6 km ≈ ■ mi

Measure the line segment to the given length. (pp. 458–461)

22. nearest centimeter; nearest millimeter

23. nearest half inch; nearest quarter inch

24. nearest centimeter; nearest millimeter

25. nearest quarter inch; nearest eighth inch

26. nearest centimeter; nearest millimeter

27. nearest inch; nearest half inch

Tell which measurement is more precise. (pp. 458–461)

28. 2 cm or 21 mm

29. 4 ft or 49 in.

30. 27 lb or 427 oz

31. 4,210 g or 4 kg

32. 4,827 mL or 5 L

33. 785 mg or 1 g

Name an appropriate customary or metric unit of measure for each item. (pp. 458–461)

34. weight (mass) of a person

35. thickness of a large phone book

36. height of a 1-story home

37. Mike is having a party. He wants each of his guests to have at least 3 cups of punch. If there are 40 guests, how many gallons of punch are needed? (pp. 452–453)

38. The newspaper reported that the new courthouse was 97 ft tall. A television reporter said the courthouse was 32 yd tall. Which measurement is more precise? (pp. 458–461)

Solve. Tell whether you need an estimate or an exact answer. (pp. 462–463)

39. Marla has 5 oz of lime juice, 16 oz of orange juice, 12 oz of white grape juice, and 6 c of pineapple juice for a punch. Does she have enough for twenty 5-oz servings?

40. Amy and 3 friends are working on a craft project. Each person needs 39 cm of ribbon. They have 1.25 m of ribbon. How much more do they need?

Decide on a plan.

See item **6.**

Use the ratio of 1 inch to the given number of centimeters. Then write a proportion using the ratio for 6 inches to an unknown number of centimeters.

Also see problem **4**, p. H63

For 1–10, choose the best answer.

1. How many feet are in 240 inches?

    A 2 ft   **C** 20 ft

    B 12 ft   **D** 24 ft

2. How many milliliters are in 8 liters?

    F 0.008 mL

    G 0.08 mL

    H 800 mL

    J 8,000 mL

3. Which measurement is the most precise?

    A $\frac{1}{2}$ ft

    B 6 in.

    C $6\frac{1}{2}$ in.

    D $6\frac{1}{4}$ in.

4. How many meters are in 12.5 kilometers?

    F 12,500 m

    G 1.25 m

    H 0.125 m

    J 0.0125 m

5. Sue bought 3 pounds of cheddar cheese and 12 ounces of Swiss cheese. Altogether, how many ounces of cheese did she buy?

    A 48 oz   **C** 68 oz

    B 60 oz   **D** Not here

6. Which proportion could be used to determine the approximate number of centimeters in 6 inches? (Use 1 in. ≈ 2.54 cm.)

    F $\frac{1}{6} = \frac{x}{2.54}$

    G $\frac{x}{6} = \frac{1}{2.54}$

    H $\frac{1}{2.54} = \frac{6}{x}$

    J $\frac{1}{x} = \frac{6}{2.54}$

7. Jeremy filled 8 2-liter bottles with water. Approximately how many quarts is that? (Use 1 qt ≈ 0.95L.)

    A 8 qt   **C** 24 qt

    B 16 qt   **D** 32 qt

8. Which term best describes the shape of a cereal box?

    F Rectangular prism

    G Triangular prism

    H Cone

    J Cylinder

9. Which symbol makes this a true number sentence?

    $$^-6 \; \bullet \; {}^+1$$

    A <   **C** =

    B >   **D** −

10. On a map, the distance between Harding and Gilford is 8 inches. In the scale, 1 inch represents 40 miles. What is the actual distance?

    F 5 mi

    G 128 mi

    H 240 mi

    J 320 mi

CHAPTER 25

Length and Perimeter

The first Ferris wheel was made for the Chicago World's Fair of 1893 by bridge builder George W. Ferris. It had 36 wooden cars that each held 60 people. Rides cost $0.50 each. The diameter of the Ferris wheel was 250 ft and the circumference was 825 ft. About how many times as great as the diameter was the circumference?

DIAMETERS AND CIRCUMFERENCES OF BIG WHEELS

Wheel	Diameter (in feet)	Circumference (in feet)
1893 Chicago Ferris Wheel	250	825
Largest carousel	27.5	86
Portable merry-go-round	20	63
2000 London Millennium Wheel	443	1,392

Check What You Know

Use this page to help you review and remember important skills needed for Chapter 25.

 Vocabulary

Choose the best term from the box.

1. The distance around a figure is the ___?___.

2. A line segment with one endpoint at the center of a circle and the other endpoint on the circle is a ___?___.

> area
> perimeter
> radius

Perimeter (See p. H26.)

Find the perimeter of the figure.

3.
8 ft
8 ft

4.
12.5 m
6 m

5.
6 cm
5 cm
9 cm

6.
0.5 m
1.3 m
1.2 m

7.
26 in.
26 in.

8.
1.4 cm
1.4 cm
1.4 cm
1.4 cm
1.4 cm

Change Units (See p. H27.)

Change to the given unit.

9. 48 in. = ▦ ft

10. 4 yd = ▦ ft

11. 12 yd = ▦ ft

12. 3 mi = ▦ yd

13. 40 mm = ▦ cm

14. 72 in. = ▦ yd

15. 3 m = ▦ mm

16. 7 m = ▦ cm

17. 5 km = ▦ m

18. 4.9 cm = ▦ mm

19. 0.09 m = ▦ km

20. 8.2 cm = ▦ m

Multiply with Fractions and Decimals (See p. H23.)

21. $3\frac{1}{7} \times 28$

22. $2\frac{2}{3} \times 112$

23. $12\frac{1}{2} \times 210$

24. 3.7×9

25. 21.06×25

26. 3.14×8

27. $3\frac{3}{4} \times 16$

28. 5.25×14

29. 0.8×92

30. $15 \times 9\frac{4}{5}$

> **LOOK AHEAD**
>
> In Chapter 25 you will
> • find the perimeter of polygons
> • find the circumference of circles

Estimate Perimeter

MATH LAB

Explore how to estimate perimeter.

You need metric ruler, string, perimeter worksheet.

Mrs. Johnson is working with the Jackson Middle School student council to clean up the shores of Pine Lake. She gives all the members drawings of the lake and asks them to estimate the perimeter. Mrs. Johnson can then assign teams of students to clean portions of the lake's shore.

You can estimate the perimeter, or distance around a figure, by using string and a ruler.

Activity

- Lay a piece of string around the perimeter of the drawing of the lake. Mark the string where it meets itself.

- Decide on a metric unit of measure. Use the ruler to measure the string from its beginning to the mark you made on the string.

- Compare your measurement with those of the other students.

Pine Lake

Think and Discuss

- **Explain** why there are different estimates for the perimeter of the drawing of the lake.

- **REASONING** Suppose the scale of the drawing is 1 cm = 100 m. How many meters would your estimate represent? How many kilometers?

Practice

Use string and a ruler to estimate the perimeter of each.

1. the outline of your hand

2. the outline of your shoe sole

CALIFORNIA STANDARDS MG 1.0 Students deepen their understanding of measurement of plane and solid shapes and use this understanding to solve problems. *also* **MR 2.4**

Perimeter

Learn how to find the perimeter of a polygon.

Activity

You need: metric ruler

- Use the rectangle below. Measure the length of each side to the nearest centimeter.

Think and Discuss
- What is the perimeter to the nearest centimeter?

- How could you get a more precise measurement for the perimeter? What would the perimeter be then?

- What is another way to find the perimeter of a rectangle?

Math Idea ▶ The perimeter, P, of a polygon is the distance around it. To find the perimeter of any polygon, you can use a formula.

EXAMPLE 1

Tony is building a brick patio with a wooden frame. The lengths of the sides are $8\frac{1}{3}$ ft, $7\frac{1}{2}$ ft, $5\frac{3}{4}$ ft, $4\frac{2}{3}$ ft, and $6\frac{1}{4}$ ft. Tony needs to find the perimeter so he will know how much wood he needs for the frame. What is the perimeter?

$P = a + b + c + d + e$ *Write a formula.*

$P = 8\frac{1}{3} + 7\frac{1}{2} + 5\frac{3}{4} + 4\frac{2}{3} + 6\frac{1}{4}$ *Replace the variables with the lengths.*

$P = \left(8\frac{1}{3} + 4\frac{2}{3}\right) + \left(5\frac{3}{4} + 6\frac{1}{4}\right) + 7\frac{1}{2}$ *Use the Commutative and Associative Properties.*

$P = 13 + 12 + 7\frac{1}{2}$ *Use mental math to add.*

$P = 32\frac{1}{2}$

So, the perimeter is $32\frac{1}{2}$ ft.

CALIFORNIA STANDARDS **AF 3.1** Use variables in expressions describing geometric quantities (e.g., $P = 2w + 2l$, $A = \frac{1}{2}bh$, $C = \pi d$, the formulas for the perimeter of a rectangle, the area of a triangle, and the circumference of a circle, respectively). **AF 3.2** Express in symbolic form simple relationships arising from geometry. *also* **MG 1.0**

Since the opposite sides of a rectangle are equal in length, you can find the perimeter by finding 2 × length and 2 × width and then adding the products. You can write a formula to find the perimeter of a rectangle. The formula is $P = 2l + 2w$.

EXAMPLE 2

Find the perimeter of Sara's rectangular yard.

$P = 2l + 2w$
$P = (2 \times 28) + (2 \times 15)$ *Write the formula.*
$P = 56 + 30$ *Replace l with 28 and w with 15.*
$P = 86$ *Add the products.*

So, the perimeter is 86 yd.

Sometimes you know the perimeter of a figure, but the length of one side is unknown.

EXAMPLE 3

The polygon at the right has a perimeter of 105 yd. Find the unknown length.

$P = a + b + c + d + e$ *Write the formula.*
$105 = a + 16 + 29 + 7 + 26$ *Use the values you know.*
$105 = a + 78$ *Add the known lengths.*
$105 - 78 = a + 78 - 78$ *Solve for a.*
$27 = a$

So, the unknown length is 27 yd.

EXAMPLE 4

The length of a rectangle is 1 cm more than 3 times the width. What is the perimeter of the rectangle if the width is 32 cm?

First find the length.
$l = 3w + 1$ *Length is 1 more than 3 times the width.*
$l = (3 \cdot 32) + 1$ *Replace w with 32.*
$l = 96 + 1$ *Multiply.*
$l = 97$ *Add.*

Find the perimeter.
$P = 2l + 2w$ *Write the formula.*
$P = 2(97) + 2(32)$ *Replace w with 32 and l with 97.*
$P = 194 + 64$
$P = 258$

So, the perimeter is 258 cm.

CHECK FOR UNDERSTANDING

Think and Discuss ▶ Look back at the lesson to answer each question.

1. **Explain** how to find the unknown length of one side of a triangle when you know the perimeter and the lengths of the other two sides.

2. **Tell** what formula you could write to find the perimeter of a regular pentagon.

Guided Practice ▶ Find the perimeter.

3.
5.56 m 3.8 m 4.21 m

4.
$4\frac{3}{4}$ in. $4\frac{3}{4}$ in. $4\frac{3}{4}$ in. $4\frac{3}{4}$ in. $4\frac{3}{4}$ in.

5.
3 ft 18 in.

PRACTICE AND PROBLEM SOLVING

Independent Practice ▶ Find the perimeter.

6.
3.8 cm 3.8 cm 3.8 cm

7.
12 m 6 m 7 m 6 m 12 m

8.
4 w + 3 w $w = 28$ in.

The perimeter is given. Find the unknown length.

9.
25 cm x 16 cm 32 cm

$P = 86.5$ cm

10.
7 mm 6 mm 2mm 2 mm 7 mm x

$P = 30$ mm

Problem Solving Applications ▶

11. Write the formula to find the perimeter of the square at the right. If $x = 31$ ft, what is the perimeter?

x ft

12. ✏ **Write a problem** that can be solved by finding perimeter.

MIXED REVIEW AND TEST PREP

13. Solve. 13.7 cm ≈ ▓ in. (p. 456)

14. Name the figure. ▱ (p. 342)

15. Solve. $316 = \frac{a}{4}$ (p. 301)

16. Write an expression for 43.8 times a number. (p. 82)

17. **TEST PREP** Mario earned $45 one week and $75 the next. He saved $30 and spent the rest. What percent of the money he earned did Mario spend? (p. 402)

A 25%   **B** $33\frac{1}{3}$%   **C** 50%   **D** 75%

Extra Practice page H56, Set A   **471**

PROBLEM SOLVING STRATEGY
Draw a Diagram

Learn how to use the strategy *draw a diagram* to find perimeter.

QUICK REVIEW

1. $(2 \times 12) + (2 \times 7)$ 2. $4\frac{3}{4} + 2\frac{1}{4} + 7\frac{1}{4}$
3. 4×9.8 4. $23 + 47 + 42$
5. $62.03 + 12.7 + 5.19 + 35.28$

Roy has a rectangular piece of land that he wants to put a barbed wire fence around. He will place a post every 9 ft along the perimeter. Each post is 5 ft tall. The land is 63 ft long and 45 ft wide. How many posts is Roy going to need ?

Analyze

What are you asked to find?

What information are you given?

Is there information you will not use? If so, what?

Choose

What strategy will you use?

Use the strategy draw a diagram to find the number of posts needed to go around the perimeter of Roy's land.

Solve

How will you solve the problem?

Draw a rectangle that is similar to the shape of Roy's land. Place marks along the perimeter of the rectangle to represent the posts.

Count the number of marks you placed around the rectangle. Each corner should have only one mark. So, Roy will need 24 posts.

Check

How can you check your answer?

What if Roy wanted to place a post every 8 ft along a rectangular perimeter 64 ft long and 48 ft wide? How many posts would he need?

CALIFORNIA STANDARDS MR 2.4 Use a variety of methods, such as words, numbers, symbols, charts, graphs, tables, diagrams, and models, to explain mathematical reasoning. AF 3.0 Students investigate geometric patterns and describe them algebraically. *also* MG 1.0, MR 3.2

PROBLEM SOLVING PRACTICE

PROBLEM SOLVING STRATEGIES

▶ **Draw a Diagram or Picture**
Make a Model
Predict and Test
Work Backward
Make an Organized List
Find a Pattern
Make a Table or Graph
Solve a Simpler Problem
Write an Equation
Use Logical Reasoning

Solve the problem by drawing a diagram.

1. Each hexagon on a quilt is made up of six congruent equilateral triangles with 11-in. sides. Each triangle shares two sides with other triangles. What is the perimeter of one hexagon?

2. Darin's pool is 4 yd wide and 10 yd long. He wants to put a cement deck around the pool. It will be $5\frac{1}{2}$ ft wide on all sides. What will be the perimeter of the deck?

For 3–4, use this information.
A window is in the shape of a square with an isosceles triangle attached at the top. The side of the triangle that is attached to the square is the same length as one side of the square.

3. Use s for the length of the side of the square and l for the length of a leg of the isosceles triangle. Which expression can you use to find the window's perimeter?

A $2s + 2l$ **C** $4s + 2l$
B $3s + 2l$ **D** $4s + 3l$

4. If the square is 2.5 m on a side and each leg of the isosceles triangle is 1.8 m, how much molding will the builder have to buy to go around the window?

F 6.8 m   **H** 11.1 m
G 8.6 m   **J** 13.6 m

MIXED STRATEGY PRACTICE

Use Data For 5–7, use the circle graph.

5. Which three categories make up about 50% of the expenses? Explain.

6. Sharon spent a total of $98 this year and a total of $81 last year for her fall party. She spent the same percent for pumpkins this year as last year. How much more did she spend on pumpkins this year?

7. What percent of the expenses was spent on decorations, entertainment, and costumes?

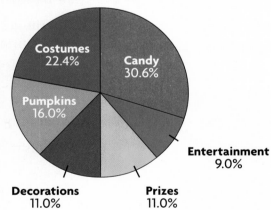

EXPENSES FOR FALL PARTY

Costumes 22.4%
Candy 30.6%
Pumpkins 16.0%
Entertainment 9.0%
Decorations 11.0%
Prizes 11.0%

8. There are 90 ft between bases on major league baseball diamonds and 60 ft between bases on Little League diamonds. How much farther does a major league player have to run if he hits a home run?

9. Kathy wants to work on her car. She needs a 25-mm wrench but her wrenches are all in customary units. To the nearest inch, what size of a customary wrench should she use?

10. Ali and Walt are taking a 50-question test. Ali answers every sixth question incorrectly and Walt answers every eighth question incorrectly. Which questions do both Ali and Walt answer incorrectly?

11. **? What's the Question?** May, Kevin, and Sean chopped wood. May chopped $\frac{2}{5}$ of the wood, Kevin chopped $\frac{1}{10}$ of it, and Sean chopped the rest. The answer is $\frac{1}{2}$.

Circumference

MATH LAB

Explore how to find the circumference of a circle.

You need compass, string, ruler, calculator.

Remember that the diameter is the distance from one point on the circle, through the center, to another point on the circle. The radius is half of the diameter.

Vocabulary

circumference

pi

The wheel on Dane's bicycle has a diameter of 26 in. Some bicycle wheels have diameters of 21 in., 24 in., and 18 in. Like bicycle wheels, circles come in different sizes.

You can find the **circumference**, or distance around a circle, by using a compass, string, and a ruler.

Activity 1

Find the circumference of a circle with a radius of 4 in.

- Open the compass to a width of 4 in. Use the compass to construct a circle with a radius of 4 in.

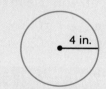
4 in.

- Lay the string around the circle. Mark the string where it meets itself.

- Use the ruler to measure the string from its beginning to the mark you made on the string.

mark

Think and Discuss

- What is the diameter of the circle?

- What is the circumference of the circle?

- Divide the circumference by the diameter. About how many times as long is the circumference?

- **REASONING**   Do all circles have the same circumference? Explain.

Practice

- Repeat the activity using four circles with different diameters.

- Compare your results with those of other students. What do you notice about the relationship between the diameter and the circumference of circles with different diameters?

CALIFORNIA STANDARDS ○━━MG 1.1 Understand the concept of a constant such as π; know the formulas for the circumference and area of a circle. **AF 3.2** Express in symbolic form simple relationships arising from geometry. **MG 1.2** Know common estimates of π (3.14; $\frac{22}{7}$) and use these values to estimate and calculate the circumference and the area of circles; compare with actual measurements. *also* **AF 3.1, MG 1.0**

The relationship between the circumference and the diameter of any circle can be written as the ratio of the circumference to the diameter, or $\frac{C}{d}$. The value of circumference divided by diameter is the same, or a "constant," for any circle. This ratio is called **pi**, or π.

In Activity 1, the value of the ratio $\frac{C}{d}$ was a little more than 3. The value of π is not a rational number but is approximately equal to 3.14 or $\frac{22}{7}$.

Because $\frac{C}{d} = \pi$, you get the formula $C = \pi d$ by multiplying each side by d. Since the diameter is twice the radius ($d = 2r$), you can also write $C = \pi(2r)$, or $C = 2\pi r$.

Activity 2

- Use a calculator with a π key, or use 3.14 or $\frac{22}{7}$ for π.

- Use the formula $C = \pi d$ and the key sequence below. To the nearest whole number, find the circumference of the circle.

 25.13274123

8 in.

$C \approx 25.13274123$, or about 25 in.

- Use the formula $C = 2\pi r$ and the key sequence below. To the nearest whole number, find the circumference of the circle.

 23.5619449

3.75 m

$C \approx 23.5619449$, or about 24 m.

TECHNOLOGY LINK
More Practice: Use E-Lab, *Circumference and Diameter*.
www.harcourtschool.com/ elab2002

Think and Discuss

- If you used 3.14 or $\frac{22}{7}$ instead of π on the calculator, how do your answers compare to the ones shown in Activity 2?

- Why do you use the formula $C = 2\pi r$ to find the circumference when you know the radius?

Practice

- Use your calculator, or 3.14 for π, to find the circumference of bicycle wheels that have diameters of 18 in., 21 in., 24 in., and 26 in.

MIXED REVIEW AND TEST PREP

1. Find the perimeter of a rectangle 26 in. long and 18 in. wide. (p. 469)

2. Write $7\frac{2}{5}$ in the form of $\frac{a}{b}$. (p. 164)

3. Solve and check. $z - 11.3 = 6.3$ (p. 294)

4. Solve. $34 \times {}^-7$ (p. 258)

5. **TEST PREP** Which shows 95°C changed to degrees Fahrenheit? (p. 304)

   A 50°F   **B** 89.6°F   **C** 94°F   **D** 203°F

Circumference

Learn how to find the circumference of a circle.

QUICK REVIEW

1. 3.14 × 8
2. 5.5 × 2
3. 21.4 ÷ 2
4. 2 × 3 × 3.14
5. 2 × 5 × 3.14

In 1962, the Seattle world's fair displayed developments in nuclear physics and featured the Space Needle. The Space Needle contains a revolving restaurant at the top.

Karen and her family eat in the restaurant at the top of the Space Needle. They ask to be seated near the window. The restaurant has a diameter of 94.5 ft and makes one revolution every hour. To the nearest foot, how far will Karen and her family travel in one revolution?

The Space Needle in Seattle, Washington, is about 605 ft high.

You need to find the circumference of the restaurant. When you know the diameter of a circle, you can use the formula $C = \pi d$.

$C = \pi d$	*Write the formula.*
$C \approx 3.14 \times 94.5$	*Replace π with 3.14 and d with 94.5.*
$C \approx 296.73$	
$C \approx 297$	*Round to the nearest foot.*

So, Karen and her family will travel about 297 ft in all.

Since the length of the diameter is twice that of the radius, use the formula $C = 2\pi r$ when you know the radius.

EXAMPLE 1

The Vienna Giant Ferris wheel was completed in 1897. It has a radius of 30.48 m. What is the circumference of the Ferris wheel? Round to the nearest hundredth.

$C = 2\pi r$	*Write the formula.*
$C \approx 2 \times 3.14 \times 30.48$	*Replace π with 3.14 and r with 30.48.*
$C \approx 191.4144$	
$C \approx 191.41$	*Round to the nearest hundredth.*

So, the circumference is about 191.41 m.

CALIFORNIA STANDARDS O⊸π **MG 1.1** Understand the concept of a constant such as π; know the formulas for the circumference and area of a circle. **AF 3.1** Use variables in expressions describing geometric quantities. **MG 1.2** Know common estimates of π (3.14; $\frac{22}{7}$) and use these values to estimate and calculate the circumference and the area of circles; compare with actual measurements. *also* **AF 3.2, MG 1.0**

Sometimes it makes sense to use the approximation $\frac{22}{7}$ for π.

EXAMPLE 2

Find the circumference of the circle at the right.

98 cm

$C = 2\pi r$ *Write the formula.*

$C \approx 2 \times \frac{22}{7} \times 98$ *98 is a multiple of 7.*
                               Replace π with $\frac{22}{7}$ and r with 98.

$C \approx \frac{2}{1} \times \frac{22}{7} \times \overset{14}{\underset{1}{\cancel{\frac{98}{1}}}}$ *Write 2 and 98 as fractions. Simplify.*

$C \approx \frac{2}{1} \times \frac{22}{1} \times \frac{14}{1}$ *Multiply.*

$C \approx \frac{616}{1}$

$C \approx 616$

So, the circumference is about 616 cm.

EXAMPLE 3

Sean has a circular outdoor wooden planter with a radius of $3\frac{1}{2}$ in. He wants to put trim around the top. What is the length of the trim he will need?

$C = 2\pi r$ *Write the formula.*

$C \approx \frac{2}{1} \times \frac{22}{7} \times 3\frac{1}{2}$ *Replace π with $\frac{22}{7}$ and r with $3\frac{1}{2}$.*

$C \approx \frac{2}{1} \times \overset{11}{\underset{1}{\cancel{\frac{22}{7}}}} \times \overset{1}{\underset{1}{\cancel{\frac{7}{2}}}}$ *Write 2 and $3\frac{1}{2}$ as fractions. Simplify.*

$C \approx \frac{2}{1} \times \frac{11}{1} \times \frac{1}{1}$ *Multiply.*

$C \approx \frac{22}{1}$

$C \approx 22$

So, Sean will need about 22 in. of trim.

You can also use the π key on a calculator. If you use a calculator, you may have to round your answer.

EXAMPLE 4

Use a calculator with a π key. Find the circumference of a circle with diameter 12.7 m. Round to the nearest tenth.

Use this key sequence.

 ✕ 12.7 ＝ | 39.8982267 |

$C \approx 39.8982267$

$C \approx 39.9$ *Round to the nearest tenth.*

So, the circumference is about 39.9 m.

Think and ▶
Discuss

Look back at the lesson to answer each question.

1. **Compare and contrast** the two formulas you can use to find the circumference of a circle.

2. **Express** $\frac{22}{7}$ as a decimal rounded to the nearest hundredth. How does it compare to 3.14?

Guided ▶
Practice

Find the circumference of the circle.
Use 3.14 or $\frac{22}{7}$ for π. Round to the nearest whole number.

3.
4.
5.

6. $r = 12.9$ mm
7. $d = 90$ yd
8. $r = 14$ mi

9. $d = 8.6$ in.
10. $r = 9.9$ m
11. $d = 10.6$ cm

Independent ▶
Practice

Find the circumference of the circle.
Use 3.14 or $\frac{22}{7}$ for π. Round to the nearest whole number.

12.
13.
14.

15.
16.
17.

18. $r = 29.62$ km
19. $d = 100.5$ in.
20. $r = 6\frac{2}{3}$ yd

21. $d = 7$ ft
22. $r = 3.5$ m
23. $d = 5.1$ cm

24. Look at the figure of a circle inside a square. The length of each side of the square is 3.5 ft. What is the circumference of the circle?

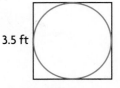

Find the radius for each circumference.

25. $C \approx 28.26$ ft
26. $C \approx 47.1$ m
27. $C \approx 58.404$ cm

Problem Solving ▶ Applications

28. A Ferris wheel at the local amusement park has a circumference of 800 ft. This is the same distance as one complete revolution. Find the radius of the Ferris wheel.

29. *REASONING* How does the circumference of two circles compare if the diameter of one is twice the diameter of the other? Give an example to explain your reasoning.

30. **? What's the Error?** Rhonda and Caren each used 3.14 to find the circumference of a circle with a radius of 34 mm. Rhonda's answer is 213.52 mm and Caren's is 106.76. Which girl made an error? What is the error?

31. George Ferris created the first Ferris wheel in 1893. The wheel had a maximum height of 264 ft. How many yards is this?

32. The cost of four pizzas was shared equally among six friends. Let c represent the total cost of the pizzas. Write an algebraic expression that shows the amount each friend paid.

MIXED REVIEW AND TEST PREP

33. Find the perimeter of a rectangle that is 46.3 m long and 28.2 m wide. (p. 469)

34. Is 131 prime or composite? (p. 148)

35. **TEST PREP** $12\frac{2}{3} \div 5\frac{3}{5}$ (p. 210)

     A $2\frac{11}{42}$ **B** $3\frac{11}{42}$ **C** $7\frac{1}{2}$ **D** $17\frac{6}{15}$

36. **TEST PREP** At 8:00 Wednesday morning the temperature was 9°F. At 4:00 that afternoon, the temperature had dropped 12°F. What was the temperature at 4:00? (p. 244)

     F ⁻21°F   **G** ⁻3°F   **H** 3°F   **J** 21°F

LiNKUP to Social Studies

Making Money The United States uses the dollar as the basic unit of money. France uses the franc and Mexico uses the peso. Artists play a big part in producing money. A new coin begins with an artist's design. The artist makes a clay model of the coin, and a mold is made from that model. A plaster cast is made from the mold. The artist then carves the details of the coin into the plaster.

1. Use a metric ruler to measure the diameter of a penny, nickel, dime, and quarter to the nearest millimeter. Find the circumference of each coin.

2. Research different countries to find the sizes of the coins they use. Compare the sizes of their coins with the ones used in the United States.

1. **VOCABULARY** The distance around a polygon is the ___?___ .
(pp. 469–471)

2. **VOCABULARY** The distance around a circle is called the ___?___ .
(pp. 476–479)

Find the perimeter. (pp. 469–471)

3. 35 yd
    19 yd

4. $9\frac{1}{4}$ ft
    5 ft $12\frac{1}{2}$ ft

5. 8.2 cm   8.2 cm
    14.7 cm   14.7 cm
    10.5 cm

6. 19 in.
    6 in.   14 in.
    24 in.

7. 5.8 m
    5.8 m   5.8 m
    5.8 m   5.8 m
    5.8 m

8. $4\frac{1}{10}$ ft $5\frac{1}{5}$ ft
    $6\frac{7}{10}$ ft $8\frac{1}{2}$ ft

The perimeter is given. Find the unknown length. (pp. 469–471)

9. 13.4 m   11.9 m
    11.06 m   11.06 m
    x

    $P = 65.7$ m

10. 13.8 cm   13.8 cm
    22.2 cm   22.2 cm
    x

    $P = 84$ cm

11. 13 mi   35 mi
    r
    29 mi

    $P = 135$ mi

Find the circumference. Use 3.14 or $\frac{22}{7}$ for π. Round to the nearest whole number. (pp. 476–479)

12. 9 in.

13. 35 ft

14. 4.3 m

15. 12.5 in.

16. 15.6 m

17. 60.8 km

Solve. (pp. 472–473)

18. Juli has a piece of fabric 32 in. long and 25 in. wide. What is its perimeter?

19. An official tournament-size dartboard has a radius of 9 in. Find its circumference to the nearest whole number.

20. Nathan is building a pen with four sides. The perimeter is 92 ft. Three of the sides measure 18 ft, 27 ft, and 21 ft. What is the length of the fourth side?

Get the information you need.
See item **7.**

You need to remember the formula for finding circumference when you know the radius of a circle. Since the choices have decimals, you should use 3.14 for pi.

Also see problem **3**, p. H63.

Choose the best answer.

1. What is the perimeter of this figure?

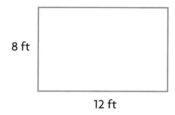

8 ft

12 ft

A 10 ft   **C** 40 ft

B 20 ft   **D** 100 ft

2. A rectangle has a perimeter of 24 ft. The length is 4 ft longer than the width. What is the length of the rectangle?

F 4 ft   **H** 28 ft

G 8 ft   **J** Not here

3. The circumference of a circle measures about 38 inches. Which is a reasonable estimate of the radius?

A 6 in.   **C** 20 in.

B 12 in.   **D** 120 in.

4. The diagram shows the dimensions of a lot. The perimeter of the lot is 211 feet. What is the unknown length?

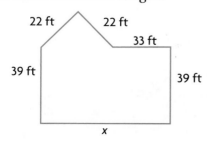

22 ft   22 ft
          33 ft
39 ft   39 ft
        x

F 23 ft   **H** 37 ft

G 35 ft   **J** Not here

5. One side of a regular pentagon measures 6.25 centimeters. What is the perimeter of the pentagon?

A 12.5 cm   **C** 30 cm

B 25 cm   **D** 31.25 cm

6. Which formula can be used to find the perimeter, P, of a rectangle?

F $P = l + w$ **H** $P = 2l \times 2w$

G $P = 2l + 2w$ **J** $P = l \times w$

7. Which is the best approximation of the circumference of a circle with a 3.5 cm radius?

A 10.9 cm   **C** 22.0 cm

B 12.3 cm   **D** 38.5 cm

8. The principal of Lincoln Middle School wants to display data about the number of absent students for each day during the past three months. Which type of graph would be best to use if she wants to identify the median number quickly?

F Histogram

G Stem-and-leaf plot

H Circle graph

J Multiple-line graph

9. Gil walks dogs to earn money. On Monday he spent $1\frac{2}{3}$ hours walking dogs. On Thursday he spent $2\frac{1}{2}$ hours walking dogs. What is the total time he spent walking dogs?

A $3\frac{1}{2}$ hr   **C** $4\frac{1}{6}$ hr

B $3\frac{3}{5}$ hr   **D** $4\frac{1}{3}$ hr

10. A survey showed that 1 out of every 25 students at Hillside High takes German. Which of the following decimals represents this ratio?

F 0.04   **H** 0.25

G 0.12   **J** 0.4

CHAPTER 26 Area

The Louvre in Paris, France, is one of the world's great art museums. In 1984, the French government modernized the Louvre under the direction of United States architect I. M. Pei. The new entrance to the Louvre is shaped like a square pyramid. The pyramid has about 150 congruent panes of glass on each of its 4 sides. Each pane is in the shape of a rhombus. About how many glass rhombuses are on all 4 sides?

Check What You Know

Use this page to help you review and remember important skills needed for Chapter 26.

✔ Vocabulary

Choose the best term from the box.

1. To multiply a number by itself is to __?__ that number.

2. A __?__ is a symbol that represents a number.

3. The polygons that form the sides of prisms and pyramids are called __?__ .

> cube
> faces
> square
> variable

✔ Find the Square of a Number (See p. H27.)

Find the square of each number.

4. 8	**5.** 12	**6.** 15	**7.** 20
8. 50	**9.** 9	**10.** 1.3	**11.** 2.4
12. $1\frac{1}{2}$	**13.** 49	**14.** 83	**15.** 129

✔ Evaluate Expressions (See p. H18.)

Evaluate each expression.

16. lw for $l = 6$ and $w = 9$

17. $\frac{1}{2}bh$ for $b = 5$ and $h = 10$

18. $2ab$ for $a = 4$ and $b = 12$

19. $3c^2$ for $c = 8$

20. $3(w + 5)$ for $w = 7$

21. $\frac{1}{2}s^2 + 2s$ for $s = 4$

✔ Faces of Prisms and Pyramids (See p. H20.)

Tell the number of faces of each figure.

22.

23.

24.

25.

> ### LOOK AHEAD
>
> **In Chapter 26 you will**
>
> - estimate and find the areas of rectangles, triangles, parallelograms, trapezoids, and circles
> - find the surface areas of prisms and pyramids

Estimate and Find Area

Learn how to estimate area and find the area of rectangles and triangles.

Vocabulary

area

Remember that area is measured in square units such as square feet (ft²) and square meters (m²).

Elena wants to put a layer of black netting on her garden to keep the weeds from growing through the mulch. She needs to know the area of her garden so she will know how much netting to buy.

It is not easy for Elena to find the area of her garden since it is not a polygon. She made a drawing of her garden on grid paper to estimate the area. Each square on the grid represents 1 ft².

The **area** of a figure is the number of square units needed to cover it. Use this drawing of Elena's garden to estimate the area.

There are 12 full red squares.

There are 6 almost-full blue squares.

There are 4 half-full orange squares. Combine the half-full squares to make 2 full squares. Do not count the squares that are less than half full.

Add the number of squares you counted.

$12 + 6 + 4 \times \frac{1}{2} = 12 + 6 + 2 = 20$

So, the area of Elena's garden is about 20 ft².

You can also use a grid to estimate the area of polygons. In the rectangle at the right, there are 28 full squares and 4 almost-full squares. There are 0 half-full squares.

$$28 + 4 = 32$$

So, the area is about 32 cm².

7.7 cm

4.2 cm

CALIFORNIA STANDARDS AF 3.1 Use variables in expressions describing geometric quantities. **MG 1.0** Students deepen their understanding of measurement of plane and solid shapes and use this understanding to solve problems. *also* ○──NS 2.0, NS 2.1, AF 3.2, MR 1.3, MR 2.5

You can use the formula $A = lw$ to find the area of a rectangle and the formula $A = s^2$ to find the area of a square.

EXAMPLE 1

Find the area of the rectangle shown at the right.

$A = lw$ *Write the formula.*

$A = 16 \times 13$ *Replace l with 16 and w with 13.*

$A = 208$ *Multiply.*

So, the area of the rectangle is 208 in.²

13 in. (width)

16 in. (length)

EXAMPLE 2

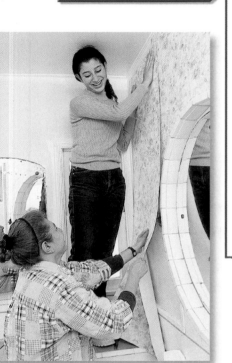

Cinzia is going to wallpaper the wall shown at the right. How many square feet of wallpaper will she need?

Think of the wall as a rectangle that is 8 ft × 6 ft and a square that is 4 ft × 4 ft.

Find the area of each figure and then find the sum.

Rectangle	Square
$A = lw$	$A = s^2$
$A = 8 \times 6$	$A = 4^2 = 4 \times 4$
$A = 48$	$A = 16$

Total area: $48 + 16 = 64$

So, Cinzia will need 64 ft² of wallpaper.

If you draw the diagonal of the rectangle at the right, you divide it into two congruent right triangles. The area of each triangle is half that of the rectangle. The area of each of the right triangles is

$\frac{1}{2}lw$, or $\frac{1}{2} \times 16 \times 13 = 104$.

In triangles, the dimensions are called the base (b) and the height (h). So, the formula for the area of a triangle is written as $A = \frac{1}{2}bh$.

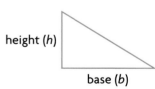

16 in.

(height or width) 13 in.

13 in.

16 in. (base or length)

height (h)

base (b)

EXAMPLE 3

Use the formula $A = \frac{1}{2}bh$ to find the area of the triangle at the right.

$A = \frac{1}{2}bh$ *Write the formula.*

$A = \frac{1}{2} \times 8 \times 5$ *Replace b with 8 and h with 5.*

$A = \frac{1}{2} \times 40$ *Multiply.*

$A = 20$ So, the area is 20 cm².

5 cm

8 cm

485

Look at triangle *ABC* at the right below.

Area of triangle *ABD*: $A = \frac{1}{2}bh = \frac{1}{2} \times 4 \times 6$.

Area of triangle *CBD*: $A = \frac{1}{2}bh = \frac{1}{2} \times 7 \times 6$.

To find the area of triangle *ABC*, find the sum of those areas.

$A = \left(\frac{1}{2} \times 4 \times 6\right) + \left(\frac{1}{2} \times 7 \times 6\right)$

$A = \left(\frac{1}{2} \times 24\right) + \left(\frac{1}{2} \times 42\right)$ *Multiply.*

$A = 12 + 21$ *Add.*

$A = 33$ So, the total area of triangle *ABC* is 33 in.²

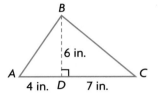

- Find the area of triangle *ABC* by using 4 in. + 7 in., or 11 in., as the base and 6 in. as the height. Compare your answer to 33 in.²

Sometimes the height is shown outside the triangle. The height of this triangle is 1.8 m.

$$A = \frac{1}{2} \times 2.8 \times 1.8 = 2.52$$

CHECK FOR UNDERSTANDING

Think and Discuss ▶ **Look back at the lesson to answer the question.**

1. **Explain** how the areas of a rectangle and a triangle with the same base and height are related.

Guided Practice ▶ **Estimate the area of the figure. Each square on the grid is 1 m².**

2.   3.   4.

Find the area.

5.   6.   7.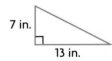

PRACTICE AND PROBLEM SOLVING

Independent Practice ▶ **Estimate the area of the figure. Each square on the grid is 1 m².**

8.   9.   10.

Find the area.

11.
20.3 mm
27.5 mm

12.
9 ft
12 ft

13.
0.5 m
0.8 m

14.
5.8 cm
9.3 cm

15.
2.5 ft
3 ft
2.5 ft
6 ft
3 ft
5 ft

16.
5 in.
5 in.
$\frac{1}{2}$ ft

Problem Solving ▶ Applications

17. The diagram shows a backyard. It will be covered with grass. How many square yards of grass are needed?

20 yd
9 yd
15 yd
3 yd
14 yd
3 yd

18. REASONING The length of this rectangle is twice the width.
  a. Find the area of the rectangle.
  b. Find the perimeter of the rectangle.

2w
w

MIXED REVIEW AND TEST PREP

19. Evaluate $\frac{1}{2} \times c$ for $c = 24$. (p. 216)

20. Evaluate $2d - 6$ for $d = {}^-6$. (p. 276)

21. Write $\frac{3}{5}$ as a percent. (p. 169)

22. TEST PREP If Rod is twice as tall as Cathy, and Rod's height is x, what expression represents Cathy's height? (p. 216)

  A $2x$ **B** x **C** $\frac{1}{2}x$ **D** $x - 2$

23. TEST PREP Which shows the circumference of a circle with a radius of 13 cm? (p. 476)

  F 23.55 cm   **G** 40.82 cm   **H** 81.64 cm   **J** 93.01 cm

---

Thinker's CORNER

Algebra   How is the area of a rectangle changed when its length and its width are each doubled?

For Rectangle 1, Area = $l \times w$.

For Rectangle 2, Area = $(2l) \times (2w)$.

$$= (2 \times 2) \times (l \times w)$$

$$= 4 \times (l \times w)$$

Rectangle 1 w
  l

Rectangle 2 $2w$
  $2l$

So, the area of a rectangle is four times as great when its length and width are each doubled.

• Use algebra to show how the area of a triangle changes when the base and height are each doubled.

ALGEBRA
Areas of Parallelograms and Trapezoids

Learn how to find the areas of parallelograms and trapezoids.

QUICK REVIEW

1. $(12 \times 9) \div 2$ 2. $(4.2 \times 6) \div 2$
3. $(9\frac{1}{2} \times 3) \div 2$ 4. $(240 \times 4) \div 2$
5. $(83.5 \times 24.2) \div 2$

A contractor has to put glass tiles on the side of an escalator. The side has the shape of a parallelogram. How many square feet of tiles will the contractor need?

Activity

You need: scissors, graph paper

* The parallelogram at the right is a drawing of the side of the escalator. Draw the parallelogram on a piece of graph paper and cut it out.

  height (*h*) 3 ft
  base (*b*) 22 ft

* Cut along the dotted line. Move the triangle to the right side to form a rectangle.

  width (*w*) 3 ft
  length (*l*) 22 ft

* What is the area of the rectangle? What is the area of the parallelogram?

* How are the dimensions and area of the parallelogram related to those of the rectangle?

* What formula can you write for the area of a parallelogram?

You can use the formula for the area of a rectangle to write a formula for the area of a parallelogram.

$A = lw$ *The length of the rectangle is the base of the parallelogram.*
   ↓ *The width of the rectangle is the height of the parallelogram.*
$A = bh$

Use $A = bh$ to find the area of the side of the escalator.

$A = bh$
$A = 22 \times 3$ *Replace b with 22 and h with 3.*
$A = 66$

So, the area of the side of the escalator is 66 ft².

CALIFORNIA STANDARDS   **AF 3.1** Use variables in expressions describing geometric quantities. **AF 3.2** Express in symbolic form simple relationships arising from geometry. **MG 1.0** Students deepen their understanding of the measurement of plane and solid shapes and use this understanding to solve problems. *also* **MR 1.3, MR 2.5, MR 2.7**

> **EXAMPLE 1**

Find the area of the parallelogram at the right.

$A = bh$ *Write the formula.*

$A = 1.5 \times 2.6$ *Replace b with 1.5 and h with 2.6.*

$A = 3.9$ *Multiply.*

So, the area is 3.9 m².

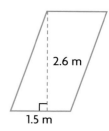

The trapezoid at the right is the shape of the glass tiles used by the contractor. You can use what you know about finding the area of a rectangle to find the area of each glass tile.

If you place two of these trapezoids end to end, you form a rectangle.

The length of the rectangle is $b_1 + b_2$, where b_1 and b_2 are the lengths of the two bases of the trapezoid. The width of the rectangle is the height of the trapezoid.

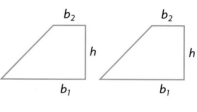

The area of the rectangle is $(b_1 + b_2) \times h$.

Since each trapezoid is half of the rectangle, you can use $\frac{1}{2} \times (b_1 + b_2) \times h$, or $\frac{1}{2}h(b_1 + b_2)$ to find the area.

> **EXAMPLE 2**

Find the area of each glass tile. Use the bases of 7 in. and 18 in. and the height of 12 in.

$A = \frac{1}{2}h(b_1 + b_2)$ *Write the formula.*

$A = \frac{1}{2} \times 12 \times (18 + 7)$ *Replace h with 12, b₁ with 18, and b₂ with 7.*

$A = \frac{1}{2} \times 12 \times 25$ *Multiply.*

$A = \frac{1}{2} \times 300$

$A = 150$

So, the area is 150 in.²

CHECK FOR UNDERSTANDING

Think and ▶ Discuss

Look back at the lesson to answer each question.

1. **Tell** how to find the area of the trapezoid at the right.

2. **Explain** which formula you would use to find the area of a rhombus.

Find the area of each figure.

3.

4.

5.

PRACTICE AND PROBLEM SOLVING

Find the area of each figure.

6.

7.

8.

9.

10.

11.

12. When triangular flaps are raised on opposite sides of the square table at the right, it has the shape of a parallelogram. The base of each triangle is 2 ft. What is the area of the table top?

13. **REASONING** Copy the parallelogram in Exercise 6 and the trapezoid in Exercise 9. Show how you can divide each figure into 2 triangles to find the area. Then find the areas and compare your answers to the answers you originally got for Exercises 6 and 9.

14. Four months ago the JB Paper Company had 1,480 employees. Three months ago it had 1,590 employees and two months ago it had 1,700. Describe the pattern. Then find the number of employees one month ago if the pattern continued.

MIXED REVIEW AND TEST PREP

15. Find the area of a triangle where $b = 4$ m and $h = 12$ m. (p. 484)

16. $7\frac{1}{6} \times \frac{3}{4}$ (p. 206)   **17.** 3 g = ■ kg (p. 454)   **18.** ⁻1,125 ÷ ⁻15 (p. 260)

19. **TEST PREP** A bakery has a 40 lb bag of flour. Six pounds of flour are used to make cakes. Half of what is left is used to make cookies. How much flour remains in the bag after the cookies are made? (p. 22)

 A 17 lb   **B** 18 lb   **C** 19 lb   **D** 20 lb

Extra Practice page H57, Set B

Area of a Circle

Explore how to find the area of a circle.

You need compass, scissors.

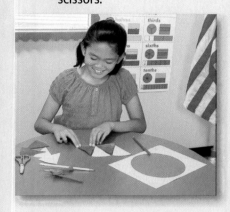

TECHNOLOGY LINK

More Practice: Use E-Lab, *Area of a Circle*. **www.harcourtschool.com/ elab2002**

Activity

To see how the area of a circle and its radius are related, you can rearrange the circle to approximate a parallelogram.

- Use the compass to construct a circle on a piece of paper.

- Cut out the circle. Fold it three times as shown.

- Unfold it, and trace the folds. Shade one half of the circle as shown.

- Cut along the folds. Fit the pieces together to make a figure that approximates a parallelogram.

Think and Discuss

Think of the figure as a parallelogram. The base and the height of the parallelogram relate to the parts of the circle.

base (b) = $\frac{1}{2}$ the circumference of the circle, or πr

height (h) = the radius of the circle, or r

- What is the formula for the area of a parallelogram?

- Use the formula for the area of a parallelogram to write a formula for the area of a circle. Use πr for the base and r for the height.

- Use your formula to find the area of a circle with a radius of 7 yd to the nearest whole number.

Remember that the formula for the circumference of a circle is $C = \pi d$ or $C = 2\pi r$.
π is approximately 3.14, or $\frac{22}{7}$.

Practice

Find the area of each circle with the given radius. Use your formula for the area of a circle, and round to the nearest meter.

1. $r = 1$ m   **2.** $r = 3$ m   **3.** $r = 7$ m   **4.** $r = 4$ m

 CALIFORNIA STANDARDS ⊶ **MG 1.1** Understand the concept of a constant such as π; know the formulas for the circumference and area of a circle. **MG 1.2** Know common estimates of π (3.14; $\frac{22}{7}$) and use these values to estimate and calculate the circumference and the area of circles; compare with actual measurements. *also* **AF 3.0, AF 3.1, AF 3.2, MR 2.4, MR 2.5, MR 2.6**

491

ALGEBRA
Areas of Circles

Learn how to use a formula to find the area of a circle.

Jackie is on the girls' wrestling team at her high school. The photo at the right shows a wrestling mat used for practice and competition. Square mats used for high school students have sides measuring 38 ft long. The circle inside the square is the wrestling circle.

EXAMPLE 1

Find the area of the wrestling circle for a radius of 14 ft. Use the formula for the area of a circle, $A = \pi r^2$. Use $\frac{22}{7}$ for π.

$A = \pi r^2$ *Write the formula.*

$A \approx \frac{22}{7} \times (14)^2$ *Replace π with $\frac{22}{7}$ and r with 14.*

$A \approx \frac{22}{7} \times 196$

$A \approx \frac{22}{7} \times \frac{\overset{28}{196}}{\underset{1}{1}}$ *Simplify.*

$A \approx \frac{22}{1} \times \frac{28}{1}$ *Multiply.*

$A \approx \frac{616}{1}$, or 616

So, the area of the wrestling circle is about 616 ft².

• **What if** you use 3.14 for π? What do you get as the area of the wrestling circle, to the nearest whole number?

Sometimes you are given the diameter of a circle.

EXAMPLE 2

The wrestling circle for sumo, a Japanese form of wrestling, has a diameter of 4.6 m. What is the area of the circle, to the nearest whole number? Use 3.14 for π.

$4.6 \div 2 = 2.3$ *Find the radius.*

$A = \pi r^2$ *Write the formula.*
$A \approx 3.14 \times (2.3)^2$ *Replace π with 3.14 and r with 2.3.*
$A \approx 3.14 \times 5.29$
$A \approx 16.6106$

So, the area of the circle is about 17 m².

CALIFORNIA STANDARDS O—¬MG 1.1 Understand the concept of a constant such as π; know the formulas for the circumference and area of a circle. **MG 1.2** Know common estimates of π (3.14; $\frac{22}{7}$) and use these values to estimate and calculate the circumference and the area of circles; compare with actual measurements. *also* O—¬NS 2.0, NS 2.1, AF 3.1, AF 3.2, MR 1.0

Think and ▶
Discuss

Look back at the lesson to answer each question.

1. **Explain** how to find the area of a circular dartboard with a diameter of $1\frac{1}{2}$ ft.

2. **Tell** which has the greater area: a circle with a radius of 5 m or a circle with a radius of 50 cm.

Guided ▶
Practice

Find the area of each circle to the nearest whole number.

3.
4.
5.
6.

Independent ▶
Practice

Find the area of each circle to the nearest whole number.

7.
8.
9.
10.

11. $r = 4$ mm   12. $r = 0.9$ m   13. $d = 66$ in.   14. $d = 100$ ft

Find the area of the semicircle or quarter circle to the nearest whole number. Use 3.14 for π.

15.

14 cm

16.

8 cm

Problem Solving ▶
Applications

17. College wrestling rules require that the mat have a wrestling circle with a minimum diameter of 32 ft. What is the minimum area of the wrestling circle to the nearest whole number?

18. **? What's the Error?** A circle has a diameter of 18 m. Sean says the area is 28 m² to the nearest meter. What error did Sean make? What is the area?

19. Find the area of a parallelogram with a base of 12.5 ft and height of 8 ft. (p. 488)

20. $^-15 \times {}^-5$ (p. 258)

21. Complete. $\frac{16}{24} = \frac{2}{\blacksquare}$ (p. 160)

22. Find the LCM of 5, 8, and 20. (p. 150)

23. **TEST PREP** Which shows the perimeter of a rectangle $12\frac{1}{2}$ ft by $18\frac{2}{3}$ ft? (p. 469)

   A $55\frac{1}{3}$ ft   **B** $60\frac{1}{2}$ ft   **C** $62\frac{1}{3}$ ft   **D** $65\frac{2}{3}$ ft

Extra Practice page H57, Set C

ALGEBRA
Surface Areas of Prisms and Pyramids

Learn how to find the surface areas of prisms and pyramids.

Sam wants to paint the box above. He wants to find the total area of the outside of the box. How many square inches will Sam have to paint?

You can use the formula for the area of a rectangle to find the surface area of a rectangular prism. The **surface area** is the sum of the areas of the faces of a solid figure.

EXAMPLE 1

Use a net to find the surface area.
Use the formula $A = lw$.

face A: $A = 11 \times 5 = 55$

face B: $A = 21 \times 11 = 231$

face C: $A = 21 \times 5 = 105$

face D: $A = 21 \times 11 = 231$

face E: $A = 21 \times 5 = 105$

face F: $A = 11 \times 5 = 55$

$55 + 231 + 105 + 231 + 105 + 55 = 782$ *Find the sum.*

So, Sam will paint 782 in.^{2}

Use the formula for the area of a rectangle to find the area of each face.

Another way to find the surface area, S, of a prism is to remember that opposite faces have the same area.

EXAMPLE 2

Find the surface area of the prism at the right.
Use the formula $A = lw$.

front and back: $(6 \times 8) \times 2 = 96$

top and bottom: $(6 \times 4) \times 2 = 48$

left and right sides: $(4 \times 8) \times 2 = 64$

$S = 96 + 48 + 64 = 208$

So, the surface area is 208 cm^2.

Multiply by 2 to include opposite faces.

Find the sum.

CALIFORNIA STANDARDS **AF 3.1** Use variables in expressions describing geometric quantities.
MG 1.0 Students deepen their understanding of the measurement of plane and solid shapes and use this understanding to solve problems. *also* **MR 1.3, MR 2.5**

Remember that a pyramid is named by the shape of its base.

To find the surface area of a pyramid, think about its net.

The surface area of a pyramid equals the sum of the areas of the triangular faces and the area of the base. This pyramid has 4 triangular faces and a square base.

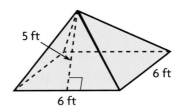

EXAMPLE 3

Find the surface area of the pyramid.

S = area of square + 4 × (area of triangle)

$S = s^2 + 4 \times \left(\frac{1}{2}bh\right)$ *Replace s with 6, b with 6, and h with 5.*

$S = 6^2 + 4 \times \left(\frac{1}{2} \times 6 \times 5\right)$

$S = 36 + 4 \times 15$

$S = 36 + 60$

$S = 96$

So, the surface area is 96 ft².

EXAMPLE 4

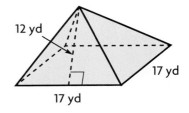

Find the surface area of the square pyramid at the right.

S = area of square base + 4 × (area of each triangular face)

$S = s^2 + 4 \times \left(\frac{1}{2} bh\right)$

$S = 17^2 + 4 \times \left(\frac{1}{2} \times 17 \times 12\right)$ *Replace s with 17, b with 17, and h with 12.*

$S = 289 + 4 \times 102$

$S = 289 + 408$

$S = 697$

So, the surface area is 697 yd².

495

Think and Discuss ▶ Look back at the lesson to answer each question.

1. **Describe** how you would find the surface area of a pentagonal prism.

2. **Describe** how to find the surface area of a square pyramid.

Guided Practice ▶ Find the surface area.

3. 6 ft 2 ft 10 ft

4. 15 cm 4 cm 8 cm

5. 8 cm 12 cm 12 cm

PRACTICE AND PROBLEM SOLVING

Independent Practice ▶ Find the surface area.

6. 3 in. 9 in. 6 in.

7. 12 cm 10 cm 10 cm

8. 6 cm 4 cm 5 cm

9. $3\frac{1}{2}$ m $3\frac{1}{2}$ m $3\frac{1}{2}$ m

10. 12 cm 14 cm 14 cm

11. 5 cm 3 cm 8 cm 4 cm

Find the surface area of each cube with the given side length, s.

12. $s = 1\frac{1}{2}$ m

13. $s = 3.4$ cm

14. $s = 28$ ft

15. Find the length, width, and height of each rectangular prism. Then find the surface area of each.

a. The length is twice the width. The height is twice the length. The width is 3 meters.

b. The width is half the length. The height is twice the width. The length is 6 feet.

c. The height is three times the length. The length is half the width. The width is 10 inches.

d. The length is four times the height. The width is one fourth the height. The height is 8 centimeters.

Problem Solving ▶
Applications

For 16–17, use the figure at the right.

16. Tamara is painting a 12 ft × 10 ft × 10 ft room. She will not paint the ceiling or the floor. How much surface area is she painting?

10 ft

10 ft

12 ft

17. One can of paint covers 350 ft². How many cans of paint will Tamara need to paint the 4 walls?

Use Data **For 18–19, use the graph at the right.**

18. About how many more square inches of cardboard are used for Box C than Box A?

19. If Box A is a cube, what is the length of each side?

20. **What's the Question?** The surface area of a cube is 96 m². The answer is 4 m.

MIXED REVIEW AND TEST PREP

21. Find the area of the circle at the right. (p. 492)

12 ft

22. Evaluate ⁻4 + 7 × ⁻4 ÷ 14 − 12. (p. 262)

23. Which is greater, $\frac{^-1}{4}$ or 0.25? (p. 234)

24. **TEST PREP** Lydia wants to put a ribbon around the edge of a circular tablecloth with a radius $1\frac{1}{2}$ yards. The ribbon costs $0.45 per yard. What is the cost of the ribbon to the nearest cent? (p. 474)

A $3.20 **B** $4.23 **C** $4.71 **D** $12.20

25. **TEST PREP** Which shows the perimeter of a hexagon with sides $9\frac{1}{2}$ ft long? (p. 469)

F 38 ft **G** $47\frac{1}{2}$ ft **H** 57 ft **J** $66\frac{1}{2}$ ft

 Thinker's CORNER

REASONING Look at the large cube at the right. It is made up of 27 unit cubes. Each unit cube is 1 in. long, 1 in. wide, and 1 in. high.

1. What is the surface area of the large cube?

2. What would happen to the surface area if you removed one cube from each of the eight corners? Explain.

3. What would happen to the surface area if you removed the cubes labeled A, B, and C? Explain.

Extra Practice page H57, Set D

Find the area. (pp. 484–490)

1.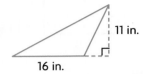
11 in.
16 in.

2.
13 yd
15 yd

3.
28 cm
34 cm

4.
$2\frac{1}{2}$ m
$6\frac{1}{2}$ m

5.
9.1 cm
6.5 cm

6.
2 ft
20 in.

7.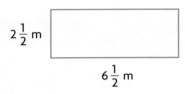
36 cm
32 cm
48 cm

8.
2 ft
1 yd

9.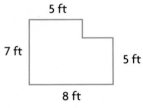
5 ft
7 ft
5 ft
8 ft

Find the area of each circle. Round to the nearest tenth. Use 3.14 or $\frac{22}{7}$ for π. (pp. 492–493)

10. $r = 6$ m

11. $r = 12$ ft

12. $d = 18$ cm

13. $d = 10.5$ m

14. $r = 1\frac{1}{2}$ yd

15. $d = 15$ in.

16. $r = 2.9$ cm

17. $d = 8.4$ m

Find the surface area. (pp. 494–497)

18.
7 yd
5 yd
5 yd

19.
1.5 m
2 m
4 m

20.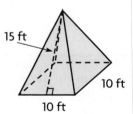
23 cm
8 cm
8 cm

21.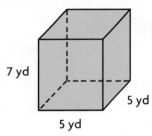
5.2 m
5.2 m
5.2 m

22.
15 ft
10 ft
10 ft

23.
3 in.
2 in.
4 in.

24. Deanna wants to plant grass in three circular plots in her garden. The three plots have diameters of 9 ft, 8 ft, and 3 ft. To the nearest square foot, how much grass must she buy? Use 3.14 for π. (pp. 492–493)

25. The area of the base of a square prism is 25 ft^2. Each of the 4 faces has a length of 6 ft. What is the surface area of the prism? Explain. (pp. 494–497)

Look for important words.
See item **7.**

Important words are *rounded to the nearest tenth*. You need to remember to round your answer.

Also see problem **2**, p. H62.

Choose the best answer.

1. The area of a rectangular closet is 42 square feet. The width of the closet is 6 feet. What is the length of the closet?

   A 7 ft   **C** 14 ft

   B 12 ft   **D** 36 ft

2. What is the area of a triangle with a base of 3.2 meters and a height of 1.5 meters?

   F 1.8 m²   **H** 2.75 m²

   G 2.4 m²   **J** 3.2 m²

3. The diameter of a circular skydiving target is 20 meters. What is the area of the target? Use 3.14 for π.

   A About 60 m²   **C** About 187 m²

   B About 122 m²   **D** About 314 m²

For 4–5, use this diagram.

4. Which numerical expression can be used to find the surface area of the figure?

   F $9 \times 4 \times 4$

   G $4(9 \times 4) + 2(4 \times 4)$

   H $2(9 \times 4) + 4(4 \times 4)$

   J $9(2 \times 4) + (9 \times 4) + (4 \times 4)$

5. What is the surface area of the solid?

   A 36 in.²   **C** 144 in.²

   B 72 in.²   **D** 176 in.²

6. What is the area of the parallelogram?

   F 37 in.

   G 74 in.

   H 150 in.²

   J 300 in.²

7. What is the area of the trapezoid, rounded to the nearest tenth?

   A 26.7 cm²

   B 67.7 cm²

   C 85.3 cm²

   D 670.3 cm²

8. The simple interest rate paid on a principal of $600 is 6%. How much interest will be earned in 1 year?

   F $3.60   **H** $360

   G $36   **J** Not here

9. The figures are similar. What is the unknown length?

   A $x = 2.4$ cm

   B $x = 12$ cm

   C $x = 20$ cm

   D $x = 24$ cm

10. What is the value of $4m - 3 \times 2$ for $m = 3$?

   F 6

   G 9

   H 18

   J Not here

Bars of gold are called
bullion. To make bullion, gold is
heated to the point at which it melts.
The liquid gold is then poured into a mold.
A typical 400-oz gold bar has the approximate
dimensions shown in the diagram below. If 3 of these
gold bars were stacked one on top of the other, how
tall would the stack of bars be?

$1\frac{1}{2}$ in.

3 in.

10 in.

Check What You Know

Use this page to help you review and remember important skills needed for Chapter 27.

☑ Vocabulary

Choose the best term from the box.

base
cube
dimensions
height
square

1. The length and width of a rectangle are the __?__ .

2. Area is measured in __?__ units.

3. A rectangular prism has length, width, and __?__ .

☑ Find the Cube of a Number (See p. H27.)

Find the value.

4. 2^3 5. 4^3 6. 5^3 7. 3^3 8. 8^3

9. 6^3 10. 1.1^3 11. 0.2^3 12. 10^3 13. 12^3

☑ Areas of Squares, Rectangles, and Triangles (See p. H28.)

Find the area of each figure.

14. 9 cm, 6 cm

15. 5 ft, 8 ft

16. 15 mi, 15 mi

17. 7 m, 6 m

18. 60.3 mm, 3.62 cm

19. $4\frac{3}{4}$ ft, $12\frac{1}{3}$ ft

20. 4 ft, $5\frac{1}{3}$ yd

21. 5.6 m, 18.7 m

☑ Areas of Circles (See p. H28.)

Find the area to the nearest centimeter.

22. 4 cm

23. 6.7 cm

24. 13 cm

25. 21 cm

LOOK AHEAD

In Chapter 27 you will
- estimate and find the volumes of prisms, pyramids, and cylinders

Estimate and Find Volume

Learn how to estimate and find the volumes of rectangular prisms and triangular prisms.

Vocabulary

volume

Jean wants to estimate how much sand will fill a sandbox. To estimate, Jean needs to think of the volume of the sandbox.

The **volume** is the number of cubic units needed to occupy a given space.

Jean will use a net to make an open box to estimate the volume.

Activity

You need: box net, scissors, tape, 30 centimeter cubes

• Cut out the net. Fold along the dashed lines, and tape the sides to make an open box.

• Estimate how many cubes will fit in the box. Put as many cubes as you can in the box.

• Is your estimate greater than or less than the actual number of cubes you put in the box?

You can visualize how many cubes will fill a prism.

EXAMPLE 1

Use the centimeter cube to estimate the volume.

About 5 cubes fit along the length and about 3 cubes along the width. The bottom layer has about 15 cubes. There are about 4 layers of about 15 cubes each.

4 layers × 15 cubes = 60 cubes

So, the volume is about 60 cubic cm, or 60 cm³.

CALIFORNIA STANDARDS MG 1.0 Students deepen their understanding of the measurement of plane and solid shapes and use this understanding to solve problems. **MG 1.3** Know and use the formulas for the volume of triangular prisms and cylinders, compare these formulas and explain the similarity between them and the formula for the volume of a rectangular solid. *also* **MR2.0, MR 2.5**

A layer of centimeter cubes has been placed on the base of the rectangular prism below. It takes 10, or 5 × 2, centimeter cubes to make the bottom layer.

The picture below shows the prism filled with centimeter cubes.

There are 3 layers of 10 cubes each. It takes 30, or 5 × 2 × 3, cubes to fill the prism.

• Look at the table below. What relationship do you see among the length, width, height, and volume?

LENGTH	WIDTH	HEIGHT	VOLUME
5	2	3	30
3	3	6	54
7	4	3	84

The relationship between the dimensions and the volume of each prism can be written as Volume = length × width × height, or $V = lwh$.

The formula $V = Bh$ can also be used to find the volume of a rectangular prism. In this formula, B is equal to $l \times w$, which is the area of the base of the prism, and h is the height of the prism.

EXAMPLE 2

Find the volume of the prism at the right.

$V = Bh$ where $B = l \times w$ *Write the formula.*

$V = (12 \times 5) \times 2$ *Replace B with 12 × 5 and h with 2.*

$V = 120$ *Multiply.*

So, the volume of the rectangular prism is 120 in.³

• **What if** you cut the rectangular prism into two congruent triangular prisms? What would be the volume of one triangular prism?

503

Remember that a triangular prism has bases that are congruent triangles.

The volume of a triangular prism is one half the volume of a rectangular prism with the same length, width, and height. To find the volume of a right triangular prism, you can use the formula $V = \frac{1}{2}lwh$. To find the volume of any triangular prism, you can use the formula $V = Bh$ where B is the area of the triangular base.

EXAMPLE 3

Find the volume of the prism at the right.

One Way

$V = \frac{1}{2}lwh$

$V = \frac{1}{2} \times 2.8 \times 4.2 \times 5$

$V = 29.4$

Another Way

$V = Bh$

$V = \left(\frac{1}{2} \times 2.8 \times 4.2\right) \times 5$

$V = 5.88 \times 5$

$V = 29.4$

So, the volume is 29.4 m³.

• **What if** the dimensions of the prism were tripled? What would the volume be?

CHECK FOR UNDERSTANDING

Think and Discuss ▶ **Look back at the lesson to answer each question.**

1. **Tell** how to find the volume of a cereal box that measures 26 cubes by 3 cubes by 18 cubes.

2. **Explain** how to find the height of a rectangular prism if you know the length, width, and volume of the prism.

Guided Practice ▶ **Find the volume.**

3. 3 in. 4 in. 2 in.

4. 6 in. 6 in. 10 in.

5. 4 cm 3 cm 8 cm

PRACTICE AND PROBLEM SOLVING

Independent Practice ▶ **Find the volume.**

6. 2 cm 5 cm 3 cm

7. 3 ft 12 ft 9 ft

8. 10 m 8 m 18 m

9. 3 ft 14 ft 7 ft

10. 7.2 cm 4.5 cm 3.7 cm

11. 4 in. 9 in. 8 in.

Find the unknown length.

12.
9 cm
x
5 cm
V = 900 cm³

13.
x
8.2 m   9 m
V = 369 m³

14. 6 cm   6 cm
x
V = 162 m³

Problem Solving ▶
Applications

15. Bobby orders sand to fill the long-jump pit. The pit is 14 ft long, 9 ft wide, and $1\frac{1}{2}$ ft deep. How much sand does Bobby need to order?

16. *REASONING* How many different rectangular prisms with whole-number dimensions have a volume of 8 ft³? Write the dimensions of each prism.

17. Juanita made a model using the scale 1 in. = 5 ft. The actual length is 200 ft. What is the model's length?

MIXED REVIEW AND TEST PREP

18. Find the surface area of a rectangular prism 8.2 ft long, 6.4 ft wide, and 4.5 ft high. (p. 494)

19. Evaluate $3y^2$ for $y = 3.9$. (p. 282)   **20.** Solve $8.1t = 49.41$. (p. 82)

21. **TEST PREP** Which type of angle is formed by two perpendicular lines? (p. 330)

   A straight   **B** obtuse   **C** right   **D** acute

22. **TEST PREP** The tax rate is 6.5%. The price is $19.99. Which shows the sales tax? (p. 408)

   F $0.13   **G** $1.30   **H** $13.00   **J** $130.00

LiNKUP to Reading

Analyze Information When you analyze details in a problem, look for information that is needed to solve the problem.

Calvin is mailing a model car to a friend. To make sure the model box does not move around in the mailing box, he must fill the empty space with packing material. How many cubic inches of packing material will he need?

10 in.
8 in.
6 in.
Model Box

12 in.
12 in.
10 in.
Mailing Box

1. What information do you need to solve the problem?

2. Solve the problem. Explain how you solved it.

3. **What if** Calvin uses a mailing box that is 1 ft by $\frac{3}{4}$ ft by $\frac{3}{4}$ ft? Will he have more or less space to fill? How much more or less?

PROBLEM SOLVING STRATEGY
Make a Model

Analyze
Choose
Solve
Check

Learn how to use the strategy *make a model* to solve problems.

A rice company uses different-sized boxes. A small box of rice is 2 in. wide, 4 in. long, and 6 in. high. How does the volume of the small box change when the length, width, and height are halved to make a sample box?

Analyze   What are you asked to find?

What information are you given?

Is there information you will not use? If so, what?

Choose   What strategy will you use?

You can use the strategy *make a model*.

Solve   How will you solve the problem?

Make a model of each box. Use a ratio to compare the volumes.

Use cubes to make a model of each box. Count the cubes to find the volume of each box.

6 in.
4 in.   2 in.

3 in.
2 in.   1 in.

The volume of the small box is 48 in.³
The volume of the sample box is 6 in.³

Now write a ratio to compare the volumes.

$$\frac{sample\ box}{small\ box} = \frac{6}{48} = \frac{1}{8}$$

So, the volume of the sample box is 6 in.³, or $\frac{1}{8}$ the volume of the small box.

Check   How can you check your answer?

What if the dimensions of the small box are doubled to make a larger box? How will the volume of the box change?

CALIFORNIA STANDARDS MG 1.0 Students deepen their understanding of the measurement of plane and solid shapes and use this understanding to solve problems. **MR 2.4** Use a variety of methods, such as words, numbers, symbols, charts, graphs, tables, diagrams, and models, to explain mathematical reasoning. *also* O⊷ NS 1.0, O⊷ NS 1.2, MR 1.0, MR 2.0

PROBLEM SOLVING PRACTICE

PROBLEM SOLVING STRATEGIES

Draw a Diagram or Picture
▶ **Make a Model**
Predict and Test
Work Backward
Make an Organized List
Find a Pattern
Make a Table or Graph
Solve a Simpler Problem
Write an Equation
Use Logical Reasoning

Solve the problem by making a model.

1. The volume of the box at the right is 240 in.^{3} Double the height of the box. How does the volume of the box change?

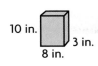

10 in. 3 in. 8 in.

2. The dimensions of the large box at the right are halved to make a smaller box. How does the volume change?

12 in. 6 in. 8 in.

For 3–4, a rectangular box is 6 cm wide, 8 cm long, and 13 cm high.

3. The length and width of the rectangular box are halved. The height remains the same. How does the volume of the smaller box compare to the volume of the larger box?

   A $\frac{1}{8}$ of the volume   **C** $\frac{1}{3}$ of the volume

   B $\frac{1}{4}$ of the volume   **D** $\frac{1}{2}$ of the volume

4. The length, width, and height of the rectangular box are doubled. How does the volume of the larger box compare to the volume of the smaller box?

   F equal to the volume

   G twice the volume

   H four times the volume

   J eight times the volume

MIXED STRATEGY PRACTICE

Use Data **For 5–6, use the line graph at the right.**

5. About how much did Howie make in sales during the six months shown on the graph?

6. At what point were sales and costs the same?

7. Grander's Clothes Shop sells shirts for $12 and shorts for $15. Terance will spend $120. How many different combinations of shirts and shorts can Terance buy?

8. Ana makes a circular pattern of stars on a piece of paper. Successive stars are the same distance apart. The sixth star is directly opposite the eighteenth star. How many stars form the circle?

10. A side of square A is 6 times the length of a side of square B. How many times as great is the area of square A than the area of square B?

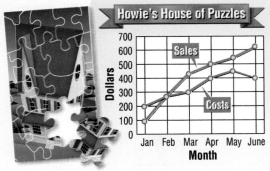

Howie's House of Puzzles

Sales

Costs

Dollars — 700, 600, 500, 400, 300, 200, 100, 0

Month — Jan Feb Mar Apr May June

9. Jasmine, Hugh, Sharon, and Mike are standing in line. A girl is not first or last in line. Mike is before Hugh. Jasmine is directly in front of Hugh. Who is first, second, third, and fourth in line?

11. **? What's the Question?** The volume of a cube is 216 cm^3. The length of each side is halved. The answer is 27 cm^3.

ALGEBRA
Volumes of Pyramids

Learn how to use a formula to find the volumes of pyramids.

$\mathbf{Q}$uang wants to use what he knows about the volume of a prism to find the volume of his model of the Great Pyramid of Khufu.

When it was built, the Great Pyramid of Khufu was 145.75 m (481 ft) high. Each side measured 229 m (751 ft) in length.

Activity

You need: prism and pyramid nets, tape, rice

- Cut out the nets for the prism and the pyramid. Fold along the dashed lines and use tape to make an open prism and an open pyramid.

- Compare the height of the prism with the height of the pyramid.

- Compare the base of the prism with the base of the pyramid.

- Fill the pyramid with rice. Pour the rice into the prism. Repeat until the prism is full.

- What is the relationship between the heights and bases of the prism and pyramid?

- What is the relationship between the volume of the pyramid and the volume of the prism?

To find the volume of a pyramid, use the formula $V = \frac{1}{3}Bh$.

EXAMPLE

Find the volume of Quang's pyramid model at the right to the nearest tenth.

$V = \frac{1}{3}Bh$, where $B = l \times w$ *Write the formula.*

$V = \frac{1}{3}(23 \times 23) \times 14.6$ *Replace B with 23 × 23 and h with 14.6.*

$V = \frac{1}{3} \times 529 \times 14.6$

$V = \frac{1}{3} \times 7{,}723.4$ *Multiply.*

$V = 2{,}574.5$ So, the volume is 2,574.5 cm³.

CALIFORNIA STANDARDS MG 1.0 Students deepen their understanding of the measurement of plane and solid shapes and use this understanding to solve problems. **MG 1.3** Know and use the formulas for the volume of triangular prisms and cylinders (area of base x height); compare these formulas and explain the similarity between them and the formula for the volume of a rectangular solid. *also* ○━┓NS 2.0, MR 2.0, MR 2.5, MR 2.7, MR 3.1

Think and ▶
Discuss

Look back at the lesson to answer the question.

1. **Compare** the formulas for the volume of a prism and the volume of a pyramid.

Guided ▶
Practice

Find the volume.

2. 6 in.
3 in.
5 in.

3. 9 m
4 m
4 m

4. 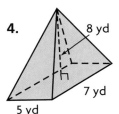 8 yd
7 yd
5 yd

Independent ▶
Practice

Find the volume.

5. 8 cm
7 cm
12 cm

6. 2 yd
5 yd
5 yd

7. 30 mm
40 mm
6 cm

8. square pyramid:
   $l = 12$ cm, $w = 12$ cm,
   $h = 25$ cm

9. rectangular pyramid:
   $l = 18$ ft, $w = 25$ ft,
   $h = 30$ ft

10. rectangular pyramid:
    $l = 10$ dm, $w = 4$ dm,
    $h = 7.5$ dm

11. rectangular pyramid: $l = 9$ ft,
    $w = 5$ yd, $h = 4\frac{4}{5}$ yd

Problem Solving ▶
Applications

12. The weight of a cubic foot of steel is 490 lb. What is the weight of a solid steel rectangular pyramid that is 2 ft by 3 ft by 18 in.?

 7 cm
6 cm
4 cm

13. ✎ **Write About It**   Tell how you would find the volume for the triangular pyramid at the right.

14. Arlena has $8.40 in nickels, dimes, and quarters. She has the same number of each coin. How many of each coin does she have?

15. $13\frac{1}{5} \times 8\frac{1}{3}$ (p. 206)

16. $4\frac{1}{2} + 3\frac{4}{5}$ (p. 186)

17. Write 79% as a decimal. (p. 60)

18. Volume = ▨ (p. 504)

 12 in.
11 in.
3 in.

19. **TEST PREP**   Which list shows $\frac{3}{4}, \frac{2}{3}$, and $\frac{5}{8}$ in order from least to greatest? (p. 166)

A $\frac{2}{3}, \frac{5}{8}, \frac{3}{4}$

B $\frac{5}{8}, \frac{3}{4}, \frac{2}{3}$

C $\frac{5}{8}, \frac{2}{3}, \frac{3}{4}$

D $\frac{2}{3}, \frac{3}{4}, \frac{5}{8}$

Volume of a Cylinder

MATH LAB

In geometry, a cylinder is a solid figure that has two congruent, parallel circles as its bases. In the activity below, you will find the volume of a cylinder.

Explore how to use manipulatives to find the volume of a cylinder.

You need 2 pieces of centimeter graph paper, scissors, tape.

Activity

- Use one piece of graph paper to cut out a rectangle 20 cm long and 8 cm wide. Without overlapping, roll the rectangle the short way and tape it to form a cylinder.

20 cm

8 cm

- Stand the cylinder on the second piece of graph paper. Trace around the base of the cylinder. Count the whole centimeter squares and the parts of centimeter squares inside the tracing.

8 cm

Think and Discuss

- About how many cubes would fit on the bottom layer of the cylinder?

- About how many layers of cubes can fit inside the cylinder?

- What is the approximate volume of the cylinder?

TECHNOLOGY LINK

More Practice: Use E-Lab, *Exploring the Volume of a Cylinder.*
www.harcourtschool.com/elab2002

Practice

Draw the following rectangles on centimeter graph paper. Cut out each one, roll it, and tape it to make a cylinder. Find the volume of each cylinder.

1.

6 cm

12 cm

2.

4 cm

18 cm

CALIFORNIA STANDARDS MG 1.0 Students deepen their understanding of the measurement of plane and solid shapes and use this understanding to solve problems. **MG 1.3** Know and use the formulas for the volume of triangular prisms and cylinders (area of base x height); compare these formulas and explain the similarity between them and the formula for the volume of a rectangular solid. *also* **MR 2.0, MR 2.5, MR 3.0, MR 3.2, MR 3.3**

Volumes of Cylinders

Learn how to find the volume of a cylinder.

You can find the volume of a rectangular prism by multiplying the area of the base by the height. You can use the same method to find the volume of a cylinder.

Math Idea ▶ Volume of a cylinder = area of base × height

$V = Bh$

$V = \pi r^2 h$ πr^2 is the area of the base, which is a circle.

EXAMPLE 1

Larisa has a cylindrical flower vase like the one shown at the right. To the nearest cubic inch, what is the volume of water the vase will hold?

4 in.

15 in.

$V = \pi r^2 h$	*Write the formula.*
$V \approx 3.14 \times 4^2 \times 15$	*Replace π with 3.14 or $\frac{22}{7}$,*
$V \approx 3.14 \times 16 \times 15$	*r with 4, and h with 15.*
$V \approx 753.6$	*Multiply.*

So, the volume of water is about 754 in.³

The volume of a cylinder can also be found if you know the diameter and the height of the cylinder.

EXAMPLE 2

Find the volume to the nearest cubic centimeter.

$5 \div 2 = 2.5$	*Find the radius.*
$V = \pi r^2 h$	*Write the formula.*
$V \approx 3.14 \times (2.5)^2 \times 15$	*Replace π with 3.14 or $\frac{22}{7}$,*
$V \approx 3.14 \times 6.25 \times 15$	*r with 2.5, and h with 15.*
$V \approx 294.375$	*Multiply.*

So, the volume is about 294 cm³.

5 cm

15 cm

• **What if** you have a cylinder with a diameter of 6.5 cm and a height of 12.5 cm? What is the volume of the cylinder?

CALIFORNIA STANDARDS MG 1.0 Students deepen their understanding of the measurement of plane and solid shapes and use this understanding to solve problems. **MG 1.3** Know and use the formulas for the volume of triangular prisms and cylinders (area of base × height); compare these formulas and explain the similarity between them and the formula for the volume of a rectangular solid. *also* O─ᴎ NS 2.0, MG 1.1, MR 2.5, MR 2.7, MR 3.1

511

Sometimes you may have to find the volume of part of a cylinder.

EXAMPLE 3

Cylinder

Look at the water tank at the right. The wall is 1.2 m thick. To the nearest meter, how much water can the inside cylinder hold?

Find the radius of the inside cylinder.

$2.7 - 1.2 = 1.5$ ← radius

Use the radius to find the volume of the inside cylinder.

$V \approx 3.14 \times (1.5)^2 \times 4.3$ *Replace π with 3.14, r with 1.5, and h with 4.3.*

$V \approx 3.14 \times 2.25 \times 4.3$

$V \approx 30.3795$ *Multiply.*

So, the water tank can hold about 30 m³ of water.

2.7 m   1.2 m

4.3 m

CHECK FOR UNDERSTANDING

Think and Discuss ▶ **Look back at the lesson to answer each question.**

1. **What if** the dimensions of Larisa's vase are doubled? What is the volume of the vase to the nearest cubic inch?

2. **Explain** how the formula for the volume of a cylinder is similar to the formula for the volume of a rectangular prism.

3. **Explain** in your own words which parts of a cylinder are represented by πr^2 and h in the formula $V = \pi r^2 h$.

Guided Practice ▶ **Find the volume. Round to the nearest whole number.**

4.

4 cm
7 cm

5. 7 in.

9 in.

6.

13 in.
8 in.

PRACTICE AND PROBLEM SOLVING

Independent Practice ▶ **Find the volume. Round to the nearest whole number.**

7.
10 in.

12 in.

8.

4.7 m
1.6 m

9.
10.3 cm
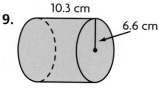
6.6 cm

Find the volume. Round to the nearest whole number.

10. $5\frac{1}{2}$ in. $3\frac{1}{2}$ in.

11. 6 cm 11 cm

12. 5 m 2.2 m

Find the volume of the inside cylinder to the nearest whole number.

13. 12 cm 6 cm 21 cm

14. 13 in. 9 in. 8 in.

15. 37 ft 17 ft 9 ft

Problem Solving ▶
Applications

16. Nicole bought a decorative outdoor planter with a diameter of 21 in. and a height of 28 in. She wants to fill it with potting soil. How much potting soil will Nicole need to the nearest cubic inch?

17. How does the volume of a cylinder with a radius of 10 ft and a height of 5 ft change when the dimensions are doubled?

18. **(?) What's the Error?** A circular swimming pool has a diameter of 20 ft and a height of 6 ft. Mr. Alvarez determined that it would take 7,536 cubic feet of water to fill the pool. What did he do wrong? What is the solution?

19. **Write a problem** about finding the volume of a real-life object shaped like a cylinder. Share with your classmates.

20. A large bag of potting soil weighs 40 lb and costs $3.79. Julio has $30 to buy as many of the large bags of potting soil as he can. How many pounds of potting soil can he buy?

MIXED REVIEW AND TEST PREP

21. Find the volume to the nearest foot of a pyramid 1.2 ft long, 2.8 ft wide, and 3.6 ft high. (p. 508)

22. Write the algebraic expression for 6.9 more than 5 times d. (p. 274)

23. Which has the greater value, $\frac{1}{10} \div \frac{1}{2}$ or $\frac{1}{2} \div \frac{1}{10}$? (p. 210)

24. $^{-}50 \times ^{-}11$ (p. 258)

25. **TEST PREP** Karl made a model, using the scale 1 in. = 4 ft. The actual length is 200 ft. Which is the length of the model? (p. 387)

    A 30 in.   **B** 40 in.   **C** 50 in.   **D** 500 in.

Extra Practice page H58, Set C

513

Find the volume. (pp. 502–505)

1.
5 in. 12 in. 7 in.

2.
15 cm 13 cm 7 cm

3.
$7\frac{1}{2}$ yd $3\frac{1}{2}$ yd 15 yd

4.
0.8 cm 0.8 cm 0.8 cm

5.
7 ft 10 ft 8 ft

6.
15 yd 5.6 m 9.3 m 7.8 m

For 7–12, find the volume of the pyramid. (pp. 508–509)

7.
8 in. 6 in. 6 in.

8.
11 ft 9 ft 27 ft

9.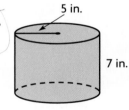
22.8 m 17 m 11.4 m

10. length = 5 m
width = 9 m
height = 10 m

11. base = 25.5 ft^2
height = 12 ft

12. base = 216 m^2
height = 24 m

Find the volume. Round to the nearest whole number. (pp. 511–513)

13.
4 m 9.6 m

14.
8 cm 3.8 cm

15.
5 in. 7 in.

Find the volume of the inside cylinder to the nearest whole number. (pp. 511–513)

16.
10 in. 3 in. 16 in.

17.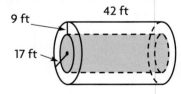
9 ft 42 ft 17 ft

18.
18 m 9 m 15 m

Solve. (pp. 506–507)

19. A box is 8 in. high, 2 in. wide, and 4 in. long. The length and the width of the box are doubled to make a larger box. How does the volume of the larger box compare to the volume of the smaller box?

20. A box is 12 cm long, 8 cm wide, and 9 cm tall. Each dimension is halved to make a smaller box. How does the volume of the smaller box compare to the volume of the larger box?

Choose the answer.

See item **10.**

You need to find the sale price, but first you need to find the amount of the discount. Be sure that the answer you choose answers the question asked.

Also see problem **6**, p. H64.

Choose the best answer.

1. Seth needs a box that has a volume of 6,000 cm^3. Which dimensions could the box have?

  A $l = 25$ cm, $w = 20$ cm, $h = 12$ cm

  B $l = 20$ cm, $w = 30$ cm, $h = 5$ cm

  C $l = 6$ cm, $w = 5$ cm, $h = 20$ cm

  D $l = 3$ cm, $w = 2$ cm, $h = 10$ cm

2. Which unit of measure below is the only one that could be used for the volume of a box?

  F in.   **H** cm^2

  G cm   **J** ft^3

3. Which is the volume of the figure?

  A 9 in.^{3} **C** 72 in.^{3}

  B 48 in.^{3} **D** 144 in.^{3}

4. What is the volume of the barrel shown below?

  F About 76 ft^3 **H** About 128 ft^3

  G About 84 ft^3 **J** About 141 ft^3

5. What is the volume of the prism shown?

  A 6.82 m^3 **C** 7.4 m^3

  B 7.3 m^3 **D** 8.36 m^3

6. Which is a reasonable estimate for the volume of the cylinder?

  F 10,000 ft^3 **H** 600 ft^3

  G 1,000 ft^3 **J** 60 ft^3

7. Which is the best estimate for the volume of a square pyramid with a height of 5 cm and a base of 4.3 cm on a side?

  A 7 cm^3 **C** 30 cm^3

  B 20 cm^3 **D** 90 cm^3

8. What is the median of the data?
30, 25, 40, 35, 25

  F 25   **H** 31

  G 30   **J** 40

9. Ian's age is 3 years greater than twice Barb's age. Ian is 27 years old. Which of the following equations can be used to find b, Barb's age?

  A $b + 5 = 27$ **C** $3 + b = 27$

  B $2b = 27$ **D** $2b + 3 = 27$

10. The regular price of a jacket is $56. If it is discounted 30%, what is the sale price?

  F $16.80   **H** $34.50

  G $22.60   **J** Not here

MATH DETECTIVE

Missing Pieces

Solve each mystery by finding the missing pieces that satisfy the given conditions.

Mystery Number 1

The volume of a rectangular prism is 60 in.³ Give two different sets of dimensions for the length, width, and height of the figure.

LENGTH	WIDTH	HEIGHT	VOLUME
			60 in.³
			60 in.³

Mystery Number 2

The volume of a cube is 64 m³. What is the surface area of the cube?

LENGTH	WIDTH	HEIGHT	VOLUME	SURFACE AREA
			64 m³	

Mystery Number 3

The surface area of a cube is 150 m². What is the volume of the cube?

LENGTH	WIDTH	HEIGHT	VOLUME	SURFACE AREA
				150 m²

Think It Over!

Write About It   Explain what happens to the volume if you double all the dimensions of either of the rectangular prisms in Mystery Number 1.

Stretch Your Thinking

A rectangular prism has three whole-number dimensions, two of which are the same. If the volume of the rectangular prism is 36 cm³, what is the greatest surface area the prism can have?

Learn how to use a network to find all possible routes and to find the shortest route.

A **network** is a graph whose edges are labeled with numbers. In the network at the right, the cities are the vertices, and the connecting segments between them are the edges. The estimated distances between the four cities are given in miles.

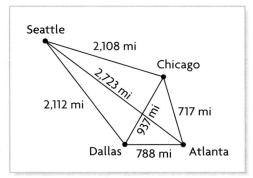

Seattle — 2,108 mi — Chicago
2,723 mi
2,112 mi
717 mi
937 mi
Dallas — 788 mi — Atlanta

You can use a network to help you find the shortest route between locations.

EXAMPLE

Use the distances shown on the network. Find the shortest route starting in Seattle that includes all the cities.

Let S = Seattle, A = Atlanta, C = Chicago, and D = Dallas.

SACD = 2,723 + 717 + 937 = 4,377   *List the different routes*
and the number of miles
SADC = 2,723 + 788 + 937 = 4,448   *for each one.*

SCAD = 2,108 + 717 + 788 = 3,613

SCDA = 2,108 + 937 + 788 = 3,833

SDAC = 2,112 + 788 + 717 = 3,617

SDCA = 2,112 + 937 + 717 = 3,766   *Compare the distances.*

The shortest route, SCAD, is 3,613 mi.

TALK ABOUT IT

How can you use a network to find the shortest route? Explain.

TRY IT

Find all the possible routes starting from A that include every vertex. Find the distance for each route. Distances are in kilometers. Figures are not drawn to scale.

1.

2.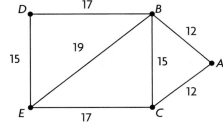

3. In Exercise 1, what is the shortest route? the longest route?

4. In Exercise 2, what is the shortest route? the longest route?

VOCABULARY

1. The sum of the areas of the faces of a solid figure is called the __?__ .
   (p. 494)

2. The __?__ is the number of cubic units needed to occupy a given space. (p. 502)

EXAMPLES	EXERCISES

Chapter 24

• **Convert units of measurement.** (pp. 452–455)

Change 20 ounces to pounds.

$\frac{1}{16} = \frac{x}{20}$ *Use the relationship 1 lb = 16 oz to write a proportion.*

$16x = 20$ *Find the cross products. Solve for x.*

$x = 1.25$, or $1\frac{1}{4}$ So, 20 oz = $1\frac{1}{4}$ lb.

Use a proportion to convert to the given unit.

3. 32 qt = ■ gal 4. 15,000 lb = ■ T

5. 15 ft = ■ in. 6. 42 m = ■ cm

7. 56 oz = ■ lb 8. 50 m = ■ km

9. 0.9 kg = ■ g 10. 60 km = ■ m

• **Convert customary and metric measurements.** (pp. 456–457)

Change 6 inches to centimeters.

$\frac{1}{2.54} = \frac{6}{x}$ *Use the relationship 1 in. ≈ 2.54 cm to write a proportion.*

$x = 6 \times 2.54$ *Find the cross products.*

$x = 15.24$ cm *Solve for x.*

So, 6 in. ≈ 15.24 cm.

Use a proportion to convert to the given unit. Use the table.

11. 4 gal ≈ ■ L

12. 20 kg ≈ ■ lb

13. 5 ft ≈ ■ cm

14. 8 mi ≈ ■ km

15. 15 m ≈ ■ yd

1 in. ≈ 2.54 cm
1 ft ≈ 30.48 cm
1 yd ≈ 0.91 m
1 mi ≈ 1.61 km
1 lb ≈ 0.454 kg
1 gal ≈ 3.79 L

Chapter 25

• **Find the perimeter of polygons.**

(pp. 469–471)

Add the lengths.

$5.6 + 3.8 + 8.7 + 4.2 = 22.3$ cm

Find the perimeter.

16.

17. a regular pentagon with sides of $12\frac{1}{4}$ in.

• **Find the circumference of circles.** (pp. 476–479)

Use the formula $C = 2\pi r$.
$C \approx 2 \times 3.14 \times 4.6$
$C \approx 28.888$ *Round to the nearest whole*
$C \approx 29$ cm *number.*

Find the circumference to the nearest whole number. Use $\pi = 3.14$ or $\pi = \frac{22}{7}$.

18.

19. $r = 42$ ft
20. $d = 125$ m

Chapter 26

• **Find the area of polygons.**
(pp. 484–490)

$A = l \times w$ *Write the formula.*

$A = 15 \times 13$ *Replace l with 15 and w with 13.*

$A = 195$ *Multiply.*

So, the area is 195 cm².

Find the area.

21.

22.

23.

24.
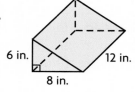

• **Find the area of circles.**
(pp. 492–493)

Use the formula $A = \pi r^2$.

$A \approx 3.14 \times 2.5 \times 2.5$

$A \approx 19.625$

$A \approx 20$ *Round to the nearest whole number.*

So, the area is about 20 cm².

Find the area to the nearest whole number. Use $\pi = 3.14$ or $\pi = \frac{22}{7}$.

25.

26.

27. $r = 26$ in.

28. $d = 42$ m

• **Find the surface area of prisms and pyramids.** (pp. 494–497)

$2 \times 12 \times 9 = 216$ *area of sides*

$2 \times 12 \times 16 = 384$ *area of front and back*

$2 \times 9 \times 16 = 288$ *area of top and bottom*

$216 + 384 + 288 = 888$

So, the surface area is 888 ft².

Find the surface area.

29.

30.

Chapter 27

• **Find the volume of prisms, pyramids, and cylinders.**
(pp. 502–513)

$V = B \times h$ *Base is a rectangle.*

$V = l \times w \times h$ *Area of a rectangle is $l \times w$.*

$V = 5.6 \times 7.4 \times 3.8$

$V = 157.472$

So, the volume is 157.472 cm³.

Find the volume.

31.

32.

33.

34.

PROBLEM SOLVING APPLICATIONS

35. Kwan framed a rectangular picture that measured 15 in. ×12 in. The framing strips were $2\frac{1}{2}$ in. wide. What was the perimeter of the frame? (pp. 472–473)

36. A box is 8 m × 6 m × 9 m. If each dimension is doubled, how does the volume change? (pp. 506–507)

Marine Mammals

Unlike fish, marine mammals are warm-blooded and are nourished by their mothers after they are born. Many species of marine mammals are found in the coastal waters off California, including seals, whales, and porpoises.

About 10,000 northern fur seals live in California. Many make their homes in Channel Islands National Park.

Use Data For 1–5, use the given measurements in the caption below. Find each northern fur seal pup measurement. Round all answers to the nearest hundredth.

1. Length in inches

2. Length in yards

3. Length in centimeters

4. Weight of female in ounces

5. Weight of male in ounces

6. Weight of female in kilograms

7. Weight of male in kilograms

Northern fur seal pups are about 1.5 ft long at birth. Males weigh about 12 lb; females weigh about 10 lb.

8. The blue whale, the world's largest animal, can sometimes be seen off the coast of California. At birth a blue whale calf weighs between 5,000 and 6,000 lb. Write the range of weights in tons and in kilograms.

9. Fur seals feed deep in the ocean. The deepest known dive of a northern fur seal is 230 m. Is this dive greater than or less than 750 ft? Explain.

10. **Stretch Your Thinking** Fur seals stay warm in cold ocean water because they are covered with thick, waterproof fur. Every square inch of fur contains an amazing number of hairs—about 370,000! How many hairs would you find on a section of fur measuring 1 ft square?

11. Groups of Dall's porpoises sometimes appear in Monterey Bay. They have been known to swim at speeds of 55 km per hour. About how many miles per hour is this? About how many miles per minute?

The Monterey Bay Aquarium

You can see more than 500 species of marine life at the Monterey Bay Aquarium, one of the largest aquariums in the world. The aquarium also features exhibits of natural communities such as tide pools and living kelp forests. At the aquarium's Outer Bay exhibit, you can watch tunas, sharks, barracudas, and stingrays through the world's largest window.

The aquarium's three-story Kelp Forest exhibit holds 335,000 gallons of water.

Use Data **For 1–5, use the drawing of the Outer Bay window.**

1. What is the perimeter of the viewing surface of the window?

2. What is the area of the viewing surface?

3. Classify the solid figure formed by the window.

4. What is the volume of the window in cubic feet? Explain.

5. ✍ **Write About It** Explain how you could find the volume of the window in cubic inches. Then find it.

56 ft

Viewing Surface

17 ft

13 in.

Use Data **For 6, use the table.**

6. You can use the formula $C = \frac{5}{9}(F - 32)$ to change Fahrenheit temperatures to Celsius. Find the Celsius equivalents of the three Fahrenheit temperatures.

TANK TEMPERATURES	
Inner Monterey Bay Exhibits	50°F
Outer Bay Exhibit	68°F
Mysteries of the Deep Exhibit	40°F

7. **What if** one of the exhibits has a tank in the shape of a rectangular prism? It measures 12 feet long by 15 feet wide by 8 feet high. There is no top to the tank. What is the surface area of the tank? Explain.

8. Suppose a tank measures 16 feet long by 12 feet wide by 10.5 feet high. How many cubic feet of water could the tank hold?

9. The aquarium pumps 2,000 gallons of sea water through its exhibits every minute. During the winter, the aquarium is open from 10 A.M. until 6 P.M. How many gallons of sea water are pumped during winter visiting hours?

10. Suppose that during the summer, the aquarium is open from 9:00 A.M. until 7:30 P.M. on Monday through Saturday, and from 10:00 A.M. until 6:00 P.M. on Sunday. How many gallons of sea water would be pumped during 1 week of summer visiting hours?

In 1992, 100 women from 20 countries set the women's skydiving record for the largest free-fall formation. Formation skydivers must plan what they want to do ahead of time. Some sketch out patterns. Others use tables to plan how many people join up at a time. They can solve equations to extend the patterns. Suppose 30 skydivers want to form linked rings. Each ring has 4 skydivers, one of whom is also part of the next ring. Complete the pattern shown in the table. Then show the pattern graphically with dots. How would you extend this pattern?

SKYDIVING LINKED RINGS						
Ring	1	2	3	4	5	6
Number of Skydivers	4	7	10	13	▪	▪

Check What You Know

Use this page to help you review and remember important skills needed for Chapter 28.

✓ Vocabulary

Choose the best term from the box.

equation
expression
variable

1. An algebraic or numerical sentence showing that two quantities are equal is a(n) __?__.

2. A mathematical phrase that combines operations, numerals, and variables to name a number is a(n) __?__.

✓ Compare Numbers (See p. H29.)

Compare. Write < or > for each ●.

3. 1.01 ● 1.1

4. $1\frac{1}{4}$ ● $1\frac{1}{2}$

5. $^-8$ ● $^-2$

6. 576 ● 479

7. 0.7 ● 0.35

8. 9.23 ● 9.31

9. 0 ● $^-2$

10. $2{,}108$ ● $2{,}110$

11. $^-3.1$ ● $^-2.9$

12. $\frac{2}{3}$ ● $\frac{3}{4}$

13. 10.1 ● 9.09

14. 0.5 ● 0.06

15. $2\frac{4}{5}$ ● $3\frac{1}{5}$

16. 3 ● $^-4.56$

17. 0.8 ● 0

18. $\frac{^-3}{4}$ ● $\frac{^-3}{8}$

✓ Input/Output Tables (See p. H29.)

Copy and complete each table.

19.

INPUT	OUTPUT
x	$x - 6$
16	■
27	■

20.

INPUT	OUTPUT
y	$y + 14.7$
12	■
12.5	■

21.

INPUT	OUTPUT
w	$w \div 4$
100	■
102	■

22.

INPUT	OUTPUT
a	$a \times 15$
10	■
23	■

✓ Evaluate Expressions (See p. H18.)

Evaluate each expression.

23. $x + 10$ for $x = 5$

24. $24 - y$ for $y = 2$

25. $20 \div z$ for $z = ^-4$

26. $8.9a$ for $a = ^-3$

27. $9s$ for $s = 7$

28. $w - 8$ for $w = 40$

29. $x^2 - {^-4}$ for $x = ^-5$

30. $(w - 7) \div 4$ for $w = ^-9$

31. $m \cdot \frac{1}{3}$ for $m = \frac{9}{14}$

32. $42 + k - 8$ for $k = 18$

33. $(w - 8) \div 2$ for $w = 14.2$

> **LOOK AHEAD**
>
> **In Chapter 28 you will**
> - identify, extend, and make number patterns in sequences
> - identify, extend, and make number patterns and write a rule for the pattern
> - identify, extend, and make patterns of geometric figures

PROBLEM SOLVING STRATEGY
Find a Pattern

Analyze
Choose
Solve
Check

Learn to solve problems by using the strategy *find a pattern*.

QUICK REVIEW

1. 6.25×4 2. 3.2×8
3. 2×7.4 4. 8.45×5
5. 6×3.86

The members of a skydiving team make formations in the air by holding on to each other. After one second, one person begins a formation. At three seconds, three others join with the first person. At five seconds, five more people join. At seven seconds, seven more join the group. If this pattern continues, how many people will be in the group after eleven seconds?

Analyze

What are you asked to find?

What information are you given?

Is there information you will not use? If so, what?

Choose

What strategy will you use?

You can use the strategy *find a pattern*.

Solve

How will you solve the problem?

Find the patterns for skydivers added and total skydivers.

Skydivers added	Total skydivers
1, 3, 5, 7, . . .	1, 4, 9, 16, . . .
Pattern: consecutive odd numbers	Pattern: squares of consecutive numbers

TECHNOLOGY LINK
More Practice: Use E-Lab, *Geometry and Number Patterns.*
www.harcourtschool.com/elab2002

Show the patterns in a table to find the number of skydivers in the formation after eleven seconds.

Time	1 sec	3 sec	5 sec	7 sec	9 sec	11 sec
Skydivers added	1	3	5	7	9	11
Total skydivers	1	4	9	16	25	36

So, there are 36 skydivers at 11 seconds.

Check

How can you check your answer?

What if there are 81 skydivers? At how many seconds are there 81?

CALIFORNIA STANDARDS O—n NS 2.0 Students calculate and solve problems involving addition, subtraction, multiplication, and division. **MR 1.1** Analyze problems by identifying relationships, distinguishing relevant from irrelevant information, identifying missing information, sequencing and prioritizing information, and observing patterns. *also* O—n NS 2.3, MR 2.4

PROBLEM SOLVING STRATEGIES

| Draw a Diagram or Picture |
| Make a Model |
| Predict and Test |
| Work Backward |
| Make an Organized List |
| ▶ **Find a Pattern** |
| Make a Table or Graph |
| Solve a Simpler Problem |
| Write an Equation |
| Use Logical Reasoning |

Solve the problem by finding a pattern.

1. Alex is beginning a new physical training program. He will do 7 push-ups the first day, 12 push-ups the second day, 17 push-ups the third day, and will continue until he does 42 push-ups in one day. If this pattern continues, on which day will Alex do 42 push-ups?

2. The school choir learns songs for a competition to be held in 7 weeks. The choir knows 5 songs after the first week, 8 after the second week, and 11 after the third week. If this pattern continues, how many songs will the choir know after 7 weeks?

For 3–4, use this information.

The air temperature decreases 8°C for every 1,500-m gain in altitude.

3. Which algebraic expression can be used to find the temperatures if the starting temperature is 20°C and n is the number of 1,500-m gains in altitude?

   A $20 - 8n$ **C** $20 + n$

   B $20 + 8n$ **D** $20 - n$

4. It was 20°C when Rachel began her ascent in a hot-air balloon. How many meters had the balloon gone up when the temperature was ⁻4°C?

   F 3 m   **H** 4,500 m

   G 3,000 m   **J** 6,000 m

MIXED STRATEGY PRACTICE

5. Paul bought previously viewed videos for $119.50. He paid $12.50 each for some and $8.00 each for others. He bought fewer than 8 videos at each price. How many videos did Paul buy for $12.50?

6. A side of square C is 8 times the length of a side of square A. How many times as great as the area of square A is the area of square C?

7. Light poles along a highway are placed at regular intervals. The distance from the first pole to the sixth pole is 300 ft. What is the distance from the twentieth pole to the thirty-first pole?

8. A card collection was divided among Michelle and her three brothers. Armand received $\frac{1}{3}$ of the cards, Ben received $\frac{1}{4}$ of the cards, and Denis received $\frac{1}{6}$ of the cards. Michelle received 345 cards. How many cards were in the collection before it was divided?

9. Marco has a garden that is 5 ft wide and 8 ft long. He decides to double the width. What is the difference in area of the gardens?

10. **?** **What's the Error?** To solve the proportion $\frac{15}{30} = \frac{x}{12}$, Sean writes the equation $15x = 12 \times 30$. What error did he make? What is the correct equation?

Patterns in Sequences

Learn how to recognize, describe, and extend patterns in sequences.

QUICK REVIEW

1. $5 + 2.1 + 6$ 2. $6 + 2.1 + 7$
3. $7 + 2.1 + 8$ 4. $8 + 2.1 + 9$
5. $9 + 2.1 + 10$

Vocabulary

triangular number

sequence

term

Mr. Fabano designed patterns for the members of his marching band to make when they perform. His patterns, shown below, are triangular arrays.

A number that can be represented by a triangular array is a **triangular number**.

triangular arrays →
triangular numbers →   1   3   6   10

The pattern of triangular numbers above can be written in a number sequence. A **sequence** is an ordered set of numbers. Each number in the sequence is called a **term**.

To get the next term in the triangular number sequence, you add 1 more than was added previously. Use this rule to find the next term in the sequence.

sequence →   1   3   6   10   ■
                 +2 +3 +4 +5

Since $10 + 5 = 15$, the next triangular number is 15.

Math Idea ▶ A sequence can have a repeating pattern that uses addition, subtraction, multiplication, division, or any combination of these operations.

EXAMPLE 1

Identify a pattern in the sequence 1, $5\frac{1}{2}$, 10, $14\frac{1}{2}$, 19, . . . and write the rule. Using your rule, find the next three terms in the sequence.

1     $5\frac{1}{2}$ 10 $14\frac{1}{2}$ 19

$+4\frac{1}{2}$ $+4\frac{1}{2}$ $+4\frac{1}{2}$ $+4\frac{1}{2}$

Look for a pattern. Compare each term with the next.

A rule is to add $4\frac{1}{2}$ to each term to get the next term.

$19 + 4\frac{1}{2} = 23\frac{1}{2}$ $23\frac{1}{2} + 4\frac{1}{2} = 28$ $28 + 4\frac{1}{2} = 32\frac{1}{2}$

So, $23\frac{1}{2}$, 28, and $32\frac{1}{2}$ are the next three terms.

CALIFORNIA STANDARDS O━┑NS 2.0 Students calculate and solve problems involving addition, subtraction, multiplication, and division. O━┑NS 2.3 Solve addition, subtraction, multiplication, and division problems, including those arising in concrete situations, that use positive and negative numbers and combinations of these operations. *also* MR 1.1, MR 2.4

Sequences that show a decreasing pattern often involve subtraction or division.

EXAMPLE 2

Find the next three possible terms in the sequence.

3,645; 1,215; 405; 135; . . .

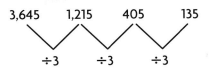

Look for a pattern. Compare each term with the next.

A rule is to divide each term by 3 to get the next term.

135 ÷ 3 = 45   *Start with 135 and divide by 3.*

45 ÷ 3 = 15

15 ÷ 3 = 5

So, 45, 15, and 5 are the next three terms.

You can use a rule to write a sequence.

EXAMPLE 3

James has $20.00. He spends $2.20 every day on his school lunch. Write a sequence to show how James will spend his $20.00. How much money will James have left after three days?

Start: $20.00

Rule: Subtract $2.20 from each term.

Day 1: $20.00 − $2.20 = $17.80   *Subtract $2.20 from each term*
Day 2: $17.80 − $2.20 = $15.60   *to find the next term.*
Day 3: $15.60 − $2.20 = $13.40

$20.00, $17.80, $15.60, $13.40, . . .   *Write the terms as a sequence.*

So, James will have $13.40 left after three days.

• **What if** James continues this pattern? How many more school lunches can he buy with the money left after three days?

CHECK FOR UNDERSTANDING

Think and Discuss ▶ **Look back at the lesson to answer each question.**

1. **Write** a rule to make the following sequence. Then find the next three terms. 8, 32, 128, 512, . . .

2. **Tell** whether the sequence 1,200; 240; 48 . . . is increasing or decreasing.

Guided Practice ▶ **Write a rule for each sequence.**

3. 5, 20, 35, 50, . . .   **4.** 0.001, 0.01, 0.1, 1, . . .

Find the next three possible terms in each sequence.

5. ⁻1, 10, 21, 32, . . . **6.** 0.2, 2, 20, 200, . . .

PRACTICE AND PROBLEM SOLVING

Independent ▶
Practice

Write a rule for each sequence.

7. 405, 135, 45, 15, . . . **8.** 17.5, 16.3, 15.1, 13.9, . . .

9. 7, 7.89, 8.78, 9.67, . . . **10.** 12, ⁻30, 75, ⁻187$\frac{1}{2}$, . . .

Find the next three possible terms in each sequence.

11. 35, 65, 125, 215, . . . **12.** 400, 200, 100, 50, . . .

13. 91, 90, 88, 85, . . . **14.** ⁻4, 40, ⁻400, 4,000, . . .

Use the rule to write a sequence.

15. Start with 9; add 3.7. **16.** Start with 5; multiply by ⁻6.

Problem Solving ▶
Applications

17. Mr. Amano bought a collector's gold coin for $138. The value of the coin increased the same amount each year he owned it. The sequence $192, $246, $300, $354, . . . shows the pattern of the increasing value. Write the rule for the pattern and find what the coin is worth the sixth year.

18. **REASONING** The sequence 1, 1, 2, 3, 5, 8, 13, . . . is called the Fibonacci Sequence. Write a rule for the pattern, and list the first 15 terms in the sequence. As the sequence continues, what other patterns do you find?

19. ✎ **Write a problem** involving a number sequence that starts with $\frac{1}{2}$ and uses multiplication to get the next number. Explain your rule.

20. The length of a rectangle is 4 more than 3 times the width. What is the perimeter of the rectangle if the width is 2 feet?

MIXED REVIEW AND TEST PREP

21. Find the volume of the cylinder to the nearest whole number. (p. 511)

22. What is the measure of ∠BEC at the right? (p. 326)

23. Evaluate $3 \times (7 + 4) \times 2^3$. (p. 44)

24. Solve. $\frac{h}{1.5} = 30$ (p. 82)

25. **TEST PREP** Jeff has $1,000 to landscape his lawn. He plans to spend 35% of it on new trees. Each tree costs $55. How many trees can he buy? (p. 402)

A 6 **B** 7 **C** 8 **D** 9

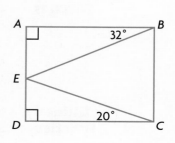

Number Patterns and Functions

Learn how to write an equation to represent a function.

Vocabulary

function

Evaluate each expression.

1. $a + 3.2$ for $a = 5.4$
2. $5.6 - r$ for $r = 1.2$
3. $0.25 \times m$ for $m = 4$
4. $q \div 0.3$ for $q = 9$
5. $y - 16$ for $y = 25.2$

Delia earns $5.25 per hour. She wants to buy a camera that costs $125. How many hours will Delia have to work to earn $125?

Use the pattern below.

HOURS WORKED	MONEY EARNED
1	$5.25
2	$10.50
3	$15.75
4	$21.00
5	$26.25

The pattern shows that for the first five hours, as the number of hours increases by one, the money earned increases by $5.25.

Delia used the pattern to write an equation in which h equals the number of hours worked and m equals the amount of money earned. She can use the equation to find how many hours she needs to work to earn $125.

TECHNOLOGY LINK

To learn more about number patterns and functions, watch the **Harcourt Math Newsroom Video** *British Code Breaker.*

$$m = 5.25h$$

$$125 = 5.25h \qquad \text{\textit{Replace m with 125.}}$$

$$\frac{125}{5.25} = \frac{5.25\,h}{5.25} \qquad \text{\textit{Solve the equation.}}$$

$$23.81 = h$$

So, Delia needs to work about 24 hr.

Math Idea ▶ A **function** is a relationship between two quantities in which one quantity depends on the other. Every input has exactly one output. Delia's equation is a function because the amount of money she earns, m, depends on the number of hours, h, that she works.

CALIFORNIA STANDARDS ○–¬NS 2.0 Students calculate and solve problems involving addition, subtraction, multiplication, and division. **MR 1.1** Analyze problems by identifying relationships, distinguishing relevant from irrelevant information, identifying missing information, sequencing and prioritizing information, and observing patterns. *also* ○–¬NS 2.3, AF 1.1, MR 1.3, MR 2.5

You can use a function table or an equation to represent some functions.

EXAMPLE 1

Mrs. Nien told her students that she used a pattern to determine the different sizes of paper to cut for them to use in class. The sizes, in inches, are shown below. Look for a pattern. Then write an equation to find the length when the width is 10 in.

width, w	3	4	5	6	7
length, l	7	9	11	13	15

Compare the length and the width.
The length is 1 in. more than 2 times the width.   ← pattern

$l = 2w + 1$ *Write the equation.*

Use the equation to find the length when the width is 10 in.

$l = (2 \times 10) + 1$ *Replace w with 10.*
$l = 20 + 1$
$l = 21$

So, the length is 21 in. when the width is 10 in.

• In Example 1, the value of l depends on what other value?

You can use an equation to find missing terms in a function table.

EXAMPLE 2

Write an equation to describe the function. Then use it to find the missing term.

x	1	2	3	4	5	6
y	1	4	9	16	25	■

THINK: Since each y-value is greater than the corresponding x-value, use either addition or multiplication. What number is each x-value multiplied by to get the y-value?

Each x-value is multiplied by itself.   ← pattern

The equation is $y = x^2$.

$y = x^2$
$y = 6^2$ *Replace x with 6.*
$y = 36$ ← $6^2 = 6 \times 6$

So, the missing term is 36.

• Find y when $x = 13$.

Think and Discuss ▶ **Look back at the lesson to answer each question.**

1. **Explain** how, in Example 1, you would find the value of w if you knew $l = 21$.

2. **Write** an equation for this function: the amount earned is $6 times the number of hours worked.

Guided Practice ▶ **Write an equation to represent the function.**

3.

a	10	9	8	7	6
b	4	3	2	1	0

4.

x	0	1	2	3	4
y	0	12	24	36	48

Independent Practice ▶ **Write an equation to represent the function.**

5.

c	1	2	3	4	5
d	2.1	3.1	4.1	5.1	6.1

6.

m	1	4	9	16	25
n	1	2	3	4	5

7.

w	1.2	1.3	1.4	1.5	1.6
l	4.8	5.2	5.6	6.0	6.4

8.

x	5	4	3	2	1
y	9	7	5	3	1

TECHNOLOGY LINK

More Practice: Use E-Lab, *Geometry and Number Patterns*.

www.harcourtschool.com/ elab2002

Write an equation to represent the function. Then use it to find the missing term.

9.

k	24	34	44	54
g	12	17	22	■

10.

s	11	14	17	20
r	22	■	28	31

11.

c	16	24	32	36
d	-4	-6	■	-9

12.

a	42	48	52	56
b	14	■	$17\frac{1}{3}$	$18\frac{2}{3}$

Problem Solving Applications ▶

13. Write an equation for each function. Tell what each variable you use represents.

   a. The length of a rectangle is 5 times its width.

   b. Distance is the product of the rate of travel and the time traveled.

   c. The cost of shipping a package is $1.25 per pound.

   d. The cost of riding in a cab is $3.00 plus $0.75 per mile.

14. Felicity wants to purchase a CD player for $115.99. She earns $5.25 per hour baby-sitting. Write an equation and find how many hours Felicity needs to baby-sit to buy the CD player.

15. **REASONING** In a function, every input has exactly one output. The equation $y = x^2$ is a function. The equation $y^2 = x$ is not a function if x is any real number. Explain.

16. Different sizes of signs are shown below. Look for a pattern. Then write an equation to find the height when the width is 13 ft.

width, w	1	3	5	7	9
height, h	5	11	17	23	29

17. Tasha sells homemade bookmarks for 75¢ each. Her sales for each of seven weeks are shown below. Write an equation to represent the function and complete the table.

n	8	15	22	29	36	43	50
t	$6	$11.25	$16.50	$21.75	$27	$32.25	▪

Use Data **For 18–19, use the graph of Vincent's budget.**

18. Vincent saved $380 to spend on his vacation. How much did he plan to pay for meals and entertainment?

VACATION BUDGET

Meals and entertainment 45%

Entrance fees 25%

Souvenirs 30%

19. How much money did Vincent plan to spend on souvenirs and entrance fees?

MIXED REVIEW AND TEST PREP

20. Using a rule, name the next three terms. ⁻0.82, ⁻0.75, ⁻0.68, ⁻0.61, . . . (p. 526)

21. $\frac{12}{13} + 1\frac{1}{8}$ (p. 182)

22. $(^-324 \div 12) + 4^3$ (p. 262)

23. **TEST PREP** Oranges are 5 for $1.80. How much will 2 dozen oranges cost? (p. 22)

A $1.92   **B** $3.20   **C** $4.32   **D** $8.64

24. **TEST PREP** Which shows $\sqrt{25} + {}^-13$? (p. 282)

F ⁻38   **G** ⁻8   **H** 12   **J** 18

Thinker's CORNER

Algebra The binary, or base-two, number system uses only the digits 0 and 1.

DECIMAL	0	1	2	3	4	5	6	7	8	9	10
BINARY	0	1	10	11	100	101	110	111	1000	1001	1010

In the decimal system, each place value is ten times the place value to the right. In the binary system, each place value is twice the place to the right. You can use powers of 2 to find the decimal equivalent of a binary number.

$$10100_{two} = (1 \times 2^4) + (0 \times 2^3) + (1 \times 2^2) + (0 \times 2^1) + (0 \times 2^0)$$
$$= (1 \times 16) + (0 \times 8) + (1 \times 4) + (0 \times 2) + (0 \times 1)$$
$$= 16 + 0 + 4 + 0 + 0$$
$$= 20$$

Find the decimal equivalent of each binary number.

1. 1111_{two} **2.** 10001_{two} **3.** 11000_{two} **4.** 11111_{two}

Geometric Patterns

Learn how to recognize, describe, and extend patterns of geometric figures.

Vocabulary

fractal

Ayita is making a wall hanging using a geometric figure. Patterns of geometric figures are based on the shape, color, size, position, or number of the figures. The pattern below is what Ayita started with for her wall hanging.

Notice that the size of the figure increases twice before the pattern repeats. The next figure in the pattern will be the large figure.

EXAMPLE 1

Ayita used this pattern for another craft project she made. Look for the pattern. Draw the next three figures.

Look for a possible pattern. The positions of the circles move clockwise around the inside of the figure, alternating one and then two.

So, the next three figures might be:

EXAMPLE 2

Look for a possible pattern. Draw the next figure.

Look for a pattern.

The figures are flipped horizontally about a vertical axis. The top left square and then the bottom left square are changed.

So, the next figure might be:

CALIFORNIA STANDARDS **MR 1.1** Analyze problems by identifying relationships, distinguishing relevant from irrelevant information, identifying missing information, sequencing and prioritizing information, and observing patterns. *also* **MR 1.2, MR 2.0**

533

You can also find patterns in three-dimensional figures.

EXAMPLE 3

Mr. Gallo is displaying boxes in his store. Look for the pattern. Draw what the next two displays will look like.

Look for a pattern. The number of boxes in the new bottom row increases by one.

So, these are the next two displays:

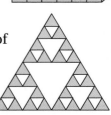

Some geometric figures contain a repeating pattern of smaller and smaller parts. This can be described as a fractal. A **fractal** has a repeating pattern containing shapes that are like the whole but of different sizes throughout.

You can build fractals from two-dimensional figures by repeating a process over and over again. This is called iterating. An iteration is a step in the process of iterating.

EXAMPLE 4

TECHNOLOGY LINK

More Practice: Use E-Lab, *Exploring Fractal Patterns.* www.harcourtschool.com/ elab2002

Build a fractal from a square by repeating this iteration process two times. Find the number of shaded squares that would be in Stage 3.

Draw a shaded square. → Reduce the entire figure to one-half in length and width. Place one copy in the upper left-hand corner and one in the lower right-hand corner.

Stage 0   Stage 1   Stage 2

Look for a pattern.

Stage 0 has 1 shaded square, Stage 1 has 2, and Stage 2 has 4.

The pattern of shaded squares is 1, 2, 4, ...

So, 8 smaller squares would be shaded in Stage 3.

CHECK FOR UNDERSTANDING

Think and ▶ Discuss

Look back at the lesson to answer the question.

1. Draw a geometric pattern that uses a triangle.

Guided ▶ Practice

Draw the next three figures in the pattern.

2.

3.

PRACTICE AND PROBLEM SOLVING

Independent ▶ Practice

Draw the next three figures in the pattern.

4.

5.

6.

7.

For 8–9, use the figures at the right.

8. The iterative process is (1) reduce the figure to one-third in length and width and (2) place five copies in the corners and center of the original square. How many yellow squares would be in Stage 3?

Stage 0   Stage 1   Stage 2

9. Use exponents to describe the pattern for the number of yellow squares at each stage.

Problem Solving ▶ Applications

10. Ayita saves part of the money made by selling her wall hangings. She saves $55 of the $450 she gets for the first hanging. She will increase the amount saved by $4 each time she sells a hanging. Write the pattern of amounts. For which hanging will she save $75?

11. The figure at the right was made with rectangular prisms. The top level has 1 prism, the next level has 4 prisms, and the bottom has 9 prisms. How many prisms would be in the next two levels added at the bottom?

12. ✎ **Write a problem** using a geometric pattern in your school. Describe it in words and with a drawing.

MIXED REVIEW AND TEST PREP

13. Solve $y = 6x + 3$ for $x = {}^-2, {}^-1, 0, 1,$ and 2. (p. 529)

14. Compare. Write $<$ or $>$ for ●.
$-\frac{1}{2}$ ● 0.50 (p. 234)

15. Write the prime factorization for 65. (p. 148)

16. Evaluate $3x + 9x - 36$ for $x = {}^-5$. (p. 276)

17. **TEST PREP** Which triangle has one right angle and no congruent sides? (p. 336)
   A an obtuse scalene triangle
   B a right isosceles triangle
   C a right scalene triangle
   D an acute scalene triangle

Extra Practice page H59, Set C

535

1. **VOCABULARY** A relationship in which one quantity depends on another is a(n) ___?___ . (p. 530)

2. **VOCABULARY** An ordered set of numbers is called a(n) ___?___ . (p. 526)

3. **VOCABULARY** Each number in a sequence is called a(n) ___?___ . (p. 526)

Write a rule for each sequence. (pp. 526–528)

4. 220, 22, 2.2, . . .

5. $^-$3, 9, $^-$27, . . . ,

6. 100, 99.1, 98.2, . . .

7. 4, $3\frac{2}{3}$, $3\frac{1}{3}$, . . .

8. 243, 81, 27, . . .

9. $\frac{7}{8}$, $3\frac{1}{8}$, $5\frac{3}{8}$, . . .

Find the next three possible terms in each sequence. (pp. 526–528)

10. 7, $5\frac{3}{4}$, $4\frac{1}{2}$, . . .

11. 90, 79, 68, . . .

12. 9, 15, 21, . . .

13. $^-$4, $^-$2, 0, . . .

14. 32, 16, 8, . . .

15. 5, 17, 29, . . .

Write an equation to represent the function. (pp. 529–532)

16.

x	9	8	7	6	5
y	0	-1	-2	-3	-4

17.

x	10	11	12	13	14
y	17	18	19	20	21

18.

x	-3	-2	-1	0	1
y	-30	-20	-10	0	10

19.

x	64	56	48	40	32
y	8	7	6	5	4

Draw the next three figures in the pattern. (pp. 533–535)

20.

21.

Draw the next two solids in the pattern. (pp. 529–532)

22.

23.

Solve. (pp. 526–528)

24. Ms. Wong received 110 orders the first month, 125 the second month, and 140 the third month. If this pattern continues, how many orders will she receive the sixth month?

25. The temperature was 21°F at 6 P.M. At 7 P.M. it was 18°F, and at 8 P.M. it was 15°F. If this pattern continues, what will the temperature be at midnight?

Check your work.

See item **2**.

You could re-write each term as a decimal to check your work. Your decimal answer should be equivalent to the fraction answer you chose.

Also see problem **7**, p. H65.

Choose the best answer.

1. What is the output when the input is 7?

Input s	Algebraic Expression $s - 3$	Output
4	$4 - 3$	1
5	$5 - 3$	2
6	$6 - 3$	3
7	$7 - 3$	■

A 4   **C** 6

B 5   **D** Not here

2. What is the next possible term in the sequence?

$$3\frac{1}{4}, 3\frac{7}{8}, 4\frac{1}{2}, 5\frac{1}{8}$$

F $5\frac{1}{4}$ **H** 6

G $5\frac{3}{4}$ **J** $6\frac{1}{4}$

3. What is the next possible term in the sequence?

89, 83.3, 77.6, 71.9

A 65.4   **C** 68.3

B 66.2   **D** 70.6

4. Alice works $7\frac{1}{2}$ hours each day. If she is paid d dollars per hour, which expression shows her earnings for 1 day?

F $d - 7\frac{1}{2}$ **H** $7\frac{1}{2} - d$

G $7\frac{1}{2} + d$ **J** $7\frac{1}{2} \times d$

5. Marisa jogs 400 meters on day 1, then 475 meters on day 2, and 550 meters on day 3. If she continues this pattern, how far will she jog on day 7?

A 700 m   **C** 850 m

B 775 m   **D** 925 m

6. Which figure comes next in the pattern?

F   **H**

G   **J**

7. Mrs. Crain filled her car with gas and then drove 261 miles. When she filled the car with gas again, she determined that the car had averaged 29 miles per gallon. Which equation could be used to find g, the number of gallons of gas it took to fill the car the second time?

A $29 \div g = 261$ **C** $29 + g = 261$

B $261 \times g = 29$ **D** $261 \div g = 29$

8. Antonio compared the lengths of four different canoes. Which lists the lengths from shortest to longest?

F 176.5 cm, 182 cm, 180.2 cm, 180.8 cm

G 182 cm, 180.8 cm, 180.2 cm, 176.5 cm

H 176.5 cm, 180.8 cm, 180.2 cm, 182 cm

J 176.5 cm, 180.2 cm, 180.8 cm, 182 cm

9. The area of a rectangular rug is 96 square feet. The rug is 8 feet wide. How long is the rug?

A 6 ft   **C** 24 ft

B 12 ft   **D** Not here

10. What is the LCM for 8 and 12?

F 2

G 4

H 24

J 96

29 Geometry and Motion

Many geometric shapes have symmetry. People and many animals also have symmetry. Many starfish, for example, have line symmetry. Look for an object in your classroom that has line symmetry. Does it have one line of symmetry or more? Sketch the object and show its line(s) of symmetry.

NUMBER OF LINES OF SYMMETRY FOR REGULAR POLYGONS

Check What You Know

Use this page to help you review and remember important skills needed for Chapter 29.

✅ Vocabulary

Choose the best term from the box.

1. Figures that are the same size and shape are __?__ .

2. The movement of a geometric figure to a new position without turning or flipping it is a __?__ .

> equivalent
> congruent
> slide

✅ Slides, Flips, and Turns (See p. H30.)

Identify the transformation as a slide, flip, or turn.

3.

4.

5.

6.

7.

8.

✅ Line Symmetry (See p. H30.)

Trace the figure. Draw the lines of symmetry.

9.

10.

11.

12.

13.

14.

15.

16.

✅ Measure Angles (See p. H31.)

Use a protractor to measure each angle.

17.

18.

19.

20.

LOOK AHEAD

In Chapter 29 you will
- make tessellations
- perform transformations of plane and solid figures
- examine symmetry in plane figures

Transformations of Plane Figures

Learn how to use translations, rotations, and reflections to transform geometric shapes.

QUICK REVIEW

Estimate the measure of each angle.

1. 2. 3.

4. 5.

A movement of a figure without changing the size or shape of the figure is a rigid **transformation**. Since the size and shape do not change, the original figure and the transformation are always congruent.

A **translation** is the movement of a figure along a straight line.

Only the location of the figure changes with translation.

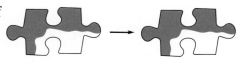

Turning a figure around a point is called a **rotation**.

Both the position and the location of the figure can change.

A point of rotation can be on or outside a figure.

point of rotation

Flipping a figure over a line is called a **reflection** about that line.

Both the position and the location of the figure change with reflection.

← line of reflection

EXAMPLE 1

Draw a 90° clockwise rotation of the figure around the given point of rotation.

Trace the figure and the point of rotation.

Place your pencil on the point of rotation.

Rotate the figure clockwise 90°.

Trace the figure in its new location.

• How would the rotation look if you turned the figure 180°?

CALIFORNIA STANDARDS MR 2.4 Use a variety of methods, such as words, numbers, symbols, charts, graphs, tables, diagrams, and models, to explain mathematical reasoning. *also* MR 2.0, MR 2.2

EXAMPLE 2

Draw a horizontal reflection and a vertical reflection of the shape of the state of California.

Trace the figure and the vertical line of reflection.

Reflect the figure horizontally over the line. This is a horizontal reflection.

Trace the figure in its new location.

Use the same process for a horizontal line of reflection.

Reflect the figure vertically over the line. This is a vertical reflection.

Trace the figure in its new location.

• Rotate the original state shape 90° clockwise about a point in its upper right corner.

CHECK FOR UNDERSTANDING

Think and ▶ Discuss

Look back at the lesson to answer each question.

1. **Explain** why a transformation is always congruent to the original figure.

2. **Describe** how the new position of a figure that has been rotated 360° compares with its original position.

Guided ▶ Practice

Tell which type or types of transformation the second figure is of the first figure. Write *translation*, *rotation*, or *reflection*.

3.   4.   5.   6.

Tell what moves were made to transform each figure into its next positions.

7.   8.

541

Independent ▶ Practice

Tell which type or types of transformation the second figure is of the first figure. Write *translation*, *rotation*, or *reflection*.

9.

10.

11.

12.

Tell what moves were made to transform each figure into its next position.

13.

14.

Trace each figure. Then draw a 90° clockwise rotation around the given point followed by a horizontal reflection.

15.

16.

17.

18.

Problem Solving ▶ Applications

19. Kathy is working on a design for a logo. She drew a profile of an eagle and wants to have two eagles facing each other. What transformation should she do to the profile?

20. The mirror image of the word AMBULANCE is written on the front of ambulances so that it can be read in a car's rearview mirror. Identify the transformation and write the word as it appears on the front of the ambulance.

21. **REASONING** Draw a figure that looks the same when translated by rotation and by reflection.

22. The lowest temperature recorded in a certain month was 48°, and the highest temperature was 64°. What was the range of temperatures for the month?

MIXED REVIEW AND TEST PREP

23. Draw the next three figures in the geometric pattern at the right. (p. 533)

24. Find the area of a circle with a radius of 4 in. (p. 492)

25. Find a rational number between $\frac{3}{8}$ and $\frac{3}{4}$. (p. 230)

26. Write the decimal for 45%. (p. 60)

27. **TEST PREP** A spinner is divided into five sections, labeled 1, 2, 3, 4, and 5. If each number is equally likely to occur, what is the probability of the spinner's landing on an odd number? (p. 418)

A $\frac{1}{5}$ **B** $\frac{2}{5}$ **C** $\frac{1}{2}$ **D** $\frac{3}{5}$

Tessellations

Vocabulary

tessellation

A repeating arrangement of shapes that completely covers a plane, with no gaps and no overlaps, is called a **tessellation** of the plane. Although most tessellations are produced by humans, a few occur in nature.

Honeycombs are naturally occurring tessellations using hexagons.

• What other natural tessellations can you think of?

In addition to being made from polygons, tessellations can be made from figures that are not polygons.

Activity 1

You need: pattern blocks, colored pencils or markers

Make a tessellation.

• Choose a pattern block to use for your tessellation shape.

• Design your tessellation. Remember that the shapes must fit together without overlapping or leaving gaps.

• Record your tessellation. Color it to make a pleasing design.

CALIFORNIA STANDARDS MR 2.4 Use a variety of methods, such as words, numbers, symbols, charts, graphs, tables, diagrams, and models, to explain mathematical reasoning. MG 2.0 Students identify and describe the properties of two-dimensional figures.

543

Math **I**dea ▶ Using regular polygons that can form tessellations, you can make other shapes that tessellate the plane.

Activity 2

You need: paper, scissors, and tape

Make a tessellation shape.

- Cut out a square that is 2 in. × 2 in.

- Use scissors to cut out a part of the square on one side.

- Translate the cutout part of the square to the opposite side of the square. Then tape it.

- Trace the new shape to form at least two rows of a tessellation. You will need to rotate, translate, or reflect the shape.

- **What if,** after you cut out part of a square, you didn't tape the cutout to the opposite side? Could the shape tessellate the plane? Explain.

TECHNOLOGY LINK

More Practice: Use **Mighty Math Cosmic Geometry**, *Tessellation Creation Station.*

CHECK FOR UNDERSTANDING

Think and ▶ Discuss

Look back at the lesson to answer each question.

1. **Explain** how you know when a pattern of shapes forms a tessellation.

2. **Give an example** of a shape that does not form a tessellation.

Guided ▶ Practice

Trace and cut out several of each polygon. Tell whether the polygon can be used repeatedly to tessellate the plane. Write *yes* or *no.*

3. 　4. 　5. 　6.

Make the tessellation shape described by each pattern. Then form two rows of a tessellation.

7.   **8.**   **9.**

PRACTICE AND PROBLEM SOLVING

Independent ▶
Practice

Trace and cut out several of each shape. Tell whether the shape can be used to form a tessellation. Write *yes* or *no*.

10.   **11.**   **12.**   **13.**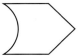

Make the tessellation shape described by each pattern. Then form two rows of a tessellation.

14.   **15.**   **16.**   **17.**

Problem Solving ▶
Applications

18. Cut out a hexagon, and then change it by cutting out a part on one side. Translate the cutout part to the opposite side of the hexagon. Can this new shape form a tessellation?

19. Fergus wants to lay floor tiles in a tessellation pattern. The floor measures 24 yd by 12 yd. What is the total area to be covered?

20. ✎ **Write About It** Explain why you can translate a cutout from the edge of a regular hexagon to the opposite side when you want to make a tessellation pattern.

MIXED REVIEW AND TEST PREP

21. Tell which type of transformation the second figure is of the first figure. (p. 540)

22. Write the ratio four to seven in three ways. (p. 374)

23. $\frac{3}{4} + \frac{1}{6}$ (p. 182)   **24.** Solve. $x - 8 = 40$ (p. 294)

25. TEST PREP Susan ran five days last week. The distances of her runs were 4.5 mi, 3 mi, 4 mi, 3.5 mi, and 5 mi. What was the mean distance Susan ran? (p. 106)

A 3.5 mi   **B** 4 mi   **C** 4.5 mi   **D** 20 mi

Extra Practice page H60, Set B

545

PROBLEM SOLVING STRATEGY
Make a Model

Learn how to solve problems that involve tessellations by making a model.

Delaney is designing a tile mosaic for the top of her table. The shape Delaney uses must tessellate a plane. She wants to use the shape at the right. Can she use this shape for her design?

Analyze   What are you asked to find?

What information are you given?

Is there information you will not use?

Choose   What strategy will you use?
You can use the strategy *make a model*.

Solve   How will you solve the problem?
Trace and cut out several copies of the shape. Use the copies to see whether the shape will tessellate a plane.

Begin by moving the shapes around to see if they fit together. The pieces must not overlap or leave any gaps when you place them together.

You may also check to see if the sum of the angles where the vertices meet is equal to 360°. If it is, the shape will tessellate a plane.

Check   How can you check your answer?

What if Delaney wants to use the figure at the right to make her mosaic? Will this figure tessellate a plane? Make a model to test your ideas.

CALIFORNIA STANDARDS **MR 1.1** Analyze problems by identifying relationships, distinguishing relevant information from irrelevant, identifying missing information, sequencing and prioritizing information, and observing patterns. **MR 3.2** Note the method of deriving the solution and demonstrate a conceptual understanding of the derivation by solving similar problems. *also* **MR 3.3**

PROBLEM SOLVING PRACTICE

Solve the problem by making a model.

PROBLEM SOLVING STRATEGIES

Draw a Diagram or Picture
▶ **Make a Model**
Predict and Test
Work Backward
Make an Organized List
Find a Pattern
Make a Table or Graph
Solve a Simpler Problem
Write an Equation
Use Logical Reasoning

1. Robert wants to make a design using a combination of two of the three shapes below. He wants the design to tessellate a plane. Which two-shape combinations can he choose from?

2. Will this shape tessellate a plane? Explain.

Foster wants to use triangular tiles in this size and shape to cover the floor of a hallway that measures 4 ft by 12 ft.

3. How many tiles does he need to cover the floor completely?

A 24   **C** 96

B 48   **D** 144

4. How many tiles would he need if the dimensions of each tile were half of those shown?

F 48   **H** 192

G 96   **J** 384

MIXED STRATEGY PRACTICE

Use Data   For 5–6, use the table.

5. Find the median temperature.

6. Make a stem-and-leaf plot of the temperature data.

HIGH TEMPERATURES IN SAN FRANCISCO						
Mon	Tue	Wed	Thurs	Fri	Sat	Sun
61	63	65	59	59	64	70

7. Shelley wants to make a quilt out of tessellating shapes. Which of the following shapes can she use?

a.   b.

8. Ed had $40 when he went to the mall. He spent $24.88 on a shirt and $8.42 on a CD. Estimate how much he had left for lunch.

9. Bernie, Nash, Lisa, and Gerta are standing in the lunch line. Bernie is not first in line. Lisa has at least two people ahead of her in line. Nash is third. Give the order of the four in line.

10. A palindrome is a number that is the same backward and forward, such as 1,221. Find a three-digit palindrome in which the product of the digits is divisible by 35.

Transformations of Solid Figures

Learn how to identify and perform transformations of solid figures.

Remember that a transformation can be a rotation, a reflection, or a translation. A solid figure and its transformation are always congruent because the size and shape do not change.

QUICK REVIEW

Name the shapes of the faces that make up each solid.

1.
2.
3.
4.
5.

You perform a transformation of a solid figure when you move your math book from your backpack to your desk.

Activity

• Place your textbook in the upper left corner of your desk, and imagine vertical and horizontal lines which divide the desk into fourths.

• Perform a 180° rotation of the textbook around the desk's vertical center line.

F = Front

B = Back

• Perform a 180° rotation around the desk's horizontal center line.

• Translate the textbook from the lower right to the lower left of the desk.

• Perform a 180° rotation around an axis that runs through the middle of the book, perpendicular to the desktop.

EXAMPLE

How many ways can you place your textbook on a piece of paper of the same size so that the paper is completely covered?

For a rectangular textbook, four positions are possible.

CALIFORNIA STANDARDS MR 2.4 Use a variety of methods, such as words, numbers, symbols, charts, graphs, tables, diagrams, and models, to explain mathematical reasoning.

Think and ▸ Discuss

Look back at the lesson to answer the question.

1. Explain whether a solid figure and its transformation are congruent.

Guided ▸ Practice

Tell how many ways you can place the solid figure on the outline.

2.

3.

4.

Independent ▸ Practice

Tell how many ways you can place the solid figure on the outline.

5.

6.

7.

8.

9.

10.

Problem Solving ▸ Applications

11. Allison hung a sign that says "Allison's Room" upside-down on her door. What transformation should she do to the sign so that it is hung correctly?

12. ✎ **Write About It** Think of transformations of solid figures that happen in your everyday life. Give several examples.

MIXED REVIEW AND TEST PREP

13. Tell which type of transformation the second figure is of the first figure. (p. 540)

14. $25 \div 5 + (7 - 3)^2$ (p. 44)

15. What is the perimeter of a square with sides 6 cm long? (p. 469)

16. Write the prime factorization of 36. (p. 148)

17. TEST PREP A computer which normally sells for $800 is on sale at 20% off. What is the discounted price? (p. 408)

    A $780   **B** $640   **C** $600   **D** $160

Extra Practice page H60, Set C

LESSON 29.5

Symmetry

Learn how to identify line symmetry and rotational symmetry.

Vocabulary

line symmetry

line of symmetry

rotational symmetry

point of rotation

QUICK REVIEW

Give the degree equivalent of each fractional portion of a full rotation.

1. $\frac{1}{4}$ 2. $\frac{1}{2}$ 3. $\frac{1}{8}$ 4. $\frac{3}{5}$ 5. $\frac{2}{3}$

Symmetry can be found all around us, both in nature and in manufactured objects. This Oregon Swallowtail butterfly has symmetry because each wing looks like a reflection of the other.

A figure has **line symmetry** if it can be folded or reflected so that the two parts of the figure match, or are congruent. The line across which the figure is symmetric is known as the **line of symmetry**.

Activity 1

You need: paper, scissors, dark crayon

- Fold the paper in half. Use the crayon to write your name in cursive along the fold line.

- Fold the paper along the fold line so your name is inside. Use the handle of the scissors to make a rubbing of your name.

- Unfold the paper. Your name appears on the other half of the paper. Your design has line symmetry.

- Where is the line of symmetry in the design you made from your name?

- How many lines of symmetry does the design have?

Some figures have several lines of symmetry.

EXAMPLE 1

Remember that regular polygons have all sides congruent and all angles congruent.

Find all the lines of symmetry in the regular hexagon.

Trace the figure and cut it out.

Fold it in half in different ways.

If the halves match, then the fold is a line of symmetry.
Count the different lines of symmetry.

The figure has six lines of symmetry.

- Do all hexagons have six lines of symmetry? Explain.

CALIFORNIA STANDARDS **MG 2.0** Students identify and describe the properties of two-dimensional figures. **MR 2.4** Use a variety of methods, such as words, numbers, symbols, charts, graphs, tables, diagrams, and models, to explain mathematical reasoning.

Letters and whole words can have lines of symmetry.

EXAMPLE 2

Do the words DECIDED and MUM have line symmetry?

-┼-DECIDED-┼-

MUM

Copy each word.

Draw the dashed lines as shown.

The two halves of each word are the same size and shape.

Since the halves formed by each line are congruent, the words have line symmetry.

A figure has **rotational symmetry** if it can be rotated less than 360° around a central point, or **point of rotation**, and still look exactly the same as the original figure.

Activity 2

You need: 4-in.-square paper, scissors

- Fold the square of paper in half and then in half again.

- Draw a petal shape on the folded paper, and cut it out so that the center is connected.

- Trace your flower on a piece of paper.

- Place the point of your pencil at the center of the flower. Match the flower with its image.

- Rotate the flower, using the point of the pencil as the point of rotation, until it matches the image again.

The figure you made has rotational symmetry.

EXAMPLE 3

Does the figure of the starfish below have rotational symmetry? If it does, what are the fraction and the angle measure of each turn?

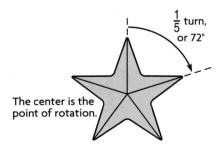

$\frac{1}{5}$ turn, or 72°

The center is the point of rotation.

Trace the figure.

Place your pencil point at the point of rotation.

Rotate the figure until it looks like the original figure.

Use a protractor to measure the degree and fraction of rotation.

The figure has $\frac{1}{5}$-turn, or 72°, rotational symmetry.

- Does the starfish figure above also have line symmetry?

Think and ▶
Discuss

Look back at the lesson to answer each question.

1. **Explain** the difference between line symmetry and rotational symmetry.

2. **Draw** a pentagon that has fewer than five lines of symmetry.

Guided ▶
Practice

Trace the figure. Draw the lines of symmetry.

3. 　　4. 　　5. 　　6.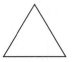

Trace the figure. Complete the other half of the figure across the line of symmetry.

7. 　　8. 　　9. 　　10.

Tell whether each figure has rotational symmetry, and, if so, identify the symmetry as a fraction of a turn and in degrees.

11. 　　12. 　　13. 　　14.

PRACTICE AND PROBLEM SOLVING

Independent ▶
Practice

Trace the figure. Draw the lines of symmetry.

15. 　　16. 　　17. 　　18.

Trace the figure. Complete the other half of the figure across the line of symmetry.

19. 　　20. 　　21. 　　22.

23. 　　24. 　　25. 　　26.

Tell whether each figure has rotational symmetry, and, if so, identify the symmetry as a fraction of a turn and in degrees.

27.   28.   29.   30.

Problem Solving ▶
Applications

31. Philip has a square cake. He cuts it along all the lines of symmetry of the square. How many pieces does he have? Are they all the same size?

32. **REASONING** Does a circle have line symmetry? rotational symmetry? Explain.

33. **(?) What's the Error?** Nick says that all figures have rotational symmetry, because when you rotate any figure 360° it matches up. What's his error?

MIXED REVIEW AND TEST PREP

34. Tell which type of transformation the second figure is of the first figure. (p. 540)

35. Find the volume of a cylinder with a diameter of 4 in. and a height of 10 in. (p. 511)

36. $\frac{2}{3} \times \frac{3}{8}$ (p. 202)

37. **TEST PREP** Which shows the number of inches in 5 ft? (p. 452)

   A 50   **B** 60   **C** 75   **D** 180

38. **TEST PREP** Andy drove 900 miles in 18 hours. At what rate was he traveling. (p. 374)

   F 45 mi per hr   **G** 50 mi per hr   **H** 55 mi per hr   **J** 60 mi per hr

---

LINKUP to Reading

Strategy • Choose Important Information A word problem may contain more information than is needed. You must decide what information you need to solve the problem.

Susan wants to use square pieces of fabric to make a quilt that will be 3 ft long and 2 ft wide. She has blue pieces that are 4 in. on a side, red pieces that are 3 in. on a side, and light and dark yellow pieces that are 6 in. on a side. How many squares will she need for a yellow quilt?

• What does the problem ask you to find?   • What information is needed to solve the problem?

• Tell the important information in this problem. Then solve.

Pauline will fold two pieces of paper along each of their lines of symmetry one time. One is a square measuring 3 in. on a side, and one is a rectangle measuring 4 in. by 8 in. How many folds must she make in all?

Extra Practice page H60, Set D

1. **VOCABULARY** A movement of a plane or solid figure without changing the figure's size or shape is a ___?___ . (p. 540)

2. **VOCABULARY** A repeating arrangement of shapes that completely covers a plane with no overlaps or gaps is called a ___?___ . (p. 543)

3. **VOCABULARY** A line across which a figure is symmetric is known as a ___?___ . (p. 550)

Tell which type or types of transformation the second figure is of the first figure. Write *translation*, *rotation*, or *reflection*. (pp. 540–542)

4.

5.

6.

7.

Tell whether each polygon can be used to form a tessellation. Write *yes* or *no*. (pp. 543–545)

8.

9.

10.

11.

Tell how many ways you can place the solid figure on the outline. (pp. 548–549)

12.

13.

14.

15.

Trace the figure. Draw the lines of symmetry. (pp. 550–553)

16.

17.

18.

19.

Tell whether each figure has rotational symmetry, and, if so, identify the symmetry as a fraction of a turn and in degrees. (pp. 550–553)

20.

21.

22.

23.

24. Richard builds a wall 6 full bricks wide, starting and ending every other layer with a half of a brick. The figure shows the first 3 layers of the wall. How many full bricks and how many half bricks does he need for 8 layers? (pp. 546–547)

25. There are 6 shelves that are 18 inches apart, and the bottom shelf is 24 inches from the floor. How far from the floor is the top shelf? (pp. 546–547)

Understand the problem.

See item **4.**

Think about the hexagon pattern block and how 6 of the triangle blocks make the same shape when you put them together. The angle measure of the rotation is related to the measure of the angle in the triangle.

Also see problem **1**, p. H62.

For 1–10, choose the best answer.

1. Which type of transformation of the first figure is shown by the second figure below?

   A Translation   **C** Reflection

   B Rotation   **D** Not here

2. Which kind of figure always has line symmetry?

   F Triangle   **H** Trapezoid

   G Circle   **J** Pentagon

3. What is the relationship of the second figure to the first?

   A Reflection   **C** Translation

   B Rotation   **D** Not here

4. A regular hexagon has rotational symmetry. What is the angle measure of each turn?

   F 30°   **H** 60°

   G 45°   **J** 90°

5. Which polygon will always tessellate a plane?

   A Regular pentagon

   B Trapezoid

   C Quadrilateral

   D Equilateral triangle

6. What is the measure of each of the angles that surround the circled vertex?

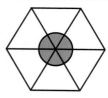

   F 30°   **H** 50°

   G 45°   **J** 60°

7. How many lines of symmetry does a regular pentagon have?

   A 0   **C** 10

   B 5   **D** 15

8. Dana's CD collection is $\frac{4}{9}$ rock and $\frac{2}{9}$ country. What part of her collection is neither rock nor country?

   F $\frac{1}{3}$

   G $\frac{2}{5}$

   H $\frac{1}{2}$

   J $\frac{2}{3}$

9. Meg sold 47 flowers. Each one was blue or white. She sold 13 more blue than white flowers. How many blue flowers did she sell?

   A 17

   B 30

   C 33

   D 34

10. The length of a rectangular prism is 8 centimeters, the width is 4 centimeters, and the height is 2 centimeters. What is the volume?

   F 32 cm^3

   G 64 cm^3

   H 112 cm^3

   J 144 cm^3

30 Graph Relationships

Before any airline flight, a flight plan is drawn on a map. The plan shows the route the airplane will fly. East-west lines of latitude and north-south lines of longitude form a coordinate grid system on the map. Pilots use these to identify places along the route. Suppose a flight is departing from Charlotte, N.C. The approximate coordinates for this city are (81,36). What are the approximate coordinates of nearby Columbia, S.C.?

Check What You Know

Use this page to help you review and remember important skills needed for Chapter 30.

Vocabulary

Choose the best term from the box.

1. Two numbers used to locate a point on a coordinate plane are called a(n) __?__.

2. A movement that does not change the size or shape of a figure is called a(n) __?__.

3. Lines that intersect at right angles to each other are __?__ lines.

> ordered pair
> origin
> parallel
> perpendicular
> transformation

Compare Numbers (See p. H29.)

Compare. Write <, >, or = for each ●.

4. $^-6$ ● $^-7$

5. $^-\frac{1}{2}$ ● $^-0.5$

6. $^-2.44$ ● 2.44

7. 3.10 ● 3.01

8. $3\frac{1}{2}$ ● $^-3.6$

9. $\frac{2}{5}$ ● 0.45

10. 0.75 ● 75%

11. $^-\frac{4}{5}$ ● $^-\frac{5}{6}$

Ordered Pairs (See p. H31.)

Write the ordered pair for each point.

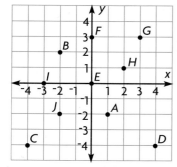

12. point A

13. point B

14. point D

15. point E

16. point G

17. point H

18. point C

19. point F

20. point I

21. point J

Transformations (See p. H30.)

Draw the reflection of the letter after it is flipped across the line.

22.

23.

24.

25.

LOOK AHEAD

In Chapter 30 you will

- solve algebraic inequalities
- use ordered pairs to describe locations on the coordinate plane and to graph functions
- solve problems by using the skill *make generalizations*
- explore linear and nonlinear relationships
- graph transformations

Graph on the Coordinate Plane

Learn how to use an ordered pair to describe a location on the coordinate plane.

Vocabulary

coordinate plane

x-axis

y-axis

axes

quadrant

origin

ordered pair

Points on a map are often located using pairs of coordinates. On the map grid shown, the town of Dexter is located at point (3, *E*). The coordinates of the location are 3 and *E*.

Use the same method to locate points on a **coordinate plane** like the one below.

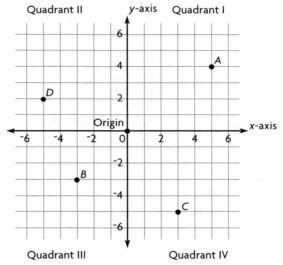

The coordinate plane is divided by a horizontal number line called the **x-axis** and a vertical number line called the **y-axis**. The two **axes** divide the plane into 4 pieces called **quadrants**. The point (0,0), where the axes intersect, is called the **origin**.

Math Idea ▶ Using an **ordered pair** of coordinates like (5,⁻2), you can locate any point on the coordinate plane. The first number of an ordered pair tells you how far to move right or left from the origin. The second number tells you how far to move up or down.

On the coordinate plane above, the coordinates of point *A* are (5,4), of point *B* are (⁻3,⁻3), of point *C* are (3,⁻5), and of point *D* are (⁻5,2). Points can be located at the origin, on the *x*- or *y*- axis, or in one of the four quadrants.

CALIFORNIA STANDARDS MR 2.0 Students use strategies, skills, and concepts in finding solutions.
also **MR 2.4**

Inequalities on a Number Line

Learn how to use inequality symbols and solve inequalities.

QUICK REVIEW

Use < or > to compare.

1. $6 + 9 \bullet 13$ **2.** $30 \div 3 \bullet 13$

3. $2 \times 6 \bullet 9$ **4.** $15 \bullet 25 - 9$

5. $68 \div 4 \bullet 16$

Vocabulary

inequality

An **inequality** is an algebraic sentence that contains the symbol $<$, $>$, $\leq$, $\geq$, or $\neq$.

These are inequalities:

$$n < 5 \qquad y > {}^-7 \qquad k \leq 3.8 + 2.7 \qquad b \geq 12 \div 4$$

If w represents the total weight the elevator can hold, then $w \leq 3{,}500$. Since w can equal any positive value less than or equal to 3,500, the inequality has many possible solutions.

ELEVATOR

MAXIMUM LOAD:

3,500 POUNDS

Math Idea ▶ You can use a number line to show all the solutions of an inequality.

EXAMPLE 1

A. Graph the solutions of the inequality $p > 3$.

The blue line shows that all values greater than 3 are included in the solutions. The blue arrow means the solutions continue to the right of 8 on the number line. The circle at 3 is not filled in because 3 is *not* a solution of the inequality.

B. Graph the solutions of the inequality $m \leq 4$.

All values less than 4 are included in the solutions. The circle at 4 is filled in because 4 *is* a solution of the inequality.

Remember that you can use these symbols when you compare numbers:

$=$	is equal to
$\neq$	is not equal to
$<$	is less than
$>$	is greater than
$\leq$	is less than or equal to
$\geq$	is greater than or equal to

Sometimes you can solve an inequality the same way you solve an equation.

EXAMPLE 2

Solve the inequality and graph the solutions on a number line.

$x - 2 < 3$

$x - 2 + 2 < 3 + 2$ *Add 2 to both sides.*

$x < 5$ *Simplify.*

CALIFORNIA STANDARDS AF 1.0 Students write verbal expressions and sentences as algebraic expressions and equations; they evaluate algebraic expressions, solve simple linear equations, and graph and interpret their results. *also* MR 1.1, MR 2.4

Think and Discuss ▶ Look back at the lesson to answer the question.

1. **Explain** whether or not the circle at ⁻3 should be filled in on the number line showing the graph of the solutions of $x \geq {}^-3$.

Guided Practice ▶ Graph the solutions of the inequality.

2. $x < 8$ 3. $x \geq {}^-7$ 4. $a \leq {}^-12$ 5. $y > 100$

PRACTICE AND PROBLEM SOLVING

Independent Practice ▶ Graph the solutions of the inequality.

6. $x \leq 0$ 7. $x \leq 8$ 8. $x \geq {}^-10$ 9. $x < 30$

Solve the inequality and graph the solutions.

10. $2 + x > 8$ 11. $a + 5 \leq 8$ 12. $y - 2 > {}^-4$ 13. $k + 3 \geq 3$

14. $b - 10 \leq 5$ 15. $n - 1 < {}^-1$ 16. $3x < 9$ 17. $\frac{c}{2} > 1$

> **TECHNOLOGY LINK**
>
> More Practice: Use **Mighty Math Astro Algebra**, Green, Level U.

For 18–19, write an inequality for the word sentence.

18. The value of c is less than or equal to two.

19. The value of p is greater than negative eleven.

Problem Solving Applications ▶ 20. **Measurement** People who fish in Big Sky Lake must throw back any trout that is less than 10 in. long. Write an inequality that represents the lengths of trout that may be kept.

Use Data For 21–22, use the data in the table.

21. Write an inequality relating a, the altitude of any U.S. location, to Mount McKinley's altitude.

22. Write an inequality relating a, the altitude of any U.S. location, to Death Valley's altitude.

UNITED STATES SUPERLATIVES	
Highest Point	**Lowest Point**
Mount McKinley, AK	Death Valley, CA
Altitude: 20,320 ft	Altitude: ⁻282 ft

23. ❓ **What's the Error?** Tyra wrote the solutions of the inequality $\frac{a}{4} < 8$ as $a < 2$. What did she do? What is the correct answer?

MIXED REVIEW AND TEST PREP

24. How many lines of symmetry does the figure at the right have? (p. 550)

Write the value in the requested form. (p. 398)

25. $\frac{3}{8}$ as a percent   26. 125% as a decimal   27. 60% as a fraction

28. **TEST PREP** Keith tossed a number cube labeled 1 to 6. He wanted to throw a 1 or a 2. What is the probability that he succeeded? (p. 418)

   A $\frac{1}{2}$ **B** $\frac{1}{3}$ **C** $\frac{1}{4}$ **D** $\frac{1}{6}$

Extra Practice page H61, Set A

EXAMPLE 1

Describe how to locate the point represented by the ordered pair (6,⁻5) on the coordinate plane.

From the origin, move 6 units to the right, since the 6 is positive. Then move 5 units down, since the 5 is negative.

• Is the point (3,4) the same as the point (4,3)? Explain.

EXAMPLE 2

Name the ordered pair and the quadrant where each point on the coordinate plane below is located.

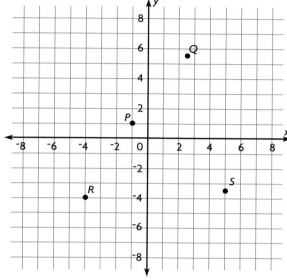

To find the ordered pair, first move right or left from the origin. Then move up or down.

Point *P* is located at (⁻1,1), in Quadrant II.

Point *Q* is located at about $(2\frac{1}{2}, 5\frac{1}{2})$, in Quadrant I.

Point *R* is located at (⁻4,⁻4), in Quadrant III.

Point *S* is located at about (5,⁻3.5), in Quadrant IV.

• Where would you find the point (0,6) on the coordinate plane? Where would you find (⁻5,0)?

You can use what you know about ordered pairs to plot points on a coordinate plane.

EXAMPLE 3

Sketch a coordinate plane, and plot the points *A*(4.5,2) and *B*(⁻3,3).

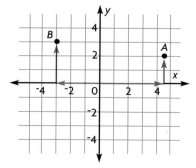

To locate point A, start at the origin and move 4.5 units to the right and 2 units up.

To locate point B, start at the origin and move 3 units to the left and 3 units up.

**Think and ▶
Discuss**

Look back at the lesson to answer each question.

1. Explain how the point (5,4) is different from the point (⁻5,⁻4).

2. Decide in which quadrant a point with two negative coordinates would appear. Explain your answer.

**Guided ▶
Practice**

Write the ordered pair for each point on the coordinate plane.

3. point *A*   **4.** point *B*

5. point *C*   **6.** point *D*

7. point *E*   **8.** point *F*

9. point *G*   **10.** point *H*

11. point *I*   **12.** point *J*

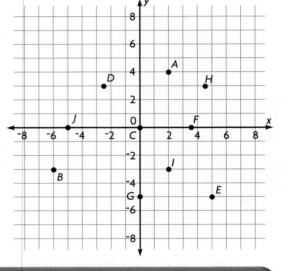

Use the coordinate plane at the right. Identify the points located in the given quadrant.

13. I   **14.** II

15. III   **16.** IV

**Independent ▶
Practice**

Write the ordered pair for each point on the coordinate plane.

17. point *A*   **18.** point *B*

19. point *C*   **20.** point *D*

21. point *E*   **22.** point *F*

23. point *G*   **24.** point *H*

25. point *I*   **26.** point *J*

27. point *K*   **28.** point *L*

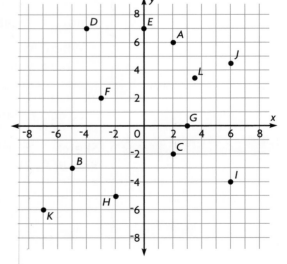

Use the coordinate plane at the right. Identify the points located in the given quadrant.

29. I   **30.** II

31. III   **32.** IV

Sketch a coordinate plane and plot the points.

33. $F(^-7,3)$ **34.** $L(^-4,6)$ **35.** $Y(^-1,3)$ **36.** $A(^-4,^-4)$

37. $K(^-2.5,^-6)$ **38.** $I(0,^-7)$ **39.** $T\left(2\frac{1}{2},^-6\frac{1}{2}\right)$ **40.** $E(6,^-3)$

Problem Solving ▶ Applications

41. Geometry Locate and connect the points (5,5), (5,‾3), and (‾2,‾3) on a coordinate plane. What kind of geometric figure do you have? What is the area of the figure?

42. Geometry Locate the points (2,3), (2,‾3), and (‾4,3) on a coordinate plane. What ordered pair tells the location of the fourth point that would make the figure a square? What is the area of the square? What is the perimeter?

43. When a point lies on the *x*-axis, what can be said about the *y*-coordinate? What is the *x*-coordinate when a point lies on the *y*-axis?

44. ✎ **Write About It** What do all of the points in Quadrant II have in common? What do all of the points in Quadrant III have in common?

MIXED REVIEW AND TEST PREP

45. Solve. $5 + y \geq {}^-2$ (p. 558)

46. A field is 249 ft long. Write the field's length in yards. (p. 452)

47. Find the circumference of a circle with a radius of 25 cm. Use 3.14 for π. (p. 476)

48. TEST PREP Find the sum of $6\frac{1}{2}$ and $3\frac{4}{5}$. (p. 186)

A $2\frac{7}{10}$ **B** $9\frac{5}{7}$ **C** $9\frac{9}{10}$ **D** $10\frac{3}{10}$

49. TEST PREP Which figure is a reflection of the figure at the right? (p. 548)

F   G   H   J

LiNKUP to Science

Astronomy Astronomers locate stars in the sky by using a coordinate grid system. Movements to the left and right are measured in hours (h). Movements up and down are measured in degrees (°).

Use the sky map of the Boötes constellation at the right.

1. Name the location of Arcturus, the brightest star in the constellation.

2. Give the letter of the star located at about $(14\frac{1}{2}\,h, 38°)$.

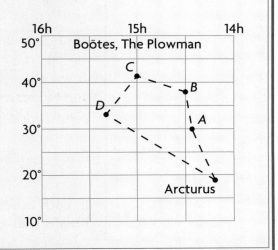

Graph Functions

Learn how to represent functions with ordered pairs, graphs, and equations.

QUICK REVIEW

Find y for $x = 2$.

1. $y = x$ **2.** $y = x + 4$ **3.** $y = x - 2$

4. $y = 8x$ **5.** $y = 3x - 1$

When you see a flash of lightning, count the seconds until you hear thunder. A count of 5 seconds means you are 1 mile from the lightning.

ELAPSED TIME (sec)	5	10	15	20	25	30
DISTANCE TO LIGHTNING (mi)	1	2	3	4	5	6

The values in the table can also be given by using the ordered pairs (5,1), (10,2), (15,3), (20,4), (25,5), and (30,6).

The ordered pairs represent a function. You can use the ordered pairs to show the function on a graph.

HOW FAR AWAY IS THE LIGHTNING?

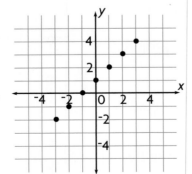

You can also use ordered pairs to graph data from a function table. Use the x and y values to form the ordered pairs.

EXAMPLE

Graph the data from the function table on a coordinate plane. Then write an equation relating y to x.

x	-3	-2	-1	0	1	2	3
y	-2	-1	0	1	2	3	4

Write the data in the table as ordered pairs:

$(^-3, ^-2), (^-2, ^-1), (^-1, 0), (0, 1), (1, 2), (2, 3), (3, 4).$

Then plot the points.

In each ordered pair, the y-value is 1 more than the x-value. This means that x and y are related by the equation $y = x + 1$.

CALIFORNIA STANDARDS O—π AF 1.1 Write and solve one-step linear equations in one variable. **MR 2.0** Students use strategies, skills, and concepts in finding solutions. *also* **MR 2.5, MR 3.2**

Think and Discuss ▶ Look back at the lesson to answer the question.

1. **Name** two ways to represent functions.

Guided Practice ▶ 2. Copy and complete the function table. Then graph the data on a coordinate plane, and write an equation relating y to x.

x	2	3	4	5
y	4	6	8	■

PRACTICE AND PROBLEM SOLVING

Independent Practice ▶ Copy and complete the function table. Then graph the data on a coordinate plane, and write an equation relating y to x.

3.
x	5	6	7	8	9
y	3	4	5	■	■

4.
x	2	1	0	⁻1	⁻2
y	6	3	0	■	■

5. Use the equation $y = 5x$ to make a function table. Use the integers from ⁻2 to 2 as values of x.

Problem Solving Applications ▶ For 6–7, Sue earns three coupons for each win in an arcade game.

6. Let w represent the number of wins. Use ordered pairs to show the number of coupons earned, c, for 7, 8, 9, and 10 wins. What equation relates c to w?

7. Let c represent the number of coupons earned. Use ordered pairs to show the number of wins, w, for 45, 48, 51, and 54 coupons. What equation relates w to c?

8. ✎ **Write a problem** involving a function. Make a function table for 5 values of x. Write the ordered pairs, and locate them on a coordinate plane.

MIXED REVIEW AND TEST PREP

9. Name the quadrant containing the ordered pairs (4,3), (2,1), and (6,5). (p. 560)

10. Tell which type of transformation the second figure is of the first figure. (p. 548)

11. A rectangle measures 8 cm by 5 cm. What is its area? (p. 484)

12. Greg ran a 5-km race. How many meters did he run? (p. 454)

13. **TEST PREP** It is 70°F in Miami, Florida, on the same day that it is ⁻10°F in Anchorage, Alaska. What is the difference between the two temperatures? (p. 250)

   A 60°F   **B** 70°F   **C** 80°F   **D** 90°F

PROBLEM SOLVING SKILL
Make Generalizations

Analyze
Choose
Solve
Check

Learn how to solve problems by making generalizations.

For a science fair project, Cara is making a display showing how popular foods such as raisins and maple syrup are made from natural ingredients. She wants to show how to calculate the amount of each type of food that can be produced from given amounts of ingredients.

It takes about 4 lb of grapes to make 1 lb of raisins. How many pounds of raisins can be made from 160 lb of grapes?

Pounds of Grapes, x	4	8	12	16	20
Pounds of Raisins, y	1	2	3	4	5

Notice that each y-value is $\frac{1}{4}$ of the corresponding x-value. So, the equation $y = \frac{1}{4}x$ describes this situation. Since the equation can be used to find the amount of raisins for any amount of grapes, it is called a generalization.

For 160 lb of grapes, $x = 160$. Then $y = \frac{1}{4} \times 160 = 40$.

So, about 40 lb of raisins can be made from 160 lb of grapes.

On average it takes about 40 gal of sap from maple trees to make 1 gal of maple syrup. How many gallons of sap does it take to make 25 gal of syrup?

Gallons of Syrup, x	1	2	3	4	5
Gallons of Sap, y	40	80	120	160	200

Notice that each y-value is 40 times the corresponding x-value. So, the equation $y = 40x$ is a generalization.

For 25 gal of syrup, $x = 25$. Then $y = 40 \times 25 = 1,000$.

So, it takes about 1,000 gal of maple sap to make 25 gal of maple syrup.

Talk About It ▶ • **What if** Cara shows in her display that about 1 qt of apple juice can be made from 60 apples? What equation could she use to find the number of quarts that can be made from a given number of apples?

PROBLEM SOLVING PRACTICE

Solve by making a generalization.

Rod spent $6 on supplies to advertise his mowing business. The table shows his profit, *y*, for several income amounts, *x*.

x	$12	$13	$14	$15	$16
y	$6	$7	$8	$9	$10

1. Which equation can be used to show Rod's profit?

A $y = 2x$ **C** $y = x - 6$

B $y = \frac{1}{2}x$ **D** $y = x + 6$

2. How much profit did Rod make if he earned $80 mowing lawns?

F $86   **H** $74

G $83   **J** $40

Liz charges $9 for each lawn that she mows.

3. What equation can Liz use to show the amount that she earns, *y*, when she mows *x* lawns?

4. How much money will Liz earn if she mows 27 lawns?

MIXED APPLICATIONS

5. Eduardo is driving from Pittsburgh to San Francisco, a distance of 2,578 mi. He drove 440 mi the first day, 380 mi the second day, and 425 mi the third day. How many miles does he have left to drive?

6. To celebrate school spirit week, Roosevelt School plans to paint every sixth locker blue and every tenth locker orange. If there are 300 lockers in the school, how many will be painted both blue and orange?

7. Mr. Wills needs to catch a flight that leaves at 1:00 P.M. It takes 45 min to get to the airport, and he wants to be there 1 hr 15 min early. At what time should Mr. Wills leave home?

8. Sandra used 60 ft of fencing to fence her garden. The garden is a rectangle that is twice as long as it is wide. What are the dimensions of the garden?

9. Use Data Use the graph at the right. The length of the main span of the Verrazano Narrows Bridge is 4,260 ft. The lengths of several other suspension bridges are shown in the graph. About how much longer than the George Washington Bridge is the Verrazano Narrows Bridge?

10. **?** **What's the Question?** Sally's Gifts sells an average of 9 beanbag animals each day. The answer is about 126 beanbag animals.

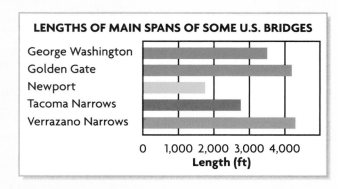

LENGTHS OF MAIN SPANS OF SOME U.S. BRIDGES

George Washington
Golden Gate
Newport
Tacoma Narrows
Verrazano Narrows

0   1,000 2,000 3,000 4,000
Length (ft)

Explore Linear and Nonlinear Relationships

MATH LAB

Explore linear and nonlinear relationships.

You need square tiles or paper squares, graph paper.

QUICK REVIEW

1. 5^2 2. 3^3
3. 2^4 4. 10^2
5. Find the value of $x + 5$ for $x = {}^-7$.

You can learn about linear relationships by looking at patterns.

Activity 1

The model at the right shows Stage 1 to Stage 3 of a pattern.

- Use square tiles or graph paper to build Stages 4, 5, and 6 of the pattern.

- Record each stage and the perimeter of each figure in a table.

- Graph the ordered pairs (x,y) from the table on a coordinate plane.

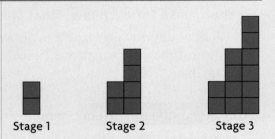

Stage 1   Stage 2   Stage 3

STAGE, x	PERIMETER, y
1	6
2	12
3	■

Think and Discuss

- Describe the pattern of the points on the graph you drew.

- Write an equation to show the relationship between each stage and its perimeter.

Practice

- Make a table that shows the stage, x, and the perimeter, y, of each figure.

- Write an equation to find the perimeter. Graph the ordered pairs (x, y).

- What pattern do you see in the graph?

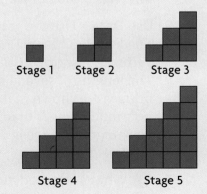

Stage 1   Stage 2   Stage 3

Stage 4   Stage 5

CALIFORNIA STANDARDS MR 1.1 Analyze problems by identifying relationships, distinguishing relevant from irrelevant information, identifying missing information, sequencing and prioritizing information, and observing patterns. **AF 1.0** Students write verbal expressions and sentences as algebraic expressions and equations; they evaluate algebraic expressions, solve simple linear equations, and graph and interpret their results. *also* O━n **AF 1.1, MR 2.2, MR 2.5**

You can also use patterns to learn about nonlinear relationships, which have graphs that do not form straight lines.

Activity 2

You need: square tiles or paper squares, graph paper

The model at the right shows Stage 1 to Stage 3 of a pattern.

Stage 1   Stage 2   Stage 3

- Use square tiles or paper squares to build Stages 4, 5, and 6 of the pattern.

- Record each stage and the area of each figure in a table.

STAGE, x	AREA, y
1	1
2	4
3	▪

- Graph the ordered pairs (x,y) from the table on a coordinate plane.

Think and Discuss
- What pattern do you see in the y-values of the ordered pairs?

- Describe the pattern of the points on the graph you drew.

- The equation for this relationship is $y = x^2$. Use the equation to find the ordered pair for Stage 8.

Practice
Use the x-values 1, 2, 3, and 4 to find ordered pairs for each equation. Then graph the equation.

1. $y = 2x^2$

2. $y = {}^-2x^2$

3. $y = 3x^2$

4. $y = \dfrac{12}{x}$

MIXED REVIEW AND TEST PREP

5. Use the table to write an equation that relates x and y. (p. 564)

x	-2	-1	0	1	2
y	6	3	0	-3	-6

Evaluate the expression for $x = 1$, 2, and 3. (p. 276)

6. $^-15 - 3x$

7. $0.25x + 6$

8. $\dfrac{^-18}{x} + 6$

9. TEST PREP A recipe for pumpkin cookies requires $1\frac{3}{4}$ c of pumpkin. Shana wants to triple the recipe. How much pumpkin does she need? (p. 206)

A $5\frac{1}{4}$ c **B** $4\frac{3}{4}$ c **C** $1\frac{5}{7}$ c **D** $1\frac{1}{4}$ c

569

Graph Transformations

Learn how to use transformations to change positions of figures on a coordinate plane.

Remember that a rigid *transformation* is a movement that doesn't change the size or shape of a figure.

Marla has chosen a card from a pack of cards with pictures of Presidents to give to Matt.

First, she *slides* the card toward Matt, face down. This is a *translation*.

Next, she *flips* the card face up. This is a *reflection*.

Finally, she *turns* the card so that Matt can see it. This is a *rotation*.

Translations, reflections, and rotations are three types of transformations. You can use what you know about the coordinate plane to explore transformations on graph paper.

Activity

You need: graph paper, scissors

- Draw and number a coordinate plane as shown.

- Plot the points (1,4), (1,1), and (3,1) and connect them to form a triangle.

- Trace the triangle on a sheet of paper. Cut it out.

- Place your cutout triangle on the triangle in the coordinate plane. Then translate it 5 units to the left, keeping the base parallel to the *x*-axis.

- Name the coordinates for the vertices of the new triangle.

- Again, place your cutout triangle on the original triangle in the coordinate plane. Translate the cutout 4 units down, keeping the left side parallel to the *y*-axis.

- Name the coordinates for the vertices of the new triangle.

CALIFORNIA STANDARDS **MR 2.5** Express the solution clearly and logically by using the appropriate mathematical notation and terms and clear language; support solutions with evidence in both verbal and symbolic work. *also* **MR 2.0, MR 2.4**

You can translate a figure on a coordinate plane by sliding it horizontally, vertically, or both.

EXAMPLE 1

Translate square *ABCD* 6 units right and 5 units down. Find the coordinates of the new square, *A'B'C'D'*.

To find the coordinates of the new square, add 6 to each *x*-coordinate and subtract 5 from each *y*-coordinate of the original square.

ABCD		*A'B'C'D'*
$A(^-4,3) \rightarrow A'(^-4 + 6, 3 - 5) \rightarrow A'(2,^-2)$		
$B(^-2,3) \rightarrow B'(^-2 + 6, 3 - 5) \rightarrow B'(4,^-2)$		
$C(^-2,1) \rightarrow C'(^-2 + 6, 1 - 5) \rightarrow C'(4,^-4)$		
$D(^-4,1) \rightarrow D'(^-4 + 6, 1 - 5) \rightarrow D'(2,^-4)$		

original

translation

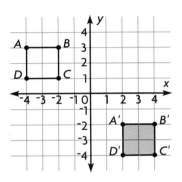

You can also draw the reflection of a figure on a coordinate plane. When you reflect a figure, you flip it across the *x*-axis or the *y*-axis.

EXAMPLE 2

Reflect △*ABC* across the *x*-axis. Find the coordinates of the new figure, △*A'B'C'*.

original

reflection

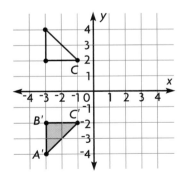

TECHNOLOGY LINK

More Practice: Use E-Lab, *Transformations and Tangrams.*
www.harcourtschool.com/elab2002

△*ABC*		△*A'B'C'*
$A(^-3,4)$	$\rightarrow$	$A'(^-3,^-4)$
$B(^-3,2)$	$\rightarrow$	$B'(^-3,^-2)$
$C(^-1,2)$	$\rightarrow$	$C'(^-1,^-2)$

You can draw a rotation of a figure about the origin on a coordinate plane.

EXAMPLE 3

Rotate trapezoid *ABCD* clockwise 90° about the origin. Find the coordinates of the new figure, *A'B'C'D'*.

original

rotation

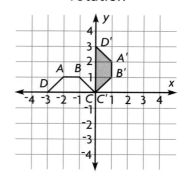

ABCD   A'B'C'D'

$A(^-2,1) \rightarrow A'(1,2)$
$B(^-1,1) \rightarrow B'(1,1)$
$C(0,0) \rightarrow C'(0,0)$
$D(^-3,0) \rightarrow D'(0,3)$

• **What if** you had rotated the trapezoid *counterclockwise* 90° about the origin? What would the coordinates of the new figure be?

CHECK FOR UNDERSTANDING

Think and ▶ Discuss

Look back at the lesson to answer each question.

1. **Compare** the *x*- and *y*-coordinates of the original figure and the new figure in Example 2. How are they related?

2. **Explain** why the coordinates of vertex *C* in Example 3 above did not change from the original figure to the new figure.

Guided ▶ Practice

Copy the figure onto a coordinate grid. Transform the figure according to the directions given. Name the new coordinates.

3. translate 2 units left

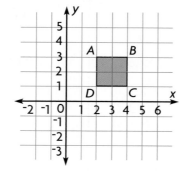

4. rotate 90° clockwise about (0,0)

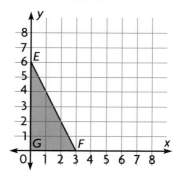

/* keep simple */

PRACTICE AND PROBLEM SOLVING

Independent ▶ Practice

Copy the figure onto a coordinate grid. Transform the figure according to the directions given. Name the new coordinates.

5.

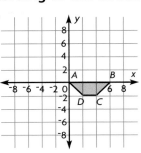

rotate 180° counterclockwise about (0,0)

6.

reflect across the *y*-axis

7.

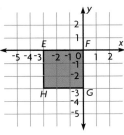

translate 3 units right and 1 unit down

Graph rectangle *EFGH* with coordinates (1,⁻2), (4,⁻2), (4,⁻4), and (1,⁻4).

8. Translate the rectangle 5 units up. What are the new coordinates of the rectangle?

9. Reflect the new rectangle in Exercise 8 across the *x*-axis. What are the new coordinates of the rectangle?

Problem Solving ▶ Applications

For 10–11, use the figure at the right. Triangle *A* has been transformed into triangle *B* by a 180° rotation about (0,0).

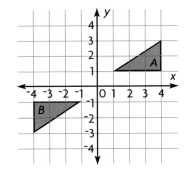

10. Could you complete the same transformation using only reflections? If so, how?

11. Could you complete the same transformation using only translations? Explain.

12. ❓ **What's the Question?** The coordinates of △*ABC* are *A*(1,⁻3), *B*(1,⁻1), and *C*(3,⁻1). The answer is that the new coordinates are *A*′(1,3), *B*′(1,1), and *C*′(3,1).

MIXED REVIEW AND TEST PREP

13. Use the table to write an equation that relates *x* and *y*. (p. 564)

x	0	1	2	3
y	0	7	14	21

14. Complete: 33, 24, 15, 6, ▨ (p. 529)   **15.** $\frac{1}{2} \times \frac{2}{3} \times \frac{3}{4}$ (p. 202)

16. TEST PREP How many lines of symmetry does an equilateral triangle have? (p. 550)

A 1   **B** 2   **C** 3   **D** 4

17. TEST PREP Find the greatest common factor of 35, 56, and 63. (p. 150)

F 6   **G** 7   **H** 8   **J** 9

Extra Practice page H61, Set D) **573**

1. **VOCABULARY** The point where the *x*-axis and *y*-axis intersect is called the __?__ . (p. 560)

2. **VOCABULARY** The *x*-axis and the *y*-axis divide the coordinate plane into four __?__ . (p. 560)

Graph the solutions of the inequality. (pp. 558–559)

3. $x < 8$ 4. $x \geq {}^-2$ 5. $x \leq {}^-4$ 6. $x > {}^-1$

Solve the inequality. (pp. 558–559)

7. $n + 5 < 8$ 8. $t - 6 > {}^-2$ 9. $5c \geq 35$

Describe how to locate the point for the ordered pair on the coordinate plane. (pp. 560–563)

10. $(2,4)$ 11. $(3,{}^-2)$ 12. $({}^-1,{}^-6)$

13. List the ordered pairs from the table. (pp. 564–565)

x	1	2	3	4
y	4	5	6	7

14. For the function table in Exercise 13, write an equation that relates *y* to *x*. (pp. 564–565)

15. For the values in the function table, write an equation that relates *b* to *a*. (pp. 564–565)

a	1	2	3	4
b	3	6	9	12

Solve. (pp. 570–573)

16. Trapezoid *DEFG* has coordinates $D(3,3)$, $E(4,3)$, $F(5,1)$, and $G(2,1)$. If it is reflected across the *x*-axis, what are the new coordinates?

17. Triangle *ABC* has coordinates $A(0,0)$, $B(3,2)$, and $C(0,2)$. What are the new coordinates after it is rotated 90° clockwise about the origin?

18. Triangle *XYZ* has coordinates $X(4,1)$, $Y(7,1)$, and $Z(7,6)$. It is translated 1 unit down and 3 units left. What are the new coordinates?

Solve. (pp. 566–567)

19. The table shows the total cost, *c*, for the number of tickets bought, *t*. Write an equation using *t* that gives the total cost, *c*.

t	1	2	3	4
c	$4	$8	$12	$16

20. Ella spends $3 on beads for each bracelet that she makes. The table shows her profit, *y*, when she sells a bracelet for price *x*. Write an equation using *x* that gives her profit, *y*.

x	$3	$4	$5	$6
y	$0	$1	$2	$3

Decide on a plan.

See item **5.**

Use the strategy *draw a diagram* to draw a coordinate plane. Plot the given coordinates and then reflect them across the *x*-axis.

Also see problem **4**, p. H63.

Choose the best answer.

1. Which algebraic inequality is represented by the graph on the number line?

A $x \le 3$ C $x < 3$

B $x \ge 3$ D $x > 3$

2. Which point represents the ordered pair ($^-$1,2) on the coordinate plane below?

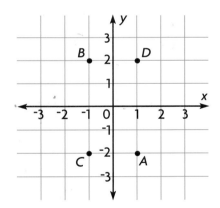

F A H C

G B J D

3. Rectangle *ABCD* has coordinates $A(0,0)$, $B(2,0)$, $C(2,1)$, and $D(0,1)$. The rectangle is rotated 90° clockwise around the origin. What are the coordinates of point B' in the new figure?

A (0,$^-$2)   C (1,0)

B (0,0)   D (2,1)

4. Triangle *ABC* has coordinates $A(1,2)$, $B(1,0)$, and $C(3,0)$. The triangle is translated 4 units left. What are the new coordinates of point C' in the new figure?

F (3,$^-$4)

G ($^-$2,$^-$3)

H ($^-$3,$^-$1)

J ($^-$1,0)

5. Rectangle *ABCD* has the coordinates ($^-$5,3), ($^-$2,3), ($^-$5,1), ($^-$2,1). What are the coordinates of A', B', C', and D' if it is reflected across the *x*-axis?

A ($^-$3,$^-$5), ($^-$3,$^-$2), ($^-$1,$^-$2), ($^-$1,$^-$5)

B ($^-$5,$^-$3), ($^-$2,$^-$3), ($^-$5,$^-$1), ($^-$2,$^-$1)

C (2,$^-$2), (5,$^-$2), (2,$^-$4), (5,$^-$4)

D (3,4), (3,1), (1,1), (1,4)

6. What is the least common multiple of 16 and 20?

F 4

G 60

H 80

J 320

7. Which can be the measure of an acute angle?

A 45°

B 90°

C 95°

D 180°

8. Bijan rides his bike for 4 hours and travels 30 miles. How many miles per hour is that?

F 15 miles per hour

G 10 miles per hour

H 7.5 miles per hour

J Not here

MATH DETECTIVE

Analyze This

For each case below, describe at least one way the sequences are the same and at least one way they are different. Then describe a rule and write the next three terms in each sequence.

Case 1

10, 20, 30, 40, . . .
10, 0, ⁻10, ⁻20, . . .

Case 2

4, 7, 10, 13, . . .
4, 12, 36, 108, . . .

Case 3

500, 50, 5, 0.5, . . .
300, 250, 200, 150 . . .

Case 4

M, A, T, H, M, A, T, H, M, A, T, H . . .
1, 2, 3, 4, 1, 2, 3, 4, 1, 2, 3, 4, . . .

Think It Over!

- 📓 **Write About It** Think of a number sequence that starts with 100 and involves division. Then describe the steps you would use to find the next term.

- **Stretch your thinking** If the sequence ↑, →, ↓, ←, ↑, →, ↓, ←, . . . is continued, what term will be in the 24th position? Explain.

Stretching Figures

Melissa is making a small quilt. She found a pattern for a star in a quilting book. She will cut out the stars she makes and sew them onto the quilt.

Here are the coordinates of the pattern:
(0,4), (1,1), (3,1), (1,0), (2,⁻2), (0,⁻1), (⁻2,⁻2), (⁻1,0), (⁻3,1), (⁻1,1)

Melissa would like the sides of the pattern to be longer, so she decides to enlarge, or stretch, them by doubling the coordinates.

When she doubles the coordinates, she gets: (0,8), (2,2), (6,2), (2,0), (4,⁻4), (0,⁻2), (⁻4,⁻4), (⁻2,0), (⁻6,2), (⁻2,2)

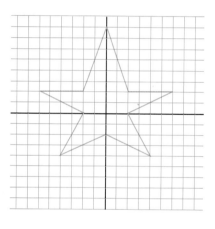

TALK ABOUT IT

What if Melissa wants to make stars of different sizes on her quilt? What coordinates should she use to draw a star if she wants the sides of the star to be five times as long as the original pattern?

TRY IT

Use the figure shown at the right. Give the new coordinates and draw the stretched figure.

1. Stretch the figure so the sides are twice as long.

2. Stretch the figure so the sides are five times as long.

3. Stretch the figure by doubling only the x-coordinates.

4. Stretch the figure by doubling only the y-coordinates.

VOCABULARY

1. Turning a figure about a point is called a __?__. (p. 540)

2. A figure has __?__ if it can be folded or reflected so that the two parts of the figure match. (p. 550)

3. The horizontal number line in the coordinate plane is called the __?__. (p. 560)

EXAMPLES

EXERCISES

Chapter 28

- **Write an equation to represent a function.** (pp. 529–532)

x	15	18	21	24
y	12	15	18	21

The x-values are 3 more than the y- values, so the equation is y = x − 3.

Write an equation to represent the function.

4.

x	15	16	17	18
y	10	11	12	13

5.

x	72	64	56	48
y	9	8	7	6

- **Extend geometric patterns.** (pp. 533–535)

What is the next figure in the pattern?

The shaded square moves clockwise 90° in these figures. So, the next figure would have the shaded square in the bottom left corner.

Draw the next two figures in the pattern.

6.

7.

Chapter 29

- **Use translations, rotations, and reflections to transform geometric shapes.** (pp. 540–542)

A *translation* moves a figure along a straight line.
A *rotation* turns a figure around a point.

A *reflection* flips a figure over a line.

Tell which type of transformation the second figure is of the first figure. Write *translation*, *rotation*, **or** *reflection*.

8.

9.

10.

11.

- **Identify rotational symmetry.** (pp. 550–553)

  How much would you have to turn the flower until it matches the original?

  $\frac{1}{4}$ turn, or 90°

Tell whether each figure has rotational symmetry. If it does, identify the symmetry as a fraction of a turn and in degrees.

12.

13.

14.

15.

Chapter 30

- **Solve algebraic inequalities.** (pp. 558–559)

$$a + 7 > 12$$
$$a + 7 - 7 > 12 - 7 \qquad \textit{Subtract 7 from both sides.}$$
$$a > 5$$

Solve the inequality.

16. $x - 9 > 14$

17. $y + 21 < 50$

18. $7z < 63$

19. $\frac{a}{5} > 14$

- **Use an ordered pair to describe a location on the coordinate plane.**

  (pp. 560–563)

  The first number of an ordered pair tells you how far right or left to move from the origin.
  The second number of an ordered pair tells you how far to move up or down.
  (⁻3,5) Move 3 units left and 5 units up.

Write the ordered pair for each point on the coordinate plane.

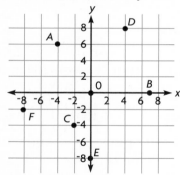

20. A

21. B

22. C

23. D

24. E

25. F

PROBLEM SOLVING APPLICATIONS

26. Nikhil wants to buy a computer that costs $850. His savings grew from $100 in January to $175 in February to $250 in March. If this pattern continues, in what month will he be able to buy the computer? (pp. 524–525)

27. A garden is being enclosed with forty-eight 1-ft sections of fencing. What is the largest possible area of the garden?

    (pp. 546–547)

579

California Connections

Patterns in Nature

Sunflowers, artichokes, pineapples, daisies, and pine cones all grow in California. These objects have something else in common, too. They exhibit one of the most famous number patterns ever discovered.

The cones of California's Coulter pine are the world's largest, weighing up to 5 pounds each.

California produces more vegetables and fruits than any other state.

1. The Fibonacci sequence is a pattern of numbers that begins 1, 1, 2, 3, 5, 8, 13, ... How is each new number in the pattern found?

2. Write the next five numbers in the Fibonacci sequence.

For 3–4, use the drawing.

3. Study the pattern of spirals on the pine cone. Start at the center. How many of the spirals turn clockwise? How many turn counterclockwise?

4. How is the pine cone related to the Fibonacci sequence?

5. Sunflowers, artichokes, pineapples, and daisies also model the Fibonacci sequence. One sunflower had 55 clockwise spirals. How many counterclockwise spirals did it have?

Use Data   For 6–8, use the diagrams.

6. Suppose square A measures 1 in. × 1 in. Find the measures of one side of squares B, C, D, and E.

7. Copy the last figure (containing square E). Then draw the next two figures in the pattern.

8. If the pattern of figures continues, what will the measurements of square M be?

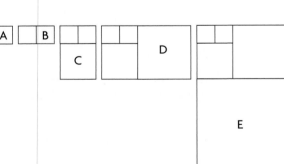

Patterns in Architecture

The J. Paul Getty Museum opened its new home in the Santa Monica Mountains above Los Angeles in 1996. Art and architecture are the main attractions at "the Getty," where even the stairwells are beautiful.

The sculptures at the Getty Museum attract visitors of all ages.

For 1–3, use the stairwell photo.

1. The photo shows a stairwell at the Getty Museum. Each flight (Line A or Line B) has 12 steps. Write a sequence showing the total number of steps in 1 flight, 2 flights, 3 flights, and so on, up to 8 flights.

2. How many flights are shown in the photograph? How many stairs are shown?

3. Write an algebraic expression you can use to find the total number of steps in n flights.

For 4–5, use the outline of the sideview of a flight of stairs.

4. Does this drawing of flight A have rotational symmetry? If so, tell the fraction and angle measure of each turn.

5. Explain how you could transform flight A into flight B by using a translation, a rotation, or a reflection.

For 6–7, use the roof photo and the table.

6. A vertical column of tiles at the Getty is outlined in the photo. Complete the table, giving the total number of tiles in the given number of columns.

NUMBER OF COLUMNS	1	2	3	4	5	6
NUMBER OF TILES	■	■	■	■	■	■

7. Use the ordered pairs in the table to draw a graph of the relation in Exercise 6.

Student Handbook

Troubleshooting H2

Prerequisite Skills Review Do you have the math skills needed to start a new chapter? Use this list of skills to review and remember your skills from last year.

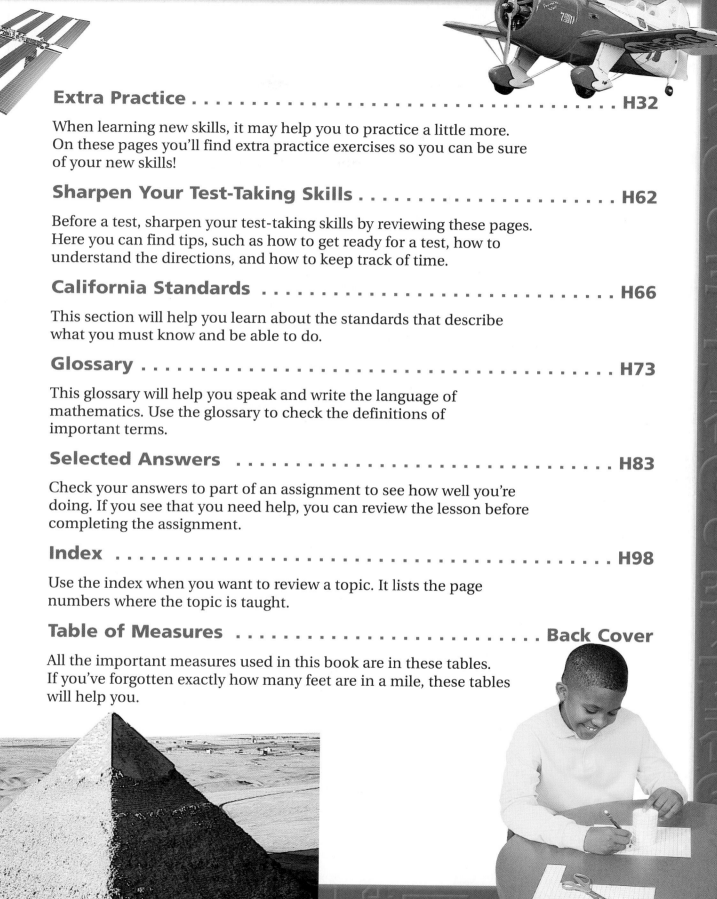

Troubleshooting

Properties

The charts list the basic properties of addition and multiplication.

Addition

PROPERTY	EXAMPLE WITH NUMBERS	EXAMPLE WITH VARIABLES
Commutative	$3 + 7 = 7 + 3$	$a + b = b + a$
Associative	$(4 + 5) + 2 = 4 + (5 + 2)$	$(a + b) + c = a + (b + c)$
Identity Property of Zero	$9 + 0 = 9$ and $0 + 9 = 9$	$a + 0 = a$ and $0 + a = a$

Multiplication

PROPERTY	EXAMPLE WITH NUMBERS	EXAMPLE WITH VARIABLES
Commutative	$8 \times 6 = 6 \times 8$	$a \times b = b \times a$
Associative	$(2 \times 9) \times 5 = 2 \times (9 \times 5)$	$(a \times b) \times c = a \times (b \times c)$
Identity Property of One	$6 \times 1 = 6$ and $1 \times 6 = 6$	$a \times 1 = a$ and $1 \times a = a$
Property of Zero	$7 \times 0 = 0$ and $0 \times 7 = 0$	$a \times 0 = 0$ and $0 \times a = 0$
Distributive	$3 \times (5 + 7) = (3 \times 5) + (3 \times 7)$	$a \times (b + c) = (a \times b) + (a \times c)$

Practice

Name the property shown.

1. $0 \times 12 = 0$

2. $(8 \times 21) \times 5 = 8 \times (21 \times 5)$

3. $72 \times 5 = 5 \times 72$

4. $4 \times (3 + 2) = (4 \times 3) + (4 \times 2)$

5. $9 + 0 = 9$

6. $k + 7 = 7 + k$

7. $(m + 4) + h = m + (4 + h)$

8. $c \times (2 + 3) = (c \times 2) + (c \times 3)$

9. $1 \times x = x$

Represent Decimals

You can use decimal squares to represent decimals. A decimal square is divided into 100 parts. Each part represents 1 hundredth of the whole, or 0.01. Count the number of shaded parts. Write this number as hundredths.

**24 out of 100 small squares are shaded.
0.24 of the whole is shaded.**

Practice

Write the decimal that is represented.

1.

2.

3.

4.

H2 Prerequisite Skills Review

Write and Read Decimals

A place-value chart can help you write and read numbers. Each three-digit group, such as ones or millions, is called a **period**.

Read: "thirty million, one hundred twenty-eight thousand, five hundred ninety-seven **and** forty-six thousandths."

MILLIONS			THOUSANDS			ONES					
Hundreds	Tens	Ones	Hundreds	Tens	Ones	Hundreds	Tens	Ones	Tenths	Hundredths	Thousandths
	3	0	1	2	8	5	9	7 .	0	4	6

Example

Name the place value of the digit, 8, in the chart.

To write the value of the 8 in the chart, multiply the digit times the value of the place-value position.

Think: $8 \times 1,000$, or 8,000.

Practice

Name the place value of the digit 4.

1. 327,489,223.78   **2.** 198,238,042.08   **3.** 149,678,935.91   **4.** 728,035.84

Give the value of the blue digit.

5. 538.92   **6.** 82,901,733.006   **7.** 29.35   **8.** 125,674.173

Compare and Order Whole Numbers

You can compare and order numbers by comparing the digits in each place-value position.

Example

Write $<$, $>$, or $=$ to compare the numbers.   14,675 ● 14,228

Step 1	Step 2	Step 3
Compare the ten thousands.	Compare the thousands.	Compare the hundreds.
14,675 ↓ same number of 14,228 ten thousands	14,675 ↓ same number of 14,228 thousands	14,675 ↓ 6 > 2 14,228 So, 14,675 > 14,228.

Practice

Write $<$, $>$, or $=$ to compare the numbers.

1. 3,919 ● 3,991   **2.** 188,937 ● 189,066   **3.** 70,001 ● 70,001

Order the numbers from greatest to least.

4. 9,774; 9,718; 9,762   **5.** 82,056; 82,856; 81,978   **6.** 23,091; 23,910; 23,109

Troubleshooting

Round Whole Numbers and Decimals

Follow these steps to round a number to a given place.

Example

Round 168,279 to the nearest ten thousand.

Step 1	**Step 2**	**Step 3**
Find the digit in the place being rounded, the ten thousands place.	Look at the next digit to the right. If it is 5 or greater, round up. Otherwise, round down.	Change to zero each digit to the right of the place being rounded.
1**6**8,279	16**8**,279 8 > 5, so round up.	168,279 → 170,000

Round to the nearest whole number.

1. 7.97 **2.** 15.58

Round to the nearest ten thousand.

5. 530,410 **6.** 12,677

Round to the nearest thousand.

3. 3,378 **4.** 7,607

Round to the nearest tenth.

7. 16.53 **8.** 2.96

Whole-Number Operations

When adding, subtracting, multiplying, and dividing whole numbers, align digits that have the same place value.

Examples

A. Find the sum. 247 + 1,496 + 89

$$\begin{array}{r} 2\,2 \\ 247 \\ 1{,}496 \\ +\ \ \ 89 \\ \hline 1{,}832 \end{array}$$

Think: 7 + 6 + 9 = 22
Regroup 22 ones as 2 tens 2 ones.
Add the other columns in a similar way.

B. Find the difference. 31 − 12

$$\begin{array}{r} 2\,11 \\ 3\!\!\!/\,1\!\!\!/ \\ -\ 1\,2 \\ \hline 1\,9 \end{array}$$

Regroup when necessary.

C. Find the product. 27 × 86

$$\begin{array}{r} 27 \\ \times\ 86 \\ \hline 162 \\ +\ 2\,160 \\ \hline 2{,}322 \end{array}$$

Multiply by the 6 ones.
Multiply by the 8 tens.
Add the products.

D. Find the quotient. 146 ÷ 9

$$\begin{array}{r} 16\ \text{r2} \\ 9{\overline{)146}} \\ -\ 9\ \ \ \\ \hline 56 \\ -\ 54 \\ \hline 2 \end{array}$$

Divide the 14 tens.
Multiply and subtract.
Bring down the 6 ones.
Divide the 56 ones.
Multiply and subtract.
Write the remainder.

Practice

Add, subtract, multiply, or divide.

1. 79 + 56 + 99 **2.** 345 − 26 **3.** 67 × 76 **4.** 376 ÷ 15

5. 700 − 388 **6.** 19 × 203 **7.** 155 + 9 + 4,823 **8.** 524 ÷ 31

Remainders

When dividing whole numbers, you can write a remainder as a whole number, as a decimal, or as a fraction.

Example

Divide: $15 \div 4$. Write the remainder as a decimal and as a fraction.

Step 1

Write the dividend as a two-place decimal.

$4\overline{)15.00}$

Step 2

Divide.

$\begin{array}{r} 3.75 \\ 4\overline{)15.00} \end{array}$

Step 3

Write the decimal as a fraction in simplest form.

$0.75 = \frac{75}{100} = \frac{75 \div 25}{100 \div 25} = \frac{3}{4}$

$15 \div 4 = 3\frac{3}{4}$

Practice

Divide. Write the remainder as a decimal and as a fraction.

1. $4\overline{)30}$
2. $8\overline{)50}$
3. $10\overline{)3,315}$
4. $12\overline{)2,442}$
5. $6\overline{)21}$
6. $8\overline{)94}$
7. $20\overline{)74}$
8. $25\overline{)210}$

Read a Table

Use the title of a table to understand what the data represent. Use the labels to understand what the items represent.

Example

How far did Angie jog on Tuesday?

DAILY JOGGING RECORD (IN MILES)							
Day	Sun	Mon	Tue	Wed	Thu	Fri	Sat
Milo	4	3	5	5	0	7	6
Angie	6	3	4	0	5	8	0

The number in the row marked "Angie" and the column marked "Tue" is 4. So, Angie jogged 4 miles on Tuesday.

Practice

Use the data in the table above to answer the questions.

1. How far did Milo jog on Friday?
2. Which person jogged 5 miles on Wednesday?
3. On which day did Angie jog 6 miles?
4. On which day did Milo and Angie jog the same distance?
5. How far did Angie jog on the two days that she jogged the same distance?
6. Who ran the greater total distance during the week?

Troubleshooting

Mean, Median, Mode, and Range

Example

Find the mean, median, mode, and range for this set of data.
6, 19, 6, 9, 11, 15

Mean Find the sum of the data items. Divide the sum by the number of items.	$6 + 19 + 6 + 9 + 11 + 15 = 66$ mean $= 66 \div 6 = 11$
Median Arrange the items from least to greatest. The median is the middle value. If there are two middle values, the median is the average of the two values.	6, 6, **9**, **11**, 15, 19 9 and 11 are the middle numbers. median $= (9 + 11) \div 2 = 10$
Mode Arrange the items from least to greatest. The mode is the value or values that repeat most often. If no value repeats, there is no mode.	**6**, **6**, 9, 11, 15, 19 mode $= 6$
Range Arrange the items from least to greatest. The range is the difference of the greatest and least values.	**6**, 6, 9, 11, 15, **19** range $= 19 - 6 = 13$

Practice

Find the mean, median, mode, and range for each set of data.

1. 2, 1, 3, 5, 9   **2.** 6, 52, 41, 21, 35   **3.** 11, 15, 6, 11, 22   **4.** 9, 5, 2, 5, 6, 1, 14

Read Bar Graphs

A **bar graph** uses bars of different lengths to show and compare data.

Example

How many books did Anita read last month?

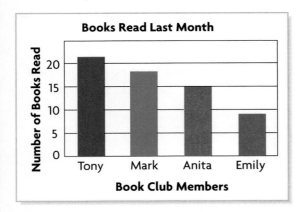

Find the bar representing Anita.

Look at the top of the bar. Read the scale at the left for the number the bar represents.

Anita read 15 books last month.

Practice

Use the bar graph to answer the questions.

1. How many books did Mark read?

2. Who read 22 books?

3. Who read the least number of books?

4. How many more books did Tony read than Anita?

Read Stem-and-Leaf Plots

In a **stem-and-leaf plot**, data are organized by "stems," or the tens digits, and "leaves," or the ones digits. The stems and leaves are listed in order.

Example

Each entry in the stem-and-leaf plot gives the number of cards that were stacked in a house of cards when it collapsed. How many times were there from 40 to 49 cards when the collapse occurred?

Stem	Leaves
1	2 5 7 8 8 9
2	0 0 1 3 5 8
3	1 2 4 7
4	2 8
5	1 6

The numbers from 40 to 49 have the stem of 4, because the tens digit is 4. There are two such numbers in the table:

Stem 4, leaf 2 shows **42**.
Stem 4, leaf 8 shows **48**.

So, twice there were from 40 to 49 cards. Once there were 42 cards, and once there were 48.

Practice

Use the stem-and-leaf plot to answer the questions.

1. How many times were there from 10 to 19 cards when the collapse occurred?

2. What was the greatest number of cards stacked? the least number?

3. How many houses of cards were built in this competition?

4. What are the mode, median, and range of the data in the table?

Prime and Composite Numbers

A **prime number** is a whole number that has exactly two factors, itself and 1. A **composite number** has more than two factors.

Examples

Decide whether the number is *prime* or *composite*.

A. 13 13 has exactly two factors, 13 and 1. These are the only whole numbers that divide 13 evenly. So, 13 is prime.

B. 12 12 has the factors 1, 2, 3, 4, 6, and 12. Each of these whole numbers divides 12 evenly. Since 12 has more than two factors, it is composite.

Practice

Decide whether the number is *prime* or *composite*.

1. 19   **2.** 6   **3.** 10   **4.** 2   **5.** 15

6. 28   **7.** 31   **8.** 99   **9.** 71   **10.** 456

Troubleshooting

Factors and Multiples

Multiples of a number are the products that result when the number is multiplied by 0, 1, 2, 3, 4, and so on.

27 is a multiple of 9 because $27 = 9 \times 3$.

Factors are numbers that divide a whole number evenly.

8 is a factor of 32 because 8 divides 32 evenly: $32 \div 8 = 4$.

Examples

A. List the next three multiples of 7: 7, 14, 21.

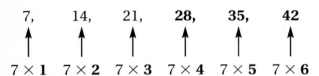

So, the next three multiples are 28, 35, and 42.

B. Find all the factors of 20.

Think: $1 \times 20 = 20$, so 1 and 20 are factors of 20. $2 \times 10 = 20$, so 2 and 10 are factors of 20. $4 \times 5 = 20$, so 4 and 5 are factors of 20.

So, the factors of 20 are 1, 2, 4, 5, 10, and 20.

Practice

List the next three multiples of the number.

1. 5: 5, 10, 15   **2.** 9: 9, 18, 27   **3.** 6: 12, 18, 24   **4.** 20: 40, 60, 80

Find all of the factors of the number.

5. 13   **6.** 12   **7.** 30   **8.** 24

9. 15   **10.** 22   **11.** 28   **12.** 36

Model Fractions

A **fraction** is a number that names a part of a whole or a part of a group.

Example

Write the fraction represented by the shaded part.

The group has 7 squares. The part that is shaded is 4 squares.

So, the fraction shaded is $\frac{4}{7}$.

Practice

Write the fraction represented by the shaded part.

1.   **2.**   **3.**   **4.**

Model Percents

Percent (%) means "per hundred." You can use a hundred square to model percents.

Example

Write the percent for the shaded part of the square.

There are 100 parts, and 75 of the parts are shaded.
So, 75% of the parts are shaded.

Practice

Write the percent for the shaded part of the square.

1.   **2.**   **3.**   **4.**

Simplify Fractions

A fraction is in **simplest form** when the numerator and denominator have no common factors other than 1.

$\frac{10}{15}$ is not in simplest form, because 10 and 15 have the common factor 5.

$\frac{7}{12}$ is in simplest form, because 7 and 12 have no common factors other than 1.

You can write a fraction in simplest form by dividing the numerator and denominator by the greatest common factor (GCF).

Example

Write the fraction in simplest form. $\frac{16}{28}$

Step 1

Find the greatest common factor of 16 and 28.

factors of 16: 1, 2, 4, 8, 16
factors of 28: 1, 2, 4, 7, 14, 28

The *greatest* common factor is 4.

Step 2

Divide the numerator and the denominator by the GCF.

$\frac{16 \div 4}{28 \div 4} = \frac{4}{7}$ Since the GCF of 4 and 7 is 1, $\frac{4}{7}$ is in simplest form.

Practice

Write the fraction in simplest form.

1. $\frac{9}{15}$ **2.** $\frac{8}{12}$ **3.** $\frac{5}{20}$ **4.** $\frac{18}{6}$ **5.** $\frac{24}{32}$ **6.** $\frac{24}{30}$

Troubleshooting

Add and Subtract Like Fractions

Like fractions are fractions that have the same denominator. To add or subtract like fractions, add or subtract the numerators. Keep the same denominator.

Example

Find the sum. $\frac{4}{9} + \frac{2}{9}$

Step 1

Add the numerators. Write the sum over the like denominator.

$$\frac{4}{9} + \frac{2}{9} = \frac{4+2}{9} = \frac{6}{9}$$

Step 2

If necessary, rewrite the fraction in simplest form.

$$\frac{6}{9} = \frac{6 \div 3}{9 \div 3} = \frac{2}{3}$$

Practice

Find the sum or difference. Write the answer in simplest form.

1. $\frac{1}{7} + \frac{4}{7}$ **2.** $\frac{4}{5} - \frac{2}{5}$ **3.** $\frac{3}{8} + \frac{3}{8}$ **4.** $\frac{3}{4} + \frac{1}{4}$ **5.** $\frac{7}{10} - \frac{5}{10}$

6. $\frac{8}{12} - \frac{4}{12}$ **7.** $\frac{15}{11} - \frac{9}{11}$ **8.** $\frac{5}{9} + \frac{8}{9}$ **9.** $\frac{2}{10} + \frac{6}{10}$ **10.** $\frac{12}{8} - \frac{4}{8}$

11. $\frac{3}{6} + \frac{1}{6}$ **12.** $\frac{5}{8} + \frac{1}{8}$ **13.** $\frac{7}{16} + \frac{5}{16}$ **14.** $\frac{17}{20} - \frac{9}{20}$ **15.** $\frac{11}{25} + \frac{4}{25}$

Round Fractions

To round a fraction to 0, $\frac{1}{2}$, or 1, compare the numerator with the denominator.

Examples

Round each fraction to 0, $\frac{1}{2}$, or 1.

A. $\frac{2}{11}$

Compared to 11, 2 is close to 0.

Round the fraction to 0.

B. $\frac{7}{15}$

7 is about half of 15.

Round the fraction to $\frac{1}{2}$.

C. $\frac{13}{16}$

13 is close to 16.

Round the fraction to 1.

Practice

Round each fraction to 0, $\frac{1}{2}$, or 1.

1. $\frac{3}{4}$ **2.** $\frac{1}{5}$ **3.** $\frac{2}{7}$ **4.** $\frac{6}{13}$ **5.** $\frac{5}{12}$

6. $\frac{14}{15}$ **7.** $\frac{4}{18}$ **8.** $\frac{3}{7}$ **9.** $\frac{4}{5}$ **10.** $\frac{12}{20}$

11. $\frac{11}{24}$ **12.** $\frac{17}{20}$ **13.** $\frac{11}{18}$ **14.** $\frac{3}{25}$ **15.** $\frac{15}{16}$

Mental Math and Equations

You can use mental math to solve equations. Try using a related equation to find the value of the variable.

Use mental math to solve the equation.

A. $k + 6 = 10$

Think: When 6 is added to k, the sum is 10. That means that $k = 10 - 6$.

So, $k = 4$. Check: $4 + 6 = 10$ ✔

B. $m - 4 = 11$

Think: When 4 is subtracted from m, the difference is 11. That means that $m = 11 + 4$.

So, $m = 15$. Check: $15 - 4 = 11$ ✔

C. $c \div 5 = 7$

Think: When c is divided by 5, the quotient is 7. That means that $c = 7 \times 5$.

So, $c = 35$. Check: $35 \div 5 = 7$ ✔

D. $h \times 4 = 36$

Think: When h is multiplied by 4, the product is 36. That means that $h = 36 \div 4$.

So, $h = 9$. Check: $9 \times 4 = 36$ ✔

Use mental math to solve the equation.

1. $g - 5 = 6$ **2.** $r \times 2 = 12$ **3.** $d \div 3 = 6$ **4.** $x + 9 = 17$

5. $v \div 6 = 5$ **6.** $y - 6 = 9$ **7.** $a + 3 = 15$ **8.** $n \times 3 = 21$

Fractions and Mixed Numbers

A **mixed number** is made up of a whole number and a fraction. You can write a mixed number as a fraction greater than 1, and a fraction greater than 1 as a mixed number.

A. Write $\frac{14}{3}$ as a mixed number.

$$3\overline{)14} \quad \text{4 r2}$$ Divide the numerator by the denominator.

$\frac{14}{3} = 4\frac{2}{3}$ Write the remainder as the numerator of a fraction.

B. Write $3\frac{4}{5}$ as a fraction.

$3 \times 5 = 15$ Multiply the denominator by the whole number.

$15 + 4 = 19$ Add the numerator.

$3\frac{4}{5} = \frac{19}{5}$ Write the sum over the denominator.

Write the fraction as a mixed number.

1. $\frac{13}{7}$ **2.** $\frac{18}{5}$ **3.** $\frac{10}{3}$ **4.** $\frac{22}{9}$ **5.** $\frac{17}{6}$

Write the mixed number as a fraction.

6. $2\frac{4}{5}$ **7.** $1\frac{3}{8}$ **8.** $3\frac{3}{4}$ **9.** $2\frac{22}{25}$ **10.** $2\frac{11}{16}$

Troubleshooting

Compare Fractions

To compare fractions with unlike denominators, rename the fractions so that they have like denominators. Then compare the numerators.

Examples

Compare. Use $<$, $>$, or $=$ for ●.

A. $\frac{5}{8}$ ● $\frac{7}{8}$

The denominators are like. Compare the numerators.

$5 < 7$, so $\frac{5}{8} < \frac{7}{8}$.

B. $\frac{7}{10}$ ● $\frac{3}{5}$

The denominators are unlike. Rename one or both fractions.

$\frac{3 \times 2}{5 \times 2} = \frac{6}{10}$

$7 > 6$, so $\frac{7}{10} > \frac{3}{5}$.

C. $2\frac{5}{12}$ ● $2\frac{1}{3}$

The whole numbers are equal. Rename one or both fractions.

$\frac{1 \times 4}{3 \times 4} = \frac{4}{12}$

$5 > 4$, so $2\frac{5}{12} > 2\frac{1}{3}$.

Practice

Compare. Use $<$, $>$, or $=$ for ●.

1. $\frac{4}{5}$ ● $\frac{3}{5}$

2. $\frac{11}{15}$ ● $\frac{13}{15}$

3. $1\frac{4}{6}$ ● $1\frac{2}{3}$

4. $2\frac{5}{8}$ ● $2\frac{1}{2}$

5. $\frac{2}{3}$ ● $\frac{5}{12}$

6. $\frac{7}{8}$ ● $\frac{3}{4}$

7. $\frac{5}{15}$ ● $\frac{2}{6}$

8. $\frac{8}{9}$ ● $\frac{2}{3}$

9. $\frac{13}{16}$ ● $\frac{7}{8}$

10. $\frac{5}{9}$ ● $\frac{14}{27}$

Understand Integers

The **integers** are the whole numbers and their opposites. Integers greater than 0 are **positive integers** and are found to the right of 0 on a number line. Integers less than 0 are **negative integers** and are found to the left of 0.

$$-6 \quad -5 \quad -4 \quad -3 \quad -2 \quad -1 \quad 0 \quad 1 \quad 2 \quad 3 \quad 4 \quad 5 \quad 6$$

Example

Name the temperatures indicated by point A and point B.

The temperature at point A is $^-35°$. The temperature at point B is $^+68°$.

Practice

Give four examples of each set of numbers.

1. whole numbers

2. negative integers

Write a positive or negative integer for each situation.

3. thermometer point C

4. thermometer point D

5. thermometer point E

6. thermometer point F

7. 35 ft above sea level

8. a loss of $5

Number Lines

You can use a number line to graph numbers and to compare and order numbers. If two numbers are graphed on a number line, the number to the right is greater.

Example

Graph ⁻3, 2, and ⁻1.5 on the number line. Then order the numbers from least to greatest.

Graph each number by placing a dot on the number line.

Notice that ⁻1.5 appears halfway between ⁻1 and ⁻2. Since values increase as you move to the right on a number line, the order of the numbers from least to greatest is ⁻3, ⁻1.5, 2.

Practice

Name the number graphed by each point.

1. A **2.** B **3.** C **4.** D **5.** E

Graph the numbers on a number line. Then order the numbers from least to greatest.

6. 4, 2, ⁻1   **7.** ⁻1, ⁻3, 0   **8.** 1.5, 2, ⁻1.5   **9.** 3, ⁻3, ⁻4   **10.** ⁻3.5, ⁻3, ⁻5

Multiplication and Division Facts

To multiply a number by a power of 10, move the decimal point one place to the right for each zero. To divide a number by a power of 10, move the decimal point one place to the left for each zero.

If you forget a multiplication fact that has 11 or 12 as a factor, think of 11 as 10 + 1, and think of 12 as 10 + 2.

Example

Find the product. 12 × 8

$12 \times 8 = (\mathbf{10} + 2) \times 8 = (\mathbf{10} \times 8) + (2 \times 8) = 80 + 16 = 96$

Practice

Find the product or quotient.

1. 56 × 10   **2.** 780 ÷ 10   **3.** 4.3 × 100   **4.** 5,280 ÷ 1,000   **5.** 2.61 × 1,000

6. 0.48 × 100   **7.** 124 ÷ 10   **8.** 5.77 ÷ 100   **9.** 11 × 10   **10.** 12 × 9

Troubleshooting

Add and Subtract Integers

You can add and subtract integers on a number line. To add a positive number on a number line, move right. To add a negative number, move left. To subtract a number on a number line, *add the opposite* of the number.

Examples

A. Add. $2 + {}^-6$

From 0, move to $^+2$. Move 6 units left to $^-4$.

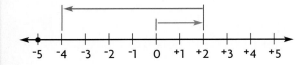

$2 + {}^-6 = {}^-4$

B. Subtract. $^-3 - {}^-7 \longrightarrow {}^-3 + {}^+7$

Add the opposite.

From 0, move to $^-3$. Move right 7 units to $^+4$.

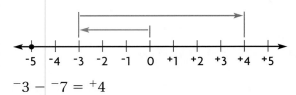

$^-3 - {}^-7 = {}^+4$

Practice

Find the sum or difference.

1. $4 - 7$ **2.** $^-3 + {}^-8$ **3.** $5 + {}^-6$ **4.** $^-8 - 3$ **5.** $7 + {}^-3$

6. $2 - {}^+5$ **7.** $9 - 4$ **8.** $11 + {}^-7$ **9.** $^-5 + {}^-6$ **10.** $2 - {}^-5$

Patterns

Look for a rule for the pattern. Then use it to extend the pattern.

Example

Find the next three numbers in the pattern.   1, 2, 4, 8

Step 1   Find the rule.

Each number in the series is 2 times the number before it.
$1 \times 2 = 2; 2 \times 2 = 4; 4 \times 2 = 8$

The rule is "multiply by 2."

Step 2   Use the rule.

Multiply each number by 2 to find the next number:
$8 \times 2 = 16, 16 \times 2 = 32, 32 \times 2 = 64$

So, the next three numbers are 16, 32, and 64.

Practice

Find the next three numbers in the pattern. Write the rule.

1. 10, 14, 18, 22, 26   **2.** 729, 243, 81, 27   **3.** 1, 4, 16

4. 63, 55, 47, 39   **5.** 49, 37, 25, 13   **6.** $2\frac{1}{2}, 1\frac{3}{4}, 1, \frac{1}{4}$

Exponents

You can use exponents to express powers of numbers. An **exponent** tells how many times the **base** is used as a factor.

$$6 \times 6 \times 6 \times 6 = 6^4$$

↓ exponent

↑ base

The exponent 4 shows that the base 6 is used 4 times as a factor.

Examples

Find the value of 5^3.

A. $5^3 = 5 \times 5 \times 5 = 125$

Write 49 by using an exponent.

B. $49 = 7 \times 7 = 7^2$

Practice

Find the value.

1. 2^4 **2.** 4^3 **3.** 5^2 **4.** 10^3 **5.** 3^4

Write by using an exponent.

6. 9   **7.** 32   **8.** 36   **9.** 100   **10.** 27

Order of Operations

When more than one operation is used in an expression, follow these rules to evaluate the expression.

RULES FOR ORDER OF OPERATIONS	
1. First, do the operations **in parentheses**.	3. Next, **multiply and divide** from left to right.
2. Next, evaluate **exponents**.	4. Finally, **add and subtract** from left to right.

Examples

Evaluate each expression.

A. $3^2 + 5 \times 2$

1. There are no parentheses.

2. Evaluate exponents. $3^2 + 5 \times 2 = 9 + 5 \times 2$

3. Multiply. $9 + 5 \times 2 = 9 + 10$

4. Add. $9 + 10 = 19$

B. $(21 - 6) \div 3$

1. Evaluate parentheses. $(21 - 6) \div 3 = 15 \div 3$

2. There are no exponents.

3. Divide. $15 \div 3 = 5$

4. There is no addition or subtraction.

Practice

Evaluate each expression.

1. $4 + 6 \times 9$ **2.** $(5 + 4) \times 3$ **3.** $25 - 12 \times 3$ **4.** $2^2 + 3^2$

5. $(2 + 3)^2$ **6.** $\frac{10 - 4}{2} \times 4^2$ **7.** $5^2 \div (1^5 + 4)$ **8.** $2 + 3 \times 3 \times 3$

Troubleshooting

Add and Subtract

To add or subtract unlike fractions, write equivalent fractions with common denominators. Then add or subtract the like fractions.

To add or subtract decimals, align the decimal points. Write equivalent decimals. Then add or subtract as you would with whole numbers.

Examples

Find the sum. $\frac{1}{6} + \frac{1}{2}$

A. $\frac{1}{6} + \frac{1}{2}$ Write equivalent fractions with a denominator of 6.

$$\frac{1 \times 3}{2 \times 3} = \frac{3}{6}$$

$\frac{1}{6} + \frac{3}{6} = \frac{4}{6}$, or $\frac{2}{3}$ Add the numerators and simplify.

Find the difference. $34.7 - 3.651$

B.
$$\begin{array}{r} 34.700 \\ -\ 3.651 \\ \hline 31.049 \end{array}$$
Align the decimal points.
Write an equivalent decimal.
Subtract as for whole numbers.

Practice

Find the sum or difference.

1. $3.12 + 2.7$ **2.** $1.197 - 1.09$ **3.** $6 - 3.4$ **4.** $2.17 + 141.8$

5. $\frac{3}{4} + \frac{1}{6}$ **6.** $\frac{5}{8} - \frac{3}{16}$ **7.** $\frac{3}{5} + \frac{1}{4}$ **8.** $\frac{1}{3} + \frac{5}{12}$

Inverse Operations

Addition and subtraction are **inverse** operations. This means that you can check a sum or difference by using the inverse operation. You can also check a product or quotient by using the inverse operation.

Examples

Use the inverse operation to check the answer.

A.
$$\begin{array}{r} 12 \\ +\ 15 \\ \hline 27 \end{array} \qquad \begin{array}{r} 27 \\ -\ 15 \\ \hline 12 \end{array}$$

B.
$$\begin{array}{r} 322 \\ -\ 180 \\ \hline 142 \end{array} \qquad \begin{array}{r} 142 \\ +\ 180 \\ \hline 322 \end{array}$$

C.
$$\begin{array}{r} 12 \\ \times\ 9 \\ \hline 108 \end{array} \qquad 9\overline{)108}$$

D.
$$14\overline{)56} \qquad \begin{array}{r} 4 \\ \times\ 4 \\ \hline 56 \end{array}$$

Practice

Use the inverse operation to check the answer.

1. $23 + 15 = 38$ **2.** $57 - 31 = 26$ **3.** $8 \times 7 = 56$ **4.** $144 \div 6 = 24$

5. $14 \times 3 = 42$ **6.** $366 - 218 = 148$ **7.** $586 + 255 = 841$ **8.** $250 \div 5 = 50$

Words for Operations

Many different words and phrases can be used in numerical and algebraic expressions to represent the operations of addition, subtraction, multiplication, and division.

Examples

Write the operation described by the phrase.

A. the sum of 6 and 7

To find a sum means to add. The operation is addition.

B. the product of k and 15

To find a product means to multiply. The operation is multiplication.

Practice

Write the operation described by the phrase.

1. k greater than 75

2. 42 less than m

3. the sum of h and 6

4. the quotient of 11 and b

5. c times w

6. the difference of a and 45

7. 9 increased by 7

8. the product of 12 and p

9. 2 reduced by m

Words and Equations

You can write equations to represent some sentences. Use a variable to represent what is unknown in the sentence. Use operation signs or other mathematical signs to represent words or phrases in the sentence.

Example

Write an equation for this sentence: A number increased by 7 is 12.

Represent "a number" by a variable such as x.
Represent "increased by" with a plus sign.
Represent "is" with an equal sign.

So, the equation is $x + 7 = 12$.

Practice

Write an equation for the sentence.

1. The product of a number and 7 is 49.

2. 8 less than a number is 50.

3. 12 and a number have a quotient of 2.

4. 13 times a number is 91.

5. 7 more than a number is 12.

6. 16 divided by a number is $\frac{4}{5}$.

7. A number reduced by 13 is 9.

8. 45 greater than a number is 26.

Troubleshooting

Evaluate Expressions

An expression that includes one or more variables is called an **algebraic expression**. To evaluate algebraic expressions, replace the variables with the given numbers. Then evaluate as you would with a numerical expression. Follow the rules for the order of operations.

Example

Evaluate $3 + m \times 2$ for $m = 6$.

$3 + m \times 2 = 3 + 6 \times 2$ Replace m with 6.

$= 3 + 12$ Multiply before adding.

$= 15$ Add.

Practice

Evaluate the expressions for the given value of the variables.

1. $4m$ for $m = 9$
2. $\frac{3}{4}p$ for $p = 8$
3. $\frac{z}{24}$ for $z = 96$

4. $t - 5$ for $t = 19$
5. $6h$ for $h = 9$
6. $5 + c$ for $c = {}^-13$

7. $72 \div k$ for $k = 6$
8. $p - 14$ for $p = 27$
9. ${}^-8f$ for $f = {}^-7$

Classify Angles

You can classify angles by their sizes.

An *acute* angle has a measure greater than 0° and less than 90°.

A *right* angle forms a square corner. It measures 90°.

An *obtuse* angle has a measure greater than 90° and less than 180°.

A *straight* angle has a measure of 180°.

A

B

C

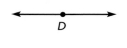
D

Practice

Classify each angle by stating whether it is *acute*, *obtuse*, *right*, or *straight*.

1.
K

2.
P

3.
Z

4.
N

5.
V

6.
M

7.
Q

8.
T

Name Angles

You can name an angle by using one letter, three letters, or a number.

Name the angle formed by the blue rays.

Use the vertex letter to name the angle. ∠G

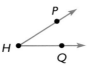

There are two angles with vertices at *N*. So, use three letters. ∠*ANV* or ∠*VNA*

Use a number. ∠5

Practice

Name the angle formed by the blue rays.

1.

2.

3. N •

4.

5.

6.

Identify Polygons

A **polygon** is a closed plane figure formed by three or more line segments. A polygon is classified by its number of sides.

Examples

State whether the figure is a polygon or not. If it is, classify it.

A. The figure is not closed, so it is not a polygon.

B. The figure is a 5-sided polygon called a pentagon.

Practice

State whether the figure is a polygon or not. If it is, classify it.

1. **2.** **3.** **4.**

Troubleshooting

Identify Solid Figures

A **polyhedron** is a solid figure with flat faces that are polygons. A **prism** is a polyhedron with two congruent, parallel **bases**. Its **lateral faces** are rectangles. A **pyramid** has a polygon for its base and triangles for its lateral faces.

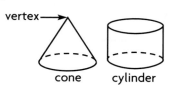

pentagonal prism   square pyramid

A **cone** has a circular base. A **cylinder** has two congruent parallel circular bases. Cones and cylinders have curved lateral surfaces.

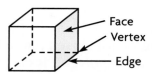

cone   cylinder

Practice

Identify each solid figure.

1.

2.

3.

4.

5.

6.

7.

8.

Faces, Edges, and Vertices

Each polygon that forms a solid figure is a **face** of the figure. A line segment where two faces meet is an **edge**. A point where three or more edges meet is a **vertex**.

Face
Vertex
Edge

Example

Tell the number of faces, vertices, and edges of the cube.

The cube has 6 faces, 8 vertices, and 12 edges.

Practice

Give the number of faces, edges, and vertices of each solid figure.

1.

2.

3.

4.

Write Equivalent Fractions

To write a fraction equivalent to a given fraction, multiply or divide the numerator and denominator by the same number.

Examples

Complete the number sentence to find an equivalent fraction.

A. $\frac{18}{24} = \frac{3}{\blacksquare}$

Think: To change 18 to 3, divide the numerator of $\frac{18}{24}$ by 6. Then divide the denominator by 6.

$\frac{18}{24} = \frac{18 \div 6}{24 \div 6} = \frac{3}{4}$

So, the equivalent fraction is $\frac{3}{4}$.

B. $\frac{3}{5} = \frac{\blacksquare}{20}$

Think: To change 5 to 20, multiply the denominator of $\frac{3}{5}$ by 4. Then multiply the numerator by 4.

$\frac{3}{5} = \frac{3 \times 4}{5 \times 4} = \frac{12}{20}$

So, the equivalent fraction is $\frac{12}{20}$.

Practice

Complete each number sentence to find an equivalent fraction.

1. $\frac{1}{2} = \frac{\blacksquare}{8}$

2. $\frac{12}{18} = \frac{2}{\blacksquare}$

3. $\frac{3}{4} = \frac{15}{\blacksquare}$

4. $\frac{\blacksquare}{6} = \frac{25}{30}$

5. $\frac{8}{24} = \frac{\blacksquare}{12}$

6. $\frac{\blacksquare}{15} = \frac{30}{45}$

7. $\frac{3}{5} = \frac{48}{\blacksquare}$

8. $\frac{\blacksquare}{40} = \frac{7}{8}$

9. $\frac{27}{36} = \frac{\blacksquare}{12}$

10. $\frac{\blacksquare}{16} = \frac{8}{32}$

11. $\frac{12}{\blacksquare} = \frac{6}{10}$

12. $\frac{\blacksquare}{100} = \frac{11}{20}$

Solve Multiplication Equations

To solve a multiplication equation, use an inverse operation. Divide both sides of the equation by the number that multiplies the variable.

Example

Solve and check. $8n = 40$

$8n = 40$

$\frac{8n}{8} = \frac{40}{8}$ Divide both sides by 8.

$n = 5$ Simplify.

Check: $8n \overset{?}{=} 40$

$8 \times 5 \overset{?}{=} 40$

$40 = 40$ ✔

Practice

Solve and check.

1. $6k = 18$

2. $3h = 27$

3. $5e = 75$

4. $10d = 700$

5. $120 = 8m$

6. $24y = 108$

7. $40 = 160b$

8. $^{-}1p = 12$

9. $15n = 135$

10. $4k = 30$

11. $1.2\,t = 96$

12. $^{-}5b = 35$

13. $198 = 18s$

14. $9 = \frac{1}{2}p$

15. $^{-}7f = {^{-}}98$

16. $^{-}84 = 14q$

Troubleshooting

Congruent and Similar Figures

Two figures are **similar** if they have the same shape. One may be an enlargement or a reduction of the other. Two figures are **congruent** if they have the same size and shape.

Examples

Tell if the figures in each pair appear to be congruent, similar, both, or neither.

A.

The figures are neither congruent or similar.

B.

The figures are similar.

C.

Both; the figures are congruent and similar.

Practice

Tell if the figures in each pair appear to be congruent, similar, both, or neither.

1.   **2.**   **3.**

Write Fractions as Decimals

To write a fraction as a decimal, divide the numerator by the denominator. Or, write an equivalent fraction with a denominator of 10, 100, or 1,000. Then rewrite the fraction as a decimal.

Example

Write $\frac{3}{4}$ as a decimal.

One Way

$$\begin{array}{r} 0.75 \\ 4\overline{)3.00} \\ -\,28 \\ \hline 20 \\ -\,20 \\ \hline 0 \end{array}$$

Divide 3 by 4. So, $\frac{3}{4} = 0.75$.

Another Way

$\frac{3}{4} = \frac{3 \times 25}{4 \times 25} = \frac{75}{100}$ Write an equivalent fraction with a denominator of 100.

$\frac{75}{100} = 0.75$ Write the equivalent fraction as a decimal.

So, $\frac{3}{4} = 0.75$.

Practice

Write the fraction as a decimal.

1. $\frac{1}{2}$ **2.** $\frac{7}{10}$ **3.** $\frac{1}{4}$ **4.** $\frac{3}{8}$ **5.** $\frac{7}{20}$

6. $\frac{2}{5}$ **7.** $\frac{17}{25}$ **8.** $\frac{43}{50}$ **9.** $\frac{48}{250}$ **10.** $\frac{19}{50}$

Multiply with Fractions and Decimals

When multiplying decimals, the number of decimal places in the product is the total of decimal places in the two factors. When multiplying fractions, write the product of the numerators over the product of the denominators.

Examples

Find the product.

A.
$$\begin{array}{r} 2.57 \\ \times\ \ 1.2 \\ \hline 514 \\ +\ 2570 \\ \hline 3.084 \end{array}$$
2.57   **2 decimal places**
× 1.2   **1 decimal place**
3.084   **2 + 1 = 3 decimal places**

B. $\frac{3}{4} \times \frac{8}{9} = \frac{3 \times 8}{4 \times 9} = \frac{24}{36}$ Multiply numerators. Multiply denominators.

$\frac{24 \div 12}{36 \div 12} = \frac{2}{3}$ Simplify.

Practice

Find the product. Write the answer in simplest form.

1. 7.2×0.9 **2.** 8.4×1.1 **3.** 112.3×0.15 **4.** 17.55×1.05 **5.** 15.82×25

6. $\frac{1}{3} \times \frac{2}{3}$ **7.** $\frac{5}{6} \times \frac{3}{4}$ **8.** $\frac{2}{3} \times \frac{5}{12}$ **9.** $\frac{5}{8} \times \frac{3}{5}$ **10.** $\frac{1}{2} \times \frac{1}{4}$

Fractions, Decimals, and Percents

You can write a percent as a decimal and a decimal as a percent.
You can write a percent as a fraction and a fraction as a percent.

Examples

A. Write 42% as a decimal.

$42\% = 42 \div 100$ Divide by 100.

So, $42\% = 0.42$.

B. Write 0.781 as a percent.

$0.781 = 0.781 \times 100$ Multiply by 100.

So, $0.781 = 78.1\%$.

C. Write 24% as a fraction.

$24\% = \frac{24}{100} = \frac{6}{25}$ Write the percent over 100. Simplify.

So, $24\% = \frac{6}{25}$.

D. Write $\frac{11}{25}$ as a percent.

$\frac{11}{25} = 11 \div 25 = 0.44$ Divide numerator by denominator.

So, $\frac{11}{25} = 44\%$.   Write the percent.

Practice

Write each decimal and each fraction as a percent.
Write each percent as a decimal and as a fraction.

1. $\frac{2}{5}$ **2.** 10%   **3.** 0.34   **4.** 25%   **5.** $\frac{1}{2}$

6. 0.81   **7.** $\frac{7}{25}$ **8.** 60%   **9.** $\frac{18}{50}$ **10.** 0.02

Troubleshooting

Certain, Impossible, Likely, Unlikely

An event is **certain** if it is sure to happen. It is **impossible** if it can never happen. It is **likely** if there is a strong chance that it will happen. It is **unlikely** if there is a strong chance that it will not happen.

Examples

Tell if the event is *certain*, *impossible*, *likely*, or *unlikely*.

A. 2 + 2 will equal 5 next Tuesday.

> 2 + 2 can never equal 5.

> So, the event is *impossible*.

B. You toss a penny 10 times and get at least 1 head.

> The probability of getting a head is $\frac{1}{2}$ each time you toss a coin.

> So, the event is *likely*.

Practice

Tell if the event is *certain*, *impossible*, *likely*, or *unlikely*.

1. Rain will fall at least once this year.

2. There will be no June next year.

3. You toss a quarter and get either a head or a tail.

4. A person who does not know you will guess your phone number.

Analyze Data

A **line graph** shows changes over time. How many people attended Game 2?

The dot on the vertical "Game 2" line is beside 140. So, 140 people attended the game.

A **circle graph** shows data as parts of a whole. How many people bought peanuts at the fair?

The section of the circle labeled "Peanuts" shows 16. So, 16 people bought peanuts.

Practice

For 1–2, use the line graph.

1. How many people attended Game 4?

2. What was the total attendance at all the games?

For 3–4, use the circle graph.

3. What was the most popular snack?

4. Popcorn and one other snack were half the sales. What was the other snack?

Customary Units

The table shows equivalents in the customary system of measurement. To change from a larger unit to a smaller one, multiply. To change from a smaller unit to a larger one, divide.

UNITS OF LENGTH
12 inches (in.) = 1 foot (ft)
3 feet = 1 yard (yd)
5,280 feet = 1 mile (mi)
1,760 yards = 1 mile

UNITS OF CAPACITY
2 cups (c) = 1 pint (pt)
2 pints = 1 quart (qt)
4 quarts = 1 gallon (gal)

UNITS OF WEIGHT
16 ounces (oz) = 1 pound (lb)
2,000 pounds = 1 ton (T)

Examples

Change to the given unit.

A. 24 in. = ▓ ft **Think:** An inch is smaller than a foot. To change from inches to feet, divide by 12.

$24 \div 12 = 2$, so 24 in. = 2 ft.

B. 4 gal = ▓ qt **Think:** A gallon is larger than a quart. To change from gallons to quarts, multiply by 4.

$4 \times 4 = 16$, so 4 gal = 16 qt.

Practice

Change to the given unit.

1. 2 yd = ▓ ft   **2.** 4 c = ▓ pt   **3.** 4 lb = ▓ oz   **4.** 2.5 gal = ▓ qt

Metric Units

The table shows how prefixes used with the basic units **meter (m)**, **liter (L)**, and **gram (g)** are related in the metric system. To change units, multiply by 10 for each place you move to the right. Divide by 10 for each place you move to the left.

kilo (k)	hecto (h)	deka (da)	BASIC UNIT	deci (d)	centi (c)	milli (m)

Examples

Change to the given unit.

A. 8 m = ▓ cm **Think:** From meter to centimeter, the move is 2 places to the right. Multiply by 10 × 10, or 100.

$8 \times 100 = 800$, so 8 m = 800 cm.

B. 4,000 mg = ▓ g **Think:** From milligrams to grams, the move is 3 places to the left. Divide by 10 × 10 × 10, or 1,000.

$4,000 \div 1,000 = 4$, so 4,000 mg = 4 g.

Practice

Change to the given unit.

1. 2 m = ▓ dm   **2.** 6 L = ▓ mL   **3.** 3,000 g = ▓ kg   **4.** 5,000 mg = ▓ cg

Troubleshooting

Solve Proportions

A **proportion** is a number sentence which states that two ratios are equal. You can solve a proportion by using the fact that in a proportion, the **cross products** are equal.

$$\frac{2}{3} = \frac{4}{6}$$ cross products

$$2 \times 6 = 3 \times 4$$

Example

Solve for n. $\frac{2}{3} = \frac{n}{12}$

$\frac{2}{3} = \frac{n}{12}$ Find the cross products.

$2 \times 12 = 3 \times n$ Use the cross products to write an equation.

$24 = 3 \times n$ Simplify.

$\frac{24}{3} = \frac{3 \times n}{3}$ Divide both sides of the equation by 3.

$8 = n$ Write the solution.

Practice

Solve for n.

1. $\frac{n}{3} = \frac{6}{9}$ **2.** $\frac{4}{n} = \frac{5}{10}$ **3.** $\frac{8}{10} = \frac{n}{15}$ **4.** $\frac{20}{15} = \frac{12}{n}$

5. $\frac{n}{6} = \frac{25}{30}$ **6.** $\frac{n}{5} = \frac{9}{15}$ **7.** $\frac{8}{n} = \frac{32}{16}$ **8.** $\frac{48}{36} = \frac{n}{9}$

Perimeter

Perimeter is the distance around a figure. To find the perimeter of a polygon, find the sum of the lengths of the sides.

Example

The opposite sides of a rectangle are congruent, so the sides of the figure that are not labeled measure 4 cm and 2 cm.

So, the perimeter is 12 cm. $P = 4\ \text{cm} + 2\ \text{cm} + 4\ \text{cm} + 2\ \text{cm}$

Practice

Find the perimeter of the figure.

1.

2.

3.

4.

5.

6.

Change Units

When you change from one unit to another, you must decide whether there are more or fewer of the new unit. If there are more units, multiply. If there are fewer units, divide.

Examples

A rug is 4 yd long. How many feet long is this?

There will be *more* feet in the measurement than there are yards. Since 1 yd = 3 ft, multiply to find the answer.

$3 \times 4 = 12$. So, the rug is 12 ft long.

Practice

Change the measurement to the given unit.

1. a 3-ft table, to in.

2. a 5-gal tank, to qt

3. a 15-ft room, to yd

4. a 1,000-m race, to cm

Find the Square and Cube of a Number

The **square** of a number is the product of the number used twice as a factor. The **cube** of a number is the product of the number used three times as a factor.

Examples

A. Find the square of 5.

$5 \times 5 = 25$ Find the product, using 5 as a factor twice.

So, 5 squared is 25, or $5^2 = 25$.

B. Find the cube of 4.

$4 \times 4 \times 4 = 64$ Find the product, using 4 as a factor three times.

So, 4 cubed is 64, or $4^3 = 64$.

Practice

Find the square of each number.

1. 7 **2.** 14 **3.** 25 **4.** 80 **5.** 5.6

Find the cube of each number.

6. 2 **7.** 3 **8.** 5 **9.** 10 **10.** 0

11. 6 **12.** 11 **13.** $\frac{1}{2}$ **14.** $\frac{3}{5}$ **15.** 0.3

Troubleshooting

Areas of Squares, Rectangles, and Triangles

The area of a rectangle or square is the product of the length and the width. The area of a triangle is *half* the product of the base and the height.

$A = l \times w$

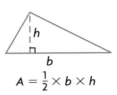

$A = \frac{1}{2} \times b \times h$

Find the area of the figure.

A.

6 ft
5 ft

$A = 6 \times 5 = 30$, or 30 ft²

B.

9 in.
9 in.

$A = 9 \times 9 = 9^2 = 81$, or 81 in.²

C.

8 cm
12 cm

$A = \frac{1}{2} \times 12 \times 8 = 48$, or 48 cm²

Practice

Find the area of the figure.

1.

13 yd
13 yd

2.

12 ft
7 ft

3.

15 cm
10.4 cm

4.

$14\frac{1}{2}$ in.
4 in.

Areas of Circles

The area of a circle is the product of π and the square of the radius. Use 3.14 for the value of π.

$A = \pi r^2$

Examples

Find the area of the circle.

A. radius = 3 cm

$\approx 3.14 \times 3^2$

$= 3.14 \times 9$

$= 28.26$ cm²

3 cm

B. The radius is half of the diameter, or 8 in.

$\approx 3.14 \times 8^2$

$= 3.14 \times 64$

$= 200.96$ in.²

16 in.

Practice

Find the area of the circle. Use 3.14 for π.

1.

10 cm

2.

12 in.

3.

18 cm

4.

42 in.

Compare Numbers

To compare integers, use a number line. The number to the right on the line is greater.	To compare decimals, align the decimal points. Then compare the digits from left to right.	To compare fractions, rename unlike fractions, then compare the numerators.

Examples

Compare. Write < or > for each ⬤.

A. 2 ⬤ ⁻3

2 is to the right of ⁻3, so 2 > ⁻3.

B. 6.25 ⬤ 6.179

6.25

6.179

0.**2** > 0.**1**, so 6.25 > 6.179.

C. $\frac{1}{2}$ ⬤ $\frac{2}{3}$

$\frac{1 \times 3}{2 \times 3} = \frac{3}{6}$ $\frac{2 \times 2}{3 \times 2} = \frac{4}{6}$

3 < 4, so $\frac{1}{2} < \frac{2}{3}$.

Practice

Compare. Write < or > for each ⬤.

1. 485 ⬤ 579   **2.** 3.03 ⬤ 3.3   **3.** $\frac{1}{2}$ ⬤ $\frac{1}{3}$ **4.** ⁻11 ⬤ ⁻3   **5.** 5 ⬤ ⁻4

6. 14.97 ⬤ 14.9   **7.** 523.6 ⬤ 532.8   **8.** 0.007 ⬤ 0.01   **9.** ⁻0.5 ⬤ ⁻0.4   **10.** $\frac{3}{4}$ ⬤ $\frac{5}{8}$

Function Tables

To find an output value in a function table, replace the variable with an input value. Then evaluate the algebraic expression.

Example

Complete the function table.

k	$k + 4$
7	▨
9	▨
11	▨
16	▨

Replace k in $k + 4$ with 7: $k + 4 = 7 + 4 = 11$.

Replace k in $k + 4$ with 9: $k + 4 = 9 + 4 = 13$.

Replace k in $k + 4$ with 11: $k + 4 = 11 + 4 = 15$.

Replace k in $k + 4$ with 16: $k + 4 = 16 + 4 = 20$.

Input Output

k	$k + 4$
7	11
9	13
11	15
16	20

Practice

Copy and complete the function table.

1.

x	$x + 7$
9	16
13	▨
18	▨
24	▨

2.

b	$b \times 16$
2	▨
5	▨
8	▨
13	▨

3.

m	$m \div 6$
192	▨
150	▨
90	▨
51	▨

4.

r	$r - 7.4$
25.8	▨
19.2	▨
13.05	▨
9.1	▨

Troubleshooting

Slides, Flips, and Turns

There are three ways that you can **transform** a figure. You can **slide,** or **translate,** the figure along a straight line. You can **flip,** or **reflect,** the figure over a line. Or you can **turn,** or **rotate,** the figure around a point.

Examples

Identify the transformation as a *slide, flip,* or *turn*.

A.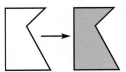

The figure has been translated along a straight line to the right. This is a *slide.*

B.

The figure has been reflected across a vertical line. This is a *flip.*

C.

The figure has been rotated clockwise around a point. This is a *turn.*

Practice

Identify the transformation as a *slide, flip,* or *turn*.

1.   **2.**   **3.**   **4.**

Line Symmetry

A figure has **line symmetry** if it can be folded or reflected so that the two parts of the figure match, or are congruent. A figure can have more than one line of symmetry.

Examples

Is the dashed line a line of symmetry? Write *yes* or *no*.

A. Yes. When the left half is folded on top of the right half, the halves match.

B. No. When the lower left is folded on top of the upper right, the halves do not match.

Practice

Is the dashed line a line of symmetry? Write *yes* or *no*.

1.   **2.**   **3.**   **4.**

Measure Angles

To measure ∠ABC with a protractor, place the center of the protractor at B.

Align the 0° mark on the protractor with one ray of the angle.

Read the angle measure where the other ray passes through the scale on the protractor.

Read the angle measure as 70°.

0° mark

Practice

Use a protractor to measure the angle.

1. **2.** **3.** **4.**

Ordered Pairs

You can use two numbers called an **ordered pair** to locate a point on a grid. The first number tells you how far to move horizontally from (0,0). The second number tells you how far to move vertically from (0,0).

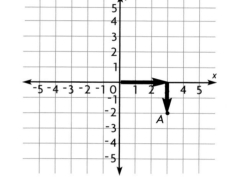

Example

Write the ordered pair for point A.

To reach point A from (0,0), go
⁺3 units right (horizontally) and
⁻2 units down (vertically).

So, the ordered pair is (3,⁻2).

Practice

Write the ordered pair for each point.

1. point A   **2.** point C

3. point D   **4.** point E

5. point F   **6.** point G

7. point I   **8.** point J

9. point B   **10.** point H

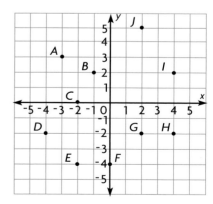

Set A (pp. 16–19)

Estimate.

1. $589 + 342$

2. $865 - 383$

3. $8,926 + 1,674$

4. $8,960$
   $9,043$
   $+8,756$

5. $3,406 - 792$

6. $4,496 \div 51$

7. 47×98

8. 321×43

9. $3,552 \div 58$

10. $6,251 \div 12$

Tell whether the estimate is an overestimate or underestimate.

11. $848 + 692 \approx 1,550$

12. $398 \times 66 \approx 28,000$

13. $317 - 67 \approx 230$

Set B (pp. 20–21)

Find the sum or difference.

1. $6,343 - 1,145$

2. $19,826 - 3,517$

3. $22,912 + 84,718$

4. $532,047 + 36,603$

5. $6,052 + 2,791 + 3,047 + 654$

6. $7,380 - 1,890 - 408 - 792$

Set C (pp. 22–25)

Multiply or divide, writing any remainder as a fraction.

1. 83×62

2. $16)\overline{869}$

3. $45)\overline{1,675}$

4. 458×194

Set D (pp. 28–29)

Write a numerical or algebraic expression for the word expression.

1. nine dollars less than twelve dollars

2. thirteen more than a number, x

3. twenty-two dogs times two bowls

4. fifty-two divided by two

Evaluate each expression.

5. $8 \times a$, for $a = 35$

6. $p - 38$, for $p = 97$

7. $175 \div k$, for $k = 5$

8. $96 + m$, for $m = 48$

Set E (pp. 30–31)

Determine which of the given values is the solution of the equation.

1. $5w = 70$; $w = 12, 13,$ or 14

2. $16 - a = 7$; $a = 8, 9,$ or 10

3. $48 \div x = 8$; $x = 6, 7,$ or 8

Solve each equation by using mental math.

4. $23 + c = 28$

5. $k \div 9 = 5$

6. $q - 25 = 25$

7. $c \times 6 = 426$

Set A (pp. 36–39)

Use mental math to find the value.

1. $26 + (4 + 18)$ **2.** $6 \times 4 \times 5$ **3.** $28 + 9 + 41$ **4.** $83 - 36$

5. $4 + 19 + 26$ **6.** 38×6 **7.** $47 - 29$ **8.** $7 + 22 + 13$

9. 51×8 **10.** $(6 \times 8) \times 5$ **11.** $19 + 31$ **12.** $92 - 29$

13. $65 + 17$ **14.** $10,000 \div 25$ **15.** $3 \times 10 \times 7$

16. $51 - 29$ **17.** $4 \times 8 \times 2$ **18.** $19 + 52 + 21$

19. Ben has three cousins under the age of 13. The product of their ages is 336. How old are they?

Set B (pp. 40–41)

Write the equal factors. Then find the value.

1. 10^4 **2.** 5^6 **3.** 2^3 **4.** 3^2 **5.** 7^4

6. 1^8 **7.** 27^2 **8.** 15^1 **9.** 12^3 **10.** 9^4

Write in exponent form.

11. $3 \times 3 \times 3 \times 3$ **12.** 10×10 **13.** $5 \times 5 \times 5 \times 5 \times 5 \times 5$

14. $12 \times 12 \times 12$ **15.** $1 \times 1 \times 1 \times 1$ **16.** $21 \times 21 \times 21 \times 21 \times 21$

17. Harold was hired to take a survey of 625 people. He wants to write this number in exponent form. If the base is 5, what is the exponent? How did you arrive at your answer?

Set C (pp. 44–45)

Give the correct order of operations.

1. $5 \times 7 + 3 \div 2$ **2.** $(6 + 15) \div 3 \times 1 - 6$

3. $12^2 \times 10 - 10^3 - 100$ **4.** $10^2 + 5^2 \div 5^2 - 5$

Evaluate the expression.

5. $7 + (2 \times 2)^4 - 9 \times 9$ **6.** $90 \times 5 - 4 \times (18 \div 6)$

7. $3^2 \times (4 + 5)^2 - 36$ **8.** $15^2 \div (4^2 + 9) + 8^1$

9. $(8^2 + 3^3) \times (9 - 5)^3$ **10.** $10^3 \div (10^2 \div 10^1) + 10^2$

11. Scott bought 5 pounds of nails that cost $2.75 per pound and 3 pounds of screws that cost $3.50 per pound. How much did Scott spend?

CHAPTER 3

Extra Practice

Set A (pp. 52–55)

Write the value of the blue digit.

1. 6.12053　　**2.** 0.0231　　**3.** 8.7　　　　**4.** 0.849

Write the number in expanded form.

5. 0.00309　　　**6.** 5.015　　　　**7.** 3.032

8. 20.0518　　　**9.** 200.05　　　**10.** 5.16

Compare the numbers. Write <, >, or = for ⬤.

11. 5.099 ⬤ 5.999　　**12.** 226.5 ⬤ 226.4　　**13.** 251.36 ⬤ 241.36

14. 18.3 ⬤ 18.30　　　**15.** 4.18 ⬤ 4.28　　　**16.** 49.089 ⬤ 49.098

Write the numbers in order from least to greatest.

17. 82.16, 82, 82.15　　**18.** 141.14, 114.41, 141.41　　**19.** 5.09, 5.49, 5.23

Set B (pp. 58–59)

Estimate.

1.　3.7　　　**2.**　62.8　　　**3.**　109.7　　　**4.**　788.3
　　3.15　　　　　× 　6　　　　− 53.622　　　　× 　92
　+2.98

5.　5.92　　　**6.**　21.513　　**7.**　5.816　　　**8.**　465.09
　　3.15　　　　　× 9.8　　　　3.215　　　　− 73.46
　+4.07　　　　　　　　　　　+1.6

9. 728 ÷ 8.1　　**10.** 8.1 − 2.456　　**11.** 20.8 ÷ 7　　**12.** 123.95 ÷ 61

Set C (pp. 60–61)

Write the decimal and percent for the shaded part.

1. 　　**2.** 　　**3.**

Write the corresponding decimal or percent.

4. 60%　　**5.** 0.9　　**6.** 39%　　　**7.** 0.04　　**8.** 0.46

9. 18%　　**10.** 0.41　　**11.** 0.38　　**12.** 7%　　**13.** 90%

Set A (pp. 66–69)

Add or subtract. Estimate to check.

1. $12.8 - 4.1$

2. $\$21.85 + \$17.48$

3. $17.3 - 16.5$

4. $8.36 + 5.216 + 0.09$

5. $8 + 7.317 + 3.06$

6. $5.08 - 2.261$

Copy the problem. Place the decimal point correctly in the answer.

7. $13.601 - 10.311 = 329$

8. $18.56 + 3.12 = 2168$

9. $5 - 3.021 = 1979$

Set B (pp. 70–73)

Tell the number of decimal places there will be in the product.

1. 21.3×18.4

2. 7.03×7.05

3. 9.2×2.13

Copy the problem. Place the decimal point in the product.

4. $360.05 \times 12.6 = 4536630$

5. $762 \times 3.285 = 2503170$

6. $10.11 \times 1.02 = 103122$

Multiply. Estimate to check.

7. 37.5×10.26

8. 6.42×9.1

9. 0.05×2.9

10. 19.3×2.41

11. 2.39×7.6

12. 2×6.005

Set C (pp. 76–79)

Rewrite the problem so that the divisor is a whole number.

1. $16.92 \div 0.12$

2. $661.44 \div 31.2$

3. $20.2 \div 0.53$

Copy the problem. Place the decimal point in the quotient.

4. $4.48 \div 2.8 = 16$

5. $28.68 \div 1.2 = 239$

6. $9.87 \div 2 = 4935$

Divide. Estimate to check.

7. $9.72 \div 1.2$

8. $25.0224 \div 3.12$

9. $20.801 \div 6.1$

10. $80.4 \div 4.8$

Set D (pp. 82–83)

Evaluate each expression.

1. $8m$ for $m = 4.2$

2. $6.3 + 5.04 + k$ for $k = 8.4$

3. $(3.4 - c) + 63$ for $c = 2.47$

Solve each equation by using mental math.

4. $p + 2.8 = 7.9$

5. $\dfrac{k}{3} = 5.6$

6. $a - 17.1 = 9.5$

7. $6s = 7.2$

8. $x + 22.6 = 30.8$

9. $15n = 4.5$

Set A (pp. 94–97)

Determine the type of sample. Write *convenience, random,* or *systematic*.

1. The teacher randomly selected a student and then surveyed every third student on his attendance list.

2. The teacher asked students in English class about their favorite books.

Set B (pp. 98–99)

A survey is to be conducted about the favorite foods of middle-school students. Tell whether the sampling method is *biased* or *unbiased*.

1. Randomly survey 10 people who get off the school bus one morning.

2. Randomly survey 1 of every 10 students in sixth, seventh, and eighth grades.

Set C (pp. 102–105)

Use the table at the right.

1. Copy and complete the table.

2. How large was the sample size?

Favorite Color	Tally	Frequency	Cumulative Frequency
Blue	ＷＷ＼	■	■
Green	Ｗ＼＼＼＼	■	■
Red	Ｗ＼＼＼	■	■
Yellow	Ｗ	■	■
Orange	＼＼	■	■

Set D (pp. 106–108)

Find the mean, median, and mode.

1. 4, 5, 7, 8, 8

2. 6.1, 8.1, 7.4, 7.2

3. 12, 14, 18, 12, 22

4. 1, 1, 7, 9, 3, 4, 2, 3

Set E (pp. 109–111)

Use the data in table.

NUMBER OF HOURS SPENT ON HOMEWORK									
Week	1	2	3	4	5	6	7	8	9	10
Hours	8	9	10	9	7	9	8	0	7	2

1. Find the mean, median, and mode with the outliers and then without the outliers.

2. How does including the outliers affect the mean? the median? the mode?

Set F (pp. 112–115)

Write *yes* or *no* to tell whether the conclusion is valid. Explain.

1. The first ten females to enter a diner say their favorite drink is orange juice. You conclude that orange juice is the favorite drink of all of the customers.

2. Two hundred students are randomly selected and asked what their favorite kind of music is. Seventy percent say pop music. You conclude that most students prefer pop music.

Set A (pp. 120–123)

1. Make a multiple-bar graph using the data in the table below.

2. Make a multiple-line graph using the data in the table below.

CLUB MEETING ATTENDANCE	Sept	Oct	Nov	Dec
Male	42	31	65	61
Female	56	47	51	60

AVERAGE RAINFALL (IN INCHES)	Jan	Feb	Mar	Apr
City A	4	5	3	5
City B	8	12	4	7

Set B (pp. 124–125)

1. Make a line graph using the data in the table at the right.

2. If the trend continues, how many miles will be traveled on the sixth day?

DISTANCE TRAVELED					
Day	1	2	3	4	5
Miles	45	85	129	175	220

Set C (pp. 126–128)

1. Make a stem-and-leaf plot of the data below. Find the mode and median.

14	22	37	41	13	18	22	29	33
36	25	21	30	48	44	19	17	28

2. Use the data in the chart below to make a histogram.

NUMBER OF HOURS OF MONTHLY EXERCISE					
0–2	**3–5**	**6–8**	**9–11**	**12–15**	**16–18**
12	15	19	22	21	6

Set D (pp. 130–131)

For 1–3, use the box-and-whisker graph. The graph shows the number of tickets sold.

1. What was the greatest number of tickets sold?

2. What is the median?

3. What are the lower and upper quartiles?

82 84 86 88 90 92 94 96

Set E (pp. 132–135)

For 1–3, use the bar graph at the right.

1. About how many times as high is the bar for Central than the bar for Lee?

2. Has Central won four times as many championships as Lee? Explain.

3. How can you change the graph so it is not misleading?

CHAPTER 7

Extra Practice

Set A (pp. 146–147)

Tell whether each number is divisible by 2, 3, 4, 5, 6, 8, 9, or 10.

1. 80   **2.** 99   **3.** 105   **4.** 126   **5.** 234

6. 370   **7.** 591   **8.** 1,620   **9.** 3,048   **10.** 8,020

Set B (pp. 148–149)

Use division or a factor tree to find the prime factorization.

1. 18   **2.** 40   **3.** 36   **4.** 27   **5.** 100

6. 150   **7.** 280   **8.** 12   **9.** 108   **10.** 420

Write the prime factorization of each number in exponent form.

11. 28   **12.** 420   **13.** 48   **14.** 81   **15.** 72

16. 80   **17.** 144   **18.** 221   **19.** 168   **20.** 1,475

Solve for *n* to complete the prime factorization.

21. $2 \times 5 \times n = 50$ **22.** $2 \times n \times 5 = 30$ **23.** $2 \times 3 \times n = 18$

24. $n \times 3 \times 3 = 27$ **25.** $2 \times 2 \times 2 \times n = 16$ **26.** $n \times 2 \times 3 = 42$

27. $2 \times 3 \times 3 \times n = 54$ **28.** $3 \times n \times 5 = 75$ **29.** $n \times n \times 3 = 12$

Set C (pp. 150–153)

List the first five multiples of each number.

1. 4   **2.** 9   **3.** 14   **4.** 5

5. 30   **6.** 8   **7.** 12   **8.** 21

Find the LCM of each set of numbers.

9. 45, 54   **10.** 10, 35   **11.** 12, 20   **12.** 18, 27

13. 150, 60   **14.** 54, 18   **15.** 55, 33, 44   **16.** 35, 25, 49

17. Jean needs eggs and muffins for the class breakfast. Eggs come in cartons of 12 and muffins come in packages of 8. What is the least number of eggs and muffins she can buy to have an equal number of each?

Find the GCF of each set of numbers.

18. 12, 108   **19.** 148, 84   **20.** 132, 108   **21.** 75, 105

22. 45, 108   **23.** 32, 128   **24.** 252, 336   **25.** 56, 280, 400

Set A (pp. 160–163)

Write the fraction in simplest form.

1. $\frac{24}{80}$
2. $\frac{14}{63}$
3. $\frac{24}{56}$
4. $\frac{15}{60}$
5. $\frac{16}{40}$
6. $\frac{50}{90}$

Complete.

7. $\frac{9}{12} = \frac{\blacksquare}{4}$
8. $\frac{7}{42} = \frac{\blacksquare}{6}$
9. $\frac{25}{\blacksquare} = \frac{5}{10}$
10. $\frac{3}{7} = \frac{21}{\blacksquare}$
11. $\frac{1}{12} = \frac{\blacksquare}{144}$
12. $\frac{9}{20} = \frac{\blacksquare}{60}$
13. $\frac{5}{\blacksquare} = \frac{15}{27}$
14. $\frac{12}{\blacksquare} = \frac{3}{8}$

Set B (pp. 164–165)

Write the fraction as a mixed number or a whole number.

1. $\frac{7}{3}$
2. $\frac{9}{2}$
3. $\frac{36}{5}$
4. $\frac{72}{8}$
5. $\frac{10}{2}$
6. $\frac{13}{4}$
7. $\frac{36}{6}$
8. $\frac{12}{9}$

Write the mixed number as a fraction.

9. $4\frac{1}{4}$
10. $5\frac{2}{3}$
11. $2\frac{1}{8}$
12. $7\frac{1}{9}$
13. $3\frac{4}{7}$
14. $1\frac{6}{11}$
15. $4\frac{3}{8}$
16. $15\frac{3}{4}$

Set C (pp. 166–167)

Compare the fractions. Write <, >, or = for each ●.

1. $\frac{5}{9}$ ● $\frac{1}{2}$
2. $\frac{3}{8}$ ● $\frac{1}{5}$
3. $\frac{7}{25}$ ● $\frac{3}{5}$
4. $\frac{3}{5}$ ● $\frac{12}{20}$

Order the fractions from least to greatest.

5. $\frac{1}{3}, \frac{1}{6}, \frac{1}{10}$
6. $\frac{4}{9}, \frac{2}{3}, \frac{3}{8}$
7. $\frac{1}{9}, \frac{3}{4}, \frac{5}{12}$

Set D (pp. 169–171)

Write the decimal as a fraction.

1. 0.9
2. 0.081
3. 0.29

Write as a decimal. Tell whether the decimal terminates or repeats.

4. $\frac{3}{8}$
5. $\frac{5}{16}$
6. $\frac{1}{3}$
7. $\frac{7}{9}$

Compare. Write <, >, or = for each ●.

8. $\frac{3}{10}$ ● 0.03
9. $\frac{2}{3}$ ● 0.7
10. 0.79 ● $\frac{5}{8}$
11. 0.15 ● $\frac{3}{20}$

Write each fraction as a percent.

12. $\frac{6}{25}$
13. $\frac{7}{10}$
14. $\frac{9}{20}$
15. $\frac{4}{5}$

Extra Practice

Set A (pp. 176–179)

Estimate the sum or difference.

1. $\frac{4}{7} - \frac{3}{8}$ 2. $3\frac{7}{9} + 1\frac{10}{11}$ 3. $5\frac{4}{5} - 3\frac{1}{2}$ 4. $\frac{8}{9} + \frac{3}{5}$

5. $\frac{4}{9} + \frac{2}{3}$ 6. $\frac{8}{9} - \frac{5}{7}$ 7. $\frac{4}{9} - \frac{1}{6}$ 8. $3\frac{10}{13} + 4\frac{1}{9}$

9. $\frac{3}{7} + \frac{1}{9}$ 10. $5\frac{3}{4} - 1\frac{1}{8}$ 11. $\frac{5}{9} - \frac{3}{7}$ 12. $8\frac{7}{8} + 5\frac{1}{2}$

13. Shawna practiced the tuba for $\frac{5}{6}$ hr on Monday, $\frac{1}{3}$ hr on Wednesday, and $\frac{1}{4}$ hr on Friday. About how many hours did Shawna practice last week?

Set B (pp. 182–185)

Write the sum or difference in simplest form. Estimate to check.

1. $\frac{3}{4} + \frac{1}{2}$ 2. $\frac{4}{7} - \frac{1}{3}$ 3. $\frac{1}{6} + \frac{5}{18}$ 4. $\frac{9}{10} - \frac{2}{5}$

5. $\frac{5}{6} + \frac{1}{4}$ 6. $\frac{6}{20} + \frac{3}{5}$ 7. $\frac{5}{9} - \frac{1}{2}$ 8. $\frac{7}{8} - \frac{1}{16}$

9. $\frac{3}{5} - \frac{1}{9}$ 10. $\frac{5}{20} + \frac{11}{20}$ 11. $\frac{11}{12} - \frac{1}{6}$ 12. $\frac{1}{10} + \frac{3}{8}$

13. Rico read $\frac{1}{5}$ of a book on Saturday and $\frac{1}{3}$ of the same book on Sunday. What portion of the book does he have left to read?

Set C (pp. 186–189)

Draw a diagram to find each sum or difference. Write the answer in simplest form.

1. $2\frac{3}{8} + 1\frac{1}{4}$ 2. $2\frac{1}{3} - 1\frac{1}{6}$ 3. $1\frac{1}{5} + 1\frac{7}{10}$ 4. $3\frac{1}{3} - 1\frac{3}{4}$

Write the sum or difference in simplest form. Estimate to check.

5. $4\frac{5}{12} - 2\frac{1}{6}$ 6. $5\frac{2}{5} - 3\frac{3}{10}$ 7. $2\frac{1}{9} + 1\frac{2}{5}$ 8. $4\frac{2}{3} + 6\frac{1}{6}$

9. Stefan bought $1\frac{1}{2}$ lb of potato salad, $2\frac{1}{4}$ lb of coleslaw, and $4\frac{3}{4}$ lb of chicken at a deli. What is the total weight of his purchase?

Set D (pp. 192–193)

Write the difference in simplest form. Estimate to check.

1. $4\frac{5}{9} - 3\frac{5}{6}$ 2. $5\frac{1}{3} - 4\frac{1}{9}$ 3. $6\frac{1}{4} - 3\frac{2}{3}$ 4. $6\frac{1}{3} - 4\frac{7}{9}$

5. $4 - 1\frac{3}{7}$ 6. $5\frac{3}{10} - \frac{2}{5}$ 7. $7\frac{1}{6} - \frac{1}{2}$ 8. $4\frac{1}{4} - 1\frac{5}{8}$

9. Mr. Norman had $51\frac{1}{3}$ ft of plastic pipe. He installed $25\frac{3}{4}$ ft in the bathroom. How much pipe does he have left?

Set A (pp. 200–201)

Estimate each product or quotient.

1. $\frac{8}{9} \times \frac{3}{5}$

2. $\frac{1}{6} \times \frac{4}{9}$

3. $5\frac{1}{2} \div \frac{5}{6}$

4. $\frac{7}{9} \div \frac{3}{4}$

5. $3\frac{1}{3} \times 5\frac{1}{5}$

6. $8\frac{3}{5} \div 2\frac{3}{4}$

7. $9\frac{7}{8} \times 10\frac{1}{2}$

8. $18\frac{3}{4} \div 5\frac{1}{8}$

9. Liz is making cookies to give to some new neighbors. The recipe calls for $1\frac{3}{4}$ cups of flour. About how much flour does Liz need if she makes $2\frac{1}{2}$ times the recipe?

Set B (pp. 202–205)

Multiply. Write the answer in simplest form.

1. $\frac{2}{4} \times \frac{3}{4}$

2. $\frac{2}{5} \times \frac{1}{4}$

3. $\frac{3}{8} \times \frac{3}{5}$

4. $3 \times \frac{7}{8}$

5. $\frac{2}{7} \times 8$

6. $\frac{5}{6} \times \frac{3}{8}$

7. $\frac{5}{9} \times \frac{1}{5}$

8. $\frac{1}{9} \times \frac{3}{4}$

9. To make a dye for art class, Jan needs $\frac{1}{4}$ tsp of red coloring and $\frac{1}{2}$ that amount of green coloring. How much green coloring does Jan need to make dye?

Set C (pp. 206–207)

Multiply. Write the answer in simplest form.

1. $2\frac{2}{3} \times 2\frac{1}{4}$

2. $2\frac{1}{2} \times \frac{1}{5}$

3. $\frac{1}{5} \times 2\frac{2}{8}$

4. $3\frac{1}{6} \times 5\frac{1}{4}$

5. $2\frac{3}{4} \times \frac{5}{6}$

6. $2\frac{2}{7} \times \frac{2}{5}$

7. $3\frac{2}{3} \times 2\frac{3}{5}$

8. $6\frac{8}{9} \times 2\frac{3}{7}$

Set D (pp. 210–213)

Find the quotient. Write the answer in simplest form.

1. $\frac{2}{9} \div \frac{8}{18}$

2. $\frac{3}{8} \div 2\frac{1}{4}$

3. $2 \div 2\frac{2}{3}$

4. $1\frac{11}{12} \div 1\frac{5}{6}$

5. $\frac{3}{4} \div \frac{1}{8}$

6. $\frac{5}{6} \div 1\frac{2}{3}$

7. $3\frac{3}{4} \div 2\frac{7}{12}$

8. $\frac{5}{7} \div \frac{10}{11}$

Set E (pp. 216–217)

Evaluate the expression.

1. $3r$ for $r = \frac{1}{2}$

2. $\frac{5}{6} + d$ for $d = \frac{3}{10}$

3. $3\frac{3}{5} \div a$ for $a = \frac{1}{2}$

4. $y + 3\frac{5}{8}$ for $y = 1\frac{1}{4}$

5. $2\frac{5}{6}k$ for $k = 1\frac{1}{3}$

6. $a \div 4\frac{1}{2}$ for $a = 3\frac{3}{5}$

Solve the equation.

7. $c + \frac{3}{4} = 1\frac{1}{4}$

8. $\frac{5}{9}s = \frac{5}{27}$

9. $x - 6\frac{1}{8} = 12\frac{1}{4}$

Set A (pp. 228–229)

Write an integer to represent each situation.

1. a temperature increase of 5°

2. the wind speed decreases by 12 mph

3. depositing $510 into a savings account

4. a temperature of 7° below zero

Write the opposite integer.

5. $^-3$
6. $^-12$
7. $^+360$
8. $^-1$

9. $^+160$
10. $^-3{,}047$
11. $^-1{,}119$
12. $^-942$

Set B (pp. 230–233)

Write each rational number in the form $\frac{a}{b}$.

1. $3\frac{1}{6}$
2. 0.5
3. 0.27
4. 13.4

5. $2\frac{2}{5}$
6. 3.18
7. 10.02
8. 300

9. 0.36
10. $5\frac{1}{3}$
11. 312
12. $4\frac{1}{4}$

Find a rational number between the two given numbers.

13. $\frac{3}{8}$ and $\frac{5}{6}$
14. $\frac{1}{8}$ and $\frac{1}{4}$
15. $\frac{5}{9}$ and $\frac{11}{15}$

16. 1.8 and 1.9
17. $^-1.5$ and $^-1.3$
18. 4.23 and 4.235

19. 3.8 and 3.82
20. $^-5$ and $^-4.9$
21. $^-12\frac{1}{4}$ and $^-12\frac{1}{3}$

Set C (pp. 234–235)

Compare. Write <, >, or = for each ●.

1. 0.4 ● 0.38
2. $\frac{2}{7}$ ● 0.25
3. $^-0.6$ ● $\frac{-2}{5}$
4. $\frac{7}{9}$ ● 0.8

5. 0.28 ● $\frac{2}{7}$
6. $\frac{3}{13}$ ● 0.13
7. $\frac{-4}{9}$ ● $^-3.1$
8. $^-3$ ● 0.31

9. $2\frac{1}{5}$ ● $2\frac{4}{13}$
10. 0.87 ● 0.868
11. $\frac{-5}{8}$ ● $\frac{5}{8}$
12. $2\frac{1}{16}$ ● $2\frac{1}{10}$

Compare the rational numbers and order them from least to greatest.

13. $\frac{1}{3}, \frac{2}{5}, 0.38, \frac{1}{2}$
14. $\frac{2}{7}, \frac{1}{3}, \frac{1}{4}, 0.26$
15. $0.92, \frac{9}{8}, \frac{8}{9}, 0.924$

16. $0.23, \frac{2}{9}, \frac{1}{4}, \frac{1}{5}$
17. $\frac{-2}{5}, \frac{-1}{3}, \frac{-1}{2}, \frac{-3}{8}$
18. $\frac{1}{10}, {}^-3, {}^-0.3$

19. $^-7, 6, 1, {}^-5$
20. $^-0.1, 0.01, 10, {}^-10$
21. $\frac{1}{2}, \frac{1}{6}, \frac{1}{9}, 0.1$

Set A (pp. 244–247)

Write the addition problem modeled on the number line.

1.

2.

Find the sum.

3. $^-3 + 9$ **4.** $1 + {}^-5$ **5.** $5 + 4$ **6.** $^-2 + 8$

7. $^-9 + 17$ **8.** $^-14 + {}^-8$ **9.** $^-19 + 8$ **10.** $22 + {}^-36$

11. $31 + 19$ **12.** $32 + {}^-45$ **13.** $63 + {}^-47$ **14.** $^-71 + 32$

$\frac{a+b}{c}$ **Algebra** **Use mental math to find the value of x.**

15. $^-2 + x = {}^-5$ **16.** $x + {}^-3 = {}^-12$ **17.** $x + {}^-12 = {}^-8$ **18.** $^-9 + x = {}^-2$

19. Walter had a trading card that was worth $15 in September. In December, the card had decreased $7 in value. Write an addition problem to find the new value.

20. At 9:00 A.M., the temperature outside was $^-9$°C. By 3:00 P.M., the temperature had risen 5°C. What was the temperature at 3:00 P.M.?

Set B (pp. 250–251)

Rewrite the subtraction problem as an addition problem.

1. $14 - 17$ **2.** $6 - {}^-8$ **3.** $^-1 - {}^-13$ **4.** $^-4 - 17$

Find the difference.

5. $2 - 9$ **6.** $^-8 - {}^-3$ **7.** $4 - {}^-2$ **8.** $4 - 9$

9. $8 - 19$ **10.** $14 - {}^-2$ **11.** $^-12 - {}^-12$ **12.** $3 - 7$

13. $23 - 32$ **14.** $14 - 29$ **15.** $^-48 - 17$ **16.** $^-24 - {}^-39$

Evaluate.

17. $^-2 - {}^-7 + {}^-9$ **18.** $8 - {}^-2 + {}^-6$ **19.** $3 - {}^-9 - 12$ **20.** $1 - {}^-7 - 15$

21. At noon in the city of Hunterville, the temperature was 5°F. By 6:00 P.M., the temperature was $^-3$°F. What was the range of temperature from noon to 6:00 P.M.?

22. The water level in the city water tower was 8 ft above normal. Three weeks later, the level was 4 ft below normal. Find the range of the water levels.

Set A (pp. 258–259)

Find the product.

1. $^-6 \times 9$ 2. $3 \times {}^-4$ 3. 8×3 4. $^-10 \times 5$

5. 11×6 6. $^-10 \times {}^-7$ 7. $^-9 \times {}^-12$ 8. $^-9 \times 4$

9. $^-7 \times 6$ 10. $^-6 \times {}^-8$ 11. $^-20 \times 4$ 12. $50 \times {}^-8$

13. 32×10 14. $40 \times {}^-9$ 15. $^-84 \times 7$ 16. $^-71 \times {}^-15$

17. The depth of a diver changed $^-5$ ft every 30 sec. How much of a depth change did the diver have after 2 min?

Algebra Use mental math to find the value of *y*.

18. $y \times {}^-5 = {}^-20$ 19. $y \times {}^-7 = 42$ 20. $^-9 \times y = 72$ 21. $^-5 \times y = {}^-25$

Set B (pp. 260–261)

Find the quotient.

1. $^-54 \div 9$ 2. $12 \div {}^-4$ 3. $24 \div 3$ 4. $^-50 \div 5$

5. $63 \div 7$ 6. $^-42 \div {}^-6$ 7. $^-96 \div {}^-12$ 8. $^-88 \div 8$

9. $^-27 \div 9$ 10. $^-64 \div {}^-8$ 11. $^-140 \div 10$ 12. $225 \div {}^-25$

13. $360 \div 15$ 14. $216 \div {}^-12$ 15. $^-980 \div {}^-14$ 16. $^-7,584 \div 24$

17. A water tower is leaking. The water level is changing $^-5$ ft per day. How many days will it take the water level to change $^-30$ ft?

Algebra Use mental math to find the value of *x*.

18. $x \div {}^-7 = {}^-4$ 19. $x \div {}^-6 = 8$ 20. $^-30 \div x = 3$ 21. $^-28 \div x = 7$

Set C (pp. 262–263)

Evaluate the expression.

1. $^-12 - 8 \div {}^-2 \times 10 + 40$ 2. $(^-8 + 6) \times 3 \div (56 - 57)$ 3. $5^3 + 14 - 200 \times 2$

4. $(^-54 + 24) \div (^-3 - {}^-8)$ 5. $15^2 \div (^-12 + 7) \times (^-6 - 2)$ 6. $5 \times (30 - 42) \div 2^2$

7. $(^-29 + 35)^2 \div (172 - 175)$ 8. $^-8 \times (6 - 11)^2 + 235$ 9. $(9 + 7 \times {}^-8) \times (^-5 + 8)^2$

Use a property to simplify the expression. Then evaluate the expression and identify the property you used.

10. $(645 + 19) + 11$ 11. $175 + 456 + 25$ 12. $^-34 + 99 + 34$

Set A (pp. 274–275)

Write an algebraic expression for the word expression.

1. 9 less than y

2. 13 more than a number, x

3. $\frac{3}{4}$ increased by y

4. 7 more than the quotient of 52 and k

5. Fred has 32 more than twice the number of baseball cards that José has. Write an algebraic expression for the number of baseball cards Fred has.

Write a word expression for each.

6. $9^2 + a$

7. $7x + 12$

8. $24 - \frac{1}{5}c$

9. $y + \frac{1}{2}x$

Set B (pp. 276–279)

Evaluate the algebraic expression for the given value of the variable.

1. $x - 6$ for $x = {}^-12$

2. $y + 13$ for $y = {}^-25$

3. $42 - k$ for $k = {}^-30$

4. ${}^-38 + p$ for $p = 22$

5. $a^2 - 25$ for $a = 3$

6. $x^3 - 19$ for $x = 2$

7. $3(a + b)^2 - c$ for $a = {}^-3$, $b = 5$, and $c = {}^-8$

8. $6xyz$ for $x = {}^-4$, $y = {}^-3$, and $z = {}^-6$

9. $3pq + r$ for $p = 2$, $q = {}^-8$, and $r = {}^-15$

Evaluate the algebraic expression for $x = 2, 3, 4,$ and 5.

10. $7x + 12$

11. ${}^-4x + 10$

12. $3x - 26$

13. ${}^-9x - 4$

14. $\frac{60}{x} - 14$

15. $\frac{{}^-60}{x} - 14$

Simplify the expression. Then evaluate the expression for the given value of the variable.

16. $5x + 3x + 12$ for $x = {}^-4$

17. $2y + 7y - 25$ for $y = 2$

18. $42z - 30z - 20$ for $z = {}^-2$

Set C (pp. 282–283)

Evaluate the expression.

1. $\sqrt{100} - 2 + 6$

2. $3 \times (9^2 - 41)$

3. $4 \times \sqrt{25} - 3 \cdot 2$

4. $\sqrt{36} \cdot \sqrt{36}$

5. $530 \times (\sqrt{4} - 2)^2$

6. $8 + \sqrt{49} - 3^2$

Evaluate the expression for $a = 16$, $b = 3$, and $c = 6$.

7. ${}^-a + (b + c)^2$

8. $\frac{c}{b} - \sqrt{a}$

9. $\sqrt{a + b + c}$

Set A (pp. 288–289)

Write an equation for the word sentence.

1. $\frac{2}{3}$ of a number is 12.

2. 1.6 more than a number is 5.

3. $1\frac{1}{2}$ less than a number is 6.

4. A number divided by $^-6$ is 30.

5. 9 times a number is 53.

6. 1.5 increased by a number is 4.2.

7. $3\frac{1}{2}$ decreased by a number is $\frac{1}{2}$.

8. The quotient of a number and 3.3 is 99.

9. The sum of 11.2 and a number is 65.07.

10. 64 divided by a number is 3.2.

11. Leroy sold 250 boxes of apples during a fund-raiser. This was 5 times as many apples as Mary sold. Write an equation that represents the situation.

Set B (pp. 291–293)

Solve and check.

1. $x + 6 = 15$

2. $15 = a + 2$

3. $11 + k = 25$

4. $z + 2.7 = 19.6$

5. $5.7 = b + 8.6$

6. $24.8 = 17.2 + c$

7. $y + 8\frac{2}{3} = 16$

8. $13\frac{1}{4} = 4\frac{4}{5} + s$

9. $13\frac{3}{8} = t + 7\frac{5}{6}$

10. $13.2 = x + 7.12$

11. $t + 7\frac{1}{3} = 10\frac{1}{12}$

12. $4.06 + r = 13.56$

13. A carpenter cut a 72-in. board into two pieces. One of the pieces is 24 in. long. How long is the second piece?

Set C (pp. 294–295)

Solve and check.

1. $x - 7 = 11$

2. $31 = a - 7$

3. $k - 10 = 42$

4. $z - 3.9 = 15.8$

5. $6.5 = b - 21.3$

6. $56.7 = c - 19.8$

7. $y - 5\frac{5}{6} = 13$

8. $22\frac{1}{3} = s - 4\frac{3}{4}$

9. $8\frac{3}{5} = t - 4\frac{5}{7}$

10. $a - 27 = 18$

11. $60.3 = b - 8.07$

12. $4\frac{7}{9} = x + 1\frac{2}{3}$

13. Reggie withdrew $175 from his checking account so he could go shopping for the new school year. His new balance was $234. How much was in the account before the withdrawal?

Set A (pp. 301–303)

Solve and check.

1. $3x = 12$
2. $7k = 56$
3. $\frac{p}{16} = {}^-3$

4. $\frac{a}{14} = 2$
5. $27 = {}^-3x$
6. $8.8 = 2.2n$

7. $4x = 24$
8. $8x = {}^-32$
9. $9 = \frac{p}{3}$

10. $56 = {}^-7p$
11. $21 = \frac{s}{3}$
12. $45 = 9n$

13. $180 = 3d$
14. $46.2 = \frac{a}{3}$
15. $1{,}486 = \frac{a}{2}$

16. $\frac{c}{1.2} = 5.6$
17. ${}^-12a = {}^-216$
18. $3.8m = 57$

For 19–20, write and solve an equation to answer the question.

19. Celia divided her baseball cards equally among 4 friends. Each friend got 23 baseball cards. How many baseball cards did Celia have originally?

20. Julie earns $6.75 per hour at her job. She wants to save $324.00. How many hours does she have to work to earn $324.00?

Set B (pp. 304–307)

Use the formula $d = rt$ to complete.

1. $d = \blacksquare$ mi
   $r = 35$ mi per hr
   $t = 4$ hr

2. $d = 1{,}600$ km
   $r = \blacksquare$ km per min
   $t = 400$ min

3. $d = 2{,}100$ km
   $r = 70$ km per sec
   $t = \blacksquare$ sec

4. $d = \blacksquare$ ft
   $r = 90.7$ ft per sec
   $t = 31$ sec

5. $d = 567$ mi
   $r = \blacksquare$ mi per hr
   $t = 17.5$ hr

6. $d = 4{,}850$ m
   $r = 250$ m per sec
   $t = \blacksquare$ sec

Convert the temperature to degrees Fahrenheit. Write your answer as a decimal.

7. 40°C
8. 2.3°C
9. 14°C
10. 20°C

Convert the temperature to degrees Celsius. Write your answer as a decimal and round to the nearest tenth of a degree.

11. 42°F
12. 47°F
13. 79°F
14. 100°F

15. The Concorde jet has a cruising speed of 1,354 mi per hr. Suppose the Concorde maintained this speed for $3\frac{1}{2}$ hr. How far would the Concorde travel?

16. The air conditioner is on only when the room temperature is greater than 75°F. Room temperature is 22°C. Is the air conditioner on? Explain.

Set A (pp. 322–323)

For 1–4, use the figure at the right. Tell how many of each you can name. Then name them.

1. points

2. line segments

3. rays

4. lines

Name the geometric figure.

5.

6.

7.

8. •F

Set B (pp. 326–329)

Tell if the angles are *vertical, adjacent, complementary, supplementary,* or *none of these.*

1. ∠CFD and ∠DFE

2. ∠AFB and ∠CFD

3. ∠EFA and ∠CFD

4. AFB and BFC

Find the unknown angle measure. The type of angle pair is given.

5. complementary

6. complementary

7. supplementary

8. supplementary

Find the measure of each angle.

9. ∠QSP

10. ∠LSR

11. ∠QSN

12. ∠RSP

13. ∠MSN

14. ∠RSN

Set C (pp. 330–331)

Use the figure.

1. Name all the lines that are parallel to $\overleftrightarrow{BE}$.

2. Name a line that is perpendicular to and intersects $\overleftrightarrow{BE}$.

3. Name all the lines that are perpendicular to $\overleftrightarrow{AD}$.

4. Name a line that is parallel to $\overleftrightarrow{AC}$.

5. Name all the lines that intersect $\overleftrightarrow{GE}$.

Set A (pp. 336–339)

Find the measure of the unknown angle and classify the triangle.

1.

2.

3.

4.

Set B (pp. 342–345)

Give the most exact name for the figure.

1.

2.

3.

4.

Complete the statement, giving the most exact name for the figure.

5. If a rhombus has four right angles, then it is a __?__.

6. If a parallelogram has four congruent sides, then it is a __?__.

7. If a polygon has four sides, then it is a __?__.

8. If a parallelogram has four right angles, then it is a __?__.

Set C (pp. 346–347)

Draw the figure. Use square dot paper or isometric dot paper.

1. a rectangle

2. a trapezoid

3. a right triangle with all sides different lengths

4. a pentagon with no congruent sides

5. a triangle with all sides the same length

6. a rhombus

Set D (pp. 348–349)

Name the given parts of the circle.

1. center

2. chords

3. radii

4. diameters

5. three arcs

6. Name a sector.

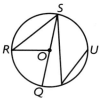

Set A (pp. 354–356)

Name the figure. Is it a polyhedron?

1.   **2.**   **3.**   **4.**

5.   **6.**   **7.**   **8.**

Write *true* or *false* for each statement. Rewrite each false statement as a true statement.

9. In a pentagonal prism, the bases are congruent.

10. A cylinder may have a polygon as its base.

11. All pyramids have rectangular bases.

12. The lateral surface of a cylinder is curved.

13. A prism with square bases may be a cube.

14. An octagonal prism has six faces.

Set B (pp. 357–359)

Name the solid figure that has the given views.

1.

2.

3.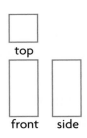

Draw the top, front, and side views for each solid.

4.

5.

6.

Set A (pp. 374–376)

Write each ratio in fraction form. Then find the unit rate.

1. $2.50 for 10

2. $2.76 for 12

3. 360 people in 3 sq mi

4. 240 miles in 6 hr

5. $2.75 for 25

6. 500 miles on 20 gallons

Set B (pp. 380–383)

The figures are similar. Write the ratio of the corresponding sides in simplest form.

1.

2.

Tell whether the figures in each pair are similar. Write *yes* or *no*. If you write *no*, explain.

3.

4.

Set C (pp. 384–386)

The figures are similar. Use a proportion to find the unknown length.

1. 24 ft, 16 ft, 12 ft, n

2. 9 m, 6 m, 12 m, n

3. 39 in., n, 15 in., 20 in.

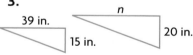

Set D (pp. 387–389)

Find the unknown dimension.

1. scale: 1 in.:5 ft
drawing length: 7 in.
actual length: ■ ft

2. scale: 1 in.:5 ft
drawing length: 4 in.
actual length: ■ ft

3. scale: 1 in.:5 ft
drawing length: ■ in.
actual length: 15 ft

Set E (pp. 390–391)

The map distance is given. Find the actual distance. The scale is 1 in. = 25 mi.

1. $1\frac{1}{2}$ in.

2. 3 in.

3. $5\frac{1}{2}$ in.

4. 7 in.

Set A (pp. 396–397)

Write the percent that is shaded.

1.

2.

3.

4.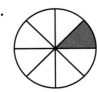

Set B (pp. 398–401)

Write as a percent.

1. 0.3
2. 0.09
3. 0.43
4. $\frac{7}{20}$
5. $\frac{3}{8}$

Set C (pp. 402–405)

Find the percent.

1. 20% of 8
2. 30% of 90
3. 45% of 75
4. 50% of 58

5. Of all the cookies Wendy baked, 40% were chocolate chip. If she baked 200 cookies, how many were chocolate chip?

Set D (pp. 408–411)

Find the sale price.

1. regular price: $31.00
   25% off

2. regular price: $65.00
   50% off

3. regular price: $42.00
   75% off

Find the regular price.

4. sale price: $45.00
   25% off

5. sale price: $24.95
   50% off

6. sale price: $14.40
   20% off

Set E (pp. 412–413)

Find the simple interest.

	Principal	Rate	Interest for 1 Year	Interest for 5 Years
1.	$65,000.00	4%	■	■
2.	$735.00	7%	■	■
3.	$1,300.00	3.9%	■	■
4.	$2,250.00	2.3%	■	■

5. Nancy put $2,500 in a savings account for 3 years at a simple interest rate of 8%. How much interest did she earn?

Set A (pp. 418–421)

For 1–5, use the spinner at the right. Find each probability. Write each answer as a fraction, a decimal, and a percent.

1. P(Dee)

2. P(Miles or Lili)

3. P(not Cara)

4. P(Marta)

5. P(Hugo, Miwa, or Chen)

A number cube is labeled 5, 10, 25, 50, 100, and 2,000. Find each probability. Write each answer as a fraction.

6. P(25)

7. P(5 or 25)

8. P(a number ending in zero)

9. P(1,000)

10. P(1,000 or 2,000)

11. P(not 500)

Cards showing pictures of team mascots are placed in a hat. There are 3 lions, 5 bears, 4 cheetahs, and 8 tigers. You choose one card without looking. Find each probability.

12. P(bear)

13. P(tiger or lion)

14. P(member of cat family)

A bag contains some buttons: 8 blue, 12 brown, 10 red, 4 green, and 6 yellow. You choose one button without looking. Find each probability.

15. P(yellow)

16. P(black)

17. P(not brown)

Set B (pp. 426–427)

A spinner is divided into 5 equal sections. Each section is labeled with one of the letters *A, E, I, O,* and *U*. Anna spins the pointer 100 times and records her results in the table below.

Letter	A	E	I	O	U
Times Landed On	15	5	30	40	10

Find the experimental probability.

1. P(*A*)

2. P(*E*)

3. P(*I*)

4. P(*O*)

5. P(*U*)

6. P(*A* or *U*)

7. P(*E, I* or *U*)

8. P(not *A*)

9. Based on her experimental results, how many times can Anna expect the pointer to land on *O* in the next 20 spins?

10. Based on her experimental results, how many times can Anna expect the pointer to land on *E* if she spins 2,000 times?

Set A (pp. 434–436)

Draw a tree diagram or make a table to find the number of possible outcomes for each situation.

1. choosing vanilla, chocolate, or strawberry yogurt, with cherry or chocolate sauce, and sprinkles or nuts

2. tossing a penny and spinning the pointer on the spinner

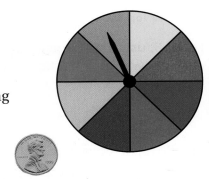

Use the Fundamental Counting Principle to find the number of outcomes for each situation.

3. a choice of pancakes, french toast, or waffles, and juice, milk, or tea

4. a choice of 6 salads and 10 dressings

5. rolling 2 number cubes labeled *A* to *F*

Set B (pp. 437–439)

Write *independent* or *dependent* to describe the events.

1. You have a bag of 6 red marbles and 4 green marbles. You draw one marble, record the color, place the marble back in the bag, and draw again.

2. Ana draws one name from a box to select the winner of a movie pass and then draws another name from the same box for the winner of a CD.

A box contains five cards labeled C, L, A, S, S. Without looking in the box, you select a card, replace it, and then select again. For 3–8, find the probability of each event. Then find the probability assuming the first card is not replaced.

3. P(C, L)

4. P(L, C)

5. P(A, S)

6. P(S, L)

7. P(C, L or S)

8. P(C or L, S)

Set C (pp. 440–441)

Seventy-five students from Park Middle School were randomly surveyed about favorite pizza toppings. The results are shown in the table.

FAVORITE PIZZA TOPPINGS	
Topping	**Number of Students**
Pepperoni	30
Black Olives	15
Sausage	12
Mushrooms	10
Other	8

1. Suppose there are 225 students who attend Park Middle School. Predict the number of students who prefer black olives as a pizza topping.

2. Suppose there are 314 students at Park Middle School. Predict the number of students who prefer sausage as a pizza topping.

Set A (pp. 452–453)

Use a proportion to convert to the given unit.

1. 15 ft = m yd

2. 12 c = r pt

3. 8 gal = h qt

4. 120 oz = x lb

5. 6 days = b hr

6. 12 in. = p yd

Set B (pp. 454–455)

Use a proportion to convert to the given unit.

1. 0.22 m = c km

2. 450 mm = h cm

3. 0.0030 kL = p L

4. 1,800 g = ■ kg

5. 10,000 cm = ■ m

6. 0.35 L = ■ mL

7. Paul buys 2 L of orange juice. He drinks 250 mL with breakfast. How many milliliters are left?

8. Tiffany runs 5,000 m. Ashley runs 3.5 km. Who runs farther and by how many meters?

Set C (pp. 456–457)

Use a proportion to convert to the given unit. Use the table on page 456. Round to the nearest hundredth if necessary.

1. 9 in. ≈ ■ cm

2. 11 yd ≈ ■ m

3. 3 mi ≈ ■ km

4. 10 L ≈ ■ gal

5. 55 cm ≈ ■ ft

6. 3.5 kg ≈ ■ lb

7. Eve weighs 51 kg. Beth weighs 116 lb. Who weighs more and by about how many pounds?

8. Zack has a board that is 6 ft long. He wants to cut a length that is 150 cm. About how many centimeters will be left?

Set D (pp. 458–461)

Measure the line segment to the given length.

1. nearest half inch; nearest inch

2. nearest centimeter; nearest millimeter

Tell which measurement is more precise.

3. 7 fl oz or 1 cup

4. 2 qt or 9 c

5. 1,245 m or 1 km

Name an appropriate customary or metric unit of measure for each item.

6. weight or mass of a bag of sugar

7. length of a shoelace

8. water in a fishbowl

Set A (pp. 469–471)

Find the perimeter.

1.

18 in.

9 in.

2.

3.3 cm 3.3 cm 3.3 cm 3.3 cm 3.3 cm 3.3 cm

3.

6.5 m · 490 cm · 9.8 m

The perimeter is given. Find the unknown length.

4.

18 cm · 41 cm · 32 cm · 52 cm · x · 52 cm

$P = 230$ cm

5.

15 m · 10 m · 9 m · 8.2 m · y

$P = 54.5$ m

6.

$13\frac{5}{6}$ ft · g · $13\frac{2}{6}$ ft · $18\frac{1}{2}$ ft

$P = 54$ ft

7. Tanya wants to put a string of lights around a rectangular window that is 40 in. wide and 48 in. high. How long will the string of lights need to be to go around the window one time?

8. Linda is building a raised flower bed $10\frac{1}{2}$ ft long and 4 ft wide. How many feet of lumber will she need to go around the flower bed?

Set B (pp. 476–479)

Find the circumference. Use 3.14 or $\frac{22}{7}$ for π. Round to the nearest whole number.

1.

7 in.

2.

12 ft

3.

49 cm

4.

$4\frac{1}{2}$ in.

5.

2.7 cm

6.

154 ft

7. $r = 16$ ft

8. $d = 23.5$ cm

9. $d = 28$ ft

10. $r = 25.9$ m

11. $d = 82.1$ mm

12. $d = 3.21$ m

13. A blue circular rug in Dominique's room has a diameter of $6\frac{1}{2}$ ft. To the nearest foot, what is the circumference of the rug?

14. A child's hat has a radius of 8 cm. To the nearest tenth of a centimeter, what is the circumference of the hat?

Set A (pp. 484–487)

Estimate the area. Each square on the grid represents 1 cm².

1.

2.

3.

Find the area.

4.
24 cm
32 cm

5.
9 ft
15 ft

6.
5 in.
8 in.

7.
9.5 m
5.2 m

Set B (pp. 488–490)

Find the area of each figure.

1.
12 yd
7 yd

2.
6 ft
4 ft
10 ft

3.
5.3 m
8.6 m

4.
18 cm
12 cm
23 cm

5. A wall plaque is shaped like a trapezoid with a height of 6 in. and bases that measure 3.5 in. and 9.5 in. Find the area.

Set C (pp. 492–493)

Find the area of each circle to the nearest whole number.

1.
9 yd

2. 12 mm

3. 7 ft

4. 4.2 in.

5. $d = 24$ cm

6. $r = 9.2$ mm

7. $d = 8$ yd

8. $d = 7\frac{1}{2}$ ft

Set D (pp. 494–497)

Find the surface area.

1.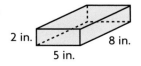
2 in.
5 in.
8 in.

2.
4 m
2 m
2 m

3.
7 m
4 m
3 m

4.
4 in.
5 in.
5 in.
9 in.
6 in.

Set A (pp. 502–505)

Find the volume.

1.

3 in.
4 in.
6 in.

2.

10.9 m
5 m
2.5 m

3

4.9 m
10.2 m
12.5 m

Find the unknown length.

4.

6 cm
14 cm
x
$V = 420 \text{ cm}^3$

5.

x
12 cm
5.5 cm
$V = 247.5 \text{ cm}^3$

6.

17 in.
x
6 in.
$V = 153 \text{ in.}^3$

Set B (pp. 508–509)

Find the volume.

1.

17 ft
8 ft
8 ft

2.

9.5 m
5 m
7 m

3.

79 in.
41 in.
42 in.

4. Find the volume of a rectangular pyramid with a length of 18 cm, a width of 16 cm, and a height of 12 cm.

5. Find the volume of a rectangular pyramid with a base of 300 ft² and a height of 50 ft.

Set C (pp. 511–513)

Find the volume. Round to the nearest whole number.

1.

17 ft
5 ft

2.

6.2 mm
4.5 mm

3.

7 cm
18 cm

Find the volume of the inside cylinder to the nearest whole number.

4.

6.7 cm
4 cm
1.5 cm

5.

3 ft
9 ft
13 ft

6.

5 mm   2 mm
9 mm

Set A (pp. 526–528)

Write a rule for each sequence.

1. 2, 6, 18, 54, . . .

2. $\frac{1}{10}, \frac{1}{5}, \frac{3}{10}, \frac{2}{5}, \ldots$

3. ⁻35, ⁻20, ⁻5, 10, . . .

4. 440, 44, 4.4, 0.44, . . .

Find the next three possible terms in each sequence.

5. $\frac{2}{5}, \frac{4}{5}, 1\frac{1}{5}, \ldots$

6. 2.3, 3.9, 5.5, . . .

7. 7, 17, 37, 67, . . .

8. 27, ⁻9, 3, ⁻1, . . .

9. Vittorio practices 6 weeks for the skateboard championship. The first 4 weeks he practices $9\frac{1}{2}$ hr, $11\frac{1}{4}$ hr, 13 hr, and $14\frac{3}{4}$ hr. Following this pattern, how many hours will Vittorio practice the sixth week?

Set B (pp. 529–532)

Write an equation to represent the function.

1.

x	0	1	2	3	4
y	3	4	5	6	7

2.

x	10	9	8	7	6
y	6	5	4	3	2

3.

x	0	1	2	3	4
y	0	8	16	24	32

4.

x	30	27	24	21	18
y	10	9	8	7	6

Set C (pp. 533–535)

Draw the next two figures in the pattern.

1.

2.

Draw the next two solids in the pattern.

3.

4.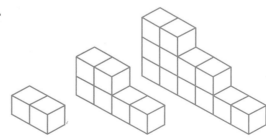

Set A (pp. 540–542)

Tell which type or types of transformation the second figure is of the first figure. Write *translation*, *rotation*, or *reflection*.

1.   **2.**   **3.**   **4.**

Set B (pp. 543–545)

Trace and cut out several of each shape. Tell whether the shape can be used repeatedly to form a tessellation. Write *yes* or *no*.

1.   **2.**   **3.**   **4.**

Set C (pp: 548–549)

Tell how many ways you can place the solid figure on the outline.

1.   **2.**   **3.**

Set D (pp. 550–553)

Trace the figure. Draw the lines of symmetry.

1.   **2.**   **3.**   **4.**

Tell whether each figure has rotational symmetry, and, if so, identify the symmetry as a fraction of a turn and in degrees.

5.   **6.**   **7.**   **8.**

Set A (pp. 558–559)

Graph the solutions of the inequality.

1. $x < 4$ **2.** $x > 9$ **3.** $x \geq 3$ **4.** $x < {}^-5$

Solve the inequality and graph the solutions on a number line.

5. $7a \leq 14$ **6.** $n + 5 > 8$ **7.** $m - 2 < 0$ **8.** $5c \geq 5$

Set B (pp. 560–563)

Write the ordered pair for each point on the coordinate plane.

1. point A **2.** point B

3. point C **4.** point D

5. point E **6.** point F

7. point G **8.** point H

9. point I **10.** point J

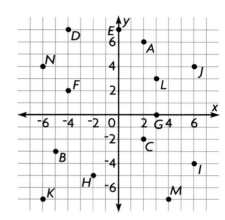

Set C (pp. 564–565)

Complete the table. Then write an equation relating *y* to *x*.

1.

x	1	2	3	4	5
y	3	4	5	■	■

2.

x	1	2	3	4	5
y	6	12	18	■	■

3.

x	8	9	10	11	12
y	1	2	3	■	■

4.

x	8	4	0	¯4	¯8
y	¯2	¯1	0	■	■

Set D (pp. 570–573)

Copy the figure onto a coordinate plane. Transform the figure according to the directions given. Name the new coordinates.

1.

2 units to the right

2.

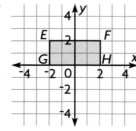

rotate 90° clockwise about (¯2,0)

3.

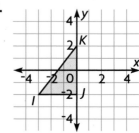

reflect across the y-axis

TIPS FOR TAKING MATH TESTS

Being a good test-taker is like being a good problem solver. When you answer test questions, you are solving problems. Remember to **ANALYZE, CHOOSE, SOLVE, and CHECK.**

Analyze

Read the problem.

- Look for math terms and recall their meanings.
- Reread the problem and think about the question.
- Use the details in the problem and the question.

---

1. The difference between two numbers is 37. Their sum is 215. What are the numbers?

A 37 and 178   **C** 89 and 126

B 83 and 120   **D** 107 and 108

TIP! **Understand the problem.** The problem requires you to find two numbers for which the difference and sum are given. Reread the problem to compare the details to the answer choices. You can use estimation instead of calculating the sum and difference of each pair of numbers. The answer is **C**.

- Each word is important. Missing a word or reading it incorrectly could cause you to get the wrong answer.
- Pay attention to words that are in **bold** type, all CAPITAL letters, or *italics* and words like *round, best,* or *least to greatest*.

---

2. Florinda took $\frac{1}{2}$ hour to complete a race. Doris took $\frac{3}{8}$ hour and Violet took $\frac{2}{3}$ hour to complete the same race. Which lists the three runners from fastest to slowest?

F Florinda, Doris, Violet

G Doris, Violet, Florinda

H Violet, Florinda, Doris

J Doris, Florinda, Violet

TIP! **Look for important words.** The words *fastest to slowest* are important. The fastest runner takes the least amount of time. Think about where each fraction would be placed on a number line between 0 and 1. Then put the times in order from least to greatest. The answer is **J**.

Think about how you can solve the problem.

- See if you can solve the problem with the information given.
- Pictures, charts, tables, and graphs may have the information you need.
- You may need to recall information not given.
- Sometimes the answer choices have information to help solve the problem.

3. Jeremy wants to make a graph to see the trend of the profit in his lawn-mowing business over the past 10 months. What type of graph would best show this?

   A circle graph   **C** line graph

   B histogram   **D** stem-and-leaf plot

 Get the information you need.
The answer choices give four different types of graphs or plots. Think about each one and the kind of data that is appropriate for it. The problem states that Jeremy wants to see the trend in his profit over the past 10 months, and line graphs show trends. The answer is **C**.

- You may need to write a number sentence and solve it to answer the question.
- Some problems have two steps or more.
- In some problems you need to look at relationships instead of computing an answer.
- If the path to the solution isn't clear, choose a problem-solving strategy and use it to solve the problem.

4. A square tile with 4-inch sides is rotated along the line below. If the tile stopped so that the letter **P** appears in an upright position, which of these distances could it have been rotated?

   P

   F 40 in.   **H** 48 in.

   G 44 in.   **J** 52 in.

Decide on a plan.
From the distances given, you must find the one that could allow for complete rotations of the square tile. Since each side of the square is 4 inches long, it moves 16 inches in one complete rotation. If you *use logical reasoning*, you see that only one of the choices is a multiple of 16. The answer is **H**.

Follow your plan, working logically and carefully.

- Estimate your answer. Compare it to the answer choices.
- Use reasoning to find the most likely choices.
- Make sure you solved all the steps needed to answer the problem.
- If your answer does not match any of the answer choices, check the numbers you used. Then check your computation.

5. Glen and Doug painted a fence. It took Glen three times as long to paint one side of the fence as it took Doug to paint the other side. If h represents the number of hours Doug painted, which expression shows the hours Glen painted?

A $3 \div h$　　　**C** $3 + h$

B $3 - h$　　　**D** $3 \times h$

TIP! Eliminate choices.
It is important to understand that h represents the hours it took Doug to paint one side of the fence. Since Glen painted three times as long as Doug, you can eliminate answers **A** and **B** because it does not make sense to divide by or subtract the hours Doug painted. Answer **C** means three more than h, so it can be eliminated. The answer is **D**.

- If your answer still does not match one of the choices, look for another form of the number, such as a decimal instead of a fraction.
- If answer choices are given as pictures, look at each one by itself while you cover the other three.
- If you do not see your answer and the answer choices include Not here, make sure your work is correct and then mark Not here.
- Read answer choices that are statements and relate them to the information in the problem one by one.

6. Tanya and her class helped plant 120 tulip and daffodil bulbs in front of the school. If 30 of the bulbs were tulips, what percent of the bulbs were daffodils?

F 90%　　　**H** 25%

G $66\frac{2}{3}$%　　　**J** Not here

TIP! Choose the answer.
Since 30 of the bulbs were tulips, 90 were daffodils. The question asks what percent are daffodils so you need to find what percent of 120 is 90 ($90 \div 120$). If your answer doesn't match one of the answer choices, check your computation. If you know your work is correct, mark the letter for Not here. The answer is **J**.

Take time to catch your mistakes.

- Be sure you answered the question asked.
- Check for important words you might have missed.
- Be sure you used all the information you needed.
- Check your computation by using a different method.

7. The temperature was 6°F today. A forecaster predicted that the temperature would drop by 5° each day for the next four days. Which sequence could be used to find the predicted temperatures?

A 6°F, 11°F, 16°F, 21°F, 26°F

B 6°F, 1°F, 6°F, 11°F, 16°F

C 6°F, 1°F, ⁻6°F, ⁻11°F, ⁻16°F

D 6°F, 1°F, ⁻4°F, ⁻9°F, ⁻14°F

TIP! **Check your work.**
You need to find the sequence that shows a drop of 5° in temperature for the next four days. Draw a thermometer or a number line to check your computation. The answer is **D**.

Don't Forget!

Before the test...

- Listen to the teacher's directions and read the instructions.
- Write down the ending time if the test is timed.
- Know where and how to mark your answers.
- Know whether you should write on the test page or use scratch paper.
- Ask any questions you have before the test begins.

During the test...

- Work quickly but carefully. If you are unsure how to answer a question, leave it blank and return to it later.
- If you cannot finish on time, look over the questions that are left. Answer the easiest ones first. Then go back to answer the others.
- Fill in each answer space carefully. Erase completely if you change an answer. Erase any stray marks.
- Check that the answer number matches the question number, especially if you skip a question.

By the end of grade six, students have mastered the four arithmetic operations with whole numbers, positive fractions, positive decimals, and positive and negative integers; they accurately compute and solve problems. They apply their knowledge to statistics and probability. Students understand the concepts of mean, median, and mode of data sets and how to calculate the range. They analyze data and sampling processes for possible bias and misleading conclusions; they use addition and multiplication of fractions routinely to calculate the probabilities for compound events. Students conceptually understand and work with ratios and proportions; they compute percentages (e.g., tax, tips, interest). Students know about π and the formulas for the circumference and area of a circle. They use letters for numbers in formulas involving geometric shapes and in ratios to represent an unknown part of an expression. They solve one-step linear equations.

Number Sense

1.0 **Students compare and order positive and negative fractions, decimals, and mixed numbers. Students solve problems involving fractions, ratios, proportions, and percentages:**

1.1 Compare and order positive and negative fractions, decimals, and mixed numbers and place them on a number line.

1.2 Interpret and use ratios in different contexts (e.g., batting averages, miles per hour) to show the relative sizes of two quantities, using appropriate notations ($\frac{a}{b}$, a to b, $a:b$).

Write the following as ratios:

1. The ratio of tricycles to tricycle wheels

2. The ratio of hands to fingers

3. If there are 6 tricycle wheels, how many tricycles are there?

4. If there are 45 fingers, how many hands are there?

1.3 Use proportions to solve problems (e.g., determine the value of N if $\frac{4}{7} = \frac{N}{21}$, find the length of a side of a polygon similar to a known polygon). Use cross-multiplication as a method for solving such problems, understanding it as the multiplication of both sides of an equation by a multiplicative inverse.

Joe can type 9 words in 8 seconds. At this rate, how many words can he type in 2 minutes?

Find n if:

1. $\frac{49}{21} = \frac{14}{n}$
2. $\frac{n}{3} = \frac{5}{7}$

(This problem also applies to Algebra and Functions Standard 1.1.)

Note: The sample problems illustrate the standards and are written to help clarify them.

The symbols 0—¬ and ● identify the Key Standards for grade six.

Number Sense (Continued)

○—ᴺ 1.4 Calculate given percentages of quantities and solve problems involving discounts at sales, interest earned, and tips.

○—ᴺ 2.0 **Students calculate and solve problems involving addition, subtraction, multiplication, and division:**

2.1 Solve problems involving addition, subtraction, multiplication, and division of positive fractions and explain why a particular operation was used for a given situation.

2.2 Explain the meaning of multiplication and division of positive fractions and perform the calculations (e.g., $\frac{5}{8} \div \frac{15}{16} = \frac{5}{8} \times \frac{16}{15} = \frac{2}{3}$).

   1. If $\frac{11}{7}$ is divided by a certain fraction $\frac{a}{b}$, the result is $\frac{3}{8}$. What is $\frac{a}{b}$?

   2. Draw a rectangle that has a perimeter of 1 and an area that is less than $\frac{1}{30}$.

○—ᴺ 2.3 Solve addition, subtraction, multiplication, and division problems, including those arising in concrete situations, that use positive and negative integers and combinations of these operations.

   Simplify to make the calculation as simple as possible and identify the properties you used at each step:

   1. $95 + 276 + 5$

   2. $^-19 + 37 + 19$

   3. $^-16\,(^-8 + 9)$

   4. $\left(\frac{-7}{8}\right)\left(\frac{17}{17}\right)$

   5. $(^-8)(^-4)(19)(6 + (^-6))$

○—ᴺ 2.4 Determine the least common multiple and the greatest common divisor of whole numbers; use them to solve problems with fractions (e.g., to find a common denominator to add two fractions or to find the reduced form for a fraction).

Algebra and Functions

1.0 **Students write verbal expressions and sentences as algebraic expressions and equations; they evaluate algebraic expressions, solve simple linear equations, and graph and interpret their results:**

○—ᴺ 1.1 Write and solve one-step linear equations in one variable.

   $6y - 2 = 10$. What is y?

Algebra and Functions (Continued)

1.2 Write and evaluate an algebraic expression for a given situation, using up to three variables.

Joe's sister Mary is twice as old as he is. Mary is 16. How old is Joe?

1.3 Apply algebraic order of operations and the commutative, associative, and distributive properties to evaluate expressions; and justify each step in the process.

1.4 Solve problems manually by using the correct order of operations or by using a scientific calculator.

2.0 **Students analyze and use tables, graphs, and rules to solve problems involving rates and proportions:**

2.1 Convert one unit of measurement to another (e.g., from feet to miles, from centimeters to inches).

Suppose that one British pound is worth $1.50. In London a magazine costs 3 pounds. In San Francisco the same magazine costs $4.25. In which city is the magazine cheaper?

When temperature is measured in both Celsius (C) and Fahrenheit (F), it is known that they are related by the following formula:

$9 \times C = (F - 32) \times 5$.

What is 50 degrees Fahrenheit in Celsius? (Note the explicit use of parentheses.)

O— **2.2** Demonstrate an understanding that *rate* is a measure of one quantity per unit value of another quantity.

2.3 Solve problems involving rates, average speed, distance, and time.

Marcus took a train from San Francisco to San Jose, a distance of 54 miles. The train took 45 minutes for the trip. What was the average speed of the train?

3.0 **Students investigate geometric patterns and describe them algebraically:**

3.1 Use variables in expressions describing geometric quantities (e.g., $P = 2w + 2l$, $A = \frac{1}{2}bh$, $C = \pi d$—the formulas for the perimeter of a rectangle, the area of a triangle, and the circumference of a circle, respectively).

A rectangle has width w. Its length is one more than 3 times its width. Find the perimeter of the rectangle. (Your answer will be expressed in terms of w.)

3.2 Express in symbolic form simple relationships arising from geometry.

Measurement and Geometry

1.0 **Students deepen their understanding of the measurement of plane and solid shapes and use this understanding to solve problems:**

○━┓ **1.1** Understand the concept of a constant such as π; know the formulas for the circumference and area of a circle.

1.2 Know common estimates of π (3.14; $\frac{22}{7}$) and use these values to estimate and calculate the circumference and the area of circles; compare with actual measurements.

1.3 Know and use the formulas for the volume of triangular prisms and cylinders (area of base × height); compare these formulas and explain the similarity between them and the formula for the volume of a rectangular solid.

Find the volumes (dimensions are in cm).

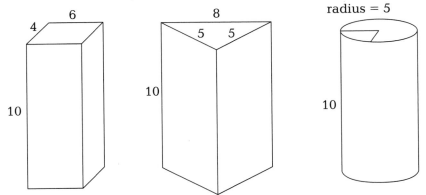

2.0 **Students identify and describe the properties of two-dimensional figures:**

2.1 Identify angles as vertical, adjacent, complementary, or supplementary and provide descriptions of these terms.

○━┓ **2.2** Use the properties of complementary and supplementary angles and the sum of the angles of a triangle to solve problems involving an unknown angle.

Find the missing angles.

2.3 Draw quadrilaterals and triangles from given information about them (e.g., a quadrilateral having equal sides but no right angles, a right isosceles triangle).

Statistics, Data Analysis, and Probability

1.0   **Students compute and analyze statistical measurements for data sets:**

      1.1   Compute the range, mean, median, and mode of data sets.

      1.2   Understand how additional data added to data sets may affect these computations of measures of central tendency.

      1.3   Understand how the inclusion or exclusion of outliers affects measures of central tendency.

      1.4   Know why a specific measure of central tendency (mean, median, mode) provides the most useful information in a given context.

2.0   **Students use data samples of a population and describe the characteristics and limitations of the samples:**

      2.1   Compare different samples of a population with the data from the entire population and identify a situation in which it makes sense to use a sample.

    2.2   Identify different ways of selecting a sample (e.g., convenience sampling, responses to a survey, random sampling) and which method makes a sample more representative for a population.

    2.3   Analyze data displays and explain why the way in which the question was asked might have influenced the results obtained and why the way in which the results were displayed might have influenced the conclusions reached.

    2.4   Identify data that represent sampling errors and explain why the sample (and the display) might be biased.

    2.5   Identify claims based on statistical data and, in simple cases, evaluate the validity of the claims.

3.0   **Students determine theoretical and experimental probabilities and use these to make predictions about events:**

    3.1   Represent all possible outcomes for compound events in an organized way (e.g., tables, grids, tree diagrams) and express the theoretical probability of each outcome.

      3.2   Use data to estimate the probability of future events (e.g., batting averages or number of accidents per mile driven).

    3.3   Represent probabilities as ratios, proportions, decimals between 0 and 1, and percentages between 0 and 100 and verify that the probabilities computed are reasonable; know that if P is the probability of an event, $1 - P$ is the probability of an event not occurring.

Statistics, Data Analysis, and Probability (Continued)

3.4 Understand that the probability of either of two disjoint events occurring is the sum of the two individual probabilities and that the probability of one event following another, in independent trials, is the product of the two probabilities.

○━┓ **3.5** Understand the difference between independent and dependent events.

Mathematical Reasoning

1.0 Students make decisions about how to approach problems:

1.1 Analyze problems by identifying relationships, distinguishing relevant from irrelevant information, identifying missing information, sequencing and prioritizing information, and observing patterns.

1.2 Formulate and justify mathematical conjectures based on a general description of the mathematical question or problem posed.

1.3 Determine when and how to break a problem into simpler parts.

2.0 Students use strategies, skills, and concepts in finding solutions:

2.1 Use estimation to verify the reasonableness of calculated results.

2.2 Apply strategies and results from simpler problems to more complex problems.

2.3 Estimate unknown quantities graphically and solve for them by using logical reasoning and arithmetic and algebraic techniques.

2.4 Use a variety of methods, such as words, numbers, symbols, charts, graphs, tables, diagrams, and models, to explain mathematical reasoning.

2.5 Express the solution clearly and logically by using the appropriate mathematical notation and terms and clear language; support solutions with evidence in both verbal and symbolic work.

2.6 Indicate the relative advantages of exact and approximate solutions to problems and give answers to a specified degree of accuracy.

2.7 Make precise calculations and check the validity of the results from the context of the problem.

Mathematical Reasoning (Continued)

3.0   Students move beyond a particular problem by generalizing to other situations:

3.1   Evaluate the reasonableness of the solution in the context of the original situation.

3.2   Note the method of deriving the solution and demonstrate a conceptual understanding of the derivation by solving similar problems.

3.3   Develop generalizations of the results obtained and the strategies used and apply them in new problem situations.

Pronunciation Key

a	add, map	f	fit, half	n	nice, tin	p	pit, stop	yo͞o fuse, few
ā	ace, rate	g	go, log	ng	ring, song	r	run, poor	v vain, eve
â(r)	care, air	h	hope, hate	o	odd, hot	s	see, pass	w win, away
ä	palm, father	i	it, give	ō	open, so	sh	sure, rush	y yet, yearn
b	bat, rub	ī	ice, write	ô	order, jaw	t	talk, sit	z zest, muse
ch	check, catch	j	joy, ledge	oi	oil, boy	th	thin, both	zh vision,
d	dog, rod	k	cool, take	ou	pout, now	th	this, bathe	pleasure
e	end, pet	l	look, rule	o͝o	took, full	u	up, done	
ē	equal, tree	m	move, seem	o͞o	pool, food	û(r)	burn, term	

ə the schwa, an unstressed vowel representing the sound spelled *a* in **above**, *e* in **sicken**, *i* in **possible**, *o* in **melon**, *u* in **circus**

Other symbols:
• separates words into syllables
' indicates stress on a syllable

absolute value [ab'sə•lo͞ot val'yo͞o] The distance of an integer from zero (p. 228)

acute angle [ə•kyo͞ot' an'gəl] an angle whose measure is greater than 0° and less than 90° (p. 324)

acute triangle [ə•kyo͞ot' trī'an•gəl] A triangle with all angles less than 90° (p. 336)
Example:

Addition Property of Equality [ə•dish'ən prä'pər•tē əv i•kwol'ə•tē] The property that states that if you add the same number to both sides of an equation, the sides remain equal (p. 294)

additive inverse [ad'ə•tiv in'vûrs] The opposite of a given number (p. 243)

adjacent angles [ə•jā'sənt an'gəlz] Angles that are side by side and have a common vertex and ray (p. 326)
Example:

∠*MRN* and ∠*NRQ* are adjacent angles.

algebraic expression [al•jə•brā'ik ik•spre'shən] An expression that includes at least one variable (p. 28)
Examples: $x + 5$, $3a - 4$

algebraic operating system [al•jə•brā'ik ä'pə•rā•ting sis'təm] A way for calculators to follow the order of operations when evaluating expressions (p. 43)

angle [an'gəl] A figure formed by two rays with a common endpoint (p. 324)
Example:

arc [ärk] A part of a circle, named by its endpoints (p. 348)
Example:

arc *AB* or $\overset{\frown}{AB}$

area [âr'ē•ə] The number of square units needed to cover a given surface (p. 484)

Associative Property [ə•sō'shē•ā•tiv prä'pər•tē] The property that states that the way addends are grouped or factors are grouped does not change the sum or the product (p. 36)
Examples: $12 + (5 + 9) = (12 + 5) + 9$
$(9 \times 8) \times 3 = 9 \times (8 \times 3)$

axes [ak′sēz] The horizontal number line (*x*-axis) and the vertical number line (*y*-axis) on the coordinate plane (p. 560)

bar graph [bär′graf] A graph that displays countable data with horizontal or vertical bars (p. 120)

base [bās] A number used as a repeated factor (p. 40)
Example: $8^3 = 8 \times 8 \times 8$; 8 is the base.

base [bās] A side of a polygon or a face of a solid figure by which the figure is measured or named (pp. 354, 485)
Examples:

biased question [bī′əst kwes′chən] A question that leads to a specific response or excludes a certain group (p. 98)

biased sample [bī′əst sam′pəl] A sample is biased if individuals or groups from the population are not represented in the sample. (p. 98)

box-and-whisker graph [bäks·ənd·hwis′kər graf] A graph that shows how far apart and how evenly data are distributed (p. 129)

Celsius [səl′sē·əs] A metric scale for measuring temperature (p. 299)

certain [sûr′tən] Sure to happen (p. 419)

chord [kôrd] A line segment with its endpoints on a circle (p. 348)
Example:

chord: $\overline{AB}$

circle [sûr′kəl] A closed plane figure with all points of the figure the same distance from the center (p. 348)
Example:

circle graph [sûr′kəl graf] A graph that lets you compare parts to the whole and to other parts (p. 122)
Example:

FAVORITE HOBBIES

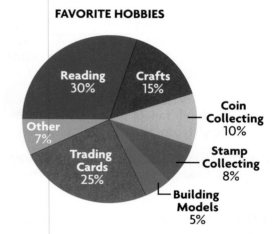

circumference [sûr·kum′fər·əns] The distance around a circle (p. 474)

clustering [klus′tər·ing] A method used to estimate a sum when all addends are about the same (p. 16)

Commutative Property [kə·myōo′tə·tiv prä′pər·tē] The property that states that if the order of addends or factors is changed, the sum or product stays the same (p. 36)
Examples: $6 + 5 + 7 = 5 + 6 + 7$
$8 \times 9 \times 3 = 3 \times 8 \times 9$

compensation [kom·pən·sā′shən] A mental math strategy for some addition and subtraction problems (p. 37)

complementary angles [kom·plə·men′tər·ē an′gəlz] Two angles whose measures have a sum of 90° (p. 327)
Example:

composite number [käm·pä′zət num′bər] A whole number greater than 1 that has more than two whole-number factors (p. 145)

compound event [käm′pound i·vent′] An event made of two or more simple events (p. 434)

congruent [kən·grōō′ənt] Having the same size and shape (p. 380)

convenience sample [kən·vēn′yənts sam′pəl] Sampling the most available subjects in the population to obtain quick results (p. 95)

coordinate plane [kō·ôr′də·nit plān] A plane formed by a horizontal line (*x*-axis) that intersects a vertical line (*y*-axis) at a point called the origin (p. 560)

corresponding angles [kôr·ə·spän′ding an′gəlz] Angles that are in the same position in different figures (p. 381)
Example:

∠A and ∠D are corresponding angles.

corresponding sides [kôr·ə·spän′ding sīdz] Sides that are in the same position in different plane figures (p. 381)
Example:

$\overline{CA}$ and $\overline{FD}$ are corresponding sides.

cube [kyōōb] A rectangular solid with six congruent faces (p. 354)
Example:

cumulative frequency [kyōō′myə·lə·tiv frē′kwən·sē] A running total of the number of subjects surveyed (p. 103)

 D

decimal [de′sə·məl] A number with one or more digits to the right of the decimal point (p. 52)

denominator [di·nä′mə·nā·tər] The part of a fraction that tells how many equal parts are in the whole (p. 159)
Example: $\frac{3}{4}$ ←denominator

dependent events [di·pen′dənt i·vənts′] Events for which the outcome of the second event depends on the outcome of the first event (p. 438)

diameter [dī·am′ə·tər] A line segment that passes through the center of a circle and has its endpoints on the circle (p. 348)
Example:

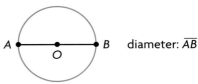

dimension [di·men′shən] The length, width, or height of a figure (p. 501)

discount [dis′kount] An amount that is subtracted from the regular price of an item (p. 408)

Distributive Property of Multiplication [di·strib′yə·tiv prä′pər·tē əv mul·tə·plə·kā′shən] The property that states that multiplying a sum by a number is the same as multiplying each addend by the number and then adding the products (p. 36)
Example: 14 × 21 = 14 × (20 + 1) = (14 × 20) + (14 × 1)

dividend [di′və·dend] The number that is to be divided in a division problem
Example: In 56 ÷ 8, 56 is the dividend.

Division Property of Equality [di·vi′zhən prä′pər·tē əv i·kwol′ə·tē] The property that states that if you divide both sides of an equation by the same nonzero number, the sides remain equal (p. 301)

divisor [di·vī′zər] The number that divides the dividend
Example: In 45 ÷ 9, 9 is the divisor.

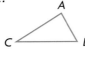 **E**

equally likely [ē′kwə·lē li′klē] Having the same chance of occurring (p. 418)

equation [i•kwā′zhən] A statement that shows that two quantities are equal (p. 30)

equilateral triangle [ē•kwə•la′tə•rəl trī′an•gəl] A triangle with three congruent sides (p. 336)
Example:

P

3 cm   3 cm

R   3 cm   Q

equivalent fractions [ē•kwiv′ə•lənt frak′shənz] Fractions that name the same amount or part (p. 160)

equivalent ratios [ē•kwiv′ə•lənt rā′shē•ōz] Ratios that name the same comparisons (p. 374)

estimate [es′tə•mit] An answer that is close to the exact answer and that is found by rounding, by clustering, or by using compatible numbers (p. 16)

evaluate [i•val′yoo•āt] Find the value of a numerical or algebraic expression (p. 28)

event [i•vent′] A set of outcomes (p. 417)

experimental probability [ik•sper•ə•men′təl prä•bə•bil′ə•tē] The ratio of the number of times an event occurs to the total number of trials or times the activity is performed (p. 426)

exponent [ik•spō′nənt] A number that tells how many times a base is used as a factor (p. 40)
Example: $2^3 = 2 \times 2 \times 2 = 8$;
3 is the exponent.

face [fās] One of the polygons of a solid figure (p. 483)
Example:

— face

factor [fak′tər] A number that is multiplied by another number to find a product

Fahrenheit [fâr′ən•hīt] A customary scale for measuring temperature (p. 305)

formula [fôr′myə•lə] A rule that is expressed with symbols (p. 304)
Example: $A = lw$

fractal [frak′təl] A figure with repeating patterns containing shapes that are like the whole but of different sizes throughout (p. 533)

frequency table [frē′kwən•sē tā′bəl] A table representing totals for individual categories or groups (p. 103)

function [funk′shən] A relationship between two quantities in which one quantity depends on the other (p. 529)

Fundamental Counting Principle [fun•də•men′təl koun′ting prin′sə•pəl] If one event has m possible outcomes and a second independent event has n possible outcomes, then there are $m \times n$ total possible outcomes. (p. 434)

greatest common factor (GCF) [grā′təst kä′mən fak′tər] The greatest factor that two or more numbers have in common (p. 151)

height [hīt] A measure of a polygon or solid figure, taken as a perpendicular from the base of the figure (p. 485)
Example:

height —

hexagon [heks′ə•gon] A six-sided polygon

histogram [his′tə•gram] A bar graph that shows the number of times data occur in certain ranges or intervals (p. 127)

Identity Property of Zero [ī•den′tə•tē prä′pər•tē əv zir′ō] The property that states that the sum of zero and any number is that number (p. H2)
Example: $25 + 0 = 25$

Identity Property of One [i•den′tə•tē prä′pər•tē əv wun] The property that states that the product of any number and 1 is that number (p. H2)
Example: 12 × 1 = 12

impossible [im•pos′ə•bəl] Never able to happen (p. 419)

independent events [in•di•pen′dənt i•vents′] Events for which the outcome of the second event does not depend on the outcome of the first event (p. 437)

indirect measurement [in•di•rekt′ mezh′ər•mənt] The technique of using similar figures and proportions to find a measure (p. 384)

inequality [in•i•kwäl′ə•tē] An algebraic or numerical sentence that contains the symbol <, >, ≤, ≥, or ≠ (p. 558)
Example: x + 3 > 5

integers [in′ti•jərz] The set of whole numbers and their opposites (p. 228)

isosceles triangle [i•sä′sə•lēz tri′an•gəl] A triangle with exactly two congruent sides (p. 335)
Example:

7 in.　7 in.

5 in.

lateral faces [lat′ər•əl fās′əz] The faces in a prism or pyramid that are not bases (p. 354)

least common denominator (LCD) [lēst kä′mən di•nä′mə•nā•tər] The least common multiple of two or more denominators (p. 182)

least common multiple (LCM) [lēst kä′mən mul′tə•pəl] The smallest number, other than zero, that is a common multiple of two or more numbers (p. 150)

like terms [lik turmz] Expressions that have the same variable with the same exponent (p. 277)

line [līn] A straight path that extends without end in opposite directions (p. 322)
Example:

⟷

line graph [līn graf] A graph that uses a line to show how data change over time (p. 121)

line of symmetry [līn əv si′mə•trē] A line that divides a figure into two congruent parts (p. 550)

line plot [līn plät] A graph that shows frequency of data along a number line (p. 102)
Example:

Miles Jogged

line segment [līn seg′mənt] A part of a line with two endpoints (p. 322)
Example:

●━━━━━━━●

line symmetry [līn si′mə•trē] A figure has line symmetry if a line can separate the figure into two congruent parts. (p. 550)

lower extreme [lō′ər ik•strēm′] The least number in a set of data (p. 129)

lower quartile [lō′ər kwôr′til] The median of the lower half of a set of data (p. 129)

mean [mēn] The sum of a group of numbers divided by the number of addends (p. 106)

median [mē′dē•ən] The middle value in a group of numbers arranged in order (p. 106)

mixed number [mikst num′bər] A number represented by a whole number and a fraction (p. 164)

mode [mōd] The number that occurs most often in a set of data (p. 106)

multiple-bar graph [mul′tə•pəl bär′graf] A bar graph that represents two or more sets of data (p. 120)

multiple-line graph [mul′tə•pəl lin′graf] A line graph that represents two or more sets of data (p. 121)

multiple [mul′tə•pəl] The product of a given whole number and another whole number (p. 145)

Multiplication Property of Equality [mul•tə•plə•kā′shən prä′pər•tē əv i•kwol′ə•tē] The property that states that if you multiply both sides of an equation by the same number, the sides remain equal (p. 302)

negative integers [ne′gə•tiv in′ti•jərz] Integers to the left of zero on the number line (p. 228)

net [net] An arrangement of two-dimensional figures that folds to form a polyhedron (p. 360)
Example:

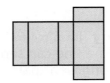

numerator [noō′mə•rā•tər] The part of a fraction that tells how many parts are being used (p. 159)
Example: $\frac{3}{4}$ ←numerator

numerical expression [noō•mâr′i•kəl ik•spre′shən] A mathematical phrase that uses only numbers and operation symbols (p. 28)

obtuse angle [äb•toōs′ an′gəl] An angle whose measure is greater than 90° and less than 180° (p. 324)
Example:

obtuse triangle [äb•toōs′ tri′an•gəl] A triangle with one angle greater than 90° (p. 336)
Example:

105°

opposites [ä′pə•zəts] Two numbers that are an equal distance from zero on the number line (p. 228)

order of operations [ôr′dər əv ä•pə•rā′shənz] The process for evaluating expressions: first perform the operations in parentheses, clear the exponents, perform all multiplication and division, and then perform all addition and subtraction (p. 42)

ordered pair [ôr′dərd pâr] A pair of numbers that can be used to locate a point on the coordinate plane (p. 560)
Examples: (0,2), (3,4), (⁻4,5)

origin [ôr′ə•jən] The point where the *x*-axis and the *y*-axis in the coordinate plane intersect, (0,0) (p. 560)

outcome [out′kəm] A possible result of a probability experiment (p. 418)

outlier [out′li•ər] A data value that stands out from others in a set; outliers can significantly affect measures of central tendency. (p. 110)

overestimate [ō•vər•es′tə•mət] An estimate that is greater than the exact answer (p. 17)

parallel lines [pâr′ə•lel linz] Lines in a plane that are always the same distance apart (p. 330)
Example:

percent (%) [pər•sent′] The ratio of a number to 100; *percent* means "per hundred." (p. 60)

perimeter [pə•ri′mə•tər] The distance around a figure (p. 467)

perpendicular lines [pər•pen•dik′yə•lər līnz] Two lines that intersect to form right, or 90°, angles (p. 330)
Example:

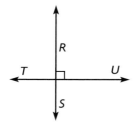

pi (π) [pī] The ratio of the circumference of a circle to its diameter; π ≈ 3.14 or $\frac{22}{7}$ (p. 475)

plane [plān] A flat surface that extends without end in all directions (p. 322)

point [point] An exact location in space, usually represented by a dot (p. 322)

point of rotation [point əv rō•tā′shən] The central point around which a figure is rotated (p. 551)

polygon [pä′lē•gän] A closed plane figure formed by three or more line segments (p. 335)

polyhedron [pä•lē•hē′drən] A solid figure with flat faces that are polygons (p. 354)
Example:

Hexagonal Prism

population [pä•pyə•lā′shən] The entire group of objects or individuals considered for a survey (p. 94)

positive integers [pä′zə•tiv in′ti•jərz] Integers to the right of zero on the number line (p. 228)

prime factorization [prīm fak•tə•ri•zā′shən] A number written as the product of all of its prime factors (p. 148)
Example: 24 = 2³ × 3

prime number [prīm num′bər] A whole number greater than 1 whose only factors are 1 and itself (p. 148)

principal [prin′sə•pəl] The amount of money borrowed or saved (p. 412)

prism [priz′əm] A solid figure that has two congruent, polygon-shaped bases, and other faces that are all rectangles (p. 354)
Example:

probability [prä•bə•bil′ə•tē] See *theoretical probability* and *experimental probability*

product [prä′dəkt] The answer in a multiplication problem (p. 15)

Property of Zero [prä′pər•tē əv zē′rō] The property that states that the product of any number and zero is zero (p. 35)

proportion [prə•pôr′shən] An equation that shows that two ratios are equal (p. 377)
Example: $\frac{1}{3} = \frac{3}{9}$

pyramid [pir′ə•mid] A solid figure with a polygon base and triangular sides that all meet at a common vertex (p. 355)
Example:

Q

quadrants [kwäd′rənts] The four regions of the coordinate plane (p. 560)

quadrilateral [kwä•drə•lat′ə•rəl] A polygon with four sides and four angles (p. 335)

quotient [kwō′shənt] The number, not including the remainder, that results from dividing (p. 23)

R

radius [rā′dē•əs] A line segment with one endpoint at the center of a circle and the other endpoint on the circle (p. 348)
Example:

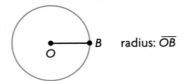

radius: $\overline{OB}$

random sample [ran′dəm sam′pəl] A sample in which each subject in the overall population has an equal chance of being selected (p. 95)

range [rānj] The difference between the greatest and least numbers in a group (p. 93)

rate [rāt] A ratio that compares two quantities having different units of measure (p. 375)

ratio [rā′shē•ō] A comparison of two numbers, a and b, written as a fraction $\frac{a}{b}$ (p. 230)

rational number [ra′shə•nəl num′bər] Any number that can be written as a ratio $\frac{a}{b}$, where a and b are integers and $b \neq 0$ (p. 230)

ray [rā] A part of a line with a single endpoint (p. 322)
Example:

ray: $\overrightarrow{JK}$

reciprocal [ri•sip′rə•kəl] Two numbers are reciprocals of each other if their product equals 1. (p. 209)

reflection [ri•flek′shən] A movement of a figure by flipping it over a line (p. 540)

regular polygon [reg′yə•lər pä′lē•gän] A polygon in which all sides are congruent and all angles are congruent (p. 340)
Example:

repeating decimal [ri•pēt′ing de′sə•məl] A decimal that doesn't end, because it shows a repeating pattern of digits after the decimal point (p. 169)

right angle [rīt an′gəl] An angle that has a measure of 90° (p. 324)
Example:

right triangle [rīt trī′an•gəl] A triangle with one right angle (p. 336)
Example:

rotation [rō•tā′shən] A movement of a figure by turning it around a fixed point (p. 540)

rotational symmetry [rō•tā′shən•əl si′mə•trē] The property of a figure that can be rotated less than 360° around a central point and still be congruent to the original figure (p. 551)

S

sales tax [sālz taks] A percent of the cost of an item, added onto the item's cost (p. 410)

sample [sam′pəl] A part of a population (p. 94)

sample space [sam′pəl spās] The set of all possible outcomes (p. 418)

scale [skāl] A ratio between two sets of measurements (p. 388)

scale drawing [skāl drô′ing] A drawing that shows a real object smaller than (a reduction) or larger than (an enlargement) the real object (p. 387)

scalene [skā′lēn] A triangle with no congruent sides (p. 335)

scatterplot [skat′ər•plät] A graph with points plotted to show a relationship between two variables (p. 139)

H80 Glossary

sector [sek′tər] A region enclosed by two radii and the arc joining their endpoints (p. 348) *Example:*

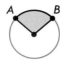

sequence [sē′kwəns] An ordered set of numbers (p. 526)

similar figures [si′mə•lər fig′yərz] Figures with the same shape but not necessarily the same size (p. 380)

simple interest [sim′pəl in′trəst] A fixed percent of the principal, paid yearly (p. 412)

simplest form [sim′pləst fôrm] The form in which the numerator and denominator of a fraction have no common factors other than 1 (p. 161)

solution [sə•loō′shən] A value that, when substituted for a variable in an equation, makes the equation true (p. 30)

square [skwâr] The product of a number and itself; a number with the exponent 2 (p. 280)

square [skwâr] A rectangle with four congruent sides (p. 501)

square root [skwâr roōt] One of two equal factors of a number (p. 281)

stem-and-leaf plot [stem ənd lēf plät] A type of graph that shows groups of data arranged by place value (p. 126)

straight angle [strāt an′gəl] An angle whose measure is 180° (p. 324) *Example:*

Subtraction Property of Equality [sub•trak′shən prä′pər•tē əv i•kwol′ə•tē] The property that states that if you subtract the same number from both sides of an equation, the sides remain equal (p. 291)

sum [sum] The answer to an addition problem (p. 15)

supplementary angles [sup•lə•men′tə•rē an′gəlz] Two angles whose measures have a sum of 180° (p. 327) *Example:*

100° 80°

surface area [sûr′fəs âr′ē•ə] The sum of the areas of the faces of a solid figure (p. 494)

survey [sûr′vā] A method of gathering information about a group (p. 94)

systematic sample [sis•tə•ma′tik sam′pəl] A sampling method in which one subject is selected at random and subsequent subjects are selected according to a pattern (p. 95)

T

term [tûrm] Each number in a sequence (p. 526)

terms [tûrmz] The parts of an expression that are separated by an addition or subtraction sign (p. 277)

terminating decimal [tûr′mə•nāt•ing de′sə•məl] A decimal that ends, having a finite number of digits after the decimal point (p. 169)

tessellation [tes•ə•lā′shən] A repeating arrangement of shapes that completely covers a plane, with no gaps and no overlaps (p. 543)

theoretical probability [thē•ə•re′ti•kəl prä•bə•bil′ə•tē] A comparison of the number of favorable outcomes to the number of possible equally likely outcomes (p. 418)

transformation [trans•fər•mā′shən] A movement that does not change the size or shape of a figure (p. 540)

translation [trans•lā′shən] A movement of a figure along a straight line (p. 540)

tree diagram [trē dī′ə•gram] A diagram that shows all possible outcomes of an event (p. 434)

triangular number [trī·an′gyə·lər num′bər] A number that can be represented by a triangular array (p. 526)

unbiased sample [un·bī′əst sam′pəl] A sample is unbiased if every individual in the population has an equal chance of being selected. (p. 98)

underestimate [un·dər·es′tə·mət] An estimate that is less than the exact answer (p. 17)

unit rate [yoo′nət rāt] A rate that has 1 unit as its second term (p. 375)
Example: $1.45 per pound

unlike fractions [un′līk frak′shənz] Fractions with different denominators (p. 180)

upper extreme [up′ər ik·strēm′] The greatest number in a set of data (p. 129)

upper quartile [up′ər kwôr′tīl] The median of the upper half of a set of data (p. 129)

variable [vâr′ē·ə·bəl] A letter or symbol that stands for one or more numbers (p. 28)

Venn diagram [ven dī′ə·gram] A diagram that shows relationships among sets of things (p. 230)

vertex [vûr′teks] The point where two or more rays meet; the point of intersection of two sides of a polygon; the point of intersection of three or more edges of a solid figure; the top point of a cone (pp. 324, 355)
Examples:

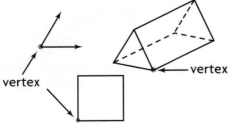

vertex — vertex

vertical angles [vûr′ti·kəl an·gəlz] A pair of opposite congruent angles formed where two lines intersect (p. 326)
Example:

∠MRP and ∠NRQ are vertical angles.

volume [väl′yəm] The number of cubic units needed to occupy a given space (p. 502)

whole number [hōl num′bər] One of the numbers 0, 1, 2, 3, 4, ... The set of whole numbers goes on without end.

***x*-axis** [eks·ak′səs] The horizontal number line on a coordinate plane (p. 560)

***x*-coordinate** [eks·kō·ôr′də·nət] The first number in an ordered pair; it tells the distance to move right or left from (0,0). (p. 560)

***y*-axis** [wī·ak′səs] The vertical number line on a coordinate plane (p. 560)

***y*-coordinate** [wī·kō·ôr′də·nət] The second number in an ordered pair, it tells the distance to move up or down from (0,0). (p. 560)

Chapter 1

Page 15

1. product   **3.** difference   **5.** sum
7. hundreds   **9.** hundred thousands
11. hundred millions   **13.** ten millions
15. 90,000   **17.** 300,000,000
19. 20,000,000,000   **21.** 2,000   **23.** 29,000
25. 35,000   **27.** 6,000   **29.** 135,000
31. 60,000   **33.** 50,000   **35.** 10,000
37. 640,000   **39.** 90,000

Pages 18–19

1. underestimate since both addends are
rounded down   **3.** 1,500   **5.** 12,000   **7.** 600
9. 24,000   **11.** 360   **13.** 1,500   **15.** 80
17. 70   **19.** 6,000   **21.** 18,000   **23.** 350
25. 14,000   **27.** 16,000   **29.** 4,000
31. 150,000   **33.** 90   **35.** 12   **37.** 3,200,000
39. over; $400 + 700$ **41.** over; 100×20
43. over; 300×30 **45.** $>$ **47.** $<$ **49.** $<$
51. about 4,500,000 books   **53.** 9,936 people
55. 235 and 345   **57.** 36 in.; 54 in.^{2} **59.** C

Page 21

3. 1,439   **5.** 26,208   **7.** 36,374   **9.** 1,111,384
11. 46,395,619   **13.** 75,125   **17.** 11,306
19. 1, 2, 3, 4, 6, 8, 12, 24   **21.** C

Pages 24–25

1. There were 102 full packages. The 18
newspapers left over were not enough to
make a full package.   **3.** 792,456   **5.**
3,553,275   **7.** 54   **9.** 38,480   **11.** 850,038
13. 56 r4   **15.** 961,184   **17.** 3,496,458   **19.**
1,745,677   **21.** 26,522   **23.** 14   **25.** 19,760
27. 26 r2   **29.** 89   **31.** 1,312,800   **33.**
1,855,998   **35.** 84   **37.** $367\frac{9}{20}$ **39.** $241\frac{11}{14}$
41. 415   **43.** \$159   **45.** 10   **47.** 64   **49.** B

Page 27

1. 13 apples, 27 oranges   **3.** B   **5.** April 26,
May 2, and May 8   **7.** 48 cards   **9.** 5 min

Page 29

1. An algebraic expression has one or more
variables.   **3.** $125 - 46$ **5.** $y \div 15$, or $\frac{y}{15}$
7. 46   **9.** 25×20 **11.** $76 - k$ **13.** 465
15. 9,000   **17.** 140   **19.** 3   **21.** $n + 12$
23. Deidre   **25.** 215   **27.** 7,643

Page 31

1. No, because $4 + 3$ equals 7; $x = 6$
3. $f = 21$ **5.** $x = 4$ **7.** $k = 4$ **9.** $x = 5$
11. $k = 18$ **13.** $m = 9$ **15.** $x = 20$
17. $k = 29$ **19.** $v = 20$ **21.** $d = 9$
23. $p = 6$ **25.** 25 students   **27.** 41
29. 4,990   **31.** D

Chapter 2

Page 35

1. 27   **3.** 64   **5.** 10,000   **7.** 512   **9.** 343
11. 256   **13.** Commutative of Addition
15. Property of Zero   **17.** Associative of
Addition   **19.** Distributive   **21.** Identity
of Multiplication   **23.** Distributive   **25.** $a =$
5   **27.** $r = 3$ **29.** $t = 8$ **31.** 20   **33.** 32
35. 8   **37.** 32   **39.** 18

Pages 38–39

3. 204   **5.** 157   **7.** 56   **9.** 145   **11.** 55
13. 142   **15.** 93   **17.** 43   **19.** 168   **21.** 495
23. 185   **25.** 108   **27.** 212   **29.** 72
31. 244   **33.** 85   **35.** 45   **37.** 1,500
39. 1,028   **41.** 160   **43.** 64   **45.** 52
49. 72 CDs   **51.** 18 CDs each   **53.** 68 more
signatures   **57.** 33,672   **59.** C

Page 41

1. 7 **3.** $2 \times 2 \times 2$; 8 **5.** $3 \times 3 \times 3 \times 3$; 81
7. $1 \times 1 \times 1 \times 1$; 1 **9.** $7 \times 7 \times 7$; 343
11. $5 \times 5 \times 5$; 125 **13.** 34×34; 1,156
15. $10 \times 10 \times 10 \times 10 \times 10 \times 10 \times 10 \times 10$;
100,000,000 **17.** $2 \times 2 \times 2 \times 2 \times 2 \times 2 \times$
$2 \times 2 \times 2 \times 2$; 1,024 **19.** 15×15; 225
21. 25; 25 **23.** 12^3 **25.** 4^4 **27.** n^2 **29.** 8^2
31. 10^4 **33.** 1947 **35.** Scott **37.** 36 in.
39. 745 r12

Pages 42–43

1. parentheses, exponent, multiplication,
addition; 154 **3.** multiplication, division,
addition; 16 **1.** No; 17 is the correct value,
not 9 **3.** $15 \div 3 + 12$ **5.** 29 **7.** 216
9. 31

Pages 44–45

1. $(420 - 100) \div 40 = 8$ **3.** 25 **5.** 28
7. 25 **9.** 68 **11.** 100 **13.** 1,218
15. 504 **17.** 60 **19.** 77
21. $\$20 - (\$3.25 + 3 \times \$1.29) = \$12.88$
23. 18,447 people **25.** 204 **27.** 32 r3

Page 47

1. Aquarium Tour, Underwater Acrobats,
Animal Acts, Whale Acts **3.** H **5.** 3 books,
4 magazines **7.** $4,700 **9.** 24 nails

Chapter 3
Page 51

1. 0.2 **3.** 0.09 **5.** 836.23 **7.** 93,450.38
9. 306,007.06 **11.** 7 **13.** 1 tenth;
hundredths **15.** < **17.** < **19.** > **21.** >
23. > **25.** > **27.** 1 **29.** 42 **31.** 72
33. 26.4 **35.** 8.6 **37.** 14.5 **39.** 55.6

Pages 54–55

1. 2.0769 **3.** 0.007 **5.** 0.06 **7.** 0.001 +
0.00003 **9.** 300 + 40 + 2 + 0.04 + 0.006
11. = **13.** 1.351, 1.361, 1.363 **15.** 0.0007
17. 0.00006 **19.** 0.03 + 0.006 + 0.0002
21. 2 + 0.4 + 0.05 + 0.006 **23.** < **25.** =
27. > **29.** < **31.** 0.405, 1.05, 1.125, 1.25,
1.45 **33.** 8.91, 9.082, 9.285, 9.82, 9.85
35. 125.4, 125.35, 125.33, 125.3 **37.** 41.01,
$14\frac{1}{10}$, $14\frac{3}{100}$, 14.01 **39.** \$94,563,020; ninety-
four million, five hundred sixty-three
thousand, twenty **43.** 228 **45.** 29

Page 57

1. C **3.** air conditioner **5.** calculator,
\$10.50; pen, \$1.50; notebook, \$2.00 **7.** 60
cards **9.** Blaster

Page 59

3. 40 **5.** 90 **7.** 210 **9.** 32 **11.** 9
13. 50 **15.** 140 **17.** 1,800 **19.** 27,000
21. \$17,000 **23.** 36 **25.** > **27.** Possible
answer: about 24 lb **31.** 2.235, 2.325, 2.523,
2.532 **33.** 1,121 r18

Page 61

1. 18% means 18 per hundred, so 18 are red.
3. 0.06, 6% **5.** 0.7 or 0.70 **7.** 0.84 **9.** 0.5,
0.50 **11.** 0.11, 11% **13.** 0.62 **15.** 0.28
17. 0.53 **19.** 85% **21.** 40% **23.** 0.79, 79%
25. 27.5, 27.8, 27.82 **27.** 15,456 **31.** B

Chapter 4
Page 65

1. 47 **3.** 44 **5.** 10 **7.** 222 **9.** 252
11. 1,444 **13.** 4,000 **15.** 3,000 **17.** 24
19. 47 **21.** 14 **23.** 29 **25.** 43 **27.** 840
29. 19,405 **31.** 8.75 **33.** 12.5 **35.** 160.8
37. $9\frac{5}{8}$ **39.** $39\frac{2}{7}$

Pages 68–69

3. $0.63   **5.** 18.585   **7.** 20.923   **9.** 37.31
11. 124.94   **13.** 87.80   **15.** 1,438.66
17. 0.6   **19.** 287.673   **21.** 488.55
23. 287.673   **25.** 257.14   **27.** 9.42   **29.** yes
31. no   **33.** 1.25   **35.** 12.3   **37.** 2.216
39. 0.090   **41.** 210; more than   **43.** 3,950;
3,905; 3,590; 3,509   **45.** 0.46

Pages 72–73

3. 0.35   **5.** 2   **7.** 3   **9.** 2.87   **11.** 1.218
13. 25.84   **15.** 0.24   **17.** 3   **19.** 6
21. 194.208   **23.** 0.410   **25.** 13.8   **27.** 36.5
29. 4.92   **31.** 0.08   **33.** 0.441   **35.** 5.5080
37. 26.8467   **39.** 0.39501   **41.** 0.8934
43. $7.31   **49.** 299 r14

Page 74

1. 1.01   **3.** 0.45

Page 75

1. 6   **3.** 3   **5.** 4   **7.** 3   **9.** 8   **11.** 0.57
13. 89 r21

Pages 78–79

1. multiply by 100   **3.** 96 ÷ 16
5. 482.4 ÷ 24   **7.** 14.86   **9.** 1.97   **11.** 4.4
13. 819 ÷ 9   **15.** 239 ÷ 5   **17.** 2335.8 ÷ 102
19. 6.1   **21.** 6.9   **23.** 9.91   **25.** 0.21
27. 21.2   **29.** 8.2   **31.** 15.05   **33.** 220
35. 2.36   **37.** 3.03   **39.** 5 weeks
41. 30 pages   **43.** greater than

Page 81

1. C   **3.** 4 magnets   **5.** 8 hr 5 min
7. 65 min, or 1 hr 5 min   **9.** 108 pieces

Page 83

1. $a \times 6.45$; 12×6.45; $45.15   **3.** 11.7
5. 5.08   **7.** $m = 19$ **9.** 4.6   **11.** 14   **13.** 3.9
15. $r = 7.7$ **17.** $a = 8.2$ **19.** $p = 17.6$
21. $n \div 6$; $n = 4.8$ **23.** 4 mi   **25.** 37
27. 3.08, 3.508, 3.58, 3.85

Chapter 5

Page 93

1. median   **3.** range   **5.** 13 more girls
7. 29 girls   **9.** 8   **11.** 8.475   **13.** 28; 32
15. 11; 10.8   **17.** 6

Pages 96–97

5. population, because there are not too
many students and they are easily available
7. systematic   **9.** convenience
11. population; There are not too many
basketball team members.   **13.** systematic
15. convenience   **21.** 9.3   **23.** 2

Pages 98–99

1. Yes. Only those in the computer club
would be represented.   **3.** biased; excludes
males   **5.** biased; excludes members who
did not use the gym in August   **7.** biased;
excludes people who are not teenagers
9. biased; excludes males   **11.** unbiased
13. biased   **17.** 4   **19.** 9,000

Page 101

1. D   **3.** 12 members   **5.** 1 3-point,
7 2-point; 3 3-point, 4 2-point; 5 3-point,
1 2-point   **7.** $42.50   **9.** 37 birds
11. Saturday

Pages 104–105

1. 1-10, 11-20; Since $2 \times 10 = 20$, you would
have 2 intervals that include 10 numbers
5. 16   **13.** 9   **15.** No, the data are categories
and not numerical   **19.** unbiased

Page 107

1. The mean would become 21.2. The mean
or median would be most useful to describe
the data.   **3.** 7 hours   **5.** 24; 23; 22
7. 15.375 points   **9.** 10 and 12   **11.** 336.8;
360; no mode   **15.** Tim should divide by 4,
not 3. 6   **17.** 71 people   **19.** 90%

Pages 110–111

1. The median and mode would not change. The mean would not increase as much.
3. 16; 14; 12 **5a.** Abe: mean 7.5, median 6, mode 7; Bart: mean 7.5, median 8.5, mode 0;
5b. Abe: mean 9.3, median 7, mode 7 and 20; Bart: mean 6.4, median 5, mode 0;
5c. Abe's: all increase; Bart's: mean/median decrease; mode doesn't change **7a.** 75
7b. No. With a perfect score of 100, the mean would be 65.7. **9.** 5.75; 5.5; 5 **11.** 9.5; 10; 10 and 12

Pages 113–115

1. No. The sample would be a convenience sample and not a random sample.
3. No. The sample is not representative.
5. Yes. The sample is random and representative; the question is unbiased.
7. No. The sample is not representative of the population. **9.** The interest in reading increased slightly. **11.** The interest in coin collecting decreased slightly. **13.** The interest in building models increased slightly.
15. increase: trading cards; decrease: coin collecting **17.** $5 **19.** 33; 31; 22 **21.** B

Chapter 6

Page 119

1. Saturn V, Ariane IV, Titan-Centaur, Titan IIIC **3.** about 55m **7.** about 15,000 sq mi
9. 92 **11.** 22 **13.** 3 or more **15.** 20

Pages 122–123

1. so you can tell the difference between the sets of data **3.** line graph **5.** bar graph
7. circle graph **11.** Each month, more comedies were rented. **13.** Comedy video rentals will increase and action video rentals will decrease. **17.** No. The sample is not representative of the student population.
19. 0.56 **21.** J

Page 125

1. Extend the graph until it intersects with 24 mi and find the corresponding time.
3. 5 hr **5.** 6 hr **7.** 8 hr **9.** about 45 mi
11. 5.6; 5; 4 **13.** 6.5; 7; 9 **15.** C

Pages 127–128

5. bar graph **7.** bar graph **9b.** 24.5 in.; 28 in. **13.** random **15.** 73%

Pages 130–131

1. A box-and-whisker graph shows the distribution of data, not all of the data
3. 11; 22; 11 **5.** 58; 67; 9 **7.** 85; 104; 19
9. 72 points; 97 points **13.** yes **15.** 0.02 or two hundredths

Pages 133–136

3. Yes. The question is biased and could lead people to choose the Mustangs as the best baseball team. **5.** No. Angel Falls is about 3,200 ft high while Tugela Falls is about 3,000 ft high. **7.** Yes. The question is biased and could lead people to choose oranges. **9.** No. Brand B costs $30 and Brand A costs $20.
13. Lin looked only at the graph and did not look at the scales. **15.** 43 and 55 **17.** B

Chapter 7

Page 145

1. prime number **3.** yes **5.** no **7.** yes
9. yes **11.** yes **13.** no **15.** yes **17.** no
19. yes **21.** no **23.** 16, 20, 24 **25.** 48, 60, 72 **27.** 20, 25, 30 **29.** 6, 12, 18, 24, 30
31. 30, 60, 90, 120, 150 **33.** 9, 18, 27, 36, 45
35. 1, 2, 4, 8 **37.** 1, 11 **39.** 1, 2, 3, 6, 9, 18, 27, 54

Page 147

3. 2, 4, 8 **5.** 2, 4, 8 **7.** 3 **9.** 2, 3, 4, 6, 9
11. 2, 4 **13.** 3, 5 **15.** 2, 3, 4, 6, 8 **17.** 3, 5
19. 2, 3, 6, 9 **21.** 3 **23.** T **25.** T
29. $28.88 **31.** 5% **33.** 16,302

Page 149

1. $2^2 \times 3 \times 13$ **3.** $2 \times 2 \times 3$ **5.** $2 \times 2 \times 2 \times 2$ **7.** 3×7 **9.** 2×127 **11.** $2 \times 2 \times 2 \times 2 \times 2 \times 2$ **13.** $2 \times 2 \times 19$ **15.** $2 \times 3 \times 3$; 2×3^2 **17.** 7×7; 7^2 **19.** $2 \times 2 \times 7 \times 19$; $2^2 \times 7 \times 19$ **21.** 2×373 **23.** $n = 2$ **25.** $n = 5$ **27.** $c = 2$ or 3 **31.** 60% **33.** 13

Pages 152–153

3. 3, 6, 9, 12, 15 **5.** 11, 22, 33, 44, 55 **7.** 21 **9.** 18 **11.** 3 **13.** 3 **15.** 4, 8, 12, 16, 20 **17.** 16, 32, 48, 64, 80 **19.** 10, 20, 30, 40, 50 **21.** 14, 28, 42, 56, 70 **23.** 24 **25.** 60 **27.** 120 **29.** 108 **31.** 2 **33.** 3 **35.** 1 **37.** 8 **39.** 9 and 12 **41. a.** 60 **b.** 4 packages of cereal samples, 3 packages of pamphlets **43.** She found the GCF instead of LCM. The LCM is 30. **45.** 124 r14

Page 155

1. October 15 **3.** H **5.** 9 in. **7.** 18 blocks

Chapter 8

Page 159

1. > **3.** denominator **5.** > **7.** < **9.** > **11.** 62,000; 61,600; 61,060 **13.** $\frac{5}{6}$ **15.** $\frac{1}{3}$ **17.** 20% **19.** 75%

Pages 162–163

1. Also multiply the denominator by 5; $\frac{10}{15}$ **3.** 12 **5.** 4 **7.** 1 **9.** 16 **11.** 1, 2, 4 **13.** 1, 2 **15.** $\frac{1}{8}$ **17.** $\frac{1}{6}$ **19.** $\frac{11}{4}$ **21.** $\frac{3}{10}$ **23.** 5 **25.** 4 **27.** 6 **29.** 8 **31.** 12 **33.** 7 **35.** 1 **37.** 1, 3 **39.** 1, 5 **41.** 1 **43.** $\frac{1}{6}$ **45.** $\frac{1}{8}$ **47.** $\frac{5}{9}$ **49.** $\frac{1}{5}$ **51.** $\frac{2}{3}$ **53.** $\frac{10}{7}$ **55.** $\frac{1}{4}$ **57.** $\frac{1}{5}$ **59.** $\frac{1}{2}$ **61.** $\frac{2}{3}$ **63.** $\frac{1}{4}$ **65.** What fractions of the muffins are bran? **67.** 24 **69.** 0.035

Page 165

1. It has a whole number part and a fraction part. **3.** $3\frac{1}{2}$ **5.** $3\frac{2}{3}$ **7.** $\frac{5}{4}$ **9.** $\frac{8}{3}$ **11.** $\frac{37}{7}$ **13.** $4\frac{1}{2}$ **15.** $5\frac{3}{4}$ **17.** $5\frac{1}{6}$ **19.** $12\frac{6}{7}$ **21.** $16\frac{2}{3}$ **23.** $3\frac{2}{3}$ **25.** $\frac{13}{2}$ **27.** $\frac{19}{10}$ **29.** $\frac{37}{4}$ **31.** $\frac{53}{11}$

33. $\frac{93}{5}$ **35.** no, only a fraction whose numerator is greater than its denominator **37.** $2^2 \times 3^2$ **39.** 23×1

Pages 166–167

1. The denominators are the same. $4 < 5$, so $\frac{4}{9} < \frac{5}{9}$ **3.** < **5.** > **7.** $\frac{1}{4}, \frac{6}{12}, \frac{2}{3}$ **9.** > **11.** = **13.** = **15.** > **17.** $\frac{2}{6}, \frac{1}{2}, \frac{9}{12}$ **19.** $\frac{1}{2}, \frac{7}{12}, \frac{5}{6}$ **21.** $\frac{3}{10}, \frac{2}{5}, \frac{1}{2}$ **23.** $\frac{1}{6}, \frac{1}{3}, \frac{1}{2}$ **25.** $\frac{1}{12}, \frac{5}{8}, \frac{3}{12}$ **27.** 3 pieces of each pizza remain. That is $\frac{3}{8}$ of the mushroom and $\frac{3}{12}$ cheese pizza. $\frac{3}{12} = \frac{1}{4} = \frac{2}{8} < \frac{3}{8}$, so more of the mushroom pizza is left. **29.** 7 **31.** 24

Page 168

1. 0.3 **3.** 0.7 **5.** 0.90 **7.** 0.4 **9.** 0.85

Pages 170–171

3. $\frac{7}{10}$ **5.** $\frac{105}{1,000}$ **7.** 0.25; T **9.** 0.66666; R **11.** > **13.** = **15.** 20% **17.** 40% **19.** $\frac{6}{100}$ **21.** $\frac{61}{100}$ **23.** $\frac{205}{1,000}$ **25.** $\frac{9}{1,000}$ **27.** 0.16666; R **29.** 0.625; T **31.** 0.3; T **33.** 0.111111; R **35.** 0.36; T **37.** 0.15; R **39.** = **41.** < **43.** < **45.** 75% **47.** 6% **49.** 50% **51.** 0.5% **53.** 0.72 **55.** Megan **57.** $\frac{9}{2}$ **59.** 76.89

Chapter 9

Page 175

1. mixed number **3.** divide **5.** $\frac{3}{4}$ **7.** $\frac{1}{3}$ **9.** $\frac{4}{3}$, or $1\frac{1}{3}$ **11.** $\frac{3}{2}$, or $1\frac{1}{2}$ **13.** $\frac{1}{2}$ **15.** $\frac{2}{3}$ **17.** $\frac{5}{3}$, or $1\frac{2}{3}$ **19.** $\frac{1}{3}$ **21.** $\frac{1}{2}$ **23.** $\frac{5}{4}$, or $1\frac{1}{4}$ **25.** $\frac{2}{3}$ **27.** $\frac{1}{2}$ **29.** $\frac{1}{2}$ **31.** $\frac{1}{3}$ **33.** $\frac{5}{6}$ **35.** 1 **37.** $\frac{2}{9}$ **39.** $\frac{5}{7}$

Pages 178–179

3. close to 1 **5.** between 0 and $\frac{1}{2}$ **7.** $1\frac{1}{2}$ **9.** $\frac{1}{2}$ **11.** $7\frac{1}{2}$ **13.** 2 **15.** close to 1 **17.** close to 0 **19.** 2 **21.** $\frac{1}{2}$ **23.** 17 **25.** 3 **27.** 0 **29.** $4\frac{1}{2}$ **31.** 9 to $9\frac{1}{2}$; $9\frac{1}{4}$ **33.** $4\frac{1}{2}$ to 5; $4\frac{3}{4}$ **35.** $5\frac{1}{2}$ to 6; $5\frac{3}{4}$ **37.** about 7 **39.** about 6 feet **41.** about 26 **43.** about 30 **47.** about $1\frac{1}{4}$ yd **49.** March 15, 24; April 2, 11, 20, 29 **51.** 45%

Page 181

1. $\frac{5}{12}$ 3. $\frac{1}{12}$ 5. $\frac{1}{10}$ 7. $\frac{1}{12}$ 9. $\frac{1}{8}$ 11. $\frac{1}{2}$
13. = 15. >

Pages 183–185

1. $\frac{5}{12}$ c more 3. $\frac{7}{10} + \frac{2}{10}$ 5. $\frac{12}{15} - \frac{5}{15}$ 7. $\frac{1}{3}$
9. $1\frac{5}{12}$ 11. $\frac{1}{15}$ 13. $\frac{7}{9}$ 15. $\frac{24}{28} - \frac{21}{28}$ 17. $\frac{5}{6}$
19. $\frac{4}{5}$ 21. $\frac{3}{14}$ 23. $\frac{1}{5}$ 25. $1\frac{3}{8}$ 27. $\frac{5}{8}$ 29. $\frac{7}{15}$
31. $\frac{2}{9}$ 33. $\frac{6}{8}$, or $\frac{3}{4}$ 35. 2 37. $\frac{5}{12}$ mile
39. $r = 1$ 41. $c = \frac{3}{5}$ 43. $m = 1$ 45. $\frac{5}{8}$ tsp
47. $\frac{2}{15}$ is left; add to find the total spent and saved; subtract to find how much is left.
49. 24 and 28 51. $\frac{4}{9}$ 53. 6,132

Pages 187–189

3. $3\frac{4}{5}$ 5. $1\frac{3}{10}$ 7. $6\frac{7}{12}$ 9. $1\frac{1}{6}$ 11. $1\frac{1}{18}$
13. $2\frac{1}{2}$ 15. $8\frac{3}{10}$ 17. $10\frac{1}{8}$ 19. $4\frac{1}{10}$ 21. $11\frac{3}{20}$
23. $8\frac{3}{14}$ 25. $28\frac{1}{8}$ 27. Commutative, $2\frac{1}{4}$
29. Associative, $1\frac{5}{6}$ 31. longer; $\frac{1}{6}$ in.; subtraction; to find the difference in length
33. $9\frac{3}{4}$ c 35. 3rd prize 37. $\frac{16}{15}$, or $1\frac{1}{15}$
39. $\frac{4}{5}, \frac{2}{3}, \frac{1}{2}$

Page 191

1. $1\frac{2}{3}$ 3. $1\frac{4}{9}$ 5. $\frac{7}{10}$ 7. $\frac{11}{12}$ 9. $3\frac{3}{4}$ 11. 21.42

Page 193

1. $3\frac{1}{3} = 2\frac{4}{3}$ 3. $2\frac{1}{12}$ 5. $3\frac{9}{10}$ 7. $3\frac{17}{18}$ 9. $1\frac{11}{12}$
11. $4\frac{7}{9}$ 13. $2\frac{7}{12}$ 15. $1\frac{3}{8}$ 17. $2\frac{7}{10}$, or 2.7
19. $2\frac{4}{5}$ 21. $\frac{1}{2}$ mi shorter 25. $7\frac{1}{3}, \frac{22}{3}$
27. 55 r5

Page 195

1. $18\frac{1}{2}$ mi 3. C 5. 12 cans 7. $4\frac{3}{4}$ mi
9. Gary

Chapter 10

Page 199

1. equation 3. 0 5. $\frac{1}{2}$ 7. $\frac{1}{2}$ 9. $\frac{1}{2}$ 11. $\frac{1}{2}$
13. $\frac{1}{2}$ 15. 0 17. $\frac{1}{2}$ 19. $C = 32$ 21. $x = 0.17$ 23. $t = 14$ 25. $r = 2.6$ 27. $1\frac{1}{6}$
29. $1\frac{1}{3}$ 31. $\frac{15}{2}$ 33. $\frac{14}{5}$

Page 201

1. about 498 million pounds 3. $\frac{1}{2}$ 5. 20
7. $63\frac{3}{4}$ 9. 118 11. $\frac{1}{2}$ 13. 1 15. 4 17. 3
19. 20 21. $\frac{1}{4}$ 23. > 25. about 7 min
29. $3\frac{4}{9}$ 31. 11

Pages 204–205

3. $\frac{3}{8}$ 5. $\frac{2}{10}$, or $\frac{1}{5}$ 7. $\frac{3}{10}$ 9. $\frac{12}{7}$, or $1\frac{5}{7}$ 11. $\frac{8}{3}$, or $2\frac{2}{3}$ 13. $\frac{5}{9}$ 15. $\frac{3}{16}$ 17. $\frac{1}{16}$ 19. $\frac{2}{9}$ 21. $\frac{2}{15}$
23. $\frac{7}{10}$ 25. $\frac{1}{10}$ 27. $\frac{4}{15}$ 29. 2 31. <
33. < 35. 108 voters 37. 6 39. $b = 12.09$

Page 207

1. $3 \times 4\frac{2}{3} = (3 \times 4) + (3 \times \frac{2}{3})$; $12 + 2 = 14$.
3. $1\frac{1}{8}$ 5. $2\frac{1}{4}$ 7. $18\frac{3}{8}$ 9. $2\frac{1}{4}$ 11. $8\frac{1}{6}$
13. 33 15. 15 17. 85 19. $7\frac{1}{5}$ 21. $22\frac{1}{2}$
23. < 25. $2\frac{1}{12}$ mi 29. $6\frac{7}{20}$ 31. 352,053

Page 208

1. 9 3. 4

Page 209

1. $n = 4$ 3. $n = \frac{3}{4}$ 5. 8 7. $\frac{3}{2}$, or $1\frac{1}{2}$ 9. $9\frac{9}{20}$
11. $606.64 > 606.074$

Pages 212–213

3. $\frac{3}{2}$ 5. $\frac{1}{7}$ 7. $\frac{3}{13}$ 9. $\frac{4}{5}$ 11. $\frac{1}{24}$ 13. 3
15. $1\frac{7}{32}$ 17. $\frac{1}{10}$ 19. $\frac{9}{2}$ 21. $\frac{7}{15}$ 23. $\frac{1}{9}$
25. $\frac{5}{13}$ 27. $1\frac{1}{6}$ 29. $9\frac{1}{3}$ 31. $2\frac{1}{4}$ 33. 5
35. $5\frac{1}{15}$ 37. $1\frac{8}{13}$ 39. 18 41. 32
43. $2\frac{5}{14}$ 45. $\frac{5}{18}$ 47. $\frac{4}{5}$ 49. 48 burgers
51. $1\frac{1}{4}$ yd 53. $46.90 55. $13\frac{3}{8}$

Page 215

1. $1\frac{5}{6}$ min; subtraction 3a. B 3b. H
5. $2\frac{2}{3}$ mi 7. $10\frac{1}{3}$ mi

Page 217

1. $b \times \frac{3}{4}$; $32 \times \frac{3}{4}$; 24 oz 3. $2\frac{1}{5}$ 5. $5\frac{5}{8}$
7. $y = \frac{3}{8}$ 9. $7\frac{5}{6}$ 11. $1\frac{1}{4}$ 13. $\frac{1}{20}$ 15. $\frac{3}{8}$
17. 1 19. $x = 2$ 21. $x = \frac{1}{4}$ 23. $x = 1\frac{1}{3}$
27. $\frac{3}{4}, \frac{3}{5}, \frac{3}{8}$ 29. $a = \frac{4}{5}$

Chapter 11

Page 227

1. negative numbers **11.** 1, 2, 3, 4
13. 1, 3, 5, 7 **15.** < **17.** = **19.** >
21. 32°F **23.** ⁻16°F **25.** 60°F

Page 229

3. ⁻350 **5.** ⁻14 **7.** ⁺25 **9.** ⁻742 **11.** ⁻5
13. ⁺12,000 **15.** ⁺2 **17.** ⁺31 **19.** ⁻207
21. 28 **23.** 660 **25.** ⁻1,310 ft **27.** $\frac{1}{5}$
29. $\frac{12}{20}$, or $\frac{3}{5}$ **31.** 13

Pages 232–233

1. All integers can be written in the form $\frac{a}{1}$
3. $\frac{-37}{100}$ **5.** $\frac{889}{1,000}$ **7.** $\frac{-22}{3}$ **9.** $-\frac{1}{2}$ **11.** 0
13. $\frac{-71}{100}$ **15.** $\frac{-21}{8}$ **17.** ⁻0.3 **19.** ⁻0.7
21. ⁻$7\frac{1}{4}$ **23.** 16.05 **25.** $\frac{-5}{8}$ **27.** yes

29. yes **31.** yes **33.** 90 steps **35.** Not all integers are whole numbers; ⁻8 is not a whole number. **37.** 0.34; $\frac{34}{100}$

Page 235

3. < **5.** < **7.** < **9.** > **11.** < **13.** >
15. > **17.** 0.4, 0.46, 0.6 **19.** $-\frac{1}{4}$, ⁻0.2, 0, $\frac{1}{4}$
21. 1°, ⁻3°, ⁻5° **23.** ⁻450 **25.** 2.096 **27.** C

Page 237

1. Jeffrey, sixth grade; Victoria, seventh grade; Arthur, eighth grade **3.** C **5.** Arlene, Kathy, Helene **7.** 15 **9.** 12 mi

Chapter 12

Page 241

1. integers **3.** ⁺62 **5.** ⁻13 **7.** ⁺12 **9.** ⁻3
11. ⁺11 **13.** ⁻8 **15.** ⁻13 **17.** ⁺4 **19.** 0

Page 242

1. ⁺13 **3.** ⁻10

Page 243

1. ⁻2 **3.** 0 **5.** ⁺7 **7.** ⁻9 **9.** ⁻6.8, ⁻6.4, ⁻6.2 **11.** ⁻3.6, ⁻$3\frac{4}{7}$, ⁻$3\frac{1}{2}$ **13.** B

Pages 246–247

1. Use sign of the addends. **3.** The sum was 12, which indicates a gain of 12 yards, not a loss of 12 yards **5.** 5 + ⁻9 = ⁻4 **7.** ⁻7
9. ⁻2 **11.** ⁻10 **13.** 0 **15.** ⁻2 + 6 = 4
17. ⁻1 **19.** 3 **21.** ⁻20 **23.** 8 **25.** ⁻28
27. ⁻12 **29.** ⁻11 **31.** 100 **33.** x = ⁻7
35. x = 5 **37.** ⁻12°F **39.** The Wildcats lost 2 yards. **41.** 12 cm **43.** greater than **45.** C

Page 249

1. ⁻11 **3.** ⁻22 **5.** 2 **7.** ⁻5 **9.** 6 **11.** ⁻8
13. D

Page 251

1. ⁻9 + 40 **3.** 3 + 6 **5.** ⁻4 + ⁻6 **7.** ⁻5
9. ⁻7 **11.** 8 + 11 **13.** ⁻9 + ⁻11 **15.** ⁻4
17. ⁻2 **19.** 0 **21.** ⁻5 **23.** ⁻4 **25.** ⁻5
27. 5 **29.** 7°F **31.** What was the range of temperatures? **33.** ⁻213 **35.** 2 × 2 × 3 × 7

Chapter 13

Page 255

1. 24 **3.** 10 **5.** 4 **7.** 72 **9.** 5 **11.** 10
13. n = 7 **15.** n = 54 **17.** n = 7 **19.** ⁻7
21. ⁻7 **23.** ⁻17 **25.** ⁻68 **27.** ⁻47 **29.** ⁻5
31. Add 1; 6, 7, 8 **33.** Divide by 2; 8, 4, 2
35. Multiply by 2; 96, 192, 384 **37.** Subtract 8; 80, 72, 64 **39.** Rotate shaded triangle 90° clockwise.

Page 257

1. 20 **3.** ⁻18 **5.** 4 **7.** 21

Page 259

1. Neither; zero is neither positive nor negative. **3.** ⁻54 **5.** ⁻16 **7.** ⁻21 **9.** ⁻90
11. ⁻40 **13.** 14 **15.** 84 **17.** 110 **19.** 288
21. 450 **23.** y = ⁻3 **25.** y = 3 **27.** y = ⁻4
29. ⁻30 **31.** The product of integers with like signs is positive; ⁻5 × ⁻2 = 10 **33.** ⁻12
35. 364

Page 261

1. The rules are the same. **3.** 6 **5.** $^-8$
7. 8 **9.** $^-22$ **11.** $^-2$ **13.** $^-9$ **15.** 9
17. $^-10$ **19.** $^-17$ **21.** $x = 18$ **23.** $x = ^-9$
25. $^-\$509$ **27.** 5 weeks **29.** $^-96$ **31.** $38\frac{1}{4}$
33. C

Pages 262–263

1. $^-2$ **3.** $^-34$ **5.** 177; Commutative
7. 576; Associative **9.** $^-30$ **11.** 120 **13.** 13
15. 421; Commutative **17.** \$6 **19.** $^-15$
21. One; 2 is prime and is an even number.
23. 7 **25.** $^-37$

Chapter 14
Page 273

1. exponent **3.** algebraic expression **5.** 32
7. 125 **9.** 16 **11.** 81 **13.** 49 **15.** 216
17. 25 **19.** 7 **21.** 2 **23.** 23 **25.** 20
27. 4 **29.** 21 **31.** 1, 2, 4, 8, 16 **33.** 1, 3,
11, 33 **35.** 1, 17 **37.** 1, 2, 3, 6, 7, 14, 21, 42
39. 1, 2, 4, 5, 8, 10, 20, 40

Page 275

3. $y \div 1.5$ **5.** 54 less than 19 times x
7. $3c + 2.9$ **9.** $h \times 4j \times k$ **11.** $6.3p - 5m$
19. $14.25 + 3.5h$ **21.** $2 + 2x$ **25.** $^-36$
27. 6 and 8

Pages 278–279

1. You will have to perform fewer computations when evaluating the expression if you simplify it first.
3. $^-16, ^-11, ^-6, ^-1$ **5.** $^-2, 6, 14, 22$ **7.** 1,
$^-2, ^-3, ^-2$ **9.** $7x - 8; ^-29$ **11.** $^-16, ^-20,$
$^-24, ^-28$ **13.** $20\frac{1}{2}, 20, 19\frac{1}{2}, 19$ **15.** $^-5, 1, 3, 4$
17. $12x - 41; ^-89$ **19.** $356 + 6a; 236$
21. $^-2$ **23.** Associative; 67 **25.** $x = 2$
27. $x = ^-1$ **31.** $^-5$

Page 280

1. 4 **3.** 49

Page 281

1. 8 **3.** 14 **5.** 22 **7.** 81; 9 **9.** $^-49$
11. $>$

Page 283

1. Operate inside the parentheses, evaluate $\sqrt{36}$, then multiply; 12 **3.** $^-33$ **5.** 24
7. 150 **9.** $^-24$ **11.** $^-10$ **13.** 77 **15.** $>$
17. 14 ft **19.** $^-2, ^-6, ^-10, ^-14$ **21.** 2, $^-4,$
$^-6, ^-7$

Chapter 15
Page 287

1. equation **3.** 37 **5.** 23 **7.** 11.37
9. 2.7 **11.** $1\frac{7}{12}$ **13.** $\frac{1}{2}$ **15.** $7\frac{11}{12}$ **17.** $2\frac{9}{28}$
19. $43 - 19 = 24$ **21.** $16 + 3 = 19$
23. $46 + 196 = 242$ **25.** $125 \div 5 = 25$
27. $2 \cdot 6 = 12$ **29.** $21 \cdot 12 = 252$
31. $25 \cdot 16 = 400$ **33.** addition
35. subtraction **37.** multiplication
39. subtraction

Page 289

1. any quantity that you do not know
3. $x + 7 = 20$ **5.** $5 \cdot m = 35$; or $5m = 35$
7. $14 = n + 12$ **9.** $n \div 2\frac{3}{4} = \frac{5}{6}$
11. $72,000 = 9e$ **15.** $8x - 15; ^-31$
17. $25 + 9z; ^-47$

Page 290

1. $x = 4$ **3.** $x = 7$ **5.** $x = 3$ **7.** $x = 1$

Pages 292–293

3. $x = 7$ **5.** $c = 2.7$ **7.** $m = 15\frac{1}{4}$ **9.** $x = 8\frac{3}{4}$
11. $k = ^-41$ **13.** $b = 2.9$ **15.** $y = 8\frac{3}{4}$
17. $t = ^-19$ **19.** $x =$ unknown length; $x + 12 + 10 = 29; x = 7$, or 7 cm **21.** $11 = n + 8$
23. 13

Page 295

1. You add the number that is being subtracted from the variable. **3.** $b = ^-6$
5. $y = 23.2$ **7.** $d = 34\frac{7}{15}$ **9.** $a = 33$
11. $z = 17.0$ **13.** $c = 23.3$ **15.** $s = 22\frac{11}{12}$
17. $m = 11.1$ **19.** $f = \frac{19}{24}$

21. s = original amount in savings; $s - (110 + 90 + 40) = 527$; $s = 8767$ **23.** $61 - 12 - 13 - 14 + 5 - 7$ **25.** $b = 4.4$ **27.** 9, 7, 5, 3

Chapter 16
Page 299

1. Celsius   **3.** $18 = n + 6$ **5.** $\frac{n}{2} = \frac{2}{3}$
7. $2x = 47$ **9.** 14   **11.** $^{-}73$ **13.** $^{-}140$
15. 9   **17.** 25   **19.** 99   **21.** 27.3   **23.** $3\frac{4}{5}$
25. $y = 5$ **27.** $a = 9$ **29.** $c = 30$
31. $q = 56$ **33.** $y = 0.04$

Page 300

1. $c = 4$ **3.** $b = 2$

Page 303

1. You use inverse operations.   **3.** $x = {^{-}}7$
5. $y = 7.2$ **7.** $k = 6$ **9.** $a = 18$ **11.** $p = 16$
13. $n = {^{-}}5$ **15.** $a = 16.48$ **17.** $w = 2.73$
19. $a = 10$ **21.** $\frac{m}{3} = 14$; $m = 42$; 42 marbles
23. $0.45 = 0.9w$; $w = 0.5$ cm   **25.** $x = 22$
27. $z = {^{-}}12$

Pages 306–307

1. The rate of speed is in feet per minute.
5. 20   **7.** 50°F   **9.** 86°F   **11.** 41°F
13. 13.9°C   **15.** 37.8°C   **17.** 36.8   **19.** 2.5
21. 98.6°F   **23.** 203°F   **25.** 0°C   **27.** 8.3°C
29. 34.4°C   **31.** 87.5 mi per hr   **33.** Yes. The shuttle would be traveling 17,550 mi per hr.
35. Earth   **39.** $y = {^{-}}6$

Page 309

1. $4x + 1 = 5$; $x = 1$ **3.** $y = 2$ **5.** Add 1 to both sides, then divide by 2; $x = 3$.   **7.** 3
9. 374

Page 311

1. 18 mi   **3.** C   **5.** 4:20 P.M.   **7.** 60 in., or 5 ft   **9.** 6 students

Chapter 17
Page 321

1. angle   **3.** line   **5.** obtuse   **7.** acute
9. straight   **11.** acute   **13.** $\angle$MNP or $\angle$PNM
15. $\angle 3$ **17.** $\angle a$ **19.** $\angle$PBF or $\angle$FBP

Pages 322–323

1. plane; point; line, line segment, or ray
3. point R **5.** point C **7.** plane DHY
9. $\overrightarrow{XY}$ **11.** $\overline{PQ}, \overline{QR}, \overline{PR}$ **13.** $\overrightarrow{PQ}, \overrightarrow{QP}, \overrightarrow{RQ}, \overrightarrow{QR}, \overrightarrow{RP}, \overrightarrow{PR}$ **15.** a point   **17.** a point
23. $n = 96$ **25.** 2

Page 325

1. 135°; obtuse   **3.** 73°; acute   **5.** 180°
7. 90°   **9.** 45°   **11.** 90°   **13.** $b = 8$ **15.** $\frac{2}{15}$

Pages 328–329

1. Vertical angles are opposite each other and do not share a common ray   **3.** $\angle AED$ and $\angle BEC$, $\angle AEB$ and $\angle DEC$ **5.** $\angle AED$ and $\angle BEC$ **7.** 51°   **9.** 33°   **11.** $\angle DOC$
13. $\angle AOB$, $\angle DOC$ **15.** 18°   **17.** 131°
19. 90°   **21.** supplementary, adjacent
23. none of these   **25.** $\angle 1, \angle 2$; $\angle 3, \angle 3$; $\angle 3, \angle 4$; $\angle 4, \angle 5$; $\angle 5, \angle 1$; They are side by side and have a common vertex.   **27.** $\angle 1, \angle 5$; $\angle 4, \angle 5$; Together they form a straight line.
29. Their measures are equal.   **31.** $\overleftrightarrow{PQ}$ or $\overleftrightarrow{QP}$ **33.** $^{-}6$

Pages 330-331

1. perpendicular and intersecting
3. intersecting / perpendicular
5. intersecting   **7.** intersecting
9. intersecting   **11.** $\overleftrightarrow{AC}, \overleftrightarrow{BD}, \overleftrightarrow{CG}, \overleftrightarrow{DH}$
13. $\overleftrightarrow{BD}, \overleftrightarrow{FH}, \overleftrightarrow{CD}, \overleftrightarrow{GH}$ **17.** Perpendicular lines always intersect but intersecting lines are not necessarily perpendicular.
19. 59.2 mi per hr   **21.** $\frac{5}{12}$

Chapter 18

Page 335

1. isosceles **3.** pentagon **5.** hexagon
7. quadrilateral **9.** triangle **15.** add 4; 20, 24, 28 **17.** subtract 6; 13, 7, 1 **19.** multiply by $\frac{1}{2}$; $\frac{1}{32}$, $\frac{1}{64}$, $\frac{1}{128}$

Pages 337–339

3. 49°; acute **5.** 51°; obtuse **7.** 61°; acute
9. 20°; obtuse **11.** 40°; acute **13.** 45°
15. 75° **17.** 30° **19.** 115° **21.** 120°
23. 55° **25.** 90° **27.** 46° **29.** 18°
31. 153° **33.** parallel **35.** 21

Page 341

1. 16, 22, and 29 dimes **3.** C
5. Silvia: blue; Rhoda: green; David: brown
7. about $4 billion **9.** 24 mi per gal

Pages 344–345

3. trapezoid **5.** square **7.** rectangle or square **9.** square **11.** rectangle
13. quadrilateral **15.** rhombus **17.** square
19. quadrilateral **21.** parallelogram, trapezoid **23.** 11 in. × 15 in. **27.** 84°
29. 0.08

Pages 346–347

13. cannot **15.** can **17.** 9 in. **19.** 55°
21. $\overrightarrow{XY}$, ray XY **23.** 0.9

Page 349

1. Divide the diameter by 2. **3.** O
7. M **9.** $\overline{MY}$, $\overline{MW}$ **11.** Possible answer: WY **13.** The length of the radius increases.
15. 25 min **17.** trapezoid **19.** 20

Chapter 19

Page 353

1. cone **3.** rectangular prism
5. triangular prism **7.** rectangular pyramid
9. 4; 4; 6 **11.** 5; 5; 8 **13.** 7; 10; 15

Page 355

1. octagonal prism; pentagonal pyramid; polyhedron **3.** cone; no **5.** square pyramid; yes **7.** cylinder; no **9.** pentagonal pyramid; yes **11.** False; a cone has one flat surface. **13.** true **15.** False; some pyramids have triangular bases **17.** hexagon
21. parallelogram, square, or rectangle
23. sector **25.** $3\frac{6}{7}$

Pages 358–359

3. triangular pyramid **5.** rectangular prism
7. rectangular pyramid **19.** a rectangle
21. triangular prism **23.** 0

Page 363

1. 20 balls of clay **3.** D **5.** Monday; 8°C
9. 9 sides

Chapter 20

Page 373

1. equivalent **3.** polygon **5.** 27 **7.** 20
9. 6 **11.** 5 **13.** 7 **15.** 6 **17.** $x = 7$
19. $x = 5$ **21.** $x = 26$ **23.** $x = 17$ **25.** $x = 240$ **27.** $x = 1$ **29.** neither **31.** similar
33. both

Pages 375–376

1. Multiply or divide both terms of a ratio by the same number. **3.** $\frac{1}{2}$; $\frac{8}{16}$ **5.** $\frac{3}{4}$; $\frac{6}{8}$
7. $\frac{150 \text{ points}}{10 \text{ games}}$; 15 points per game **9.** $\frac{90 \text{ words}}{2 \text{ min}}$; 45 words per min **11.** $\frac{\$15}{6 \text{ lb}}$; $2.50 **13.** $\frac{3}{6}$
15. $\frac{2}{6}$ **17.** $\frac{12}{32}$ **19.** $\frac{15}{27}$ **21.** $\frac{\$15}{5 \text{ tapes}}$; $3 per tape
23. $\frac{60 \text{ mi}}{3 \text{ gal}}$; 20 mi per gal **25.** $\frac{\$2.10}{6 \text{ fish}}$; $0.35 per fish **27.** $\frac{4}{2}$ **29.** 15 **31.** 60 **33.** 4:1
35. Both have 3 boxes, so look for lowest cost. **37.** $2.10 **39.** squares of 1, 2, 3, 4…; 25
41. $2 \times 3 \times 3 \times 3$

Page 379

1. $27\frac{1}{2}$ lb **3.** 81 mi **5.** J **7.** 600 in.^{2}
9. 98 **11.** 19 salespeople **13.** 60
15. He misinterpreted the labels. Four people sold between 21–30 cars each.

Pages 382–383

1. Yes, compare the small triangle with the large triangle: $\frac{20}{30} = \frac{2}{3}$, $\frac{12}{18} = \frac{2}{3}$, and $\frac{22}{33} = \frac{2}{3}$.
3. both **5.** both **9.** neither **13.** No; ratios are not equivalent. **15.** Yes
17. $\angle T = 70°$; $\angle Y = 110°$; ST = 3.5 m
23. 45 ft $\times$ 60 ft **27.** $y = 3$

Pages 385–386

1. The ratios of corresponding sides of similar figures are equal **3.** 12 in.
5. $x = 39$ ft **7.** $\frac{n}{8} = \frac{3}{8}$; $n = 3$ in. **9.** $\frac{30}{n} = \frac{18}{24}$; $n = 40$ cm **11.** $\frac{n}{6.2} = \frac{2.5}{5}$; $n = 3.1$ cm
13. 42 in. **15.** No. Corresponding sides do not have the same ratio. **17.** 19

Pages 388–389

1. The real bike is 36 times the bike in the drawing. **5.** 28 **7.** 30 **9.** 20 **11.** 4
13. 96 boxes **15.** Each length of 1 cm on the drawing represents a length of 2 m on the actual object. **17.** false **19.** 30

Pages 390–391

1. 20 mi **3.** 50 mi **5.** 425 mi **7.** 240 mi
9. 420 mi **11.** 1,440 mi **13.** 370 mi
15. 6 in. **17.** $2\frac{1}{2}$ in. **19.** about $2\frac{1}{2}$ hr
21. the auto club map **23.** 4 in. **25.** 9

Chapter 21
Page 395

1. ratio **3.** 0.25 **5.** 0.1 **7.** 0.8 **9.** 0.2
11. 0.81 **13.** 17.28 **15.** 26.65 **17.** 28.8
19. 34.2 **21.** $\frac{1}{6}$ **23.** $\frac{2}{21}$ **25.** $\frac{3}{5}$ **27.** $\frac{16}{21}$
29. 0.45 **31.** 14% **33.** 0.4 **35.** 0.53
37. 1 **39.** 0.03

Page 397

1. 63% **3.** 16% **5.** 31% **7.** 60% **9.** >
11. = **13.** 8% **15.** 18% **17.** 64% **19.** <
21. > **23.** 64% **25.** $2\frac{1}{2}$ in. **27.** 13.015

Pages 400–401

1. move left 2 places, move right two places
3. 25% **5.** 60% **7.** $\frac{1}{2}$ **9.** $\frac{3}{4}$ **11.** $1\frac{4}{5}$
13. 0.08 **15.** 2.4 **17.** 15% **19.** 9%
21. 0.7% **23.** 250% **25.** 12.5% **27.** $\frac{22}{25}$
29. $\frac{1}{25}$ **31.** $\frac{1}{3}$ **33.** 0.07 **35.** 2.2
37. > **39.** < **41.** 20% **43.** What percent of the instruments are woodwinds? **45.** 26 pg per hr **49.** $2 \cdot 3 \cdot 3 \cdot 5$

Pages 404–405

1. Simply double the value of 25% of the number. **3.** 6 **5.** 36 **7.** 7.8 **9.** 21
11. 20 **13.** 44 **15.** 1.08 **17.** 59.64
19. 18.48 **21.** 42.4 **23.** 15.3 **25.** 123
27. 3.85 **29.** 57 **31.** $1.20 **33.** $2.37
35. $4.83 **37.** $5 **39.** $0.80 **41.** 25%; 15
43. 10%; 0.6 **45.** 140,000,000 mi^2 **49.** 0.1%
51. similar

Pages 410–411

3. $60 **5.** $76.80 **7.** $100 **9.** $33.53
11. $46.62 **13.** $38.08 **15.** $30.72
17. $4.65 **19.** $46.20 **21.** $61.51
23. $0.68 per lb **27.** $b = 17$

Page 413

1. It is a formula to find simple interest. Interest equals principal · rate · time. **3.** $4
5. $1,728 **7.** $208 **9.** $1.40, $7.00
11. $82.80, $414.00 **13.** Bank B. At Bank A, he will owe $1,255 while at Bank B he will owe $1,200. **15.** He used 0.6 for 6%. The correct simple interest is $36. **17.** 5 in.
19. 10

Chapter 22
Page 417

1. impossible **3.** equally likely **5.** 0.5, 50%
7. 0.75, 75% **9.** 0.1, 10% **11.** 0.6, 60%
13. 0.72, 72% **15.** 0.12, 12% **17.** 0.95, 95%
19. $\frac{1}{3}$ **21.** $\frac{1}{2}$ **23.** $\frac{2}{3}$ **25.** $\frac{2}{5}$ **27.** $\frac{7}{8}$ **29.** $\frac{1}{3}$
31. $\frac{7}{12}$ **33.** $\frac{1}{4}$ **35.** certain **37.** impossible

Pages 420–421

1. 1, 2, 3, 4, 5, 6 **3.** $\frac{1}{4}$, 0.25, 25% **5.** $\frac{1}{4}$, 0.25, 25% **7.** $\frac{3}{4}$, 0.75, 75% **9.** $\frac{1}{4}$, 0.25, 25% **11.** $\frac{0}{8}$, 0, 0% **13.** $\frac{1}{2}$, 0.50, 50% **15.** $\frac{1}{6}$ **17.** $\frac{1}{3}$ **19.** $\frac{1}{2}$ **21.** $\frac{1}{3}$ **23.** 1 **25.** = **27.** > **29.** < **31.** 0.6 **33.** $\frac{5}{12}$ **35.** 77% **37.** the probability the event will not occur **41.** 9 blue; 15 red **43.** ⁻7.75, $7\frac{1}{4}$, $7\frac{3}{8}$, 7.5

Page 423

1. *too much*; $\frac{4}{7}$ **3.** *too little*; need the 1970 land speed record **5.** C **7.** 4 hr **9.** 8,528 steps

Page 425

1. $\frac{5}{8}$ **3.** 89 **5.** A

Page 427

3. $\frac{1}{5}$ **5.** $\frac{1}{10}$ **7.** $\frac{2}{15}$ **9.** $\frac{1}{4}$ **11.** $\frac{2}{5}$ **13.** $\frac{2}{5}$ **15.** 15 times **17.** number of times 4 lands ÷ total number of rolls **19.** $\frac{4}{7}$ **21.** 9.75 **23.** B

Chapter 23
Page 431

1. sample space **3.** outcome **5.** $\frac{1}{3}$ **7.** $\frac{3}{16}$ **9.** $\frac{2}{5}$ **11.** $\frac{1}{8}$ **13.** $\frac{1}{25}$ **15.** about 26 in. **17.** 100 students **19.** 16 students

Page 433

1. 12 choices **3.** B **5.** 1,870,737 **7.** 4 groups of 10; 3 groups of 8

Pages 435–436

1. 24 **3.** 9 outcomes **5.** 36 outcomes **7.** 16 outcomes **9.** 18 outcomes **11.** 216 outcomes **13.** yes; 12 · 4 · 8 = 384, and 384 > 365 **15.** $\frac{1}{5}$ **17.** 163°

Pages 438–439

1. 16%; 36%; yes, because there are more odd numbers on the spinner than even numbers, and the probability of two odd numbers is greater. **3.** dependent **5.** $\frac{1}{36}$ **7.** 0 **9.** $\frac{1}{20}$ **11.** $\frac{1}{5}$ **13.** independent **15.** $\frac{1}{36}$; $\frac{1}{30}$ **17.** $\frac{1}{6}$; $\frac{1}{5}$ **19.** $\frac{1}{9}$; $\frac{2}{15}$ **21.** $\frac{1}{27}$; 0 **23.** $\frac{1}{108}$; 0 **27.** 15 outcomes **29.** 18 outcomes

Page 441

1. Write and solve the proportion $\frac{5}{75} = \frac{n}{210}$ **3.** $\frac{7}{20}$, 0.35, or 35% **5.** about 153 sixth graders **7.** about 600 cars **9.** $\frac{5}{36}$ **11.** $\frac{1}{3}$ **13.** $\frac{1}{2}$

Chapter 24
Page 451

1. feet **3.** multiply **5.** 3 **7.** 568 **9.** 4 **11.** 24 **13.** 12 **15.** 5 **17.** 1,000 **19.** 10 **21.** 1,000 **23.** 2 **25.** 4,000 **27.** 9 **29.** $n = 45$ **31.** $n = 9$ **33.** $n = 10$

Page 453

1. Use the ratio $\frac{4\ qt}{1\ gal}$ on one side of the proportion and the number of quarts over x gallons on the other side. **3.** 4 **5.** 128 **7.** 12 **9.** $\frac{1}{2}$ **11.** $1\frac{1}{2}$ **13.** 60 **15.** $2\frac{1}{4}$ **17.** $3\frac{1}{4}$ **19.** = **21.** $4\frac{1}{2}$ yd **25.** $1\frac{13}{55}$ **27.** 2.361

Page 455

1. Use the ratio $\frac{10dm}{1m}$ on one side of the proportion and the number of decimeters over x meters on the other side. **3.** 0.005 **5.** 9,000 **7.** 200,000 **9.** 0.5 **11.** 1.2 **13.** 440,000 **15.** 18,000 **17.** 425 **19.** > **21.** = **25.** $\frac{1}{3}$ **27.** 40% **29.** ⁻9

Page 457

1. 1 mi; it takes 1.61 km to make 1 mi. **3.** 25.4 **5.** 427.7 **7.** 10 **9.** 76.2 **11.** 4.75 **13.** 26.37 **15.** < **17.** > **19.** 1.38 in. **21.** 18°C **23.** ⁻10 **25.** $\frac{7}{6}$

Pages 460–461

1. gram; because the gram is a smaller unit than the kilogram **3.** 2 cm; 23 mm **5.** 85 in. **7.** 8 oz **9.** meter, yard, or foot **11.** gram or ounce **13.** $1\frac{1}{2}$ in.; $1\frac{3}{4}$ in. **15.** $\frac{3}{4}$ in.; $\frac{7}{8}$ in. **17.** 8 ft **19.** 8 oz **21.** millimeter or part of an inch **23.** = **25.** = **27.** To the nearest millimeter because a millimeter is a tenth of a centimeter, which is smaller than a half centimeter. **31.** 17.055 L **33.** $^{-}96$

Page 463

1. estimate; no **3.** estimate; no **5.** C **7.** $9\frac{3}{20}$ in. **9.** $2\frac{3}{8}$ in. below **11.** 23 years old

Chapter 25
Page 467

1. perimeter **3.** 32 ft **5.** 20 cm **7.** 104 in. **9.** 4 **11.** 36 **13.** 4 **15.** 3,000 **17.** 5,000 **19.** 0.40009 **21.** 88 **23.** 2,625 **25.** 526.5 **27.** 60 **29.** 73.6

Page 471

3. 13.57 m **5.** 9 ft or 108 in. **7.** 43 m **9.** $x = 13.5$ cm **13.** 5.40 **15.** $a = 1,264$

Page 473

1. 66 in. **3.** B **7.** 42.4% **9.** 1 in.

Page 475

1. 88 in. **3.** $z = 17.6$ **5.** D

Pages 478–479

1. They both use π; one uses the diameter and the other uses 2 times the radius, which is equal to the diameter. **3.** 16 m **5.** 12 cm **7.** 283 yd **9.** 27 in. **11.** 33 cm **13.** 20 in. **15.** 9 yd **17.** 220 cm **19.** 316 in. **21.** 22 ft **23.** 16 cm **25.** 4.5 ft **27.** 9.3 cm **29.** The circumference is twice as long. **31.** 88 yd **33.** 149 m

Chapter 26
Page 483

1. square **3.** faces **5.** 144 **7.** 400 **9.** 81 **11.** 5.76 **13.** 2,401 **15.** 16,641 **17.** 25 **19.** 192 **21.** 16 **23.** 5 faces **25.** 4 faces

Pages 486–487

1. The area of the triangle is $\frac{1}{2}$ the area of the rectangle. **3.** about 20 m^2 **5.** 117 in.^{2} **7.** 45.5 in.^{2} **9.** about 6 m^2 **11.** 558.25 mm^2 **13.** 0.2 m^2 **15.** 22.5 ft^2 **17.** 273 yd^2 **19.** 12 **21.** 60%

Pages 489–490

1. $A = \frac{1}{2}h(b_1 + b_2) = \frac{1}{2} \times 4 \times (4.2 + 6.5) = 21.4$; 21.4 m^2 **3.** 10 ft^2 **5.** 102.3 m^2 **7.** 136.95 m^2 **9.** 40 ft^2 **11.** 1.008 cm^2 **15.** 24 m^2 **17.** 0.003

Page 491

1. 3 m^2 **3.** 154 m^2

Page 493

3. 28 cm^2 **5.** 1,809 ft^2 **7.** 50 yd^2 **9.** 7 m^2 **11.** 50 mm^2 **13.** 3,420 in.^{2} **15.** 77 cm^2 **17.** 804 ft^2 **19.** 100 ft^2 **21.** 3

Pages 496–497

1. Find the area of each pentagonal face and the area of the five rectangular faces and add. **3.** 184 ft^2 **5.** 336 cm^2 **7.** 340 m^2 **9.** 73.5 m^2 **11.** 108 cm^2 **13.** 69.36 cm^2 **15a.** $3 \times 6 \times 12$; 252 m^2 **15b.** $3 \times 6 \times 6$; 144 ft^2 **15c.** $10 \times 5 \times 15$; 550 in.^{2} **15d.** $2 \times 8 \times 32$; 672 cm^2 **17.** 2 cans **19.** 5 in. **21.** 113 ft^2 **23.** 0.25

Chapter 27
Page 501

1. dimensions **3.** height **5.** 64 **7.** 27 **9.** 216 **11.** 0.008 **13.** 1,728 **15.** 20 ft^2 **17.** 21 m^2 **19.** $58\frac{7}{12}$ ft^2 **21.** 52.36 m^2 **23.** 141 cm^2 **25.** 346 cm^2

Pages 504–505

1. Find $26 \times 3 \times 18$, or 1,404 cubes.
3. 24 in.^{3} **5.** 48 cm^3 **7.** 324 ft^3
9. 294 ft^3 **11.** 288 in.^{3} **13.** $x = 10$ m
15. 189 ft^3 **17.** 40 in. **19.** 45.63

Page 507

1. twice the volume of the original container
3. B **7.** 10 shirts, 0 shorts; 0 shirts, 8 shorts;
5 shirts, 4 shorts **9.** Mike, Sharon, Jasmine,
Hugh

Page 509

1. Both include the area of the base times the
height; prism: $V = bh$; pyramid: $V = \frac{1}{3}Bh$
3. 48 m^3 **5.** 224 cm^3 **7.** 24 cm^3, or 24,000
mm^3 **9.** 4,500 ft^3 **11.** 24 yd^3, or 648 ft^3
15. 110 **17.** 0.79

Page 510

1. about 75 cm^3

Page 512–513

1. 6,029 in.^{3} **3.** πr^2 represents the area of
the base and h represents the height.
5. about 346 in.^{3} **7.** about 942 in.^{3}
9. about 1,409 cm^3 **11.** about 311 cm^3
13. about 2,374 cm^3 **15.** about 7,436 ft^3
17. The volume is about eight times as large.
21. about 4 ft^3 **23.** $\frac{1}{2} \div \frac{1}{10}$

Chapter 28
Page 523

1. equation **3.** < **5.** < **7.** > **9.** >
11. < **13.** > **15.** < **17.** > **19.** 10; 21
21. 25; 25.5 **23.** 15 **25.** $^-$5 **27.** 63
29. 29 **31.** $\frac{3}{14}$ **33.** 3.1

Page 525

1. the eighth day **3.** A **5.** 7 videos
7. 660 ft **9.** 40 ft^2

Pages 527–528

1. Multiply by 4; 2,048; 8,192; 32,768 **3.** Add
15 to each successive term **5.** 43, 54, 65
7. Divide each term by 3. **9.** Add 0.89 to
each term. **11.** 335, 485, 665 **13.** 81, 76, 70
15. 9, 12.7, 16.4, 20.1, . . . **17.** add $54; $462
21. 754 ft^3 **23.** 264

Pages 531–532

3. $b = a - 6$ **5.** $d = c + 1.1$ **7.** $l = 4w$
9. $g = k \div 2$; 27 **11.** $d = c \div {}^-4$; $^-8$
15. in $y^2 = x$, if $x = 1$, $y = 1$ or $y = {}^-1$
17. $t = 0.75n$; $37.50 **19.** $209 **21.** $2\frac{5}{104}$

Page 535

9. $5^0, 5^1, 5^2, 5^3, \ldots$ **11.** 16 prisms; 25 prisms
13. $y = {}^-9, {}^-3, 3, 9, 15$ **15.** 5×13

Chapter 29
Page 539

1. congruent **3.** slide **5.** slide **7.** turn
17. 25° **19.** 135°

Pages 541–542

1. The figure does not change in size or
shape. **3.** rotation or reflection
5. reflection **7.** rotation, reflection
9. rotation or translation **11.** rotation
13. translation, reflection, reflection
19. horizontal reflection **25.** $\frac{5}{8}$

Pages 544–545

3. yes **5.** no **11.** yes **13.** no
19. 288 sq yd. **23.** $\frac{11}{12}$

Page 547

1. triangle and square; square and octagon
3. C **5.** 63° **7.** a **9.** 757 and 575

Page 549

1. Yes, the figure's size and shape do not change.   3. 8 ways   5. 6 ways   7. 8 ways
9. 12 ways   11. rotate it 180°   13. reflection
15. 24 cm

Pages 552–553

1. Line symmetry means that the figure can be folded into two congruent halves that are mirror images. Rotational symmetry means that the figure matches itself when rotated less than 360°.   11. yes; $\frac{1}{4}$; 90°   13. yes; $\frac{1}{2}$; 180°   27. yes; $\frac{1}{5}$; 72°   29. no   31. 8 pieces; yes   33. To have rotational symmetry, it must match up with a rotation of less than 360°   35. 125.66 in^3

Chapter 30
Page 557

1. ordered pair   3. perpendicular   5. =
7. >   9. <   11. >   13. ($^-$2,2)   15. (0,0)
17. (2,1)   19. (0,3)   21. ($^-$2,$^-$2)

Page 559

1. Yes.   19. $p > {}^-11$ 21. $a < 20{,}320$
25. 37.5%   27. $\frac{3}{5}$

Pages 562–563

1. (5,4) is above the x-axis. (5,$^-$4) is below the x-axis.   3. (2,4)   5. (0,0)   7. (5,$^-$5)
9. (0,$^-$5)   11. (2,$^-$3)   13. A, H   15. B
17. (2,6)   19. (2,$^-$2)   21. (0,7)   23. (3,0)
25. (6,$^-$4)   27. ($^-$7,$^-$6)   29. A, L, J
31. B, H, K   41. triangle; 28 sq units
43. It is 0; it is 0.   45. $y \geq {}^-7$ 47. 157 cm

Page 565

1. a table, a graph of ordered pairs.
7. (45,15), (48,16), (51,17), (54,18); $w = \frac{1}{3}c$
9. Quadrant I   11. 40 cm^2

Page 567

1. C   3. $y = 9x$ 5. 1,333 mi   7. 11:00 A.M.
9. about 800

Pages 572–573

3. A'(0,3), B'(2,3), C'(2,1), D'(0,1)   5. A'(0,0), B'($^-$6,0), C'($^-$4,2), D'($^-$2,2)   7. E'(0,$^-$1), F'(3,$^-$1), G'(3,$^-$4), H'(0,$^-$4)   9. E'(1,$^-$3), F'(4,$^-$3), G'(4,$^-$1), H'(1,$^-$1)   11. No.
13. $y = 7x$ 15. $\frac{1}{4}$

Index

D

E

Photo Credits Key: (t) top; (b) bottom; (l) left; (r) right; (c) center; (bg) background; (i) inset.

Page: iv NASA; v (t) Lori Adamski Peek/Stone; vii (b) Bob Daemmrich Photography; ix (t) Mark E. Gibson; ix (b) David Young-Wolff/PhotoEdit; xi (b) NASA; xiii (b) Bob Daemmrich Photography ; xv (t) Mark E. Gibson; xv (b) Mark Kelley/ Stock, Boston; xvi Ned Haines/Photo Researchers; xvii (t) Index Stock Photography; xviii Myrleen Ferguson/PhotoEdit; xix (b) Bob Daemmrich Photography; xx John Henley/The Stock Market; xxi Ralph H. Wetmore II/Stone; 2 William R. Sallaz/Duomo Photography; 6 Mark E. Gibson; 14 John Frassanito/NASA/Science Photo Library/Photo Researchers; 16 Myrleen Ferguson/ PhotoEdit; 17 Paula Lerner/Woodfin Camp & Associates; 19 Amy Etra/ PhotoEdit; 20 NASA; 21 NASA/Science Photo Library/Photo Researchers; 22 Mark E. Gibson; 23 Tim Davis/Photo Researchers; 25 (t) Tony Freeman/ PhotoEdit; 25 (b) Jon Feingersh/The Stock Market; 28 Bob Daemmrich Photography; 29 Bob Daemmrich Photography; 30 Joseph Pobereskin/Stone; 31 Phil McCarten/PhotoEdit; 34 UNEP/Peter Arnold, Inc.; 36 Nicholas Veasey/ Stone; 37 Mark E. Gibson; 40 Baron Wolman/Stone; 45 Foodpix; 46 (t) Blair Seitz/Photo Researchers; 50 FoodPix; 52 Spencer Grant/Stock, Boston; 53 John Coletti/Stock, Boston; 55 Raymond Barnes/Stone; 56 Lori Adamski Peek/Stone; 58 Milinda Ketkar/ Dinodia Picture Agency; 64 Andrew Brookes Creative/The Stock Market; 66 Daryl Balfour/Stone; 69 (t) Mark E. Gibson; 69 (b) Bob Daemmrich Photography; 71 Mark E. Gibson; 73,76 David Young-Wolff/ PhotoEdit; 79 Roy Morsch/The Stock Market; 80 Michael Newman/ PhotoEdit; 81 Mark E. Gibson; 87 Mark Wagner/Stone; 90 Bernard Boutrit/Woodfin Camp & Associates; 91 Robert Frerck/Woodfin Camp & Associates; 92 Bernhard Edmaier/Science Photo Library/Photo Researchers; 94 (t) Wally McNamee/Woodfin Camp & Associates; 96 Index Stock Photography; 101 Stan Osolinski/The Stock Market; 102 Tom Walker/Stock, Boston; 105 Carolyn A. McKeone/Photo Researchers; 106 Tony Reynolds; 107 Tom & Pat Leeson/Photo Researchers; 109 Marina Bakhtadze; 110 (t) Mark E. Gibson; 110 (b) Russell R. Grundke; 111 Ellen B. Senisi/Photo Researchers; 112 Bob Daemmrich Photography; 113 Jeff Greenberg/Photo Researchers; 114 Mark E. Gibson; 115 S. Feld/H. Armstrong Roberts, Inc.; 118 Airworks/Tom Stack & Associates; 120 (t) Francois Gohier/ Photo Researchers; 120 (b) Michael Sewell/Peter Arnold, Inc.; 124 Bob Daemmrich Photography; 125 Brian Bailey/Stone; 126 Bryan Berg; 127 Bob Daemmrich Photography; 130 (t) David A. Hardy/Science Photo Library/Photo Researchers; 130 (b) Frank Zullo/Photo Researchers; 132 David Ball/Stone; 133 David R. Frazier; 134 James Marshall/ The Stock Market; 135 David R. Frazier; 142 (tl) David Young-Wolff/PhotoEdit; 142 (tr), (br) Mark E. Gibson; 142 (bl) Davis Barber/PhotoEdit; 143 Philip Wallick/Gnass Photo Images; 147 Mugshots/The Stock Market; 149 Richard Hutchings/PhotoEdit; 150 Bob Daemmrich Photography; 153 (b) I. M. House/Stone; 153 (t) Michael Newman/ PhotoEdit; 154 Index Stock Photography; 158 Manoj Shah/Stone; 161 (t),(b) Nancy Dudley/Stock, Boston; 163 Pekka Parviainen/Science Photo Library/ Photo Researchers; 164 Jerry Schad/Photo Researchers; 165 George Olson/ Woodfin Camp & Associates; 166 Baldwin H. Ward/Corbis; 167 David R. Frazier; 169 R.L. Dunne/Bruce Coleman, Inc.; 170 (tl) Mark Newman/Bruce Coleman, Inc.; 170 (tr) Carolyn A. McKeone/Photo Researchers; 170 (b) Index Stock Photography; 171 J. Gerald Smith/Photo Researchers; 174 Nathan Benn/Woodfin Camp & Associates; 176 Mark E. Gibson; 177 David Young-Wolff/PhotoEdit; 179, 186 Mark E. Gibson; 187 Terry Whittaker/Photo Researchers; 188 E.R. Degginger/Photo Researchers; 189 (t) Mark Newman/Bruce Coleman, Inc.; 189 (b) David R. Frazier; 192 Bill Swersey/ Gamma Liaison; 193 Richard Pasley/Stock, Boston; 194 Brad Rickerby/Bruce Coleman, Inc.; 198 Doug Pensinger/Allsport Photography; 200 Bruce Coleman, Inc.; 201 Stephen Kline/Bruce Coleman, Inc.; 202 Scott Barrow, Inc.; 203 Art Montes De Oca/FPG International; 205 Bob Daemmrich Photography; 206 Dugald Bremner/Dugald Bremner Studio; 207 Myrleen Cate/Stone; 213 (t) Michael Newman/PhotoEdit; 213 (b) Arthur Tilley/FPG International; 214 Bob Daemmrich Photography; 216 Zig Leszczynski/Animals Animals/Earth Scenes; 217 Sharon Cummings/Dembinsky Photo Associates; 224 Larry Ulrich Photography; 225 Tony Freeman/PhotoEdit; 226 Scott Darsney/Alaska Stock Images; 228 Carr Clifton/Minden Pictures; 229 Mary Clay/Dembinsky Photo Associates; 231 David Madison/Bruce Coleman, Inc.; 232 David Madison Photography; 234 Patrick Endres/Alaska Stock Images; 235 John Terence Turner/ FPG International; 237 Larry Voigt/Photo Researchers; 240 Julian Baum/Science Photo Library/Photo Researchers; 245 Mark E. Gibson; 254 Tomas Utsi; 258 Frank Rossotto/The Stock Market; 259 Alan L. Detrick/Photo Researchers; 260 Don Pitcher/Alaska Stock Images; 261 Cheryl Hogue/Photo Researchers; 263 Jon Riley/Stone; 264 Mark Kelley/Stock, Boston; 270 Bob Rowan; Progressive Image/Corbis; 271 (l) Bernard Boutrit/Woodfin Camp & Associates; 271 (r) Galen Rowell/Corbis; 272 Buddy Mays/Travel Stock Photography; 274 Bob Daemmrich Photography; 275 Jilly Wendell/Stone; 282 Myrleen Ferguson/PhotoEdit; 286 Richard Megna/Fundamental Photographs; 288 Dennis Flaherty/Photo Researchers; 289 NASA; 291 Ralf Laue; 293 Tony Freeman/PhotoEdit; 298 Glen Allison/Stone; 301 Paul McKelvey/Stone; 303 Index Stock Photography; 304 David R. Frazier; 305,307 Index Stock Photography; 311 Bob Daemmrich Photography; 318 Photo Select(Moldav)/ Stone; 319 Chuck Mitchell; 320 David Barnes/The Stock Market; 322 Charles D. Winters/Photo Researchers; 323 Donovan Reese/Stone; 326 (l) Index Stock Photography; 326 (r) Robin Prange/The Stock Market; 327 Don Mason/The Stock Market; 329 Bob Daemmrich Photography; 330 Jake Rajs/Stone; 331 David Young-Wolff/Stone; 334 Andy Sacks/Stone; 341 Superstock; 343 Spencer Swanger/Tom Stack & Associates; 344 Joseph Nettis/Photo Researchers; 345 PhotoEdit; 348 Gerald Zanetti/The Stock Market; 352 Peter Mauss/Esto; 357 Soho Imageworks, Inc./Phototake; 367 Bob Daemmrich Photography; 370 (t) Myrleen Ferguson/PhotoEdit; 370 (b) Audrey Gibson/Mark E. Gibson Photography; 371 (l),(c) Lee Foster/Foster Travel Publishing; 371 (r) Bob Ecker; 372 Johan W. Elzenga/Stone; 374 Leif Skoogfors/Woodfin Camp & Associates; 378 NASA/Peter Arnold, Inc.; 383 Michael Busselle/Stone; 384 (t),(i) Michael Busselle/Stone; 384 (b) Jeff Greenberg/Photo Researchers; 384 (b) Charlie Ott Photography/Photo Researchers; 385 Ned Haines/Photo Researchers; 386 Bridgeman/Art Resource NY; 388 Robert Brons/BPS/Stone; 394 Joseph Nettis/ Stone; 396 David Young-Wolff/PhotoEdit; 399 Micheal Simpson/FPG International; 401 Chuck Savage/The Stock Market; 402 David Madison/Stone; 403 Shilo Sports/FPG International; 405 Bruce Forster/Stone; 409 (t) David Young-Wolff/PhotoEdit; 409 (b) Paul Barton/The Stock Market; 410 Mark Gamba/The Stock Market; 412 Henley & Savage/The Stock Market; 413 David Young-Wolff/PhotoEdit; 416 Index stock Photography; 421 Bruce Hands/Stock, Boston; 422 Robert Ginn/ PhotoEdit; 423 David R. Frazier; 426 Bruce Coleman, Inc.; 427 Ray Coleman/ Photo Researchers; 430 (i) Lawrence Migdale/Stock, Boston; 432 Michelle D. Bridwell/PhotoEdit; 433 J. Wright/Bruce Coleman, Inc.; 436 Index Stock Photography; 437 Arthur Beck/The Stock Market; 439 Index Stock Photography; 440 Rolf Kopfle/Bruce Coleman, Inc.; 441 Ken Lax; 445 (t) Jonathan Nourek/ PhotoEdit; 445 (b) Thomas D. Mangelsen/Peter Arnold, Inc.; 448 Mark E. Gibson; 449 ChromoSohm/The Stock Market; 450 Rudi VonBriel/PhotoEdit; 452 Allsport Photography USA; 453 Michael Kevin Daly/The Stock Market; 454 Allsport Photography USA; 455 Richard Hutchings/Photo Researchers; 456 J. Messerschmidt/Bruce Coleman, Inc.; 457 Philip Jon Bailey/Stock, Boston; 458 David Young-Wolff/PhotoEdit; 458 (i) William Johnson/Stock, Boston; 460 Index Stock Photography; 461 Spectrum Stock; 462 Spencer Grant/PhotoEdit; 463 Mug Shots by Gabe Palmer/The Stock Market; 466 Mark E. Gibson; 469 Phil Degginger/Color-Pic; 472 David R. Frazier; 476 F. Stuart Westmorland/Photo Researchers; 477 Zefa/Index Stock Photography; 479 (t) Archive Photos; 479 (b) U.S. Mint/Department of the Treasury; 482 Will & Deni McIntyre/Stone; 484,485 David Young-Wolff/ PhotoEdit; 488 Thomas Kitchin/Tom Stack & Associates; 489 E. R. Degginger/ Color-Pic; 492 Alese/MortPechter/The Stock Market; 495 Sylvain Grandadam/Stone; 496 Pete Saloutos/The Stock Market; 497 Michael Newman/ PhotoEdit; 500 Lloyd Sutton/Masterfile; 502 Index Stock Photography; 503 IFA Bildertamm/Leo de Wys, Inc.; 505 Bob Daemmrich Photography; 508 Hugh Sitton/Stone; 513 Michael Boys/Corbis; 517 Mark Richards/PhotoEdit; 520 (t) Gerd Ludwig/Woodfin Camp & Associates; 520 (b) Stuart Westmorland/Corbis; 521 Kunio Owak: The Stock Market; 522 Index Stock Photography; 524 Ken Fisher/Stone; 525 Debra P. Hershkowitz/Bruce Coleman, Inc.; 526 Philip Gould/Corbis; 528 Tony Freeman/PhotoEdit; 529 (t) John Henley/The Stock Market; 529 (b) Tony Freeman/PhotoEdit; 531 Laura Dwight/Peter Arnold, Inc.; 532 Michelle Bridwell/ PhotoEdit; 533 Chuck Place/Stock, Boston; 538 Art Wolfe/Stone; 543(t) Brownie Harris/The Stock Market; 545 Wes Thompson/The Stock Market; 550 Gary Vestal/Stone; 551 Fred Bavendam/Peter Arnold, Inc.; 556 Chad Slatery/Stone; 559 Peter Beck/The Stock Market; 561 Jeff Greenberg/Photo Reseaerchers; 563 Joe Sohm/The Stock Market; 564 Ralph H. Wetmore II/Stone; 566 (t) J. C. Carton/Bruce Coleman, Inc.; 566 (b) William H. Johnson/Stock, Boston; 567 Bob Daemmrich Photography; 570 (t),(b) Library of Congress; 572 Ross Harrison Koty/Stone; 580 (l) Dan Suzio/Photo Researchers; 580 (r),(c) Mark E. Gibson; 581 (t) Lee Foster/Foster Travel Publishing; 581 (c) J. Paul Getty Trust, Photo Scott/Esto Photographics; 581 (b) Mark E. Gibson; Hi (tl) NASA /Science Photo Library/Photo Researchers; (tr) Russell R. Grundke; (bl) Sylvain Grandadam/Stone.

All other photos by Harcourt photographers: Weronica Ankarorn, Victoria Bowen, Eric Camden, Greg Leary, Ken Kinzie, Ron Kunzman, Ed McDonald, Sheri O'Neal, and Terry Sinclair.

Table of Measures

METRIC | CUSTOMARY

Length

METRIC	CUSTOMARY
1 millimeter (mm) = 0.001 meter (m)	1 foot (ft) = 12 inches (in.)
1 centimeter (cm) = 0.01 meter	1 yard (yd) = 36 inches
1 decimeter (dm) = 0.1 meter	1 yard = 3 feet
1 kilometer (km) = 1,000 meters	1 mile (mi) = 5,280 feet
	1 mile = 1,760 yards
	1 nautical mile = 6,076.115 feet

Capacity

METRIC	CUSTOMARY
1 milliliter (mL) = 0.001 liter (L)	1 teaspoon (tsp) = $\frac{1}{6}$ fluid ounce (fl oz)
1 centiliter (cL) = 0.01 liter	1 tablespoon (tbsp) = $\frac{1}{2}$ fluid ounce
1 deciliter (dL) = 0.1 liter	1 cup (c) = 8 fluid ounces
1 kiloliter (kL) = 1,000 liters	1 pint (pt) = 2 cups
	1 quart (qt) = 2 pints
	1 quart (qt) = 4 cups
	1 gallon (gal) = 4 quarts

Mass/Weight

METRIC	CUSTOMARY
1 milligram (mg) = 0.001 gram (g)	1 pound (lb) = 16 ounces (oz)
1 centigram (cg) = 0.01 gram	1 ton (T) = 2,000 pounds
1 decigram (dg) = 0.1 gram	
1 kilogram (kg) = 1,000 grams	
1 metric ton (t) = 1,000 kilograms	

Volume/Capacity/Mass for Water

1 cubic centimeter (cm^3) → 1 milliliter → 1 gram

1,000 cubic centimeters → 1 liter → 1 kilogram

TIME

1 minute (min) = 60 seconds (sec)	1 year (yr) = 12 months (mo), or about 52 weeks
1 hour (hr) = 60 minutes	1 year = 365 days
1 day = 24 hours	1 leap year = 366 days
1 week (wk) = 7 days	